Biopsy Pathology of the Thyroid and Parathyroid

OTTO LJUNGBERG

Associate Professor of Pathology, University of Lund and
Chief Pathologist and Director of the Department of Pathology and Cytology
Central Hospital
Kristianstad
Sweden

CHAPMAN & HALL MEDICAL
London · Glasgow · New York · Tokyo · Melbourne · Madras

Published by Chapman & Hall, 2–6 Boundary Row, London SE1 8HN

Chapman & Hall, 2–6 Boundary Row, London SE1 8HN, UK

Blackie Academic & Professional, Wester Cleddens Road, Bishopbriggs, Glasgow G64 2NZ, UK

Chapman & Hall, 29 West 35th Street, New York NY10001, USA

Chapman & Hall Japan, Thomson Publishing Japan, Hirakawacho Nemoto Building, 6F, 1–7–11 Hirakawa-cho, Chiyoda-ku, Tokyo 102, Japan

Chapman & Hall Australia, Thomas Nelson Australia, 102 Dodds Street, South Melbourne, Victoria 3205, Australia

Chapman & Hall India, R. Seshadri, 32 Second Main Road, CIT East, Madras 600 035, India

First edition 1992

© 1992 Otto Ljungberg

Typeset in 10/12 Palatino by Mews Photosetting, Beckenham, Kent
Printed in Great Britain at the University Press, Cambridge

ISBN 0 412 34890 X

A catalogue record for this book is available from the British Library

Library of Congress Cataloging-in-Publication data available

Contents

Preface

The purpose of this book is a guide for the histopathological assessment of thyroid and parathyroid lesions. Its main readership target is practising histopathologists and trainees in surgical pathology, but it is hoped that it may also be of interest to clinical colleagues from other disciplines involved in the management and treatment of thyroid and parathyroid disorders.

Knowledge in the field of thyroid and parathyroid pathology has been growing rapidly during recent years. New developments of special histological techniques, mainly in the field of immunohisto-chemistry, have not only enabled a more accurate interpretation of the lesions, but also provided new information of their nature. This is especially relevant with respect to thyroid tumours and has affected our view on their current histological classification. New histological techniques have also been introduced which allow a more accurate intraoperative diagnosis of parathyroid lesions in primary hyper-parathyroidism, resulting in a more rational surgical approach in the treatment of this condition.

The surgical treatment of thyroid and parathyroid lesions is usually aimed at cure. Consequently, the majority of specimens submitted to the pathologist include the entire lesion, and usually represent the gland as a whole or, at least, a major part of it. Excisional biopsies, performed for diagnostic purposes only, are rare. Therefore, this edition will not deal solely with biopsy pathology, but rather will have the character of a treatise on surgical pathology, emphasizing the diagnostic aspects of thyroid and parathyroid pathology in general.

On the other hand, this book is not intended to be a textbook covering every aspect of thyroid and parathyroid pathology. Emphasis has been put on those structural and histochemical features which are relevant for a correct histopathological diagnosis. For the same purpose, some reference has also been devoted to recent advances in the fields of cell and molecular biology and pathogenesis, which have provided new insights into the nature of the disorders. Clinical presentation and behaviour of the lesions, especially when relevant for the diagnosis, have been briefly dealt with. With regard to the tumours, special attention has been paid to histopathological parameters which are considered important for prognostication and clinical management. Besides typical 'textbook cases', the book also illustrates some variants and atypical examples of lesions that the pathologist may encounter in his day-to-day practice. The material that forms the basis of this book is mainly derived from the files of the Department of Pathology at Malmö General Hospital. This hospital serves a city with about 240 000 inhabitants. During the last 15 years the author has personally handled all cases within the area of endocrine pathology at the department, comprising about 900 thyroid cases (including about 150 malignant tumours) and about 400 cases of primary hyperparathyroidism. The author has also acquired valuable experience from a large amount of material sent for consultation from laboratories of other hospitals in the country and abroad, often raising problems in differential diagnosis which may be of interest to the practising histopathologist and which, in various ways, have been conveyed to the reader.

Fine needle aspiration cytology has become an important tool for the clinical assessment of the majority of disease processes of the thyroid gland and for the selection of cases which require surgical intervention. However, the cytological diagnosis of aspirated material from thyroid lesions does not fall within the scope of this book, and will only be referred to sporadically. Several excellent books on this subject are available. Space only allows reference to be made to the following three: *Cytological Aspiration Biopsy of the Thyroid Gland* by Manfred Droese (F.K. Schattauer Verlag, Stuttgart, New York, 1980), *Aspiration Biopsy Cytology of the Thyroid Gland* by Torsten Löwhagen and Joseph A. Linsk in *Clinical Aspiration Cytology* (eds Joseph A. Linsk and Sixten Franzén, J.B. Lippincott Company, Philadelphia, 1983) and finally, *Manual and Atlas of Fine Needle Aspiration Cytology*, 2nd edn by Svante R. Orell, Gregory F. Sterrett, Max N.-I. Walters and Darrel Witaker (Churchill-Livingstone, Edinburgh, London, Melbourne and New York, 1992).

I would like to thank Mrs Leni Grubb, Birgitta Kjellander, Elise Nilsson and Ewa Askerlund, who have, with mastery, performed the immunohistochemical and electron microscopical preparations and Mrs Maria Vincze for invaluable photographic assistance.

Many pathologists have contributed valuable case material that has been used for illustrations in the book. They include Associate Professor Jan Alumets, Lund, Associate Professor Thorbjörn Berge, Taif, Saudi Arabia, Dr Kaj Bjelkenkrantz, Örebro, Dr Poul Boiesen, Jönköping, Associate Professor Stefan Cajander, Umeå, Dr Finn Dahl, Örebro, Dr Barbara Du Rietz, Stockholm, Dr Sune Eriksson, Trollhättan, Dr Ingemar Glifberg, Gävle, Professor Lars Grimelius,

Uppsala, Associate Professor Sven-Olof Hjertquist, Sundsvall, Dr Erik Holm, Karlskrona, Dr Aziz Hussein, Örebro, Dr Sven Johansson, Växjö, Dr Sven Lundberg, Jönköping, Dr Per Magnusson, Växjö, Dr Björn Risberg, Örebro, Dr Bengt Sandstedt, Stockholm, Associate Professor Torkel Wahlin, Trollhättan, Associate Professor Helena Willén, Lund, Associate Professor Roger Willén, Lund. I am most grateful for their help.

I am indebted to Axel Johnson Lab System AB, Stockholm and Olympus Europe, Hamburg for financial support to defray the costs for printing the colour photographs.

Finally my thanks go to Professor Munro Neville for his constant encouragement, valuable suggestions and comments on the text, and for revising my English. I also owe my sincere gratitude and appreciation to the editorial staff at Chapman & Hall for their great patience and invaluable support and assistance.

PART ONE
The Thyroid Gland

1 The normal thyroid gland

1.1 Anatomy

The normal adult thyroid gland consists of two lateral lobes, measuring
5–8 cm in length. They are connected by the isthmus, which is a band
of variable size lying close to the ventral aspect of the trachea, covering the 2nd, 3rd and 4th tracheal rings. In about 40% of cases a
pyramidal lobe extends upward from the upper margin of the isthmus
toward the hyoid bone. The two lobes of the gland embrace the ventral
and lateral aspects of the trachea, usually just below the cricoid
cartilage.

The adult gland weighs approximately 20–25 g. However, there is
considerable individual variation in size and shape, which are also
dependent on a variety of environmental and physiological factors,
e.g. age, sex, iodine content of diet and pregnancy. The right lobe
is often somewhat larger than the left. The gland has a reddish-brown
glistening, finely lobulated cut surface.

The gland is covered by a thin fibrous capsule which sends thin
interlobular septa into the deeper parts of the parenchyma.
Microscopically, the capsule is frequently found to have small defects,
through which irregular nodules of follicles may be seen to extend
into adjacent perithyroidal fat or fibromuscular tissue (Komorowski
and Hanson, 1988). Even where the capsule is intact, minute nodules
of thyroid tissue may be found outside the gland. It is important to
be aware of this anatomical aberration, especially in cases of thyroid
tumours, in order not to misinterpret such collections of follicles as
extrathyroidal tumour infiltration or metastases. Extrathyroidal follicles
are most commonly seen at the poles of the gland and in the area of
the isthmus and pyramidal lobe. Since extraglandular collections of
follicles are common in normal adults, it is easy to understand why,
in cases of nodular goitre, it would be possible for hyperplastic nodules
to develop outside the gland. (see p. 93)

The rich blood supply to the gland comes from the paired superior
and inferior thyroid artiers and an inconstant a. thyreoidea ima. The
blood is drained via a venous plexus in the fibrous capsule into the

internal jugular and brachiocephalic veins. Lymphatics drain mainly to deep cervical nodes, but also to pretracheal and retrosternal nodes.

Very rarely, the entire thyroid gland may show a coal-black pigmentation. The black thyroid is due to a melanin-like pigment, which, like lipofuscin, appears to accumulate in the follicular cells with advancing age. The most excessive deposits have been seen in patients receiving chronic minocycline therapy (Reid, 1983; Alexander *et al.*, 1985; Landas *et al.*, 1986).

1.2   Histology

The gland consists of two functionally and morphologically distinct endocrine systems. The principal components of the gland are the follicles, made up of the follicular cell proper, the thyreocyte. The follicles are the smallest complete functional units of this component, capable of synthesis, storage and release of the thyroid hormones, tetraiodothyronine (T_4 or thyroxine) and triiodothyronine (T_3).

The second minor endocrine component of the gland, known only during the last two decades, is represented by the parafollicular cells, also referred to as the C-cells, which belong to a large family of peptide hormone-producing neuroendocrine cells which are widely distributed in many endocrine and non-endocrine tissues of the body.

In addition to these two components, there is also a third anatomical structure of unknown physiological significance. This was described at the end of the nineteenth century but has not received much attention until recently. It has been referred to as the so-called solid cell nests (SCN) or ultimobranchial remnants and constitutes an inconspicuous, often multifocal epithelial ingredient of the thyroid gland.

1.2.1   The follicles

The functional unit of the main endocrine system in the thyroid is the follicle, a closed spheroid structure lined by a single layer of epithelial cells (Fig. 1.1). It is filled with colloid which is largely made up of proteins, especially the iodinated glycoprotein thyroglobulin, the secretory product of the follicular cells. The follicles vary considerably in size with an average diameter of about 200 μm. Their walls consist of a single layer of flat or cuboidal epithelial cells with their apices directed towards the lumen of the follicle.

The follicles are separated from each other by delicate septa of connective tissue. Thyroid lobules, the thyromeres, are made up of

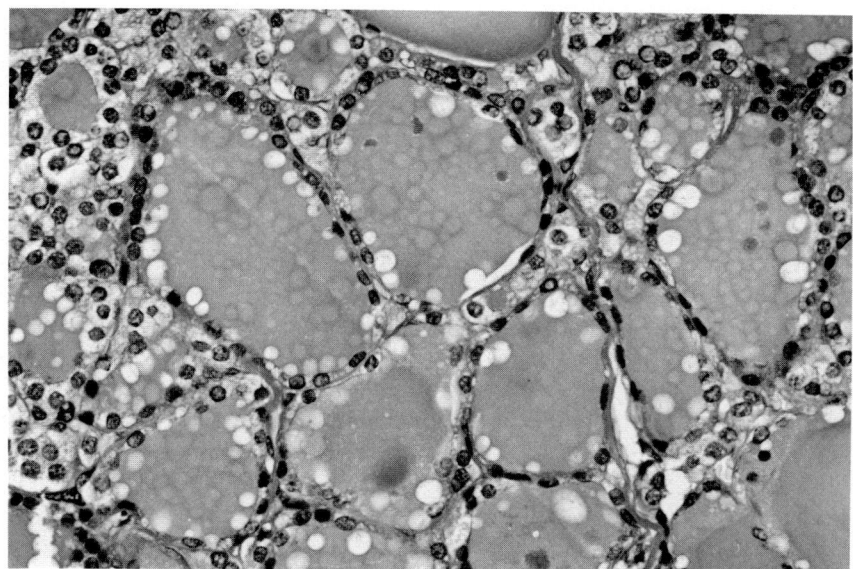

Figure 1.1 Parenchyma of normal adult thyroid gland. There are colloid-filled follicles with marginal vacuoles and lined by cuboidal follicular cells (H&E, × 300).

aggregates of 20–40 follicles bound together by a sheath of connective tissue and supplied by an end artery. A rich capillary network and a venous and lymphatic plexus surround each follicle. Many capillaries lie directly adjacent to the follicular walls with only a thin layer of epithelial cytoplasm separating the endothelium from the colloid. Follicular cells vary in size, usually being low cuboidal (height equal to width) or even flat. Columnar cells are rare. In a given follicle, the cells are similar in size.

Depending on age and functional status, the microscopic picture of the gland is very variable. In the adult gland, neoformation of follicles may occur physiologically by budding off from pre-existing follicles as well as by fusion of smaller follicles (Sugiyama, 1967). Aggregates of small follicles bulging into the lumen of larger follicles, sometimes refered to as Sanderson's polsters, are frequently seen in glands with increased hormonal activity. During various disease processes in the gland the structure may be affected by proliferative and regressive alterations.

Ultrastructural studies (Heimann, 1966; Klinck *et al.*, 1970) show that the fine structure of human thyroid tissue is essentially the same as in other mammals, although in humans, there is a greater degree of variation between different areas in a single gland and also between individuals. The basement membrane runs along the bases of the follicular cells, surrounding the entire follicle, but does not usually extend in between the lateral aspects of the cells. The microvilli, protruding from the apical surfaces of the cells into the follicular lumen vary in size, shape and number. This variability is partly related to the shape of the cell, villi of flat cells tending to be short, stubby and few and those of cylindrical cells more numerous, long and slender, even finger-like, although they seldom measure more than $0.2 \mu m$ in length.

In the resting cell the nucleus is usually located centrally. It is round or oval in shape. The nucleolus is usually eccentric and not very prominent. The mitochondria are few and evenly distributed in the whole cytoplasm. The amount of endoplasmic reticulum varies from cell to cell. In flat cells it forms only a few small, elongated cisternae whereas in cuboidal and cylindrical cells it is conspicuous and widely distributed. It is mainly of the granular type. In the basal regions of such cells the cisternae are large, elongated or irregularly formed, whereas in the apical parts they are much smaller, rounded or oval. Cisternae tend to lie close to the mitochondria and may almost completely encircle them. Occasional membrane-bound vesicles, including exocytotic vesicles, colloid droplets and other endocytotic structures, and lysosomes involved in synthesis, transport and degradation of hormonal products are also present.

By immunoelectron microscopy, thyroglobulin immunoreactivity has been localized on rough endocytoplastic reticulum, Golgi apparatus, exocytotic vesicles, luminal colloid, colloid droplets and lysosomes, whereas labelling for the hormones, T_3 and T_4, has been located on luminal colloid, colloid droplets, and lysosomes (Ring and Johanson, 1987). These findings are in keeping with the current model of thyroglobulin being synthesized in the rough endoplasmic reticulum, transported to the Golgi apparatus and then discharged into the follicle lumen via exocytotic vesicles. Thyroglobulin is iodinated in the follicular lumen, at the apical plasma membrane. This process, catalysed by thyroperoxidase present in the plasma membrane, results in iodine binding to tyrosyl residues in thyroglobulin. Monoiodotyrosines (MIT) and diiodotyrosines (DIT) are formed, some of which are coupled, yielding the thyroid hormones triodothyronine (T_3) and tetraiodothyronine (T_4). When thyroid hormones are secreted, hormone-containing thyroglobulin is brought into the follicle cell by endocytosis. Endocytotic vesicles fuse with lysosomes, thyroglobulin is hydrolysed and the liberated hormones are released into the blood and lymph vessels (for review see Ericson, 1983). However, the mechanism by which the release from the cell into the circulation takes place is unknown.

During long-standing stimulation to increased activity, as seen in toxic goitres, the follicles become smaller, the amount of colloid diminishes, the follicular cells become tall columnar and the nuclei enlarge and approach the bases of the cells. Ribosomes and endoplasmic reticulum increase and Golgi complexes hypertrophy. The microvilli increase in number and length. Colloid-containing lumina with numerous microvilli occur within the follicular cells. Intracytoplasmic vesicles, colloid droplets and dense bodies as well as mitochondria are increased in number. In the inactive state, the follicles enlarge and accumulate more colloid, the follicular cells become flat or low cuboidal and the number and size of the cytoplasmic organelles are reduced (Heimann, 1966).

The follicular colloid is pale staining and vacuoles are often observed at its margin giving it scalloped borders (Fig. 1.1). Such vacuoles are not found at the ultrastructural level. Immunohistochemically, these vacuoles are frequently found to contain thyroglobulin-immunoreactive material (Fig. 1.2). In follicles with active secretory function the colloid is densely eosinophilic. In old individuals, it tends to be broken up into globular formations and may even contain calcium salt deposits which at the light microscopic level may take the form of birefringent oxalate crystals (Reid *et al.*, 1987) or basophil microcalcifications. True psammoma bodies, i.e. concentrically laminated calcospherites, have not been reported to occur normally in the thyroid gland. Colloid is variably stainable with periodic acid Schiff (PAS) and Alcian blue,

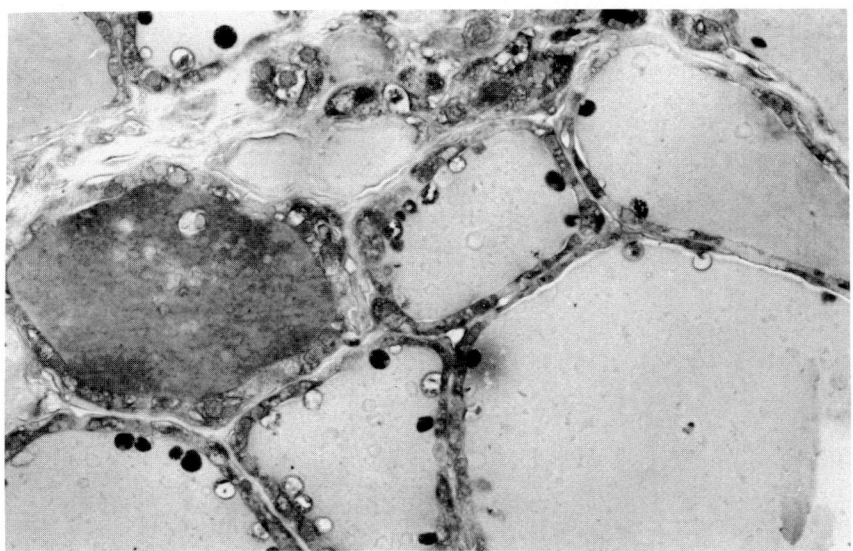

Figure 1.2 Normal thyroid parenchyma. Many marginal vacuoles contain thyroglobulin immunoreactive material (ABC technique with antibodies against human thyroglobulin, × 300).

depending on variations in the types and amounts of carbohydrate substances present (Rigaud and Bogomoletz, 1987). In addition, Harach (1985b) has described a rare, second kind of follicle, mainly composed of calcitonin-immunoreactive cells, sometimes intermingled with alcianophil mucinous cells and containing strongly Alcian blue-positive material believed to represent acid mucins. He reported follicles of this type to be most easily found in the vicinity of ultimobranchial remnants and suggested them to be derivatives of ultimobranchial tissue.

In various pathological neoplastic and non-neoplastic conditions of the thyroid gland, and rarely also in otherwise normal thyroid tissue, altered follicular cells can be found which have a characteristic granular eosinophil cytoplasm (for review, see Roediger, 1975). This type of cell was originally described by Askanazy in 1898. It is a polygonal cell, larger than ordinary follicular and parafollicular cells, whose cytoplasm is filled with strongly eosinophilic granules found to represent mitochondria on the ultrastructural level. Its nucleus is large, sometimes binucleate, often with polyploid DNA content. In the same year that Askanazy described this peculiar cell in the thyroid, Welsh (1898) reported on cells with similar microscopic appearance in the parathyroid glands. he named them 'oxyphil cells'. In 1932 Eisenberg and Wallerstein suggested the term 'oxyphil cells' to be used also for the cells in the thyroid gland. Unfortumately, Ewing (1919) introduced the erroneous name 'Hürthle cells' for the thyroid oxyphil cells, believing that Hürthle had been the first to describe these cells. However, the cell Hürthle actually reported on in 1894 was not the oxyphil cell but the canine C-cell. Hamperl (1936, 1950, 1962) objected to the term 'Hürthle cell' and suggested 'oncocyte' as a more correct term. Nevertheless, the misnomer 'Hürthle cell' has become inseparably connected with the thyroid oxyphil cells and is still flourishing in current thyroid literature. More recently the term 'eosinophil large cell' has been proposed by Krayenbühl and Hedinger (1985), a name already suggested in 1907 by Langhans, to distinguish this cell type from another variant of follicular cells which is also eosinophilic, due to abnormal increase of rough endoplasmic reticulum, but of smaller size (Böcker et al., 1978).

Oxyphil cells are most often seen in Hashimoto's thyroiditis (see p. 55) and in certain neoplasms (see p. 190, 211 and p. 243). Likewise, they may occasionally be encountered in Graves' disease and non-specific inflammatory conditions of the gland, and, very rarely, in small numbers in nodular goitre.

The nature of the thyroid oxyphil cell is still enigmatic. In addition to its richness in mitochondria it is also considered to have a marked reduction in the amount of cytoplasmic thyroglobulin and organelles related to the synthesis and transport of this protein (Böcker *et al.*, 1978).

1.2.2 The parafollicular cells

The parafollicular cells represent a minor endocrine constituent of the thyroid gland. They have been found to be responsible for the production of the peptide hormone calcitonin (Bussolati and Pearse, 1967) which was discovered in 1962 by Copp *et al.*

Most of the earlier information about these cells was derived from the study of animal thyroid glands. Baber (1876) was the first to describe epithelial cells which differed in size and topography from follicular cells in the dog thyroid gland. They were larger than follicular cells, had a large nucleus and light, faintly granulated cytoplasm. He called them 'parenchymatous cells'. Nonidez (1931/32) developed an argyrophil staining technique for selective staining of Baber's parenchymatous cells, and because of their localization at the periphery of the follicles, he suggested they be called 'parafollicular cells'. The term 'parafollicular cells' however, has been considered inadequate, since these cells were found to have other topographical relationships to the follicles as well, and they were also demonstrated in the parathyroid glands of certain mammals. Furthermore, in avian and amphibian species, these cells were found to be chiefly incorporated within the ultimobranchial body, which is a separate structure lying outside the thyroid in these animals. The term 'C-cells', introduced by Pearse (1966) defines these cells according to their function of storing and secreting calcitonin, and has since become the term preferred by most workers in the field.

In recent years, however, cells belonging to this system have also been found to elaborate other hormonal peptides (see below). Since it cannot be excluded that cells containing such peptides may, at least in part, be different from the C-cells, constituting subpopulations of the system, the broader term 'parafollicular cells' may still be useful to designate this system as a whole.

The parafollicular cells can be selectively visualized in the thyroid glands of many animals by well established histological and histochemical techniques, developed during the late 1960s (for review, see Lietz, 1971, Today, however, these methods have been largely replaced by more specific immunohistochemical methods which define the cells according to their hormonal secretory products. Yet, the argyrophil reaction of Grimelius (de Grandi, 1970) is still used as a simple screening method for the parafollicular cells and their tumours. The parafollicular cells belong to the diffuse (neuro)endocrine cell system, or APUD cell system according to Pearse (1968), whose cellular constituents have a number of structural, functional and histochemical features in common. Like the other members of this system, parafollicular cells contain secretory granules of neuroendocrine type which are the main characteristic at the ultrastructural level. These are membrane-bound, dense-core granules, ranging in size from about 150 nm to 300 nm, which are the site of hormonal peptide storage within the cells (DeLellis *et al.*, 1978). Besides calcitonin, parafollicular cells have been found to exhibit immunoreactivity for other peptides, such as somatostatin (Van Norden *et al.*, 1977; Yamada *et al.*, 1977; Kusumoto, 1980; Kameda *et al.*, 1982), bombesin, a gastrin-releasing peptide (Kameya *et al.*, 1983), substance P (Kakudo and Vacca, 1983) and helodermin, which is a VIP/secretin-like peptide (Sundler *et al.*, 1988). In addition, these cells have been found to produce peptides closely related to calcitonin, such as the so-called calcitonin gene-related peptide (CGRP) (Sabate *et al.*, 1985) and a cleavage product of pro-calcitonin, called katacalin (Ali-Rachedi *et al.*, 1983).

Experimental studies in animals have shown that the basic function of calcitonin is the regulation of plasma calcium by a feedback mechanism. This is brought about by the inhibition of bone resorption, thus reducing the amount of circulating calcium. The calcium ion stimulates secretion of calcitonin and degranulates the C-cells (for review, see Copp, 1970). Thyroid stimulating hormone (TSH) may also control calcitonin synthesis. Laboratory animals receiving an iodine-deficient diet or propylthiouracil develop high serum calcitonin levels, suggesting that TSH stimulates both follicular and parafollicular cells (Clark *et al.*, 1978; Peng *et al.*, 1978). However, in man the physiological role of calcitonin is still doubtful. As yet, no disease process in man is known which can be ascribed to excessive or insufficient production of this hormone. The physiological significance of the other hormonal peptides found in the parafollicular cells is likewise largely unknown, although a paracrine role has been suggested for the thyroid somatostatin, in inhibiting T_3 and/or T_4 production and release from thyroid follicular cells (Dhillon *et al.*, 1982).

Another major feature of the parafollicular cells, which they share with the other members of the diffuse (neuro)endocrine system, has been known as the APUD-mechanism (Pearse, 1969), that is, their ability to take up precursors of certain monoamines, such as DOPA, and 5-hydroxytryptophan, to decarboxylate these precursors, and to store and release the amines thus formed. In some species, including sheep, goat and horse, such amines are spontaneously formed and stored in amounts that can be demonstrated histochemically (Falck *et al.*, 1964; Solcia and Sampietro, 1968). Monoamines can not be demonstrated in normal human parafollicular cells, although they have been found in medullary carcinoma (Falck *et al.*, 1968). England *et al.*

(1972), however, studying surgically removed goitrous glands, showed that after the glands had been perfused *ex vivo* with L-DOPA, the parafollicular cells contained dopamine, as demonstrated with the formaldehyde-induced fluorescence method. The monoamine mechanisms in parafollicular cells have received little attention in recent years and their physiological role is still unknown. Melander *et al.* (1971) and Atack *et al.* (1972) demonstrated in animal experiments that calcium infusion and vitamin D_2 treatment cause the release of both calcitonin and amine stores through a degranulation of the C-cells.

Parafollicular cells of the human thyroid gland, being much rarer than in other mammals, were not unequivocally identified until the late 1960s. Sugiyama (1967), in a study of 326 human thyroid glands, could detect these cells only in two cases. Teitelbaum *et al.* (1971) found areas with parafollicular cells when more than 200 random sections from one thyroid lobe were examined histologically. In sections stained with standard techniques, the parafollicular cells are difficult or impossible to distinguish from follicular cells without comparison with specially stained sections. But with the latter, these cells usually appear larger and paler than follicular cells, they are polygonal or spindle-shaped, have a light, faintly granular or even foamy cytoplasm and a large, sometimes eccentric, faintly granular nucleus with a small, distinct nucleolus (Fig. 1.3).

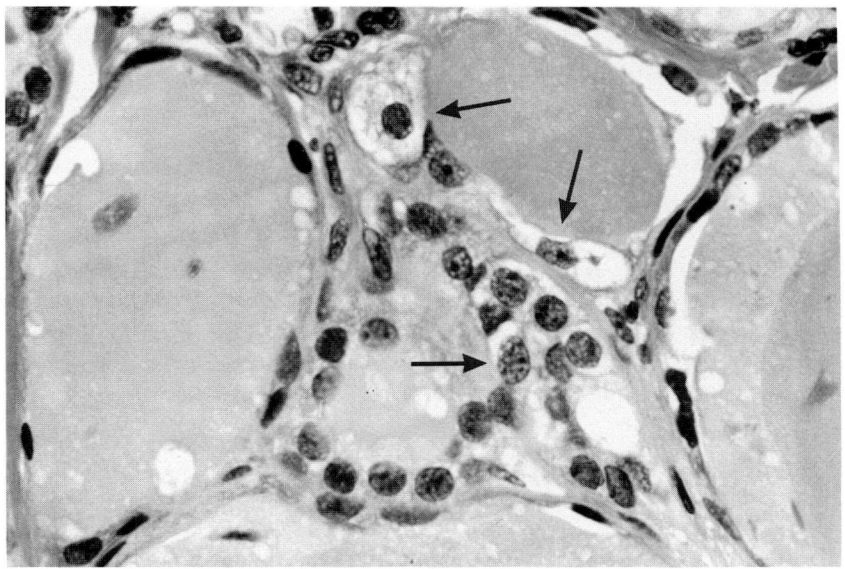

Figure 1.3 Normal thyroid parenchyma with scattered parafollicular cells (arrows). They have abundant light and granulated, sometimes foamy cytoplasm and large nuclei with a loose, granular chromatin pattern and distinct nucleoli (H&E, × 650).

The parafollicular cells, as defined by their argyrophilia or immuno-reactivity with antibodies against calcitonin, are mainly confined to a restricted zone within axial parts of the middle or upper third of each lateral lobe (Wolfe *et al.*, 1974), whereas few, if any, such cells are found in the inferior poles, the isthmus region or pyramidal lobe. Morphometric analyses have revealed a significant increase in the density of these cells in the vicinity of the so-called solid cell nests, regarded as ultimobranchial epithelial remnants (Janzer *et al.*, 1979; see below). Characteristic for the parafollicular cells is their topographical arrangement within the thyroid follicle and surround-ing stroma (Fig. 1.4). The intrafollicular position is the most typical in many mammals, including man. This means that they are located at the periphery of the follicular wall, within its basement membrane, but do not contact the colloid from which they are separated by thyreocytes or parts thereof. Histologically, human parafollicular cells may also appear as if they were situated in the stroma between follicles, but this has never been shown by electron microscopy (Teitelbaum *et al.*, 1971). Serial sections of such interfollicular groups of cells have indicated that at least some represent tangential cuts of parafollicular cell containing follicles (DeLellis *et al.*, 1977).

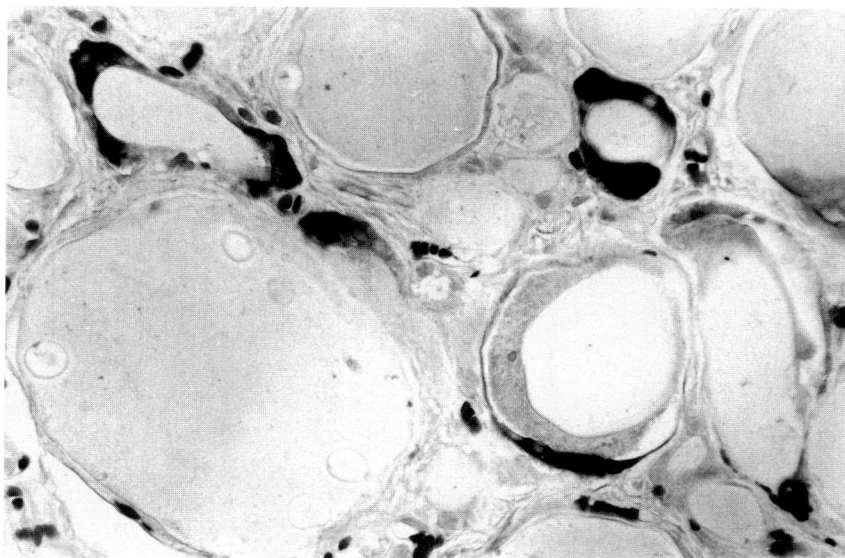

Figure 1.4 Thyroid parafollicular (C) cells demonstrated by immunoperoxidase staining for calcitonin. They are lying within the follicular wall, as single elements or in groups of a few cells, usually close to the basement membrane and have no contact with follicular lumen (PAP technique with antibodies against human calcitonin × 300).

A second type of parafollicular cell has also been described, identified by argentaffin rather than argyrophil cytoplasmic granulation and a spider-like configuration with long cytoplasmic processes and occurring in non-tumorous thyroid tissue, as well as in tumour tissue in glands containing medullary carcinoma (Ljungberg, 1970, 1972).

Studies on thyroid glands from autopsy cases have indicated that the number of parafollicular cells tends to be related to age, being more numerous in the neonate and in the child (up to 75 cells per low power field) than in adults (maximum 10 cells per low power field) (Wolfe et al. 1975). O'Toole et al. (1985), in an autopsy study comprising 60 patients ranging from 16 to 89 years, found that the number of calcitonin immunoreactive cells was essentially similar for the young and middle-aged groups (approximately one cell per mm^2). In the elderly group however, the number of such cells was very variable between individuals (from 0.1 to 12.5 cells per mm^2). An increased number of cells with age was suggested, but the values obtained were not statistically significant. In cases where the cells were most numerous large clusters of up to 20 or more cells were observed. Similar findings were reported by Gibson et al. (1982). Komorowski and Hanson (1988) found such hyperplastic C-cell clusters in 31 of 138 autopsy cases, aged 20–40 years and with no known clinical or laboratory evidence of thyroid disease. Interestingly, 25 of the 31 cases were males.

There seems to be a great variation in parafollicular cell density within a given gland. The cells are most numerous in the vicinity of the ultimobranchial remnants (see Fig. 1.10) where they may even develop into hyperplastic nodules. The lack of disturbances in calcium metabolism in cases with such C-cell nodules, as well as absence of family histories of medullary carcinoma in these cases, suggest that these hyperplastic changes are not premalignant in nature, but rather a normal phenomenon (Janzer et al., 1979).

1.2.3 The solid cell nests (ultimobranchial remnants)

Single or multiple, benign-looking formations of solid cell nests (SCN), sometimes also including small cysts or tubules, are occasionally encountered as incidental findings in surgical specimens of the thyroid gland (Figs. 1.5, 1.6 and 1.7). Janzer et al. (1979) have given a thorough historical review of the literature dealing with the problem of SCN in the thyroid. Solid cell clusters were first described in the human gland by Wölfer in 1880. They were long regarded as embryonic remnants of immature thyroid follicles or cell clusters budding out from follicular walls during neoformation of follicles. Erdheim (1904) and Getzowa (1907, 1911) however, also describing solid and

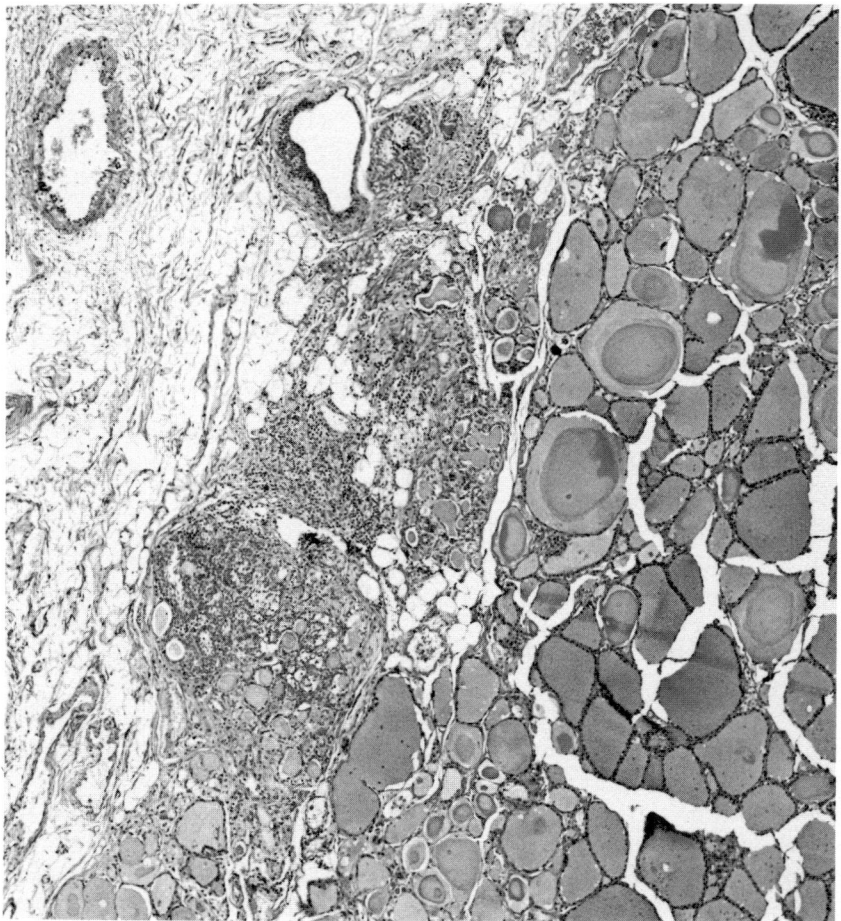

Figure 1.5 Ultimobranchial remnant tissue found incidentally in a thyroid lobe removed because of a follicular adenoma. The remnant is located in the border zone between the normal thyroid parenchyma and the extrathyroidal supporting tissue (H&E, × 50).

occasional cystic clusters of epithelial cells, regarded their cell nests as being of branchiogenic or ultimobranchial origin because both authors had found similar epithelial structures in cases of thyroid gland aplasia. Squamous differentiation, even with frank keratinization, has also been noticed in these epithelial nests. This was originally described by Kloeppel (1910). The nature of these structures in man has since been the subject of much controversy. They have been interpreted as nests of squamous metaplasia of follicular cells (for review see LiVolsi

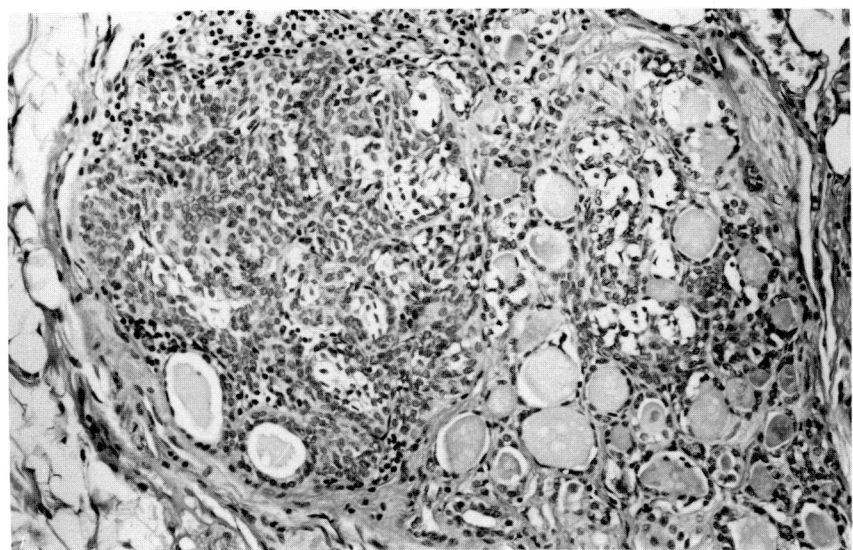

Figure 1.6 Detail from the remnant illustrated in Fig. 1.5, showing a solid epithelial cell nest (on the left), and a cluster of small immature-looking thyroid follicles with colloid (on the right). To the lower left are two small ultimobranchial tubules partly lined by a double-layered epithelium (H&E, ×165).

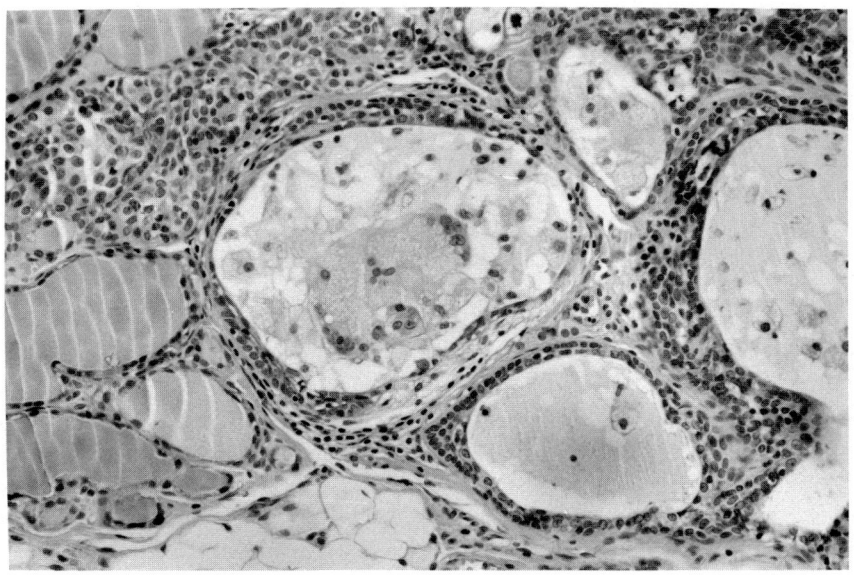

Figure 1.7 A group of small ultimobranchial cysts lined by flattened epithelial cells in several layers and containing detritus and degenerated cells. To the upper left is a small solid cell nest. Same case as in Fig. 1.5 (H&E, × 165).

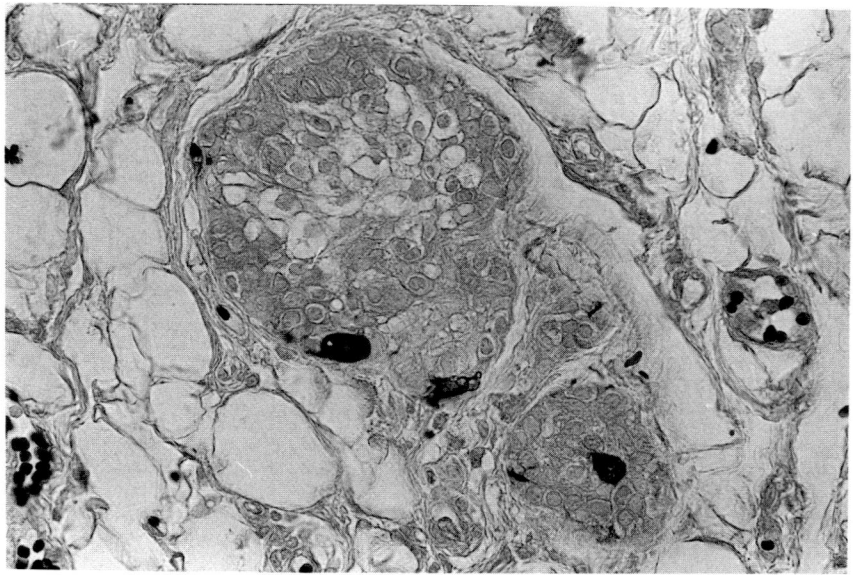

Figure 1.8 Ultimobranchial cell nests with occasional calcitonin immunoreactive cells (PAP technique with antibodies against human calcitonin, × 340)

and Merino, 1978), which might be the origin of epidermoid carcinoma of the thyroid; Fukunaga and Lockett (1971) and Autelitano *et al.* (1987) considered them to be composed of parafollicular cells and possibly to represent latent medullary carcinoma; and Gibson *et al.* (1981) thought them to be nests of parafollicular cell hyperplasia.

Most modern authors, however, have come to the conclusion that these epithelial nests are probably embryonic remnants of ultimobranchial derivation. Some of the nests, but not all of them, have been found to contain occasional calcitonin-immunoreactive cells (Fig. 1.8) (Nadig *et al.*, 1978; Janzer *et al.*, 1979; Harach 1985a; Autelitano *et al.*, 1987). Occasionally, small immature follicles with colloid can be seen within or adjacent to the solid nests (Fig. 1.6). Thyroglobulin-immunoreactive cells have also been reported in these nests, although such cells are usually absent. The majority of the cells within the nests show no hormonal immunoreactivity, but some nests may contain material reactive with antibodies against carcinoembryonic antigen (CEA), cytokeratin, and/or epidermal keratin (Fig. 1.9) (Harach, 1985a; Autelitano *et al.*, 1987; Cameselle *et al.*, 1989). Ultrastructurally, human solid cell nests contain cells with secretory granules similar to those in thyroidal parafollicular cells, as well as cells characterized by

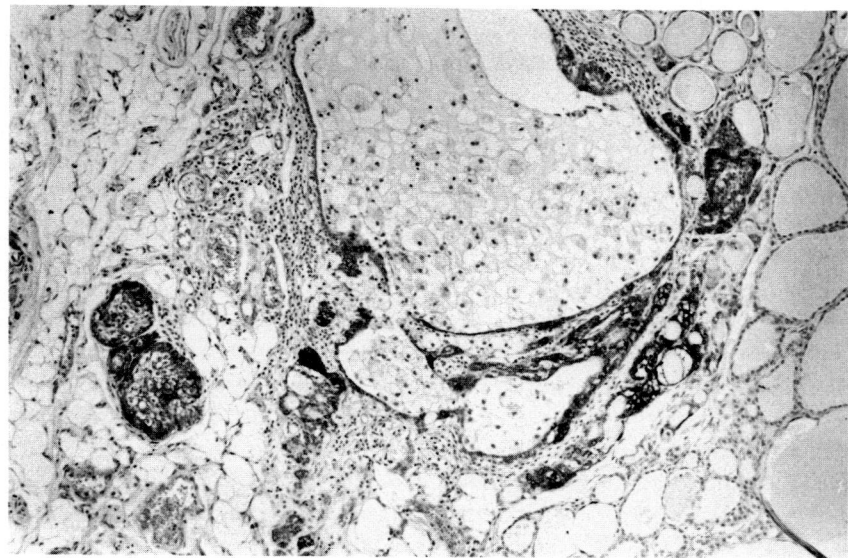

Figure 1.9 Ultimobranchial remnant showing immunoreactivity for epidermal keratin, present in some of the cells lining the cysts and tubules as well as in cells of the solid nests (PAP technique, × 85).

abundant tonofibrils (Yamaoka, 1973). The light microscopic and ultrastructural features of the human solid cell nests are similar to those found in ultimobranchial bodies described in the rat (Calvert and Isler, 1970; Nève and Wollman, 1971) and those found in hyperplastic and neoplastic lesions of ultimobranchial remnants in bulls (Jubb and McEntee, 1959; Black et al., 1973; Ljungberg and Nilsson, 1985). Interestingly, the same epithelial arrangement may be seen in the so-called intermediate type of thyroid carcinoma, which is believed to represent an ultimobranchial derivative and the human counterpart to the bull tumours (see p. 238).

In systematic studies on autopsy material solid cell nests have been found at a prevalence rate of 7–61%, this variability probably being largely dependent on the number of histological sections studied. Their prevalence is higher in males than in females; the significance of this difference is unknown (Yamaoka, 1973; Janzer et al., 1979; Harach, 1986). They occur singly or as multiple structures and chiefly in the region of the central vertical axis of the lateral lobes, with a slight tendency to a dorsocranial position. This is the same area in which most parafollicular cells occur. Interestingly, parafollicular cells, as defined by their calcitonin immunoreactivity, tend to accumulate in the vicinity of these solid cell nests (Fig. 1.10), as first noticed by Janzer *et al.* (1979). These authors point out that this phenomenon raises the possibility that solid cell nests may be a probable source of postnatal proliferation of parafollicular cells.

The solid cell nests are embedded in a connective tissue stroma of varying amounts and are more or less clearly demarcated from the surrounding thyroid follicles. They most frequently occur as solid rounded nests of polygonal or elongated-oval cells with homogeneous, basophil cytoplasm and usually oval or spindle-shaped nuclei. Sometimes there is also an admixture of cells with clear or vacuolated, occasionally finely granular, cytoplasm. Some nests may contain groups of small follicles lined by low cuboidal cells, enclosing colloid-like material and appearing as immature, newly formed thyroid follicles (Fig. 1.6). In these areas the nests tend to blend diffusely with the surrounding thyroidal parenchyma. In addition to solid nests there are sometimes also varied-sized cysts (Fig. 1.7), sometimes lined by flattened epithelial cells arranged in several layers and resembling squamous cells. Other cysts are lined by cells resembling ordinary

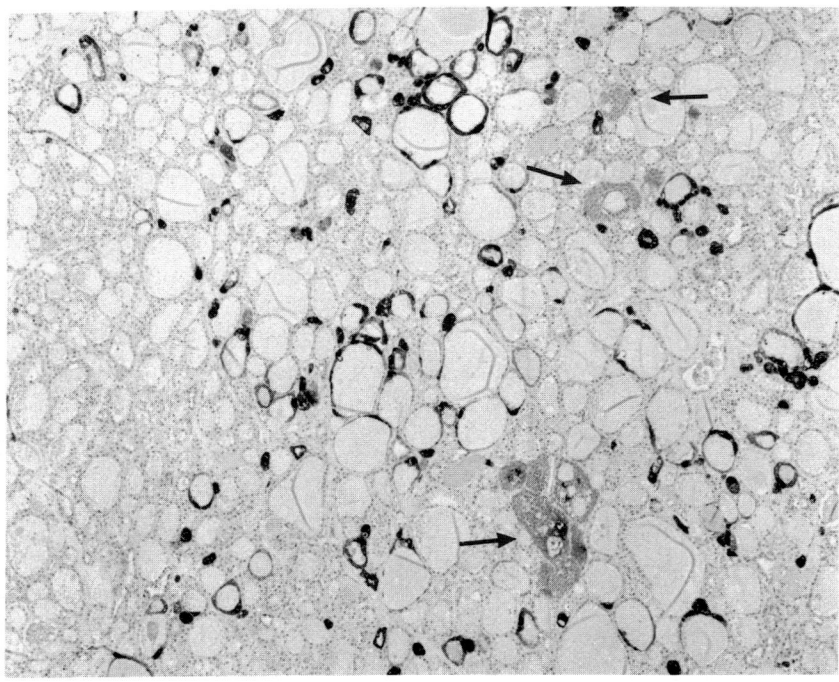

Figure 1.10 Calcitonin-immunoreactive parafollicular cells tend to concentrate in the neighbourhood of ultimobranchial remnants (arrows) (PAP technique with antibodies against human calcitonin and weak haematoxylin counterstain, × 50).

follicular cells, still others are mixed forms, lined on one side by multilayered flattened cells of squamous type, and on the other side by a single layer of cuboidal follicular cells. Occasional cysts and tubules are lined by tall, basophil cylindrical cells. All these different kinds of cysts may appear empty or contain degenerating cells or basophil granules, often floating in a background of more or less homogeneous, eosinophil material.

1.3 Development

In vertebrates, the development of the thyroid glands commences as a median entodermal downgrowth from the primitive pharynx. This median thyroid anlage forms a hollow diverticulum which descends into the anterior neck, maintaining its connection with the tongue by a narrow tube, the thyroglossal duct. The diverticulum becomes solid as it migrates caudally and gives rise to the pyramidal lobe, the isthmus and parts of the two lateral lobes of the gland, reaching its final position anterior to the trachea at about 7 weeks of gestation. The thyroglossal duct usually disappears at this stage, but remnants of the duct may persist at any level along its course and may later develop ectopic thyroid tissue and median cysts. The original opening of the thyroglossal duct is represented by a vestigial pit, the *foramen caecum* of the tongue.

In addition, branchiogenic tissue, in the form of the so-called ultimobranchial bodies, fuses with the median thyroid anlage during the early embryonic period and becomes part of the thyroid, contributing to the lateral lobes. The ultimobranchial bodies, thus forming the so-called lateral thyroid anlage, arise from the ventral aspects of the fourth (last) pharyngeal pouches. They migrate downwards, on each side of the neck together with the parathyroid IV primordia. Reaching their final position, they separate from the proximal parathyroids and fuse with the dorsolateral aspects of the median thyroid anlage and become incorporated into the developing lateral lobes (Lietz et al., 1971). The ultimobranchial bodies are considered to be the source of the parafollicular cells, which during embryonic life migrate out into the surrounding thyroid parenchyma (Chan and Conen, 1971). This was suggested in 1937 by Godwin who states that in the dog, parafollicular cells arise from the ultimobranchial body and that follicular cells have a double origin. Based on ultrastructural findings in the rat, Calvert (1972) came to the conclusion that follicles differentiating from the median thyroid primordium do not have parafollicular cells in their walls, whereas follicles carrying such cells were regarded as ultimobranchial derivatives. Kemeda et al. (1980) reached the same opinion, from studies on fetal dog thyroids. This view is supported by the finding that calcitonin-immunoreactive cells were completely absent in a human series of surgical specimens consisting of thyroglossal duct remnants and cysts with adjacent thyroid tissue and also in the thyroid tissue of a case of lingual thyroid (Ljungberg, unpublished observations). It is further supported by the fact that in man, parafollicular cells and follicles carrying such cells in their walls, are especially numerous in the vicinity of the ultimobranchial remnants in the adult gland (Fig. 1.10). The mucin-containing follicles with many C-cells, described by Harach (1985b), possibly belong to the same category.

Follicle formation and colloid production commence by 10–13 weeks of gestation. At this time, trapping and concentration of iodine and production of thyroid hormones have also been shown, and TSH has been demonstrated in the primitive pituitary gland (Fisher and Klein, 1981).

References

Alexander, C.B., Herrera, G.A., Jaffe, K. and Yu, H. (1985) Black thyroid: clinical manifestations, ultrastructural findings, and possible mechanisms. *Hum. Pathol.*, **16**, 72–8.

Ali-Rachedi, A., Varndell, I.M., Facer, P. *et al.* (1983) Immunocytochemical localization of katacalcin, a calcium-lowering hormone cleaved from the human calcitonin precursor. *J. Clin. Endocrinol. Metab.*, **57**, 680–2.

Askanazy, M. (1898) Patologisch-anatomische Beiträge zur Kentniss der Morbus Basedowii, inbesondere über die dabei auftretende Muskelerkrankung. *Dtsch. Arch. Klin. Med.*, **61**, 118–86.

Atack, C.V., Ericson, L.E. and Melander, A. (1972) Intracellular distribution of amines and calcitonin in the sheep thyroid gland. *J. Ultrastruct. Res.*, **41**, 484–92.

Autelitano, F., Santeusanio, G., Di Tonto, U. *et al.* (1987) Immunohistochemical study of solid cell nests of the thyroid gland found from an autopsy study. *Cancer*, **59**, 477–83.

Baber, E.C. (1876) Contributions to the minute anatomy of the thyroid gland of the dog. *Proc. R. Soc. Lond. [Biol]*, **24**, 240–1.

Black, H.E., Capen, C.C. and Young, D.M. (1973) Ultimobranchial thyroid neoplasms in bulls. *Cancer*, **32**, 865–78.

Böcker, W., Dralle, H., Koch, G. *et al.*(1978) Immunohistochemical and electron microscopic analysis of adenomas of the thyroid gland: II. Adenomas with specific cytological differentiation. *Virchows Arch. [A]*, **380**, 205–20.

Bussolati, G. and Pearse, A.G.E. (1967) Immunofluorescent localization of calcitonin in the 'C' cells of the pig and dog thyroid. *J. Endocrinol.*, **37**, 205–9.

Calvert, R. (1972) Electron microscopic observation on the contribution of the ultimobranchial bodies to thyroid histogenesis in the rat. *Am. J. Anat.*, **133**, 269–90.

Calvert, R. and Isler, H. (1970) Fine structure of a third epithelial component of the thyroid gland of the rat. *Anat. Rec.*, **168**, 23–41.

Cameselle, J., Sambade, C., Villaneuva, J.P. *et al.* (1989) Immunocytochemical study of solid cell nest (SCN) of the thyroid. *Pathol. Res. Pract.*, **185**, 26–7.

Chan, A.O. and Conen, P.E (1971) Ultrastructural observations on cyto-differentiation of parafollicular cells in the human fetal thyroid. *Lab. Invest.*, **25**, 249–59.

Clark, O.H., Rehfeld, S.J., Castner, B. *et al.* (1978) Iodine deficiency produces hypercalcemia and hypercalcitoninemia in rats. *Surgery*, **83**, 626–32.

Copp, D.H. (1970) Endocrine regulation of calcium metabolism. *Ann. Rev. Physiol.*, **32**, 61–86.

Copp, D.H., Cameron, E.C, Cheney, B.A. *et al.* (1962) Evidence for calcitonin – a new hormone from the parathyroid that lowers blood calcium. *Endocrinology*, **70**, 638–49.

de Grandi, P. (1970) The routine demonstration of C-cells in human and animal thyroid glands. Value of a simple silver stain. *Virchows Arch. [B]*, **6**, 137–50.

DeLellis, R.A., May, L., Tashjian A.H. Jr *et al.* (1978) C-cell granule heterogeneity in man. An ultrastructural immunocytochemical study. *Lab. Invest.*, **38**, 263–9.

DeLellis, R.A., Nunnenmacher, G. and Wolfe H.J. (1977) C-cell hyperplasia. An ultrastructural analysis. *Lab. Invest.*, **36**, 237–48.

Dhillon, A.P., Rode, J., Leathem, A. and Papadaki, L. (1982) Somatostatin: a paracrine contribution to hypothyroidism in Hashimoto's thyroiditis. *J. Clin. Pathol.*, **35**, 764–70.

Eisenberg, A.A. and Wallerstien, H. (1932) Hürthle cell tumour. *Arch. Pathol.*, **13**, 716–24.

Englund, N.E., Nilsson, G., Owman, C. *et al.* (1972) Human thyroid C cells: occurrence and amine formation studied by perfusion of surgically removed goitrous glands. *J. Clin. Endocrinol. Metab.*, **35**, 90–6.

Erdheim, J. (1904) I. Über Schilddrüsenaplasie. II. Geschwülste des Ductus Thyreoglossus. III. Über einige menschliche Kiemenderivate. *Beitr. Pathol. Anat.*, **35**, 366–433.

Ericson, L.E. (1983) Ultrastructural aspects on secretion and iodination in the thyroid gland. *J. Endocrinol. Invest.*, **6**, 311–24.

Ewing, J. (1919) *Neoplastic Diseases – A Textbook on Tumors*, 1st edn. W.B. Saunders, Philadelphia, p. 879.

Falck, B., Larsson, B., von Mecklenburg, C. *et al.* (1964) On the presence of a second specific cell system in mammalian thyroid gland. *Acta Physiol. Scand.*, **62**, 491–2.

Falck, B., Ljungberg, O. and Rosengren, E. (1968) On the occurrence of monoamines and related substances in familial medullary thyroid carcinoma with phaeochromocytoma. *Acta Pathol. Microbiol. Scand.*, **74**, 1–10.

Fisher, D.A. and Klein, A.H. (1981) Thyroid development and disorders of thyroid function in the newborn. *N. Engl. J. Med.*, **304**, 702–12.

Fukunaga, F.H. and Lockett, L.J. (1971) Thyroid carcinoma in the Japanese in Hawaii. *Arch. Pathol.*, **92**, 6–13.

Getzowa, S. (1907) Über die Glandula parathyreoidea, intrathyreoidale Zellhaufen derselben und Reste des postbranchialen Körpers. *Virchows Arch. [A]*, **188**, 181–235.

Getzowa, S. (1911) Zur Kenntnis des postbranchialen Körpers und der branchialen Kanälchen des Menschen. *Virchows Arch. [A]*, **205**, 208–57.

Gibson, W.C.H., Peng, T-Ch. and Croker, B.P. (1981) C-cell nodules in adult human thyroid. A common autopsy finding. *Am. J. Clin. Pathol.*, **75**, 347–50.

Gibson, W.C.H., Peng, T-Ch. and Croker, B.P. (1982) Age-associated C-cell hyperplasia in the human thyroid. *Am. J. Pathol.*, **106**, 388–93.

Godwin, M.C., (1937) Complex IV in the dog with special emphasis on the relation of the ultimobranchial body to the interfollicular cells in the postnatal thyroid gland. *Am. J. Anat.*, **60**, 299–339.

Hamperl, H. (1936) Über das Vorkommen von Onkozyten in verschiedenen Organen und ihren Geschwülsten. *Virchows Arch. [A]*, **298**, 327–75.

Hamperl, H. (1950) Oncocytes and the so-called Hürthle cell tumor. *Arch. Pathol.*, **49**, 563–7.

Hamperl, H. (1962) Benign and malignant oncocytoma. *Cancer*, **15**, 1019–27.

Harach, H.R. (1985a) Histological markers of solid cell nests of the thyroid. With some emphasis on their expression in thyroid ultimobranchial-related tumors. *Acta. Anat.*, **124**, 111–6.

Harach, H.R. (1985b) Thyroid follicles with acid mucins in man: a second kind of follicles? *Cell Tissue Res.*, **242**, 211–5.

Harach, H.R. (1986) Solid cell nests of the human thyroid in early stages of postnatal life. Systematic study. *Acta Anat.*, **127**, 262–4.

Heimann, P. (1966) Ultrastructure of human thyroid. A study of normal thyroid, untreated and treated diffuse goitre. *Acta Endocrinol.*, **53** (Suppl. 110), 1–102.

Hürthle, K. (1894) Beiträge zur Kenntnis der Secretionsvorgangs in der Schilddrüse. *Arch. Gesamte Physiol.*, **56**, 1–44.

Janzer, R.C., Weber, E. and Hedinger, Chr. (1979) The relation between solid cell nests and C cells of the thyroid gland. *Cell Tissue Res.*, **197**, 295–312.

Jubb, K.V. and McEntee, K. (1959) The relationship of ultimobranchial remnants and derivatives to tumor of the thyroid gland in cattle. *Cornell Vet.*, **49**, 41–69.

Kakudo, K.and Vacca, L.L. (1983) Immunohistochemical study of substance P-like immunoreactivity in human thyroid and medullary carcinoma of the thyroid. *J. Submicrosc. Cytol.*, **15**, 563–8.

Kameda, Y., Oyama, H., Endoh, M. *et al.* (1982) Somatostatin immunoreactive C cells in thyroid glands from various mammalian species. *Anat. Rec.*, **204**, 161–70.

Kameda, Y., Shigemoto, H. and Ikeda, A. (1980) Development and cyto-differentiation of C cell complex in dog fetal thyroids. *Cell Tissue Res.*, **206**, 403–15.

Kameya, T., Bessho, T., Tsumuraya, M. *et al.* (1983) Production of gastrin releasing peptide by medullary carcinoma of the thyroid. An immunohistochemical study. *Virchows Arch. [A]*, **401**, 99–108.

Klinck, G.H., Oertel, J.E. and Winship, T.H. (1970) Ultrastructure of normal human thyroid. *Lab. Invest.*, **22**, 2–22.

Kloeppel, F.C. (1910) Vergleichende Untersuchung über Gebirgsland- und Tieflandschilddrüsen. *Beitr. Pathol. Anat.*, **49**, 579–94.

Komorowski, R.A. and Hanson, G.A., (1988) Occult thyroid pathology in the young adult: an autopsy study of 138 patients without clinical thyroid disease. *Hum. Pathol.*, **19**, 689–96.

Krayenbühl, J. Ch. and Hedinger, Chr. (1985) Grosszellig-eosinophile Tumoren der Schilddrüse. *Schweiz. Med. Wochenschr.*, **115**, 512–22.

Kusumoto, Y. (1980) Calcitonin and somatostatin are localized in different cells in the canine thyroid gland. *Biomed. Res.*, **1**, 237–41.

Landas, S.K., Schelper, R.L., Tio, F.O. *et al.* (1986) Black thyroid syndrome: exaggeration of a normal process? *Am. J. Clin. Pathol.*, **85**, 411–8.

Langhans, Th. (1907) Über die epithelialen Formen der malignen Struma. *Virchows Arch. [A]*, **189**, 69–188.

Lietz, H. (1971) C-Cells: source of calcitonin. A morphological study. *Curr. Top. Pathol.*, **55**, 109–46.

Lietz, H., Wöhler, J. and Pomp, H. (1971) Zur Entwicklung und Ultrastruktur der embryonalen Schilddrüse des Menschen. *Z. Zellforsch.*, **113**, 94–110.

LiVolsi, V.A. and Merino, M.J. (1978) Squamous cells in the human thyroid gland. *Am. J. Surg. Pathol.*, **2**, 133–40.

Ljungberg, O. (1970) Two cell types in in familial medullary thyroid carcinoma: a histochemical study. *Virchows Arch. [A]*, **349**, 312–22.

Ljungberg, O. (1972) Argentaffin cells in human thyroid and parathyroid and their relationship to C cells and medullary carcinoma. *Acta Pathol. Microbiol. Scand. [A]*, **80**, 589–99.

Ljungberg, O. and Nilsson, P.-O. (1985) Hyperplastic and neoplastic changes in ultimobranchial remnants and in parafollicular (C) cells in bulls: a histologic and immunohistochemical study. *Vet. Pathol.*, **22**, 95–103.

Melander, A., Owman, Ch. and Sundler, F. (1971) Effect of vitamin D_2 stimulation on amine stores and secretory granules in the calcitonin cells of the mouse. *Histochemie*, **25**, 32–44.

Nadig, J., Weber, E. and Hedinger, Chr. (1978) C-cells in vestiges of the ultimobranchial body in human thyroid glands. *Virchows Arch. [B]*, **27**, 189–91.

Nève, P. and Wollman, S.H. (1971) Fine structure of ultimobranchial follicles in the thyroid gland of the rat. *Anat. Rec.*, **171**, 259–72.

Nonidez, J.F. (1931/1932) The origin of the 'parafollicular' cell, a second epithelial component of the thyroid gland of the dog. *Am. J. Anat.*, **49**, 479–505.

O'Toole, K., Fenoglio-Preisler, C. and Pushparaj, N. (1985) Endocrine changes associated with the human aging process: III. Effect of age on the number of calcitonin immunoreactive cells in the thyroid gland. *Hum. Pathol.*, **16**, 991–1000.

Pearse, A.G.E. (1966) The cytochemistry of the thyroid C-cells and their relationship to calcitonin. *Proc. R. Soc. Lond. [Biol]*, **164**, 478–87.

Pearse, A.G.E. (1968) Common cytochemical and ultrastructural characteristics of cells producing polypeptide hormones (the APUD series) and their relevance to thyroid and ultimobranchial C-cells and calcitonin. *Proc. R. Soc. Lond. [Biol]*, **170**, 71–80.

Pearse, A.G.E. (1969) The cytochemistry and ultrastructure of polypeptide hormone-producing cells of the APUD series and the embryologic, physiologic and pathologic implications of the concept. *J. Histochem. Cytochem.*, **17**, 303–13.

Peng, T.C., Cooper, C.W., Garner, S.C. *et al.* (1978) Hypercalcitonism and C-cell hyperplasia in rats with goiters produced by a low iodine diet or propylthiouracil. *J. Pharmacol. Exp. Ther.*, **206**, 710–7.

Reid, J.D. (1983) The black thyroid associated with minocycline therapy. *Am. J. Clin. Pathol.*, **79**, 738–46.

Reid, J.D. , Choi, C-H. and Oldroyd, N.O. (1987) Calcium oxalate crystals in the thyroid. Their identification, prevalence, origin, and possible significance. *Am. J. Clin. Pathol.*, **87**, 443–54.

Rigaud, C. and Bogomoletz, W.V. (1987) 'Mucin secreting' and 'mucinous' primary thyroid carcinomas. Pitfalls in mucin histochemistry applied to thyroid tumors. *J. Clin. Pathol.*, **40**, 890–5.

Ring, P. and Johanson, V. (1987) Immunoelectron microscopic demonstration of thyroglobulin and thyroid hormones in the rat thyroid gland. *J. Histochem. Cytochem.*, **35**, 1095–104.

Roediger, W.E.W. (1975) The oxyphil and C-cells of the human thyroid gland. A cytochemical and histopathological review. *Cancer*, **36**, 1758–70.

Sabate, M.I., Stolarsky, L.S., Polak, J.M. *et al.* (1985) Regulation of neuroendocrine gene expression by alternative RNA processing: colocalization of calcitonin and calcitonin gene-related peptide in thyroid C-cells. *J. Biol. Chem.*, **260**, 2589–92.

Solcia, E. and Sampietro,. R. (1968) New methods for staining secretory granules and 5-hydroxytryptamine in the thyroid C cells. In *Calcitonin. Proceedings of the Symposium on Thyrocalcitonin and the C Cells*, William Heinemann Medical Books Ltd, London, p. 127.

Sugiyama, S. (1967) Histological studies of the human thyroid gland observed from the viewpoint of its postnatal development. *Ergeb. Anat. Entwickl.-Gesch.*, **39**, 1–72.

Sundler, F., Robberecht, P., Yanaihara, N. *et al.* (1988) Is helodormin produced by medullary thyroid carcinoma cells and normal C-cells? Immunocytochemical evidence. *Regul. Pept.* **20**, 83–9.

Teitelbaum, S.L., Moore, K.E. and Shieber, W. (1971) Parafollicular cells in the normal human thyroid. *Nature*, **230**, 334–5.

Van Norden, S., Polak, J.M. and Pearse, A.G.E. (1977) Single cellular origin of somatostatin and calcitonin in the rat thyroid gland. *Histochemistry*, **53**, 243–7.

Welsh, D.A. (1898) Concerning the parathyroid glands –a critical anatomical and experimental study. *J. Anat. Physiol.*, **32**, 292–307.

Wolfe, H.J., DeLellis, R.A., Voelkel, E.F. *et al.* (1975) Distribution of calcitonin-containing cells in the normal neonatal human thyroid gland: a correlation of morphology with peptide content. *J. Clin. Endocrinol. Metab.*, **41**, 1076–81.

Wolfe, H.J., Voelkel, E.F. and Tashjian, A.H. Jr (1974) Distribution of CT-containing cells in the normal adult human thyroid gland: a correlation of morphology with peptide content. *J. Clin. Endocrinol. Metab.*, **38**, 688–94.

Wölfer (1880) Über die Entwicklung und den Bau der Shilddrüse mit Rücksicht auf die Entwicklung der Kröpfe. Berlin (cited by Janzer *et al.* 1979).

Yamada, Y., Ito, S., Matsubara, Y. and Kobayashi, S. (1977) Immunohistochemical demonstration of somatostatin-containing cells in the human, dog and rat thyroid. *Tohoku J. Exp. Med.*, **122**, 87–92.

Yamaoka, Y. (1973) Solid cell nest (SCN) of the human thyroid gland. *Acta Pathol. Jpn.*, **23**, 493–506.

2 Pathological examination

2.1 Handling and analysis of specimens

Surgical specimens from the thyroid gland submitted to the pathological department include the entire organ (total thyroidectomy), the major part of the gland leaving minor posterior parts of both or either lobe(s) (subtotal thyroidectomy), a lobe (hemithyroidectomy or lobectomy), sometimes with the attached isthmus, and infrequently, only a portion of a lobe (lobe resection) or a small excisional biopsy. In some cases the thyroid specimen may be accompanied by cervical lymph nodes. It is advisable that the specimen be marked by the surgeon, preferably with a suture, to facilitate orientation. This is especially helpful when the gland is much deformed because it may be impossible to reconstruct its proper orientation and to identify its topographic relationships once it has been removed from its anatomical location. Sites of special interest (for example areas where extra-thyroidal tumour growth are suspected, attached lymph nodes etc.) should also be tagged. The surgical specimen should not be sliced in the operating theatre merely to satisfy an immediate curiosity. This may cause distortion of the specimen during fixation and make its orientation unnecessarily difficult for the pathologist.

If practically possible, it is preferable to have the specimens submitted unfixed as soon as possible to the pathologist. If not, the tissue should be immediately immersed in fixative, and sent to the pathological department. Neutral buffered formalin appears to be the best fixative and is also satisfactory for most immunohistochemical studies relevant to the diagnostic pathology of thyroid lesions (see below). Large specimens may need to be sliced to facilitate penetration of the fixative. This must be done with great care in order to secure proper orientation and sampling for the histological examination. When the thyroid specimens are very friable, it is often better to fix them before sectioning (Schmidt, 1983).

The thyroid specimen should be weighed and measured, if possible in the unfixed state. The external surface should be carefully inspected and attached structures, such as lymph nodes, parathyroid glands and muscle should be noted. Then the specimen may be cut, and the cut surfaces carefully inspected. In cases of tumours and nodular goitre, the first cut should be made through the largest lesion and if possible should also include surrounding normal parenchyma and its relationship to the thyroid capsule. If possible, the following sections should be made in a manner that allows the largest possible surfaces to be exposed for inspection and tumour areas to be related to the normal surrounding parenchyma. All lesions should be recorded and measured. The two largest diameters, perpendicular to each other, should be recorded. Involvement of the thyroid capsule or extra-thyroidal infiltration and its location, when present, should also be identified and recorded.

Blocks for histological examination should be taken from the various lesions and separately numbered. Macroscopically normal thyroid tissue should also be carefully cut into slices, no more than 2 mm in thickness, and examined for minute lesions. Discrete white spots found on the cut surfaces will not infrequently turn out to be papillary macrocarcinomas. At least one block including an adequate sample of the macroscopically normal parenchyma from each lobe should always be taken. If a primary lesion cannot be found when known metastases are present, it is necessary to embed all sections for histological examination (Kennedy, 1977). In tumour cases all resection lines, when present, must be included in the histological sections. All lymph nodes as well as parathyroid glands and muscle which may be attached should of course also be taken for histological examination.

In cases of thyrotoxicosis where no focal lesions are found, one or two blocks from each lobe, and one from the isthmus may be satisfactory.

Encapsulated tumours are usually follicular in type. In such cases the diagnosis of malignancy is wholly dependent on the demonstration of invasion of blood vessels and/or infiltration into the surrounding parenchyma. Such invasion, when present, takes place in the capsular region. It may be merely microscopic and may occur at a single site only. Therefore, adequate sampling from these parts of the lesion is essential (Lang et al., 1980; Franssila et al., 1985). Recommendations of how many blocks should be taken vary. Obviously, the number is dependent on the size of the lesion, but most tumours in this category are small enough for the entire capsular area to be included in less than four to six blocks. In the case of larger tumours up to six blocks should be taken initially from the capsular area. Additional blocks are required if the initial histological examination indicates the possibility of malignancy, such as thick hyalinized capsule, marked cellularity, atypical structure or features that might indicate invasive growth into the capsule and/or incipient invaginations into blood vessels. In large tumours, a few separate blocks from the central parts of the tumour may be needed to demonstrate any special structural or cytological features which may be of interest.

In papillary carcinoma, which is often an unencapsulated tumour with irregular outline, vascular invasion is not a distinctive feature. Instead, this tumour tends to invade lymph vessels and the interstitial tissue of the surrounding parenchyma, and may grow extensively into other parts of the gland. It commonly appears as multiple growths. In these instances it is, therefore, important to take blocks not only from grossly evident tumour areas but also to sample macroscopically normal-looking parts of the thyroid gland. From a prognostic point of view it is essential to determine the size of the tumour and especially to determine whether or not the tumour has extended beyond the thyroid capsule. Blocks taken for histological examination should cover adequately the outlines of the gland and especially account for those areas where there may be macroscopic suspicion of penetration of the glandular capsule.

In cases of medullary carcinoma, it is usually necessary to search for evidence of hyperplasia of the parafollicular cell system in the non-neoplastic parts of the parenchyma, suggesting a genetic basis of the lesion (Ljungberg, 1972; Wade and Williams, 1985). In these instances it is important to examine the whole gland. Since the parafollicular cells are mainly concentrated in the region of the perpendicular, central axis of the lobe and chiefly in its middle and upper thirds, it is preferable to cut each lobe into transverse slices (about 2 mm in thickness) from above downwards, and to label each block in order (Plate 1). The same procedure applies for known familial cases of medullary carcinoma. To-day, familial medullary carcinoma is usually diagnosed at a very early stage by means of family screening programmes using provocative tests for calcitonin secretion from the tumours. The tumours present in these glands are typically multiple and bilateral and always associated with diffuse and/or multifocal hyperplasia of the parafollicular cell system. The tumours may be very small, sometimes hardly visible to the naked eye.

2.2 Histological and histochemical methods

2.2.1 Histological stains

Haematoxylin and eosin is usually sufficient for the histological examination of submitted sections. A stain for elastic tissue (e.g. orcein), or a connective tissue stain such as van Gieson–Hansen or Mallory's trichrome, may be valuable for visualizing vascular and/or capsular tumour invasion.

Amyloid deposits in medullary carcinoma are most easily demonstrated with the alkaline Congo red method (Puchtler *et al.*, 1962). Congo red is a fluorochrome which fluoresces with a bright red–orange colour when excited with light at 365 nm (Puchtler and Sweat, 1965). This property is especially valuable for the identification of very small amounts of amyloid, such as the minute intracellular clumps sometimes present in medullary carcinoma cells, which are difficult to demonstrate by conventional microscopy (Ljungberg, unpublished observation). The same sections used for conventional microscopy can be examined in the fluorescence microscope, equipped for rhodamine fluorescence.

The silver impregnation technique, such as the Grimelius' procedure (Grimelius, 1968), has long been used as a valuable method for identifying the parafollicular cells and medullary carcinoma, reflecting neuroendocrine secretory granules within the cell cytoplasm. Argyrophilia may be regarded as a characteristic of neuroendocrine cells in general, and thus is not specific for the parafollicular cells. The silver stains need proper fixation to give reliable results. Grimelius' procedure for instance, requires fixation in Bouin's solution, or alternatively, postfixation of formalin-fixed sections in this fixative. The chromogranins, which are acidic proteins originally found in the chromaffin granules of the bovine adrenal medulla (Winkler *et al.*, 1986), have recently been introduced as another group of general markers for neuroendocrine cells, including the parafollicular cells of the thyroid. The demonstration of chromogranin A by immunolocalization techniques (see below) has proved to be a very reliable and practically useful method to visualize the parafollicular cells and their tumours and requires no special fixatives (Schmid *et al.*, 1987).

2.2.2 *Immunohistochemical techniques*

Thyroid pathology has benefited considerably from the application of immunohistochemical techniques, especially in the study of thyroid tumours. These methods, demonstrating specific markers for follicular and parafollicular cells and their tumours, have been used increasingly during the last decade. They have proved to be of great diagnostic value in defining more clearly the spectrum of thyroid neoplasia, and have facilitated the distinction between tumours of different histogenesis. They have also greatly helped in establishing the nature of distant metastases from thyroid carcinomas, whose site of origin could not easily be determined merely on the basis of their structural features. Finally they have facilitated the differential diagnosis between primary thyroid tumours and metastatic thyroidal disease, such as between primary clear cell tumours and metastases of renal clear cell carcinoma.

A thorough description of the principles of immunohistochemistry as applied in diagnostic pathology is given by DeLellis and Dayal (1988). The most widely used methods include the immunoperoxidase procedures, especially the unlabelled antibody peroxidase–antiperoxidase complex (PAP) method (Sternberger, 1979) and the avidin–biotin–peroxidase complex (ABC) method (Hsu *et al.*, 1981), and more recently, the alkaline phosphatase–anti-alkaline phosphatase (APAAP) method, which has also been adapted for use with murine monoclonal antibodies (Cordell *et al.*, 1984).

All these procedures give comparable results. Alkaline phosphatase would appear to be advantageous compared to peroxidase as a label in thyroid tissue, which is rich in endogenous peroxidase activity. Such activity however, is seldom a problem when staining paraffin embedded tissue; it is chiefly destroyed during the fixation and embedding procedures. On the other hand, when cryostat sections are used, the APAAP procedure is probably the technique of choice.

The monoclonal APAAP procedure can also be used in conjunction with conventional immunoperoxidase methods (such as the PAP technique) for double immunostaining of two antigens, one of which is detected by a monoclonal mouse antibody and the other by a polyclonal antibody raised in a different species (Cordell *et al.*, 1984).

It should be pointed out that immunolocalization techniques can also be applied on fine-needle aspirates. This may be of great help especially in cases where invasive diagnostic measures are not possible and may considerably improve the planning of therapeutic strategies (Walts *et al.*, 1985; Rastad *et al.*, 1987; Davila *et al.*, 1988).

The older immunofluorescence methods give comparable results but are less sensitive, usually require a higher concentration of the primary antibody, and do not allow permanent tissue sections. However, they are still valuable in double immunostaining in order to reveal the presence and distribution of two antigens (such as thyroglobulin and calcitonin) in the same section, especially in cases where they may occur at the same site (cf. intermediate type of thyroid carcinoma. p. 232).

Although the application of immunohistochemistry probably represents the most important advance in the field of diagnostic pathology that has been made during the last 30 years, a word of caution would be appropriate at this time. Immunohistochemical techniques are subject to a variety of pitfalls, some of which are methodological, whereas others are related to the nature of the lesion. Immunological preparations must always be examined with great care.

In order to minimize false reactions due to methodological factors it is crucial that immunohistochemical work is always accompanied by the appropriate negative and positive controls, recommended for the technique under study. It is also important to be aware of insidious immunological cross-reactions. It is necessary to have complete information about the immunogen used to raise the antibodies and which may have been bound to various large molecules in order to enhance antigenicity. This is especially true when working with thyroid specimens, since ironically, certain commercial antisera against some of the peptides present in thyroid tissues, such as somatostatin and neurotensin, may have been raised against antigens that have been coupled with thyroglobulin. Such antisera may cause serious misinterpretation of the staining, unless they are absorbed with thyroglobulin. Under these circumstances it is advisable only to use antibodies which have been raised against antigens bound to other large molecules.

The infiltrating nature of a neoplasm may cause entrapment of residual normal structures as components of the lesion. This is a common

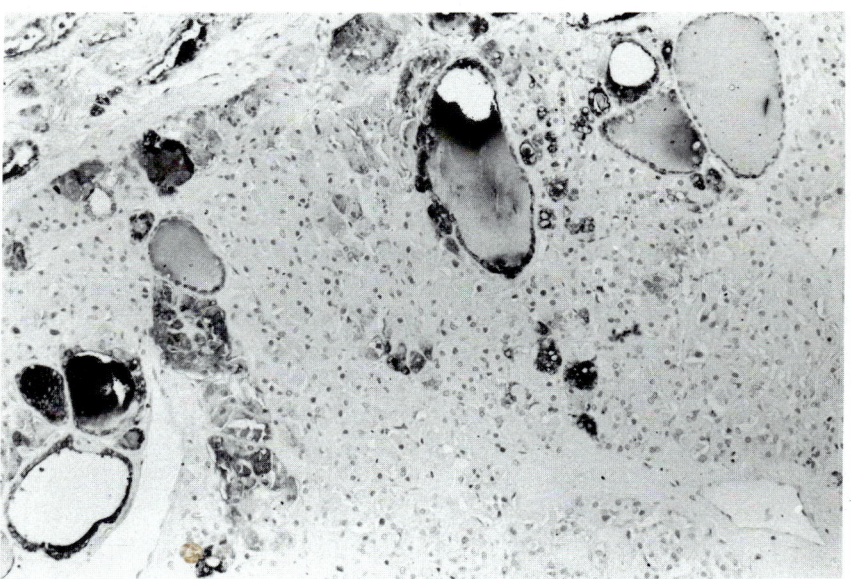

Figure 2.1 Entrapped normal follicles in the peripheral parts of a medullary carcinoma, showing positive immunostaining for thyroglobulin. Among these follicles, there are also small groups of thyroglobulin-positive cells intermingled with the tumour cells, interpreted as residual elements from disintegrated follicles (ABC technique with antibodies against human thyroglobulin and weak haematoxylin counterstain, × 100).

cause of misinterpretation when examining immunohistochemical preparations. In medullary carcinoma for instance, entrapped, normal-looking follicles are frequently seen at the periphery of the tumour showing strong thyroglobulin immunoreactivity (Fig. 2.1). Sometimes we have also encountered clustered or scattered thyroglobulin-positive cells in the tumour tissue, even in its central part, which may be extremely difficult to interpret. Unless such cells show unequivocally malignant features, we are inclined to regard them as residual normal elements.

Diffusion of the antigen or the substrate used in the staining procedure into the interstitium or into neighbouring cells, or a possible secondary uptake by such cells, is another problem. This may again be illustrated with thyroglobulin immunoreactivity in a case of metastatic renal clear cell carcinoma, where carcinoma cells around entrapped benign thyroid follicles are diffusely stained (Fig. 2.2). In this case, as has also been shown in medullary carcinomas, thyroglobulin immunoreactivity was frequently present in the plasma of blood vessels within the tumour. In some places

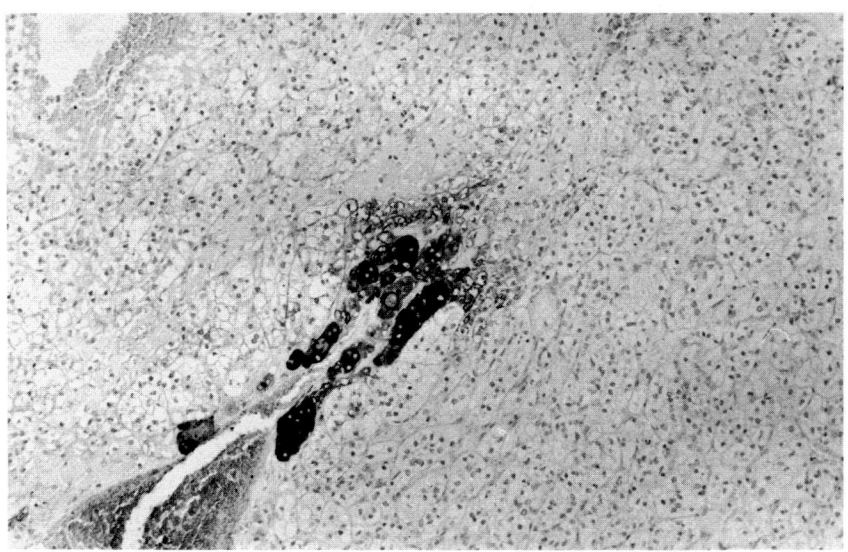

Figure 2.2 Metastasis of renal clear cell carcinoma to the thyroid gland. A strong thyroglobulin immunoreactivity is seen in normal follicles entrapped by the tumour, and there is some diffusion of the chromogen into the surrounding structures with some non-specific staining of adjacent stroma and tumour cells (ABC technique. Antibodies against human thyroglobulin and 3-amino-9-ethylcarbazole as chromogen, and weak haematoxylin counterstain, × 100).

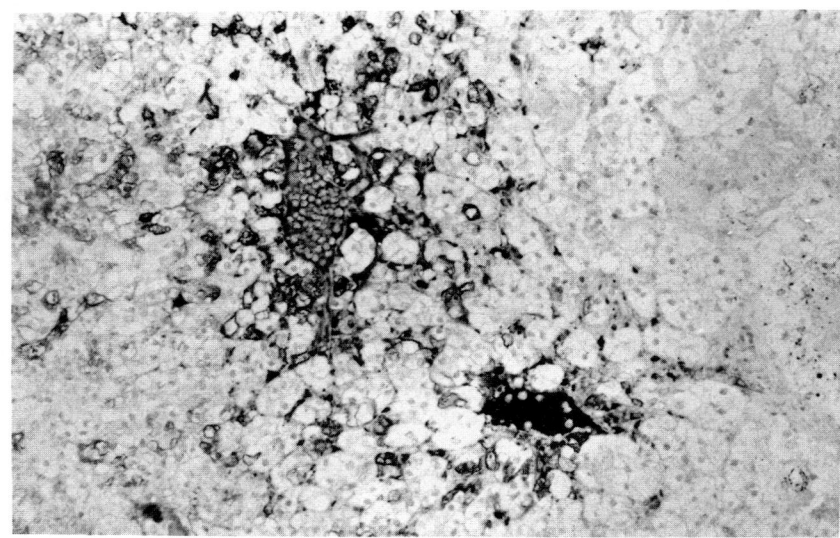

Figure 2.3 Same case as in Fig. 2.2. Strong immunoreaction for thyroglobulin is seen in the plasma of blood vessels within the metastatic tumour. There is a marked diffusion of the stain into the vessel walls and adjacent tumour stroma with non-specific staining also of some neighbouring tumour cells (same staining procedure as in Fig. 2.2, × 160).

there was a diffuse staining of the vessel wall and sometimes even of tumour cells lying next to the vessel (Fig. 2.3).

2.2.3 Immunohistochemical markers

With the above-mentioned reservations, thyroglobulin is a reliable marker for follicular cells and tumours related to these cells, i.e. follicular adenomas and carcinomas as well as papillary carcinomas (Albores-Saavedra et al., 1983). It can be demonstrated in formalin-fixed paraffin-embedded tissue, with any of the above-mentioned immunolocalization techniques. The demonstration of thyroglobulin is especially helpful in the classification of metastatic lesions according to their site of origin (Plate 2). Logmans and Jobsis (1984) studied 15 cases of metastatic thyroid carcinoma and found thyroglobulin immunoreactivity in the metastases of all cases. In 53 cases with metastases of non-thyroidal origin, however, the metastases were invariably unreactive. Likewise, thyroglobulin antibodies may also be useful in deciding whether certain thyroid neoplasms are primary lesions or metastatic, masquerading as primaries. This problem is most often encountered in cases of clear cell tumours which structurally may be difficult to distinguish from metastases of renal clear cell carcinoma (see Fig. 12.2), and in some cases of papillary carcinoma with diffuse growth pattern which may be imitated by a metastatic papillary carcinoma, especially of the lung (see also p. 284).

Immunohistochemical demonstration of ceruloplasmin and lactoferrin, which are structurally related glycoproteins, have been claimed to be of value in the differential diagnosis of thyroid tumours. These markers have both been reported to occur in follicular and papillary carcinomas as well as in oxyphil tumours, but not in follicular adenomas and medullary carcinomas (Tuccari and Barresi, 1985, 1987). However, further studies are needed to establish their role in this context.

The role of cytokeratin antibodies in thyroid pathology is not fully established. The use of various polyclonal antibodies against high-molecular-weight keratins has yielded inconsistent staining patterns and appears to have limited value in the differential diagnosis between various benign and malignant epithelial lesions of the thyroid. Schelfhout et al. (1989) reported that the monoclonal antibody LP2K (Lane et al., 1985) against the low-molecular-weight keratin Moll no. 19 (Moll et al., 1982) may be useful in distinguishing papillary carcinoma from benign and malignant follicular neoplasms. However, this method requires frozen sections which limits its value in routine pathology practice. A new monoclonal antibody, NCL-5D3 (Milab AB, Malmö) also recognizes low-molecular-weight cytokeratins, in particular Moll no. 8, but also nos. 18 and 19 (Angus et al., 1987). It may be practically useful in identifying papillary carcinomas on paraffin sections from routinely fixed tissue (see also p. 158 and Figs 9.17 and 9.18 and Plate 16). The demonstration of epithelial membrane antigen as well as cytokeratins may also be useful in distinguishing between undifferentiated thyroid carcinomas and non-epithelial tumours such as sarcoma and malignant lymphoma (Burt et al., 1985; Carcangiu et al., 1985).

Calcitonin immunolocalization is most widely used to identify the parafollicular (C) cells and their tumours (DeLellis and Wolfe, 1981; Fig. 9.64). Calcitonin immunohistochemistry is superior to previous histochemical methods such as so-called masked metachromasia, using basic dyes after acid hydrolysis, or silver techniques which give inconsistent results on routinely fixed tissues. Calcitonin can easily be visualized with any of the afore-mentioned techniques. Calcitonin immunohistochemistry plays an important role in excluding or recognizing parafollicular cell hyperplasia in association with medullary carcinoma, and may, therefore, help identifying those patients with probable familial disease (Fig 9.66 and Plate 24). Needless to say, calcitonin immunoreactivity is of great help in making the definitive diagnosis of medullary carcinoma. In fact, this diagnosis relies on the demonstration of calcitonin in tumour cells. As pointed out earlier in this chapter, chromogranin A is a very reliable general marker of neuroendocrine cells and their tumours, and consequently, can also be used to identify parafollicular cells and medullary carcinoma.

The number of neurohormonal peptides and other hormonal and non-hormonal substances which may be elaborated by medullary carcinoma is extensive. In addition to calcitonin, peptides such as the calcitonin gene-related peptide (CGRP), somatostatin, neurotensin, ACTH, substance P, human pancreatic protein, and several others can be demonstrated immunohistochemically in this tumour. Most of these, however, are not invariably present or occur only in minority cell populations. These peptides are valuable in identifying the very rare forms of atypical parafollicular cell tumours which are calcitonin-negative, and tumour forms with traits of both follicular and parafollicular cell carcinoma (so-called intermediate type), which may exhibit immunoreactivity for somatostatin and neurotensin (Fig. 9.76), but they are of limited value as diagnostic tools in common clinical practice.

Carcinoembryonic antigen (CEA) is a useful non-specific marker of medullary carcinoma (Wilson et al., 1986). Its value for distinguishing parafollicular cell hyperplasia from early stages of medullary carcinoma in cases with the familial disease, however, appears uncertain (Dasovic-Knezevic et al., 1989; Mendelsohn, 1987).

Lymphoid cell marker evaluation is of great importance in the diagnosis of cases of small cell tumours of the thyroid. Previously, such tumours were frequently misdiagnosed as small cell carcinomas, but in recent years the application of immunohistochemistry has shown that the majority of such tumours are malignant lymphomas (Burt et al., 1985). Malignant lymphoma of the thyroid usually shows immunoreactivity with antibodies against leukocyte common antigen or similar antigens (Salter et al., 1985; Fig. 10.2). Plasmacytoma can be evaluated with antibodies against human kappa and lambda light chains (Aozasa et al., 1986).

2.3 Electron microscopy

Electron microscopy has considerably broadened our understanding of the nature of thyroid tumours, especially in conjunction with the knowledge gained from immunohistochemical studies. Its value in the differential diagnosis of differentiated thyroid neoplasms and between benign and malignant tumours, however, is limited. Here light microscopy, with the adjunct of immunohistochemical techniques, is still the main tool (Johannessen *et al.*, 1978; Rosai and Carcangiu, 1984).

Ultrastructural studies together with immunohistochemical investigation may be helpful in identifying the various types of cells occurring in the rare intermediate (mixed) types of thyroid carcinoma which have been reported recently (Figs 9.77 and 9.78; Ljungberg and Nilsson, 1991). Electron microscopic investigation is also particularly important for the identification of the rare forms of parafollicular cell carcinoma which are found to be calcitonin-negative, and for the differential diagnosis of poorly differentiated medullary carcinomas. The demonstration of intracytoplasmic secretory granules at the ultrastructural level helps to reveal the neuroendocrine nature of such neoplasms. The electron microscope may be useful, as in any other organ, in distinguishing epithelial from non-epithelial lesions. In the thyroid, this is especially true for the differential diagnosis between undifferentiated spindle or giant cell carcinoma and sarcoma (Sobrinho-Simões *et al.*, 1985).

2.4 Frozen section diagnosis

It is not possible to define dogmatically the indications for frozen section. These will depend on the surgical strategy in use at a particular institution, as well as on the resources of the pathological department. The more conservative attitude in the surgical treatment of the common thyroid lesions attained in recent years (Schroder *et al.*, 1986; Starnes *et al.*, 1988) has reduced the need for frozen section diagnosis, and in some institutions, as is the case in Malmö, this is only occasionally performed today (Hamburger and Hamburger, 1986). The increased accuracy of preoperative fine-needle aspiration diagnosis has further contributed to this development. With respect to medullary carcinoma, the diagnosis has mostly been established preoperatively by fine-needle biopsy and serum calcitonin measurements, and frozen section diagnosis is then superfluous. In certain instances however, such as in cases with a suspicion of carcinomatous extrathyroidal infiltration or lymph node metastases, the pathologist may be asked to perform a frozen section of specimens taken from these sites during the exploration of a thyroid lump. In these cases, the diagnosis is usually easy for experienced pathologists, although sequestered nodules in benign nodular goitre, lying in the lateral region of the neck, and the presence of extrathyroidal hyperplastic thyroid tissue in cases of Graves' disease or Riedel's thyroiditis may be treacherous diagnostic pitfalls (Rosai and Carcangiu, 1987).

When a thyroid specimen (usually lobectomy done for an intra-thyroidal nodule) is sent in with a request for frozen section diagnosis, the differential diagnosis in the majority of cases will be between a benign lesion, such as the dominating nodule in a nodular goitre or a follicular adenoma and a carcinoma. In cases of undifferentiated or poorly differentiated carcinomas as well as in cases of widely invasive follicular carcinomas the experienced pathologist can usually make a correct diagnosis without problems. This also applies to classical medullary carcinoma.

In cases of papillary carcinoma, however, frozen section diagnosis is unreliable. The papillary structures, and especially the cytological features, including ground glass appearance of nuclei and nuclear grooving, on which the diagnosis relies, may not be evident at all. The newly defined follicular and encapsulated variant of papillary carcinoma is very difficult or impossible to identify on frozen sections. The most difficult problem is to distinguish between a follicular adenoma and a minimally invasive follicular carcinoma, and their oxyphil counterparts (Nakazawa et al., 1968; Meissner, 1977; Kraemer, 1987). In such cases the diagnosis of follicular carcinoma is seldom possible on frozen section. Here the diagnosis is entirely based on the identification of capsular and/or blood vessel invasion, which may be merely focal and may not be evident in the samples taken for frozen sections. The frozen section technique, therefore, is unreliable under these circumstances. However, this methodological deficiency is of little therapeutical significance. Regardless of whether or not the tumour eventually was found to be malignant on the basis of permanent sections, and provided the tumour was completely removed, the operation (usually a lobectomy or subtotal thyroidectomy) is generally considered adequate in these cases.

References

Albores-Saavedra, J., Nadji, M., Civantos, F. and Morales, A.R. (1983) Thyroglobulin in carcinoma of the thyroid: an immunohistochemical study. *Hum. Pathol.*, **14**, 62–6.

Angus, B., Purvis, J., Stock, D. *et al.* (1987) NCL-5D3: a new monoclonal antibody recognizing low molecular weight cytokeratins effective for immunohistochemistry using fixed paraffin-embedded tissue. *J. Pathol.*, **153**, 377–84.

Aozasa, K., Inoue, A., Yoshimura, H. *et al.* (1986) Plasmacytoma of the thyroid gland. *Cancer*, **58**, 105–10.

Burt, A.D., Kerr, D.J., Brown, I.L. and Boyle, P. (1985) Lymphoid and epithelial markers in small cell anaplastic thyroid tumours. *J. Clin. Pathol.*, **38**, 893–6.

Carcangiu, M.L., Steeper, T., Zampi, G. and Rosai, J. (1985) Anaplastic thyroid carcinoma. A study of 70 cases. *Am. J. Clin. Pathol.*, **83**, 135–58.

Cordell, L.J., Falini, B., Erber, W.N. *et al.* (1984) Immunoenzymatic labeling of monoclonal antibodies using immune complexes of alkaline phosphatase and monoclonal anti-alkaline phosphatase (APAAP complexes). *J. Histochem. Cytochem.*, **32**, 219–29.

Dasovic-Knezevic, M., Bormer, O., Holm, R. *et al.* (1989) Carcinoembryonic antigen in medullary thyroid carcinoma: an immunohistochemical study applying six novel monoclonal antibodies. *Mod. Pathol.*, **2**, 610–7.

Davila, R.M., Bedrossian, C.W.M. and Silverberg, A.B. (1988) Immuno-cytochemistry of the thyroid in surgical and cytologic specimens. *Arch. Pathol. Lab. Med.*, **112**, 51–6.

DeLellis, R.A., and Dayal, Y. (1988) Principles of immunohistochemistry as applied to the surgical pathology of neoplasms. In *Pathology of Human Neoplasms* (ed. H.A. Azar), Raven Press, New York, pp. 131–81.

DeLellis, R.A. and Wolfe, H.J. (1981) The pathobiology of the human calcitonin (C)-cell: a review. *Pathol. Annu.*, **16** (2), 25–52.

Franssila, K.O., Ackerman, L.V., Brown, C.L. and Hedinger, Chr. E. (1985) Session II: follicular carcinoma. *Semin. Diagn. Pathol.*, **2**, 101–22.

Grimelius, L. (1968) A silver nitrate stain for alpha-2-cells in human pancreatic islets. *Acta Soc. Med. Upsal.*, **73**, 243–70.

Hamburger, J.I. and Hamburger, S.W. (1986) Declining role of frozen section in surgical planning for thyroid nodules. *Surgery*, **98**, 307–12.

Hsu, S.M., Raine, L. and Fanger, H. (1981) Use of avidine–biotin–peroxidase complex (ABC) in immunoperoxidase techniques: a comparison between ABC and unlabeled antibody (PAP) procedures. *J. Histochem. Cytochem.*, **29**, 577–80.

Johannessen, J.V., Gold, V.E. and Jao, W. (1978) The fine structure of human thyroid cancer. *Hum. Pathol.*, **9**, 385–400.

Kennedy, A. (1977) *Basic Techniques in Diagnostic Histopathology*, Churchill Livingstone, Edinburgh, London and New York, p. 92.

Kraemer, B.B. (1987) Frozen section diagnosis and the thyroid. *Semin. Diagn. Pathol.*, **4**, 169–89.

Lane, E.B., Bartek, J., Purkis, P.E. and Leigh, I.M. (1985) Keratin antigens in differentiating skin. *Ann. NY Acad. Sci.*, **455**, 241–58.

Lang, W., Georgii, A., Stauch, G. and Kienzle, E. (1980) The differentiation of atypical adenomas and encapsulated follicular carcinomas in the thyroid gland. *Virchows Arch. [A]*, **385**, 125–41.

Ljungberg, O. (1972) On medullary carcinoma of the thyroid. *Acta Pathol. Microbiol. Scand. [A]*, Suppl. 231, 1–57.

Ljungberg, O. and Nilsson, P-O. (1991) Intermediate thyroid carcinoma in man and ultimobranchial tumors in bulls. A comparative morphologic and immunohistochemical study. *Endocr. Pathol.*, **2**, 24–39.

Logmans, S.C. and Jobsis, A.C. (1984) Thyroid-associated antigens in routinely embedded carcinomas. *Cancer*, **54**, 274–9.

Meissner, W.A. (1977) Frozen section diagnosis. Follicular carcinoma of the thyroid. *Am. J. Surg. Pathol.*, **1**, 171–3.

Mendelsohn, G. (1987) Diagnostic histochemistry and immunohistochemistry of the thyroid, parathyroid, and adrenal glands. In *Histochemistry in Pathologic Diagnosis* (ed. S.S. Spicer), Marcel Dekker, New York and Basel, pp. 648–59.

Moll, R., Franke, W.W., Schiller, D.L. *et al.* (1982) The catalog of human cytokeratins: patterns of expression in normal epithelia, tumors and cultured cells. *Cell*, **31**, 11–24.

Nakazawa, H., Rosen, P., Lane, N. and Lattes, R. (1968) Frozen section experience in 3000 cases. Accuracy, limitations and value in residency

training. *Am. J. Clin., Pathol.*, **49**, 41–51.
Puchtler, H. and Sweat, F. (1965) Congo red as a stain for fluorescence microscopy of amyloid. *J. Histochem. Cytochem.*, **13**, 693–4.
Puchtler, H., Sweat, F. and Levine, M. (1962) On the binding of Congo red by amyloid. *J. Histochem. Cytochem.*, **10**, 355–64.
Rastad, J., Wilander, E., Lindgren, P-G. *et al.* (1987) Cytologic diagnosis of a medullary carcinoma of the thyroid by Sevier–Munger silver staining and calcitonin immunocytochemistry. *Acta Cytol.*, **31**, 45–7.
Rosai, J. and Carcangiu, M .L. (1984) Pathology of thyroid tumors: some recent and old questions. *Hum. Pathol.*, **15**, 1008–12.
Rosai, J. and Carcangiu, J.L. (1987) Pitfalls in the diagnosis of thyroid neoplasms. *Pathol. Res. Pract.*, **182**, 169–79.
Salter, D.M., Krajewski, A.S. and Dewar, A.E. (1985) Immunohistochemical staining of non-Hodgkin's lymphoma with monoclonal antibodies specific for the leucocyte common antigen. *J. Pathol.*, **146**, 345–53.
Schelfhout, L.J.D.M., Van Muijen, G.N.P. and Fleuren, G.J. (1989) Expression of keratin 19 distinguishes papillary thyroid carcinoma from follicular carcinoma and follicular thyroid adenoma. *Am. J. Clin. Pathol.*, **92**, 654–8.
Schmid, K.W., Fischer-Colbrie, R., Hagn, C. *et al.* (1987) Chromogranin A and B and secretogranin II in medullary carcinomas of the thyroid. *Am. J. Surg. Pathol.*, **11**, 551–6.
Schmidt, W.A. (1983) *Principles and Techniques of Surgical Pathology*, Addison-Wesley, Menlo Park, CA, pp. 173–81.
Schroder, D.M., Chambors, A. and France, C.H.J. (1986) Operative strategy for thyroid cancer. Is total thyroidectomy worth the price? *Cancer*, **58**, 2320–8.
Sobrinho-Simões, M.A., Nesland, J.M., Holm, R. and Johannessen, J.V. (1985) Transmission electron microscopy and immunocytochemistry in the diagnosis of thyroid tumors. *Ultrastruct. Pathol.*, **9**, 255–75.
Starnes, H.F., Brooks, D.C., Pinkuss, G.S. *et al.* (1988) Surgery for thyroid carcinoma. *Cancer*, **55**, 1376–81.
Sternberger, L.A. (1979) *Immunocytochemistry*, 2nd edn, Wiley, New York.
Tuccari, G. and Barresi, G. (1985) Immunohistochemical demonstration of lactoferrin in follicular adenomas and thyroid carcinomas.*Virchows Arch. [A]*, **406**, 67–74.
Tuccari, G. and Barresi, G. (1987) Immunohistochemical demonstration of ceruloplasmin in follicular adenomas and thyroid carcinomas. *Histopathology*, **11**, 723–31.
Wade, J.H.S. and Williams, E.D. (1985) Surgical pathology of the thyroid gland. In *Pathology in Surgical Practice* (eds G.J. Hadfield, M. Hobsley and B.C. Morson), Edward Arnold, London, Melbourne and Baltimore, pp. 165–76.
Walts, A.E., Said, J.W., Shintaku, I.P. and Lloyd, R.V. (1985) Chromogranin as a marker of neuroendocrine cells in cytologic material – an immunocytochemical study. *Am. J. Clin. Pathol.*, **84**, 273–7.
Wilson, N.W., Pambakian, H., Richardson, T.C. *et al.* (1986) Epithelial markers in thyroid carcinoma: an immunoperoxidase study. *Histopathology*, **10**, 815–29.
Winkler, H., Apps, D.K. and Fischer-Colbrie, R. (1986) The molecular function of adrenal chromaffin granules: established facts and unresolved topics. *Neuroscience*, **18**, 261–90.

3 Developmental abnormalities

Thyroid tissue can occur remote from the normal location of the gland. Modern scintigraphic techniques have revealed such tissue frequently to occur as aberrant nests within the gland's normal embryonic sphere of distribution, i.e. in association with remnants of the thyroglossal duct, along its course from the foramen caecum of the tongue to the isthmus of the gland, and adjacent to the lateral lobes (Sauk, 1970; see also p. 19). Heterotopic thyroid tissue, located outside this sphere, is much rarer. Such locations may be the larynx or trachea, the anterior or posterior mediastinum, the heart, the breast and even the porta hepatis. The presence of thyroid tissue in the lateral regions of the neck, including the occurrence within cervical lymph nodes, and often referred to as lateral aberrant thyroid, has been a matter of much debate in the literature.

3.1 Thyroglossal duct cysts and sinuses

Remnants of the thyroglossal duct may persist anywhere along the course of the original duct, from the region of the *foramen caecum* of the tongue to the level of the postnatal thyroid gland. According to Ellis and Van Norstrand (1977), such remnants occur in about 7% of adult larynges.

   The most common anomaly is a median cyst (Allard, 1982; Dalgaard and Witteland, 1956). Thyroglossal duct cysts usually occur in children and young adults, but may be seen at all ages. They develop in the anterior part of the neck, in the midline and usually below the level of the hyoid bone. About one-fifth are located in a suprahyoid position, and some are found in the region of the foramen caecum of the tongue. The chief complaint is usually a palpable, painless mass of several years' duration. Typically, the cyst is associated with a duct remnant. A sinus tract is frequently identified in its vicinity and may sometimes be traced for various distances cranially or caudally. Multiple cysts along the course of the tract or branched thyroglossal sinuses have also been reported (Tovi and Eyal, 1985). Nests of thyroid tissue are frequently found along these ducts and within or close to

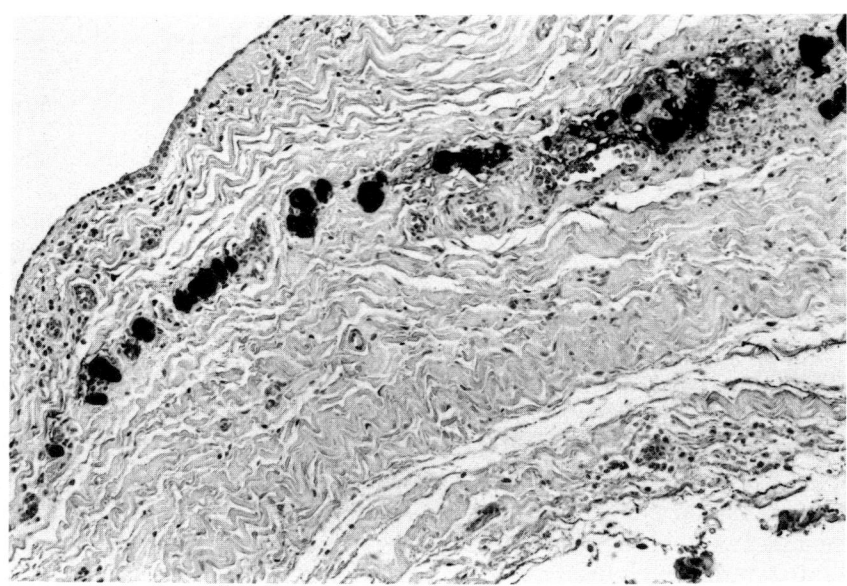

Figure 3.1 Thyroglossal duct cyst lined by flattened epithelium of squamous type. In the fibrous cyst wall, there are groups of thyroid follicles showing a strong immunoreaction for thyroglobulin (PAP technique with antibodies against human thyroglobulin and weak haematoxylin counterstain, × 130).

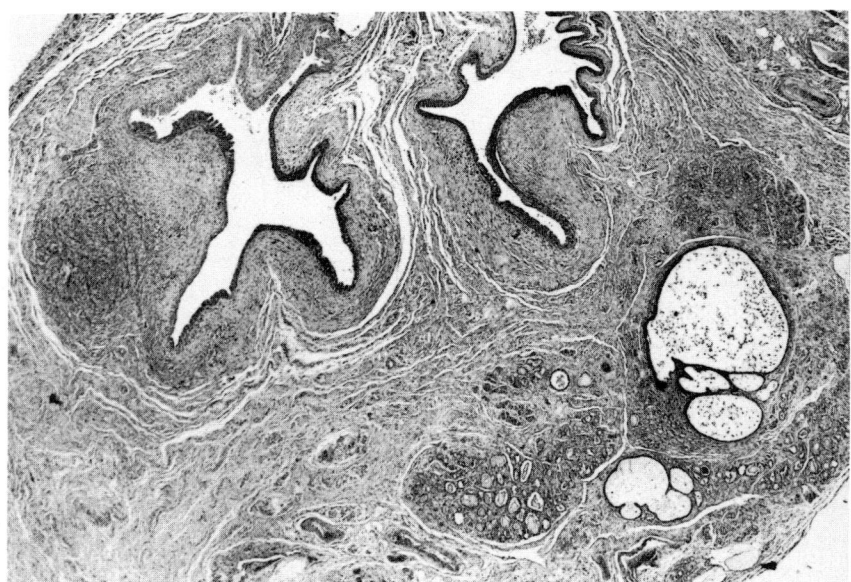

Figure 3.2 Group of thyroglossal duct sinuses with lymphoid tissue in their walls; to the lower right there are small cysts containing detritus material, and foci of thyroid tissue may be seen (H&E, × 22).

the cyst walls (Figs 3.1 and 3.2). Incidence figures reported for the presence of adjacent thyroid tissue vary from about 25 to 65%, being relative to the number of histological sections examined (LiVolsi *et al.*, 1974).

The typical cyst measures 1–4 cm, occasionally up to 10 cm, and contains a pale mucoid fluid. Usually, cysts do not communicate with the exterior. When infection occurs or there has been a previous inadequate surgical interference, fistulous openings in the skin may develop. The cysts and ducts are lined by epithelium of respiratory or squamous type, but sometimes transitional or mixed epithelial forms are also encountered (Fig. 3.3). Usually, there is a varying amount of lymphoid tissue in the cyst wall which may contain germinal centres. There are no thyroglobulin-positive follicular cells within the lining epithelium (Fig. 3.4). Non-specific inflammation and fibrosis are often also seen. The presence of associated thyroid tissue distinguishes thyroglossal duct cysts from branchial cleft cysts, the latter occurring in the anterolateral regions of the neck. It may be necessary to perform extensive histological examination in order to identify thyroid tissue remnants, which may occur merely as inconspicuous groups of small follicles. Parafollicular cells have not been found within these nests of thyroid tissue. In a study of median anlage anomalies, including

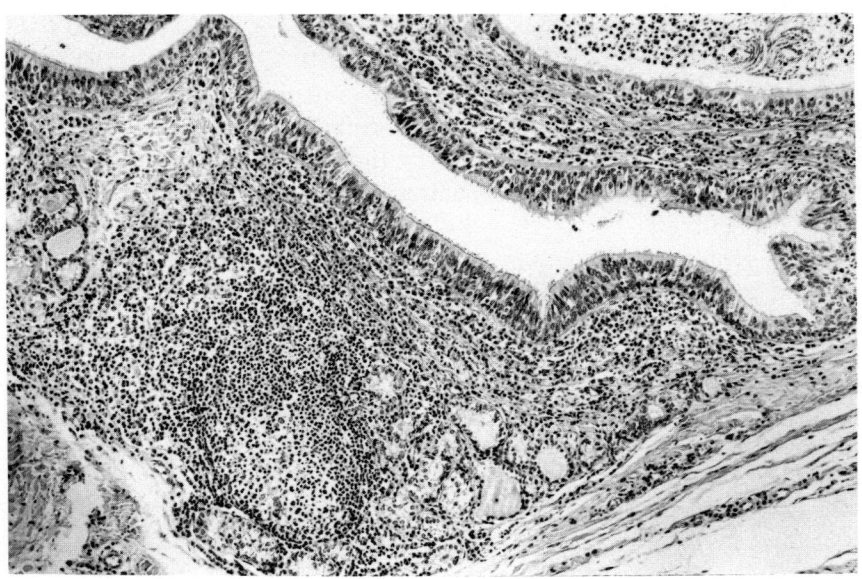

Figure 3.3 Thyroglossal duct sinus lined by epithelium of respiratory type. Lymphoid tissue with germinal centres and scattered thyroidal follicles are seen in the wall (H&E, × 85).

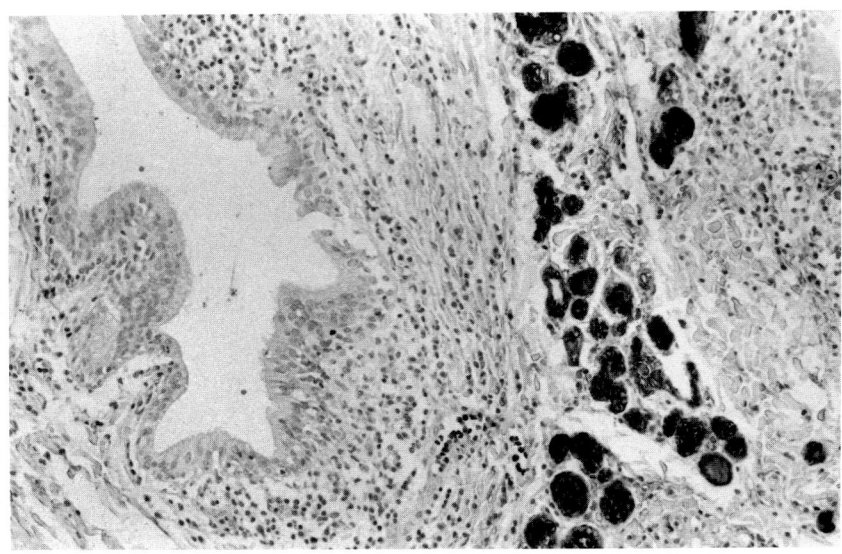

Figure 3.4 Thyroglossal duct sinus surrounded by lymphoid cells and a cluster of follicles showing a strong immunoreaction for thyroglobulin. There are no thyroglobulin-immunoreactive cells within the epithelium lining the sinus (PAP technique with antibodies against human thyroglobulin and weak haematoxylin counterstain, × 85).

23 cases of thyroglossal cysts with adjacent thyroid tissue and one case of lingual thyroid, not a single calcitonin-immunoreactive cell was found, in either the thyroid tissue or the epithelium lining the cysts (Ljungberg, unpublished observation). This finding is in agreement with the current belief that parafollicular cells are not embryo-logically related to the median anlage, but are derivatives of the lateral thyroid primordium (section 1.3). It should also be mentioned here that branchial cleft derivatives may occasionally be found within the substance of the thyroid gland. These include intrathyroidal thymic tissue or parathyroid glands. Very rarely, intrathyroidal ultimo-branchial cysts, lined by mucinous or squamous epithelium, with intermingled neuroendocrine (parafollicular) cells have also been recorded (Roediger *et al.*, 1977).

Carcinoma arising in the thyroglossal duct or cyst is rare. Little more than one hundred cases have been reported (Bosch *et al.*, 1986). Of these more than 90% were of the papillary type, with the same microscopic features as its counterpart in the thyroid gland (Figs 3.5–3.7). They have usually been diagnosed postoperatively, as an incidental finding in a thyroglossal cyst. In recent years, fine-needle

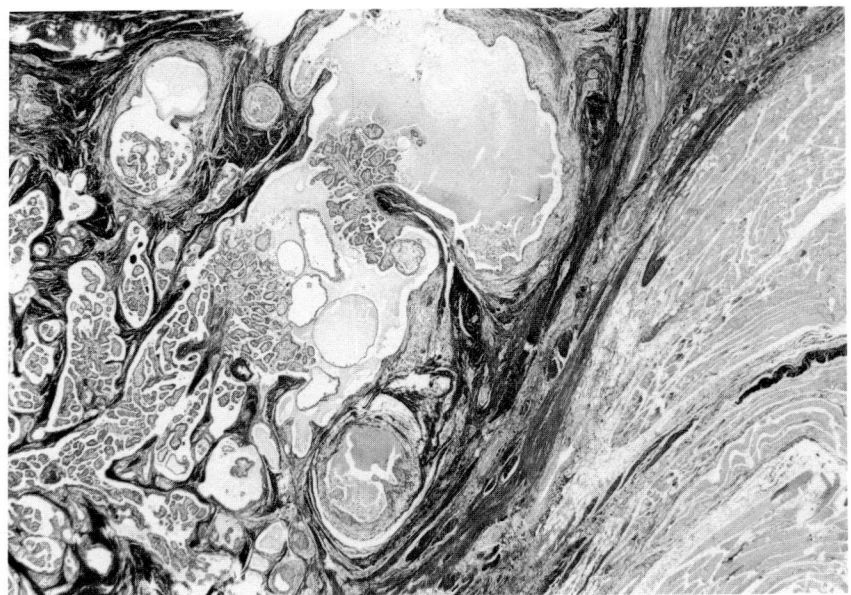

Figure 3.5 Papillary carcinoma originating in thyroglossal duct cysts. There are irregularly formed sinuses and cysts partly lined by papillary excrescences (Van Gieson-Hansen stain, × 17).

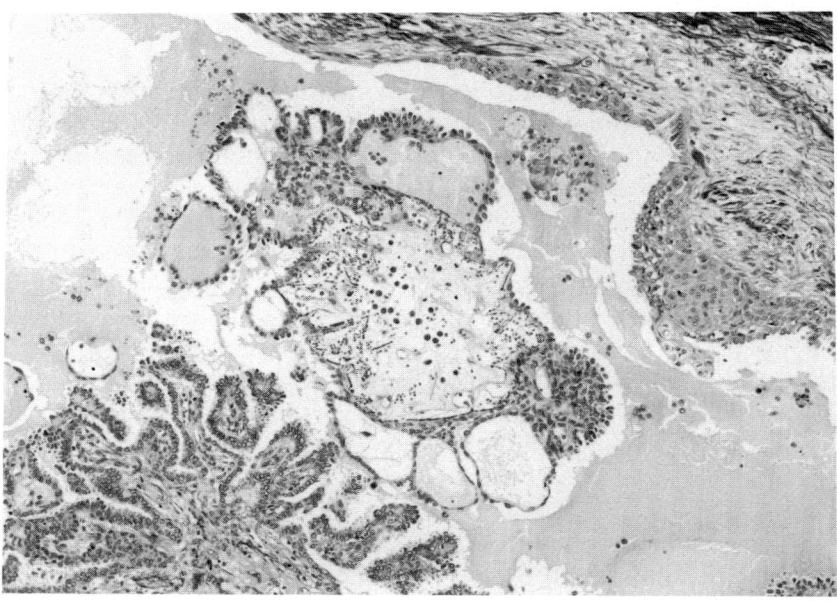

Figure 3.6 Same case as shown in Fig. 3.5. Thyroglossal cyst is partly lined by squamous epithelium, and partly by carcinomatous papillary formations (× 80).

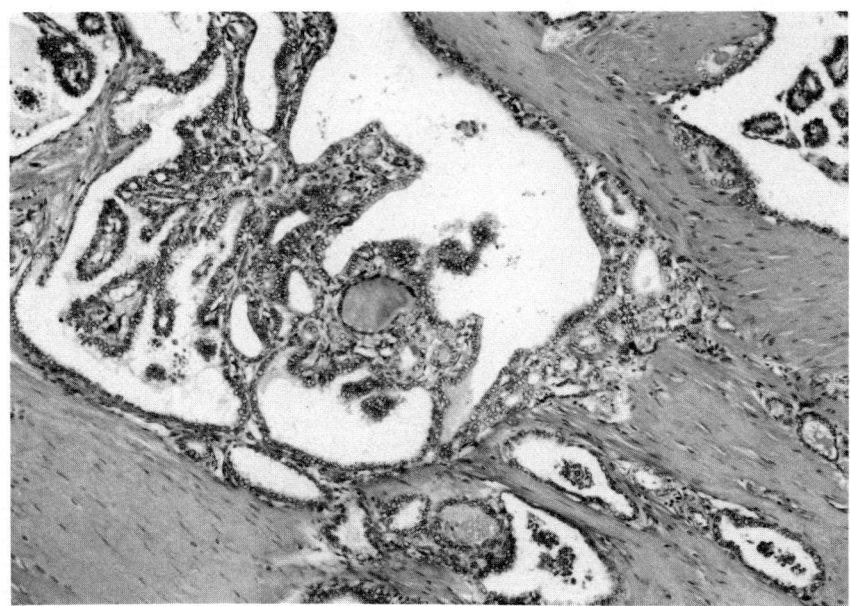

Figure 3.7 Same case as shown in Fig. 3.5. Papillary structures are covered by cylindrical cells with ground glass nuclei characteristic of papillary thyroid carcinoma. Carcinomatous follicles are seen infiltrating the stroma (× 80).

aspiration biopsy has proved to be a useful diagnostic technique for establishing a preoperative diagnosis (Widström *et al.*, 1976; Berridge and Webb, 1986). In most cases, these tumours have been confined to the thyroglossal duct anomaly; distant metastases have been very rare and the majority of cases have had a favourable prognosis. Rarely, the tumour has been of squamous type (Mobini *et al.*, 1974) and occasionally, anaplastic carcinoma has also been reported, with a highly malignant course (Nussbaum *et al.*, 1981). Interestingly, pure follicular carcinoma has been encountered only exceptionally, and recently, a case of oxyphil cell adenoma of a thyroglossal duct remnant has also been recorded (Tovi *et al.*, 1988). As would be expected, primary medullary carcinoma is unknown in this location.

3.2 Lingual thyroid

Thyroid tissue within the base of the tongue has been discovered in 10% of individuals, with equal sex distribution, as judged from autopsy studies (Sauk, 1970). The development of a true lingual thyroid gland, due to complete failure of descent to the normal postnatal location in the neck, however, is a very rare event. Ulrich (1932) found an incidence of two cases among 4000 thyroid patients. It may give rise to dysphagia, dysphonia or respiratory obstruction which may even be life threatening. In a number of cases, haemorrhage has also been recorded, occasionally causing death. Symptomatic lesions dominate among females. The condition may also be associated with hypothyreosis. About 70% of the patients with a grossly evident lingual thyroid gland have no glandular tissue at the normal site. It is important to realize that under such circumstances, removal of the ectopic tissue will render the patient athyreotic, unless substitutional medical treatment is given postoperatively (LiVolsi et al., 1974).

Lingual thyroid tissue may be subjected to the same pathological changes as the normal gland. Atoxic goitre may develop, being most frequent in females and usually becoming clinically evident during periods of increased endocrine activity, e.g. puberty, adolescence, pregnancy or the menopause (Baughman and Gainesville, 1972; Weider and Parker, 1977). The histological picture is usually that of a colloid and/or hyperplastic goitre, sometimes with areas of microfollicular or fetal patterns. Tumours are very rare (Smithers, 1970). The most common carcinoma developing from the lingual thyroid tissue is the papillary type, with a generally favourable prognosis, similar to that of its counterpart in the normal gland (Joseph and Komorowski, 1975).

3.3   Lateral aberrant thyroid

This term refers to the presence of thyroid tissue lateral to the jugular veins. Usually the thyroid tissue is included within a lymph node. In the 1930s and early 1940s, this condition was believed to represent developmental inclusions of thyroid tissue (Wozencraft et al., 1948). Although microscopic nests of benign-looking follicles in the marginal sinus of a lymph node may be a result of anomalous development (Meyer and Steinberg, 1969), it is now generally agreed that, in most instances, the presence of thyroid tissue, especially when replacing a larger part of the node and/or occurring in several nodes, is a manifestation of a metastasis of an occult papillary thyroid carcinoma (Butler et al., 1967; Sampson et al., 1970). It is important to examine such tissue carefully with many histological sections. Even if most of it is composed of benign-looking follicles with flat epithelium devoid of specific nuclear features (Fig. 3.8), not infrequently, careful study will identify a minute focus of papillary structures with typical ground glass nuclei, revealing the malignant nature of the lesion (Fig. 3.9).

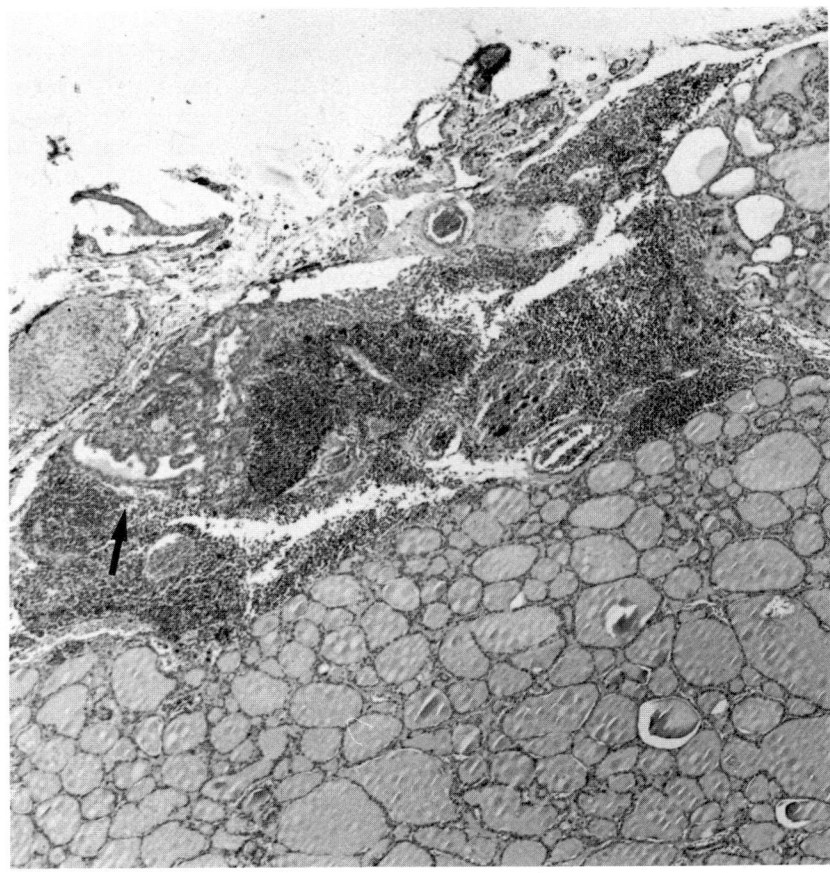

Figure 3.8 So-called lateral aberrant thyroid. Part of a cervical lymph node is shown, containing mostly normal-looking thyroid follicles; however, there is a tiny focus of papillary structures, lying in the marginal sinus (arrow) (H&E, × 50).

On the other hand, it is important to realize that implantation of apparently benign thyroid tissue in the lateral parts of the neck can, very rarely, occur after thyroid surgery or accidental trauma (Moses *et al.*, 1976). Thyroid tissue may also occur in the form of nodules which have been sequestered from a nodular goitre, separated from the gland and displaced laterally (Sisson *et al.*, 1964; see also Chapter 5, p. 93). Although such thyroid nodules tend to be located in the region of the cervical lymph nodes, they are separated from these nodes, usually being demarcated by fibrous capsules of their own. In addition, they lack a marginal sinus and other structural features typical of lymph

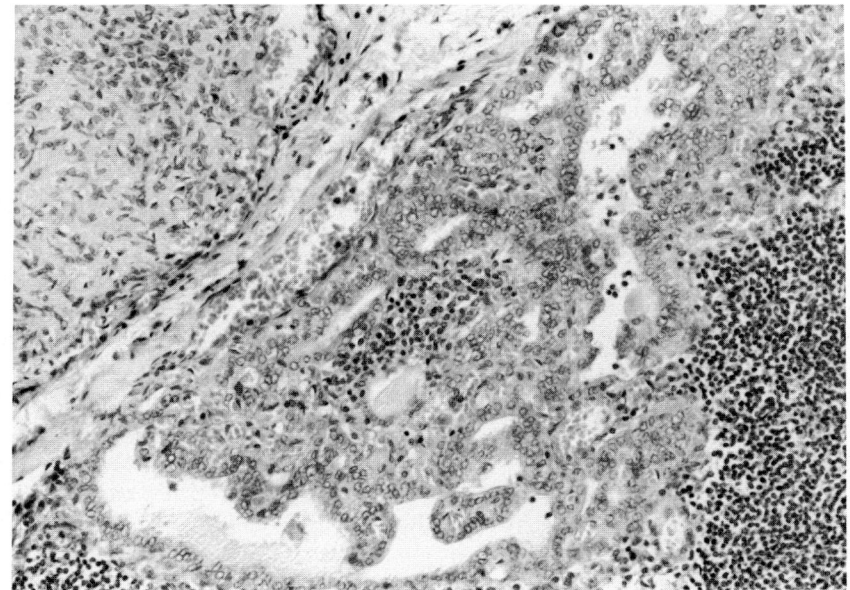

Figure 3.9 Close-up view of the small focus of papillary structures arrowed in Fig. 3.8. The papillae are lined by atypical epithelial cells with ground glass nuclei, revealing the malignant nature of the lesion. The patient was found to have a papillary microcarcinoma in the ipsilateral thyroid lobe (H&E, × 350).

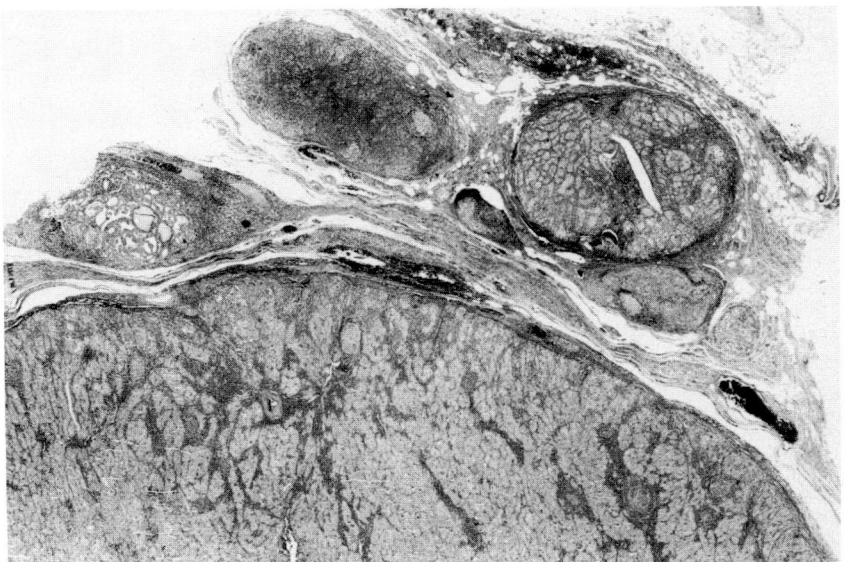

Figure 3.10 Nodular Hashimoto's thyroiditis. Some nodules have been extruded and separated from the surface of the gland, and lie in the extrathyroidal tissue and simulate lymph node metastases of thyroid carcinoma (H&E, × 11).

nodes. Occasionally, sequestered nodules may be derived from nodular Hashimoto's thyroiditis. Such nodules may simulate lymph node metastases because of their lymphoid stromal component, but they show histological features similar to the thyroid gland proper (Fig. 3.10). Closer examination will again reveal the lack of a typical lymph node capsule with marginal sinus, which may help to distinguish the lesion from a metastasis. Finally, in florid cases of Graves' disease, hyperplastic thyroid tissue may be encountered outside the confines of the gland, often in the surrounding skeletal muscles. It is not known whether this is a result of active infiltration of hyperplastic thyroid tissue from the gland proper or represents pre-existing extrathyroidal nests of follicles (see p. 106), which have become especially prominent as a result of the hyperplastic process. The presence of clinical manifestations of thyrotoxicosis in these cases helps to discriminate such foci from malignant deposits.

3.4   Struma ovarii

Struma ovarii is not a developmental anomaly of the thyroid gland but is to be regarded as a teratoma. Aggregates of thyroid follicles may occur as a component of a benign ovarian cystic teratoma (dermoid) in 5–20% of cases (Blackwell et al., 1946; Scully, 1970). By definition, struma ovarii is present when the thyroid tissue comprises the whole or the major part of the teratoma. Struma ovarii has been reported to occur in 1–2% of ovarian teratomas (Marcus and Marcus, 1961). Rarely, struma ovarii may be hyperfunctioning and cause clinical hyperthyroidism (Pantoja et al., 1975).

Grossly, struma ovarii is a circumscribed brownish nodule. It sometimes occurs in the wall of a cystic teratoma (dermoid). Microscopically, the structure is that of ordinary thyroid tissue. In some cases the struma ovarii shows the features of a multinodular goitre with colloid nodules intermingled with nodules showing hyperplastic changes. Such nodules may be extruded from the ovarian lesion and seeded on the peritoneum and omentum. This condition, which is referred to as benign ovarian strumatosis, may erroneously be interpreted as a manifestation of a malignant ovarian struma (see also Chapter 5, p. 95).

Malignant ovarian struma is uncommon, and the diagnosis has most often been based on microscopic examination, rather than on clinical grounds. Occasional examples of so-called strumal carcinoid have been reported in recent years (Kimura et al., 1986; Snyder and Tavassoli, 1986). This is a biphasic neoplasm composed of trabeculae and nests of neuroendocrine cells, intermingled with follicles built up of thyroglobulin-immunoreactive follicular epithelial cells. In addition this tumour also contains cells of intermediate type, i.e. with microscopic, immunohistochemical and ultrastructural features of both cell types. The tumour is of great interest from a histogenetic point of view, in that these findings suggest a common progenitor for both cell types. A counterpart to the strumal carcinoid has also been described in the thyroid gland (intermediate type of thyroid carcinoma, see p. 227).

References

Allard, R.H.B. (1982) The thyroglossal cyst. *Head Neck Surg.*, **5**, 134–46.

Baughman, R.A. and Gainsville, M.S.D. (1972) Lingual thyroid and lingual thyroglossal tract remnants. *Oral Surg.*, **34**, 781–99.

Berridge, D.C. and Webb, A.J. (1986) Cervical thymus cyst and thyroglossal cyst carcinoma. *Br. J. Surg.*, **73**, 44.

Blackwell, W.J., Dockerty, M.B., Masson, J.C. *et al.* (1946) Dermoid cysts of the ovary: their clinical and pathological significance. *Am. J. Obstet. Gynecol.*, **51**, 151–72.

Bosch, J.L.H.R., Kummer, E.W. and Hohmann, F.R. (1986) Carcinoma of the thyroglossal duct. *Neth. J. Surg.*, **38**, 36–40.

Butler, J.J., Tulinius, H., Ibanez, M.L. *et al.* (1967) Significance of thyroid tissue in lymph nodes associated with carcinoma of the head, neck or lung. *Cancer*, **20**, 103–12.

Dalgaard, J.B. and Witteland, P. (1956) Thyroglossal anomalies. A follow-up study of 58 cases. *Acta Chir. Scand.*, **111**, 444–55.

Ellis, P. and Van Norstrand, A.W. (1977) The applied anatomy of thyroglossal tract remnants. *Laryngoscope*, **87**, 765–70.

Joseph, T.J. and Komorowski, R.A. (1975) Thyroglossal duct carcinoma. *Hum. Pathol.*, **6**, 717–29.

Kimura, N., Sasano, N. and Namiki, T. (1986) Evidence of hybrid cell of thyroid follicular cell and carcinoid cell in strumal carcinoid. *Int. J. Gynecol. Pathol.*, **5**, 269–77.

LiVolsi, V.A., Perzin, K.H. and Savetsy, L. (1974) Carcinoma arising in median ectopic thyroid (including thyroglossal duct tissue). *Cancer*, **34**, 1303–15.

Marcus, C.C. and Marcus, S.L. (1961) Struma ovarii. A report of 7 cases and a review of the subject. *Am. J. Obstet. Gynecol.*, **81**, 752–62.

Meyer, J.S. and Steinberg, L.S. (1969) Microscopically benign thyroid follicles in cervical lymph nodes. *Cancer*, **24**, 302–11.

Mobini, J., Krouse, T.B. and Klinghoffer, J.F. (1974) Squamous cell carcinoma arising in a thyroglossal duct cyst. *Am. Surg.*, **40**, 290–4.

Moses, D.C., Thompson, N.W., Nishiyama, R.H. and Sisson, J.C. (1976) Ectopic thyroid tissue in the neck. Benign or malignant? *Cancer*, **38**, 361–5.

Nussbaum, M., Buchwald, R.P., Ribner, A. *et al.* (1981) Anaplastic carcinoma arising from median ectopic thyroid (thyroglossal duct remnant). *Cancer*, **48**, 2724–8.

Pantoja, E., Noy, M.A., Axtmayer, R.W. *et al.* (1975) Ovarian dermoids and their complications. Comprehensive historical review. *Obstet. Gynecol. Surv.*, **30**, 1–20.

Roediger, W.E.W., Kalk, F., Spitz, L. *et al.* (1977) Congenital thyroid cyst of ultimobranchial gland origin. *J. Pediatr. Surg.*, **12**, 575–6.

Sampson, R.J., Oka, H., Key, Ch. R. *et al.* (1970) Metastases from occult thyroid carcinoma. An autopsy study from Hiroshima and Nagasaki, Japan. *Cancer*, **25**, 803–11.

Sauk, J.J., Jr (1970) Ectopic lingual thyroid. *J. Pathol.*, **102**, 239–43.

Scully, R.E. (1970) Recent progress in ovarian cancer. *Hum. Pathol.*, **1**, 73–98.

Sisson, J.C., Schmidt, R.W. and Beierwaltes, W.H. (1964) Sequestered nodular goiter. *N. Engl. J. Med.*, **270**, 927–32.

Smithers, D.W. (1970) Carcinoma associated with thyroglossal duct anomalies. In *Tumours of the Thyroid Gland* (ed. D.W. Smithers), E. and S. Livingstone, Edinburgh and London, pp. 155–61.

Snyder, R.R. and Tavassoli, F.A. (1986) Ovarian strumal carcinoid: immunohistochemical, ultrastructural, and clinicopathologic observations. *Int. J. Gynecol. Pathol.*, **5**, 187–201.

Tovi, F. and Eyal, A. (1985) Branched and polycystic thyroglossal duct anomaly. *J. Laryngol. Otol.*, **99**, 1179–82.

Tovi, F., Fliss, D. and Inbar-Yanai, I. (1988) Hürthle cell adenoma of the thyroglossal duct. *Head Neck Surg.*, **10**, 346–9.

Ulrich, H.F. (1932) Lingual thyroid. *Ann. Surg.*, **95**, 503–7.

Weider, D.J. and Parker, W. (1977) Lingual thyroid: review, case reports, and therapeutic guidelines. *Ann. Otol. Rhinol. Laryngol.*, **86**, 841–8.

Widström, A., Magnusson, P., Hallberg, O. *et al.* (1976) Adenocarcinoma originating in the thyroglossal duct. *Ann. Otol. Rhinol. Laryngol.*, **85**, 286–90.

Wozencraft, P., Foote, F.W., Jr and Frazell, E.L. (1948) Occult carcinoma of the thyroid. *Cancer*, **1**, 574–83.

4 Thyroiditis

Surgery, as a rule, is not the treatment of choice for thyroiditis. The majority of cases are treated medically, although some come to surgery for cosmetic reasons or obstructive symptoms. In addition thyroid tissue is sometimes excised for diagnostic purposes, because the clinical condition may mimic a neoplastic process. Clinical and gross features such as rapid asymmetric enlargement, nodularity, firmness and even fixation to surrounding structures may occur in both inflammation and neoplasia. Without biopsy, the distinction between these two conditions may be difficult or impossible. In most cases with suspected lesions in the thyroid gland, a fine-needle biopsy is done, and where thyroiditis is suspected, a definite diagnosis can often be established with this method. However, thyroiditis may cause destruction of parenchyma, which may evoke a compensatory hyperplasia, sometimes being focal in distribution and even giving rise to nodules. In addition, certain forms of thyroiditis are accompanied by thyroid-stimulating autoantibodies producing hyperplastic changes in the thyroid parenchyma. Fine-needle biopsies from such hyperplastic areas may show very cellular, and even atypical, patterns which are difficult to distinguish from a neoplastic process. Sometimes, carcinoma or malignant lymphoma occurs concurrently, lesions which may be difficult to interpret cytologically in the background of thyroiditis (Willems and Löwhagen, 1981; Kini, 1987).

4.1 Lymphocytic thyroiditis

The term 'lymphocytic thyroiditis' embraces a heterogeneous group of thyroiditides. Since the aetiology and pathogenesis of the various types of thyroiditis with lymphocytic infiltration are often uncertain, they are difficult to classify and explain. The majority of cases of chronic lymphocytic thyroiditis are now considered to be various clinicopathological manifestations of an autoimmune disease. This is clinically the most important group. However, there are also forms of less clinical significance which are quite often seen by the pathologist, and which constitute a heterogeneous group of lesions of uncertain nature. These thyroiditides, often classified under the neutral term 'non-specific lymphocytic thyroiditis', are probably related to a great variety of factors noxious to the thyroid parenchyma, such as virus infection, trauma, drugs and other chemicals, radiation, and intrathyroidal neoplastic processes.

4.1.1   Classic Hashimoto's disease

Autoimmune thyroiditis, chronic lymphocytic thyroiditis, struma lymphomatosa, and lymphadenoid goitre are the most commonly used terms for this condition, described by Hashimoto in 1912. Three major varieties of this thyroiditis are known – the classic form, the fibrosing type, and the type occurring in children and adolescents (juvenile autoimmune thyroiditis). Graves' disease is also closely linked to this group of diseases, all of which are considered to be different clinicopathological expressions of the same basic process (Volpé et al., 1974). Therefore, Graves' disease is included by some together with the others under the general term 'autoimmune thyroid disease' (Volpé, 1978a).

(a)   Clinical features

The classic form is characterized by a diffuse goitre with or without nodularity. It is seen more commonly in women; fewer than 5% of cases occur in males (Doniach et al., 1979). It is most common in the 30–60 years age group, but no age is exempt. The patient usually presents with a small or moderate-sized, painless goitre with an average duration of two to four years and has thyroid function tests which are normal or borderline hypothyroid. It is common that the patient initially has only traces of thyroglobulin antibodies in serum. However, a test for microsomal antibodies is invariably positive. When hypothyreosis develops, the titres in both tests usually increase to high levels. It has been estimated that more than half the patients will progress to hypothyroidism (Lamberg, 1979). A transient phase of mild hyperthyroidism may occur early in the disease, sometimes referred to as 'Hashitoxicosis' (Volpé et al., 1974). Thyrotoxicosis, which has also been described as a late complication, may be due to the development of thyroid-stimulating antibodies (Gavras and Thomson, 1972). Conversely, cases have also been encountered where a goitre has started to grow again several years after subtotal thyroidectomy for Graves' disease and histology has shown typical Hashimoto's thyroiditis (Doniach et al., 1979). It is estimated that about 10% of thyrotoxics finally develop spontaneous myxoedema, although they usually do not have a goitre. Some patients have a past history of goitre during adolescence, followed by a remission and long

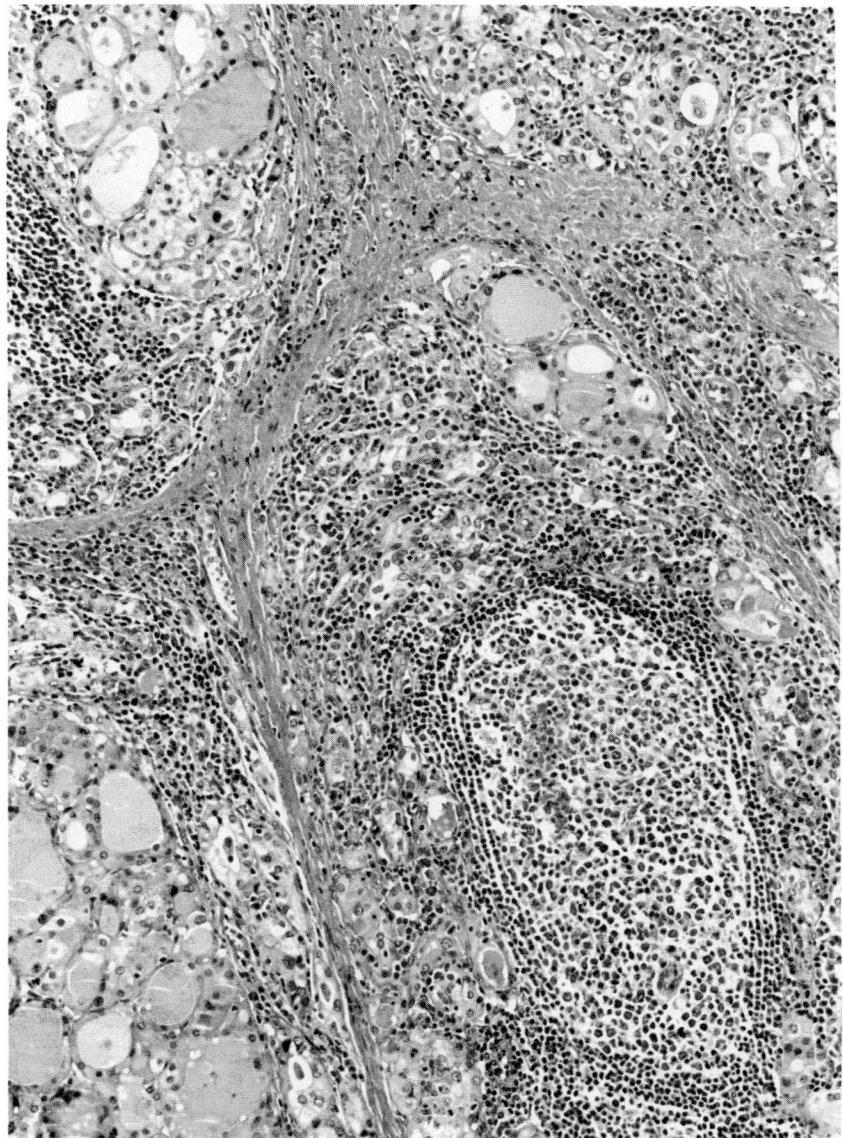

Figure 4.1 Classic Hashimoto's thyroiditis. Lymphoplasmacytic stromal infiltration, with germinal centres, atrophy of follicles, oxyphil change of follicular cells, and lobular fibrosis are the cardinal features (H&E, × 125).

symptomless interval, with recurrence of a slowly progressive goitre during the menopause. Occasionally, however, the goitre has a more rapid onset and may grow to many times its normal size in less than a year. Male patients usually have myxoedema or pressure symptoms from their goitre when first seen and their antibody titres are usually very high.

(b) Pathology
Most authorities are of the opinion that the diagnosis of Hashimoto's thyroiditis should be founded on the typical clinical, immunological and histopathological features of the disease. It should not be loosely applied to any form of chronic thyroiditis with lymphocytic stromal infiltration.

Grossly, the typical case shows a diffuse enlargement of the gland. Sometimes one lobe is more enlarged than the other. The gland is firm but not hard and usually does not adhere to surrounding tissues. The cut surfaces have a slightly lobulated or nodular, fleshy appearance with a pale tan or grey–white colour. In some cases nodules of hyperplastic tissue may replace all the gland and may grossly mimic carcinoma. Nodules composed entirely of oxyphil cells typically have homogeneous, tan-coloured cut surfaces. Peripherally located nodules may project as buds from the surface of the gland, and even lose their connection with it and be displaced laterally in the neck (so-called sequestered nodules, see Fig. 3.10). Such nodules, containing a mixture of thyroid follicles and lymphoid tissue, may easily be misdiagnosed as lymph nodes containing metastatic thyroid carcinoma (see also p. 48).

Histologically (Fig. 4.1), the three major diagnostic criteria are:

1. Lymphoplasmacytic stromal infiltration;
2. Follicular destruction and atrophy with interstitial fibrosis;
3. Oxyphil change of the remaining follicular cells.

The intensity of the histological abnormality varies from one part to another. The lobular architecture, however, is usually preserved, which is an important detail in the distinction from carcinoma. The lymphocytic infiltrate contains large lymphoid follicles with prominent germinal centres. The plasma cell population has been found to be polyclonal (Ben-Ezra et al., 1988). There is an admixture of histiocytes and occasional multinucleate giant cells may also be seen. The thyroid follicles show varying degrees of destruction. The follicles that remain may show focally a marked degree of hyperplasia with high columnar epithelium that may even be heaped up into pseudopapillary formations (Fig. 4.2). Such features may be misinterpreted as a neoplastic change were it not for the preserved lobular architecture of the tissue. Minor

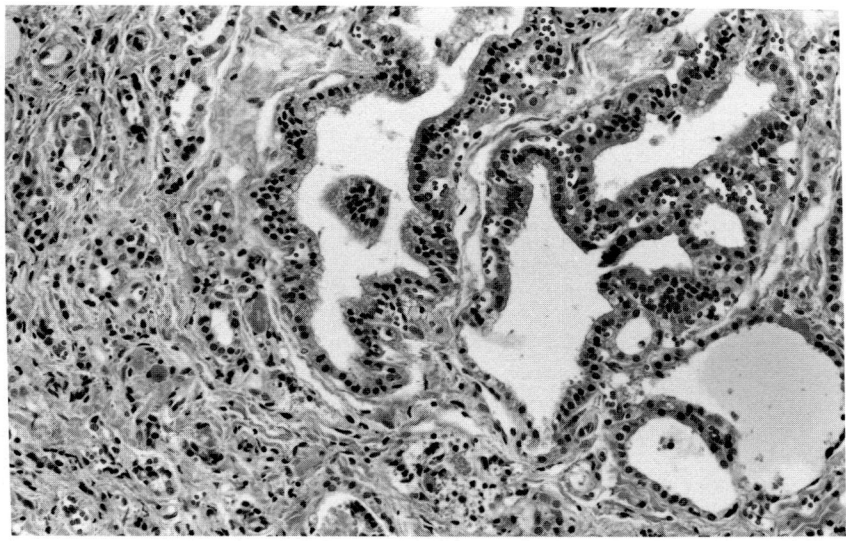

Figure 4.2 Hashimoto's thyroiditis. A cluster of preserved follicles showing hyperplastic changes is seen together with tall cylindrical cells forming papillary structures that may imitate those of papillary carcinoma. The nuclear features of papillary carcinoma are lacking (H&E, × 170).

nodules often composed of follicles or solid accumulations of oxyphil cells may be seen (Fig 4.3), often also containing patchy aggregates of lymphoid cells. Rarely, the nodularity may be more pronounced, with the development of multiple hyperplastic nodules, sometimes surrounded by fibrous bands and consisting of macro- or microfollicular or even pseudopapillary and solid structures of follicular cells, which may also show oxyphil changes (Plate 3). The recognition of an early oxyphil neoplasm or papillary carcinoma in such areas may be problematic (Rosai, 1989). A carcinoma, however, is usually evident by the irregular growth of the lesion, and by infiltration into the adjacent thyroid tissue or invasion of blood vessels.

Follicles in the process of disintegration are small, atrophic and contain little colloid. In severe cases there may be few follicles left. The epithelial cells show a prominent oxyphil metaplasia (Fig. 4.4); they are also known as Askanazy cells, and sometimes, erroneously, as Hürthle cells (see also p. 8). These cells may show prominent cytological polymorphism with irregular, hyperchromatic nuclei and atypia, which must not be mistaken for malignancy. The Askanazy cells show an intracytoplasmic accumulation of mitochondria and are functionally insufficient with respect to the production of thyroglobulin,

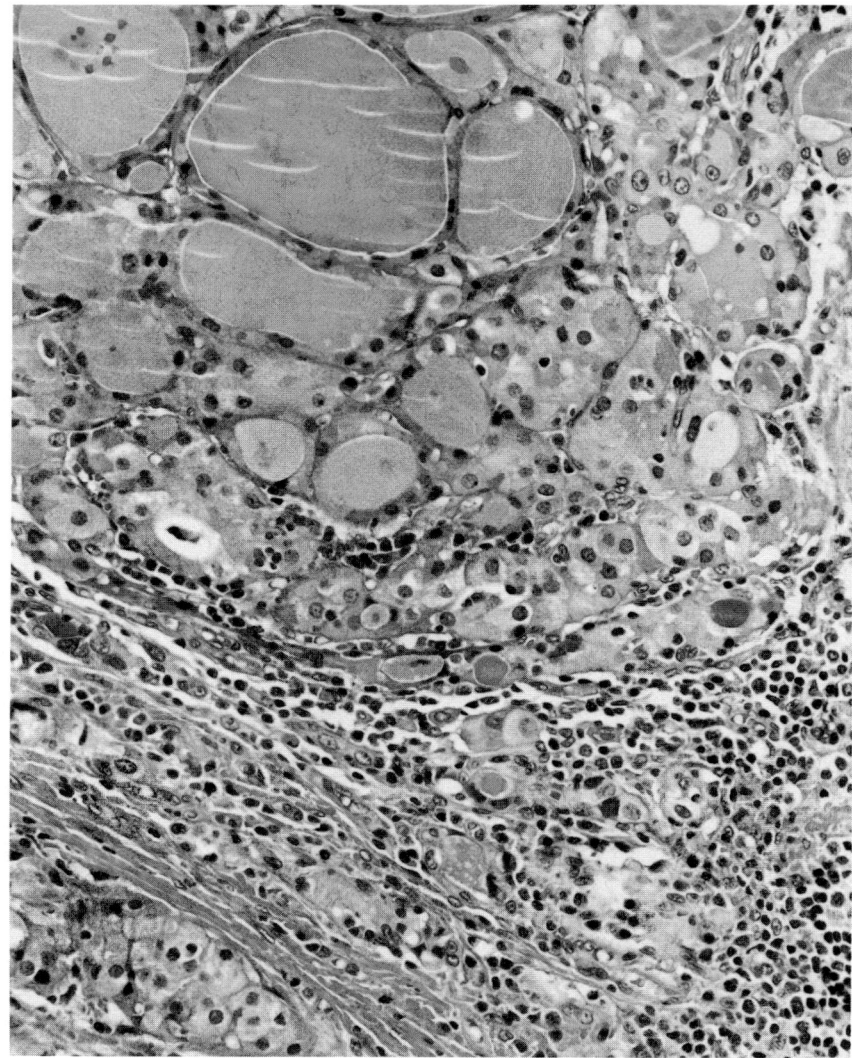

Figure 4.3 Hashimoto's thyroiditis is illustrated with part of a small nodule composed of follicles and small nests of follicular cells with varying degrees of oxyphil change (H&E, × 250).

T_3 and T_4. Some oxyphil cells may show a gradual change towards cells with light, foamy cytoplasm gradually losing their oxyphilia, probably due to swelling and degeneration of mitochondria (Fig. 4.5). The Askanazy cells are not pathognomonic for Hashimoto's thyroiditis but can occur in other thyroid lesions, including thyroid atrophy,

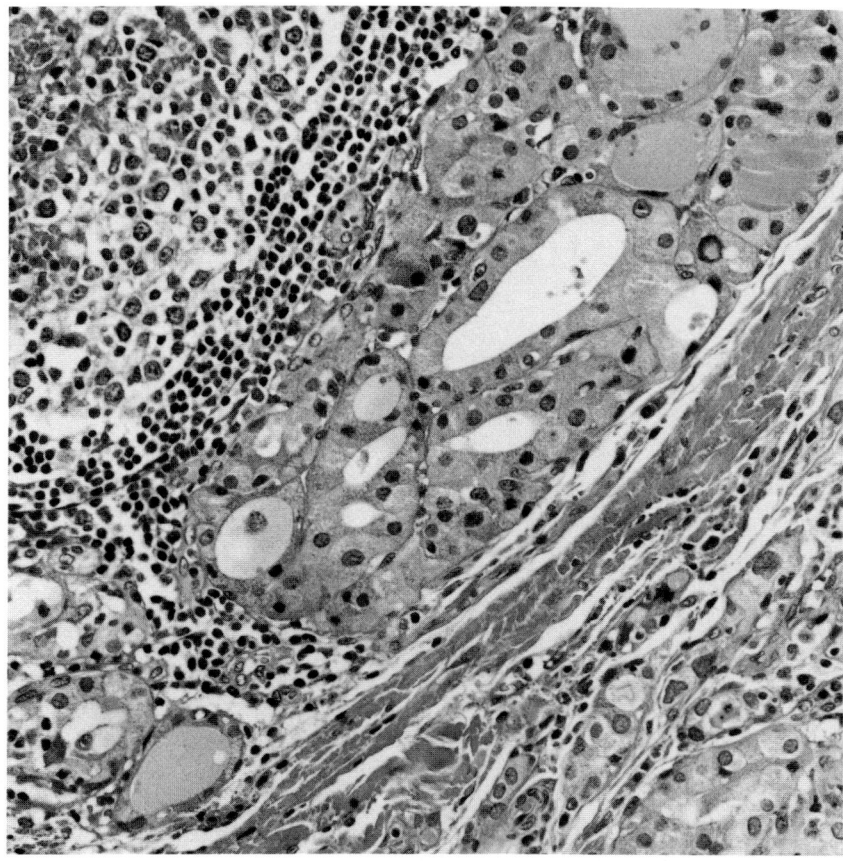

Figure 4.4 Hashimoto's thyroiditis showing a focus of pronounced oxyphil change of the follicular cells (H&E, × 250).

multinodular goitre and thyroid tumours. Another type of oxyphil follicular cell may also be seen in Hashimoto's thyroiditis, which is smaller than the Askanazy cell. Ultrastructurally, it is found to contain an increased amount of ergastoplasmic cisternae with a highly developed Golgi apparatus, and increased amounts of colloid droplets mixed with dense bodies. These are features which are characteristic of the stimulated cell seen in hyperthyroidism (Volpé, 1978a). Squamous metaplasia of follicular cells may occur, forming small epithelial nests of squamous cells and transitional elements between follicular and squamous cells. This phenomenon however, is more prominent in the fibrous variant of autoimmune thyroiditis.

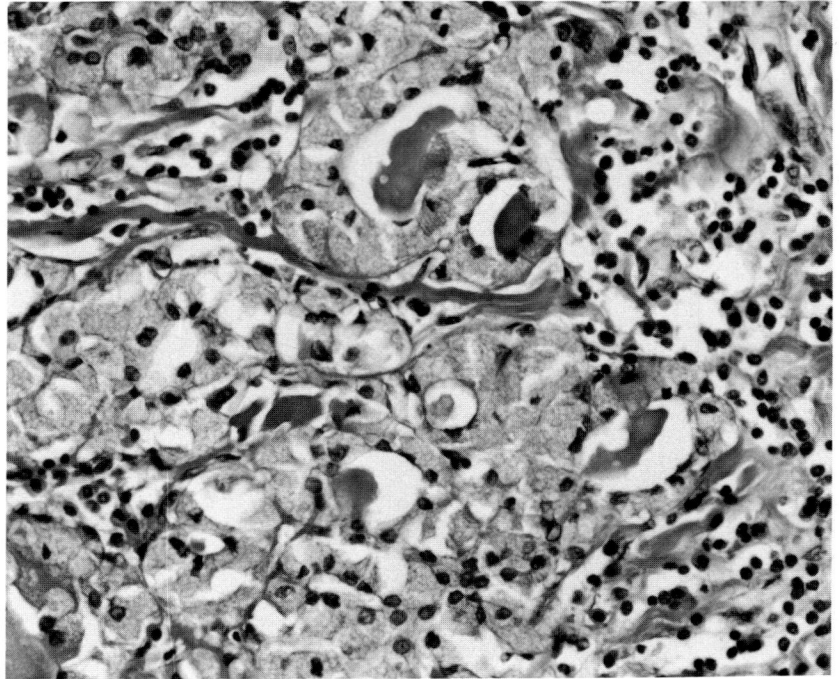

Figure 4.5 Hashimoto's thyroiditis with foamy follicular cells (H&E, × 330).

(c) Aetiology and pathogenesis

Autoimmune thyroiditis tends to be familial. It may coexist with other autoimmune disorders in the same patient or that individual's family (Volpé, 1977; Brown *et al.*, 1978; Strakosch *et al.*, 1982). These include so-called autoimmune lymphocytic adrenalitis (Schmidt's syndrome) (Carpenter *et al.*, 1964), as well as autoimmune disorders in other endocrine organs, which sometimes lead to polyendocrine insufficiency syndromes (Irvine, 1975), chronic liver diseases (Doniach, 1972), pernicious anaemia, and myasthenia gravis (Segal and Weintraub, 1976).

The basic underlying defect in autoimmune thyroid disease appears to be an inherited deficiency in antigen-specific suppressor T lymphocytes (Strakosch *et al.*, 1982; Aichinger *et al.*, 1985; Burek and Rose, 1986). This specific defect may allow effector T lymphocytes to attack the thyroid cells and allow helper T lymphocytes to induce plasma cells to produce autoantibodies against components and secretory products of the thyroid cells. Autoimmune thyroid disease appears to be more common in individuals with certain HLA genotypes. It has been proposed that individuals with HLA-DR5 are predisposed to the production of thyroid antibodies (Strakosch *et al.*, 1982). Patients with Hashimoto's thyroiditis tend to develop high serum titres of autoantibodies to several thyroid antigens, such as thyroglobulin, other colloid components, microsomes, nuclear components, the thyroid hormones and the TSH receptor. The level of circulating auto-antibodies, generally, parallels the intensity of the pathological process in the gland. Little is known, however, about the mechanism(s) by which the disease process is precipitated and the sequence of events involved in the pathogenesis of the disease. Recent studies (Aichinger *et al.*, 1985) suggest that humoral autoimmune mechanisms, in the form of complement-fixing immune complexes, are involved in the early stages of the disease, which are followed at a later stage by cell-mediated immunopathological effects, most likely brought about by antibody-dependent T cell cytotoxicity.

(d) Neuroendocrine cells in Hashimoto's thyroiditis
Increased numbers of somatostatin and/or calcitonin immunoreactive parafollicular cells have been reported in cases of Hashimoto's thyroiditis (Dhillon *et al.*, 1982; Albores-Saavedra *et al.*, 1988; Libbey *et al.*, 1989; Biddinger *et al.*, 1991), and also in cases of Hashimoto's thyroiditis accompanied by malignant lymphoma of the thyroid (Baschieri *et al.*, 1989). The significance of this finding is difficult to evaluate, because under normal conditions, the density of parafollicular cells may vary considerably (see p. 13). Secondly, it cannot be excluded that, at least in some of the reported cases, the increase of parafollicular cells may not be absolute but merely spurious, since in Hashimoto's thyroiditis the destructive process may be expected to be primarily directed against the follicular cell component and not against the parafollicular cells. In the two cases described by Biddinger *et al.* and Libbey *et al.*, however, the hyperplasia was marked and beyond doubt, and in addition, both patients had increased serum calcitonin levels, which returned to normal after total thyroidectomy. Interestingly, a similar marked hyperplasia of parafollicular cells has also been recorded in the normal-appearing peritumorous thyroid tissue in cases of papillary carcinoma and follicular tumours (Albores-Saavedra *et al.*, 1988), and one patient, in whom preoperative serum calcitonin measurements had been done, showed elevated values. These findings are remarkable in that hypercalcitoninaemia in the presence of a thyroid lump has currently been considered almost diagnostic for medullary carcinoma. The pathogenesis of the parafollicular cell hyperplasia in these cases is enigmatic. It may be speculated that the destruction of normal parenchyma brought about by the thyroiditis or the tumour

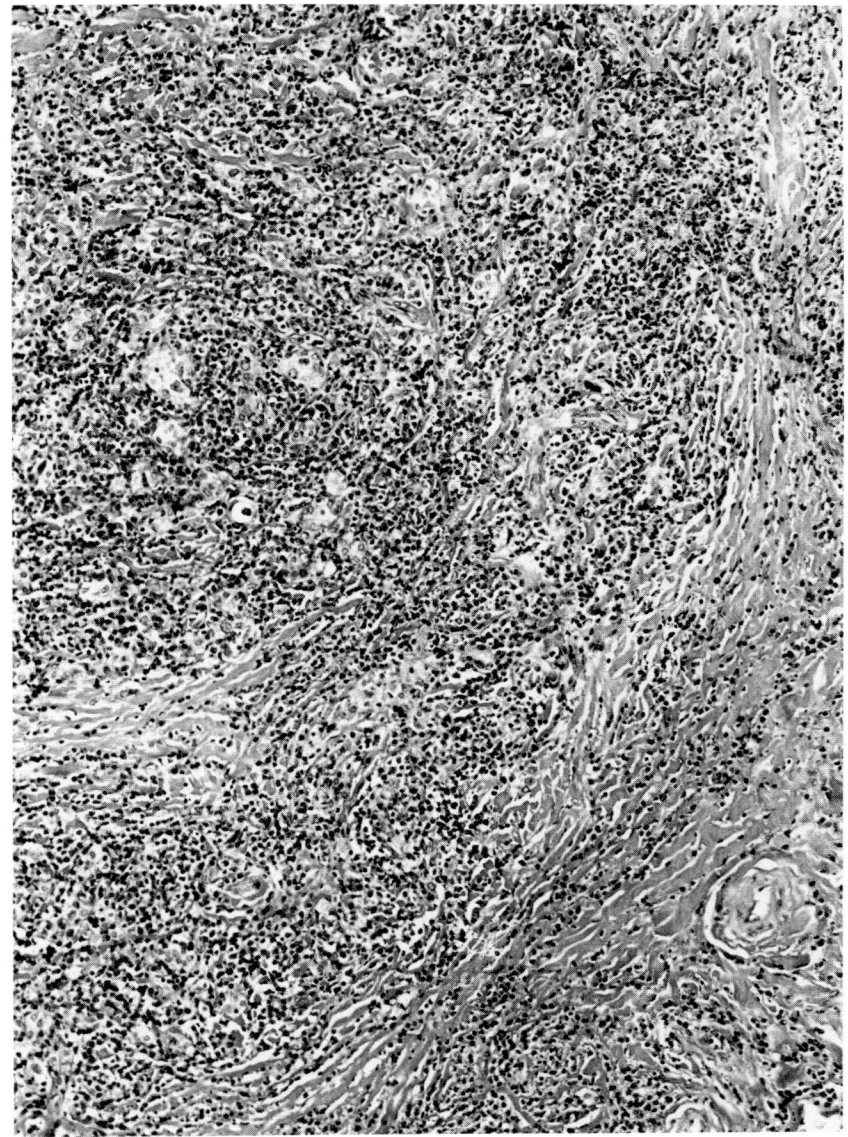

Figure 4.6 Fibrous variant of Hashimoto's thyroiditis. There is extensive parenchymal atrophy and pronounced fibrosis (H&E, × 125).

leads to chronic TSH overstimulation and hyperplasia of parafollicular cells. As pointed out earlier (see p. 10), there is evidence that TSH may stimulate not only the follicular, but also the parafollicular cells. The author agrees with Albores-Saavedra *et al*. that further studies are needed to correlate serum calcitonin levels with morphological changes of the parafollicular cells in thyroid lesions other than medullary carcinoma, in order to clarify the significance of a possible accompanying hyperplasia of the parafollicular cells.

4.1.2 Fibrous variant of Hashimoto's disease

The fibrosing form tends to affect elderly individuals and is frequently associated with hypothyroidism and very high titres of circulating antibodies against thyroglobulin. It accounts for about 12% of all cases of Hashimoto's thyroiditis (Katz and Vickery, 1974). The patients typically present with severe pressure symptoms and a very firm gland often with a rapidly enlarging goitre. An important feature is the frequent clinical confusion with thyroid carcinoma. This suspicion, as well as unrelieved neck pressure symptoms, and a persistently large goitre, are the main reasons why thyroidectomy is not infrequently performed in these cases.

Histologically (Figs. 4.6 and 4.7), this variant has a less prominent lymphoplasmacytic infiltrate and a more extensive parenchymal atrophy associated with broad bands of dense fibrosis, with marked lobulation of the tissue (Katz and Vickery, 1974). Typically, the capsule of the gland is intact without any adhesions to surrounding structures (Fig. 4.8). This is in contrast to Riedel's thyroiditis, where extensive fibrosis also involves extrathyroidal structures. There may be marked squamous metaplasia of remaining epithelial islands (Fig. 4.9), with features of a pseudoinvasive pattern of squamous cells in the collagen-rich fibrous tissue (LiVolsi and Merino, 1978). Such a pattern can cause confusion with a squamous cell carcinoma which may be highly differentiated and grow in a stroma with chronic inflammation (Fig. 9.85). However, in carcinoma, the squamous cells usually reveal cytological atypia and frequent frank keratinization, as well as unequivocal signs of infiltrative growth, often extending beyond the confinements of the gland. In addition, squamous thyroid carcinoma runs a highly malignant clinical course, resembling that of undifferentiated forms.

Fibrosing Hashimoto's thyroiditis is possibly related to atrophic thyroiditis, or so-called idiopathic myxoedema. Both conditions show similar histology and have many clinical features in common, although in the latter, the gland is quite small. The nature of such a relationship however, remains speculative (Katz and Vickery, 1974; LiVolsi, 1988).

4.1.3 Juvenile forms of autoimmune thyroiditis

Lymphocytic thyroiditis is a frequent cause of goitre in children and adolescents (Rallison et al., 1975). The affected children usually present

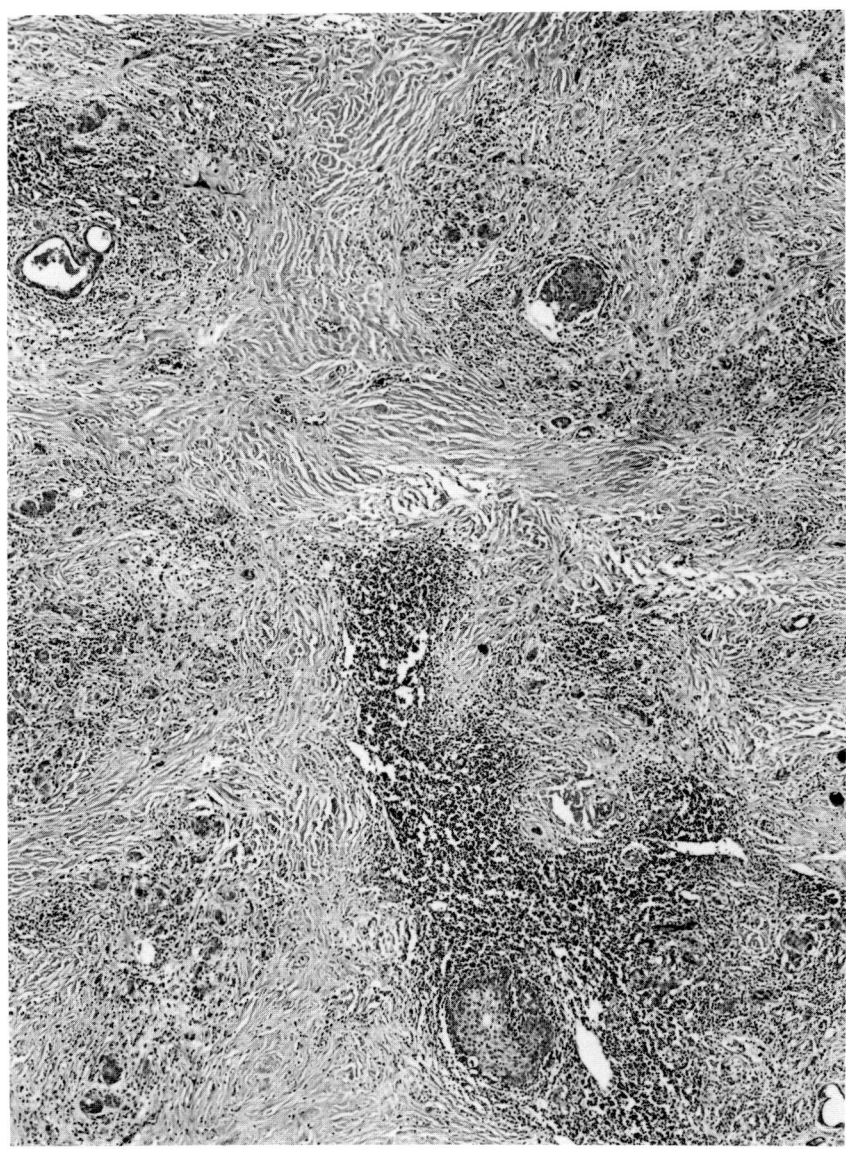

Figure 4.7 Fibrous variant of Hashimoto's thyroiditis. Advanced changes with broad bands of fibrous tissue replacing disintegrated parenchyma and marked atrophy of remaining follicles are seen (H&E, × 50).

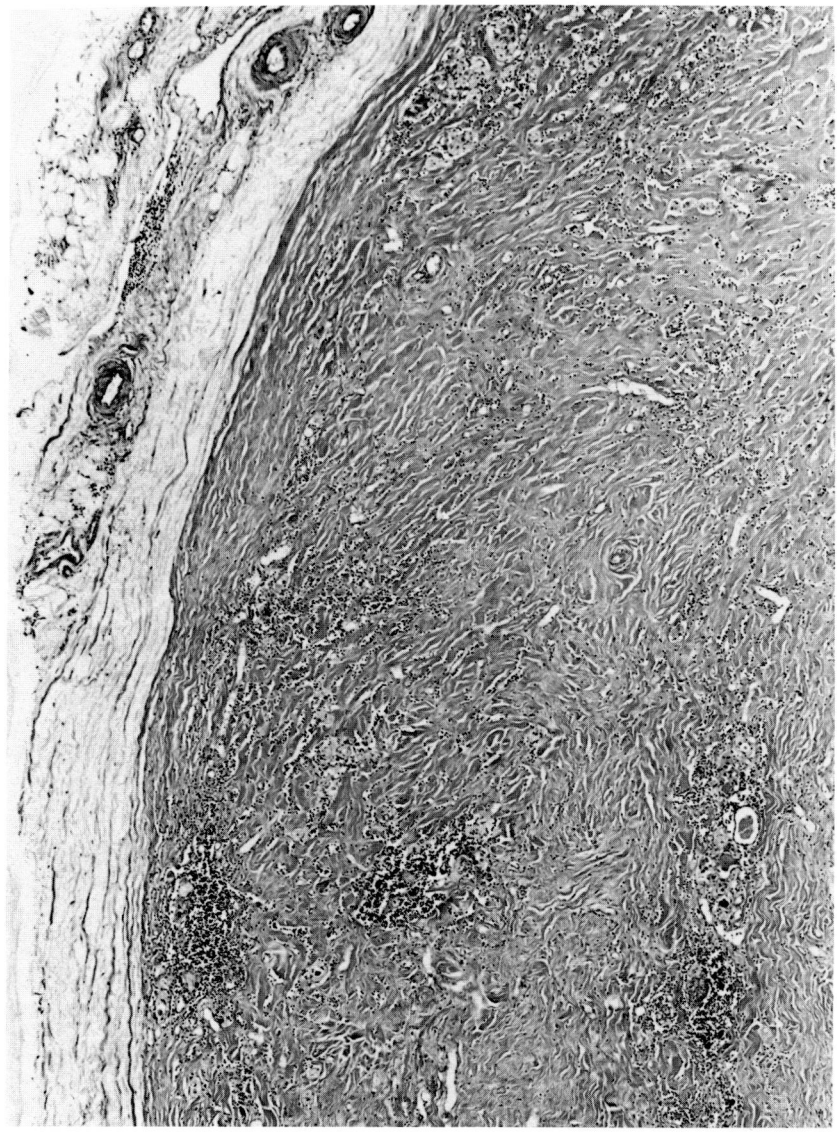

Figure 4.8 Fibrous variant of Hashimoto's thyroiditis with extreme fibrosis and loss of parenchyma. Typically, the process arrests at the capsule of the gland, leaving the extrathyroidal structures uninvolved. This is in contrast to Riedel's thyroiditis which also affects the perithyroidal soft tissues (H&E, × 50).

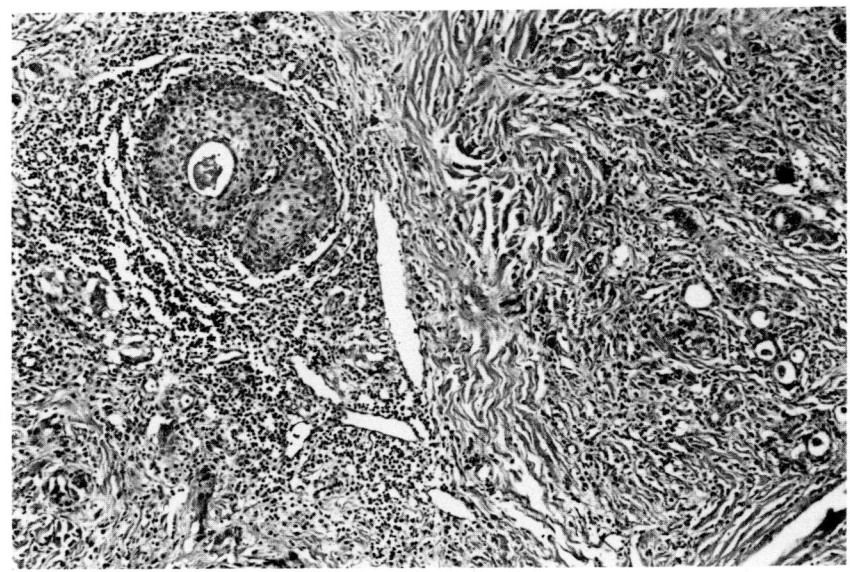

Figure 4.9 Fibrous variant of Hashimioto's thyroiditis with an epithelial nest showing squamous metaplasia (H&E, × 85).

with asymptomatic goitre, but some may have transient hyper-thyroidism. The clinical course often includes phases of remission and exacerbation. Macroscopically, the gland is usually diffusely enlarged and firm. The lymphocytic infiltration is marked, although lymphoid follicles are reported to be rare. Oxyphil metaplasia of the follicular epithelial cells is also rare or absent (Volpé, 1978a).

4.1.4 Non-specific lymphocytic thyroiditis

This is the most common type of thyroiditis in general hospital pathological material, although of less clinical interest than the other forms of chronic thyroiditis. It is also referred to as simple chronic thyroiditis or focal lymphocytic thyroiditis. The lesion is usually an incidental finding in thyroid tissue surgically removed because of a mass in the gland. The microscopic picture is quite different from that seen in Hashimoto's thyroiditis. There are multifocal collections of lymphocytes, sometimes with admixture of plasma cells, but germinal centres are rare and oxyphil change is usually absent. The infiltrate tends to spare the follicles and is mainly located in the fibrous septa (Fig. 4.10).

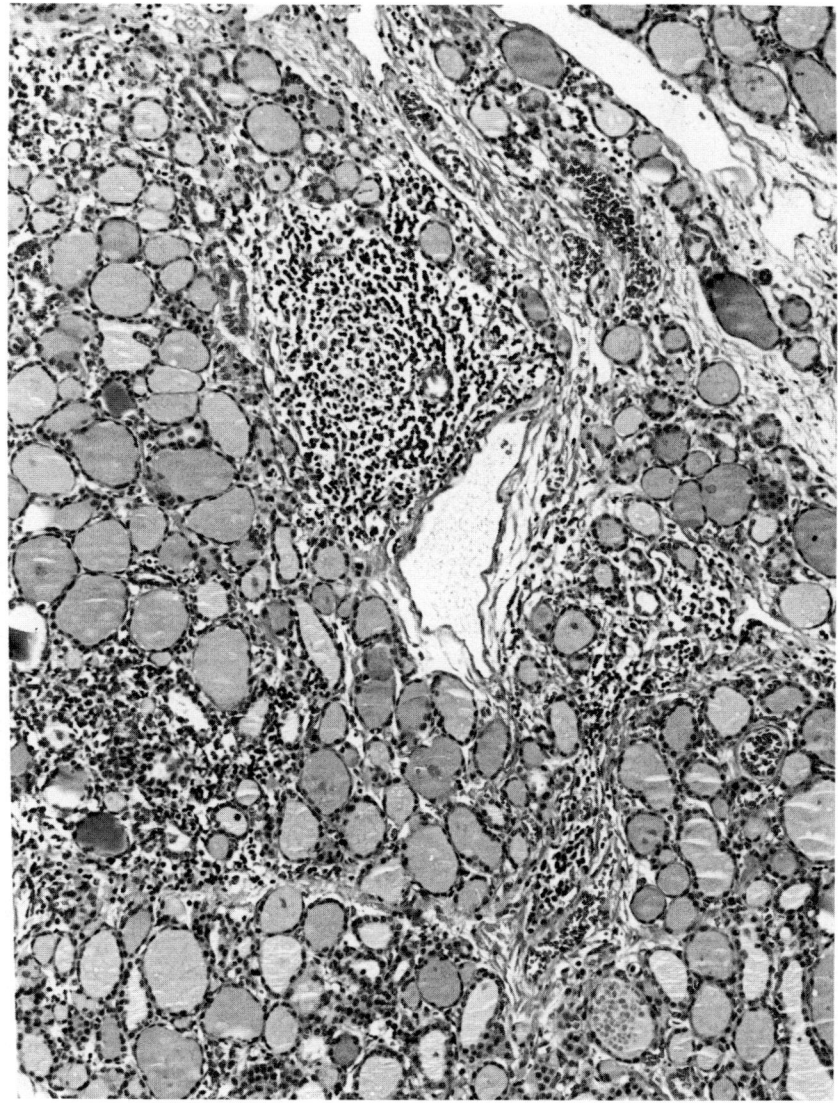

Figure 4.10 Non-specific lymphocytic thyroiditis. There are interstitial collections of lymphocytes with little involvement of the follicles. Oxyphil cell change is absent (H&E, × 125).

This form of thyroiditis is probably not a nosological entity but represents a heterogeneous group of conditions with different aetiology. It probably includes reactions to various insults to the thyroid gland. A considerable proportion of these non-specific thyroiditides may represent low-grade autoimmune thyroiditis. It has been noted that the incidence of focal lymphoid thyroiditis parallels that of thyroid antibodies in the general population and increases with advancing age (Williams and Doniach, 1962). Focal lymphocytic thyroiditis may also occur in multinodular atoxic goitres and these patients may have low titres of thyroid antibodies in the serum. The presence of lymphocytic thyroiditis in such goitres appears to be associated with an increased risk of postoperative hypothyroidism (Berglund *et al.*, 1991). Almost all thyrotoxic glands removed at operation also contain small lymphocytic foci.

Lymphoid infiltrates, often associated with plasma cells and even fibrosis, are frequently seen around neoplasms, especially papillary carcinoma. Whether this peritumour thyroiditis represents a reaction to the tumour or a pre-existing lesion, and, in the case of the latter, whether it may have premalignant significance, is a matter of controversy (LiVolsi and Merino, 1981). A neoplasm arising in a thyroid affected by Hashimoto's thyroiditis, is more likely to be a malignant lymphoma than carcinoma (Compagno and Oertel, 1980; Holm *et al.*, 1985; Aozasa *et al.*, 1987).

A variety of aetiological factors, such as viral infections, dietary agents, and drugs, can cause inflammatory disorders of the thyroid. Drugs in particular may induce autoimmune cell-mediated damage and cause chronic thyroiditis, as has been suggested in so-called lithium-associated thyroiditis (Kontozoglou and Mambo, 1983).

4.2   Subacute thyroiditis (de Quervain's thyroiditis)

Subacute thyroiditis, also called granulomatous thyroiditis, is a rare inflammatory disease of the thyroid of unknown but probably viral aetiology. The condition was described in 1904 by de Quervain, who later clearly separated it from other forms of thyroiditis (de Quervain and Giordanengo, 1936). It would appear, however, that it may take acute, subacute or chronic courses. Typically, it affects adult women (aged 20–60 years) who present with fever, tender thyromegaly, and a sore throat. In about half the patients there may initially be signs of a mild hyperthyroidism with elevated serum values of T_4 and T_3 in combination with suppression of radioiodine uptake (Volpé, 1979). There is usually a striking elevation of the sedimentation rate. Since the clinical pictures and laboratory data are characteristic in the majority of cases of subacute thyroiditis, diagnostic biopsies are not routinely used to confirm the diagnosis. The patient usually recovers spontaneously within several weeks to a few months. The pathologist, therefore, only rarely encounters this disorder in his diagnostic

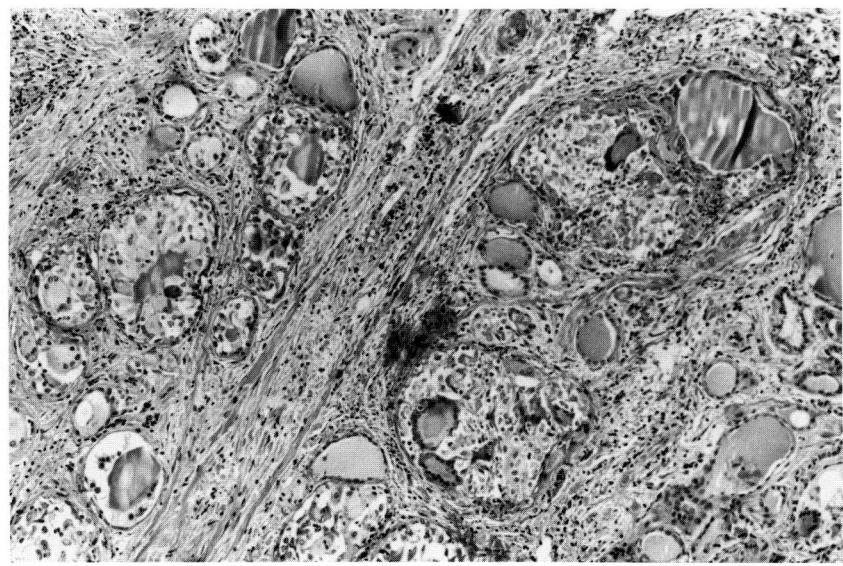

Figure 4.11 Subacute thyroiditis (de Quervain's thyroiditis). Various degrees of follicular destruction and interstitial fibrosis can be seen. The involved follicles are infiltrated by large, pale histiocytes and multinucleate giant cells, forming a granulomatous reaction to the colloid (H&E, × 100).

practice. However, asymmetric involvement of the gland is common, giving the impression of a focal process. Under such circumstances the condition may be clinically confused with a neoplasm and the lesion removed for diagnosis.

Macroscopically, the process is seen to involve the entire gland, although often asymmetrically. The gland is usually slightly or moderately enlarged and in advanced stages often rather firm. Fine adhesions may occur to adjacent structures.

Microscopic examination (Figs 4.11 and 4.12) shows an irregular pattern of involvement with various stages of destruction in different parts of the same gland (Meachim and Young, 1963; Greene, 1971). Initially, there seems to be infiltration of the follicle with large mononuclear cells, lymphocytes and neutrophils. The follicles are markedly enlarged and show hyperplasia and disruption of the lining epithelium. A progressive disruption of the basement membranes occurs, with collapse of follicles. Multinucleate giant cells accumulate in groups of five to ten which are arranged peripherally around remaining masses of colloid. A granulomatous reaction to spilled colloid is the most important histological hallmark. In later stages, there

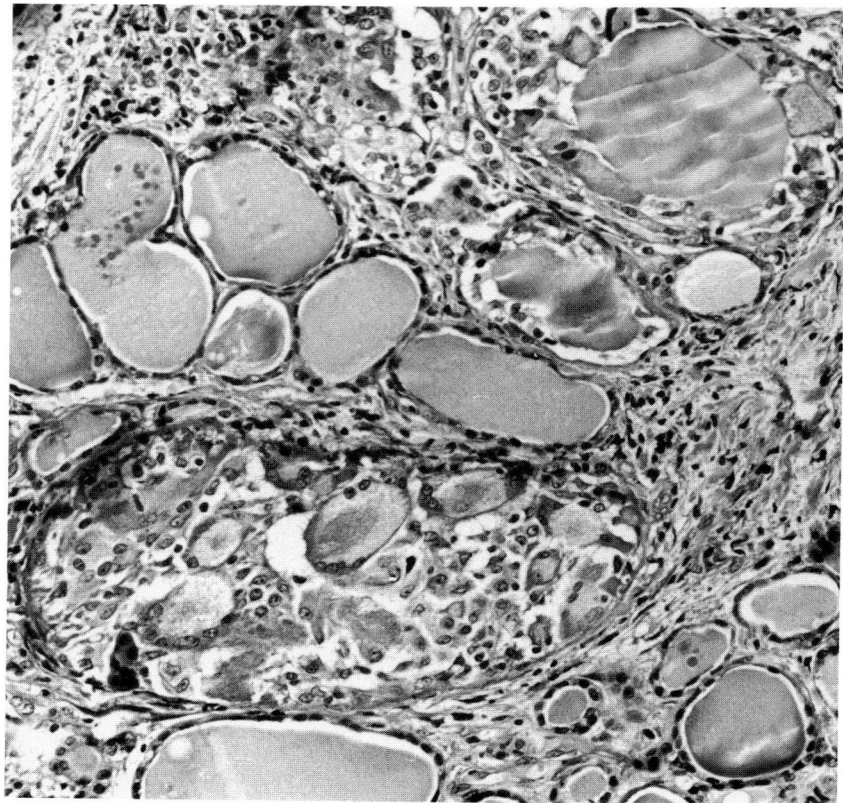

Figure 4.12 Same case as shown in Fig. 4.11. Some follicles are dilated and more or less filled with histiocytes and multinucleate giant cells with foamy cytoplasm (H&E, × 215).

are signs of interfollicular fibrosis and follicular regeneration. Although a transient hypothyreosis may occur during the course of the disease, full restitution of thyroid function is the rule. Very rarely, however, patients can progress to permanent myxoedema, in some instances due to a failure to restore the parenchyma (Bastenie and Ermans, 1972).

The aetiology of this disorder is uncertain. The disease often follows an upper respiratory tract infection. Accumulated cases have been recorded during the course of a mumps epidemic. Furthermore, it has been reported in association with several other virus infections, such as measles, Coxsackie virus and influenza (Volpé, 1979). The disease is currently believed, therefore, to have a virus aetiology, although this remains to be proved.

4.2.1   Painless thyroiditis

The majority of patients with so-called painless thyroiditis have had a goitre and have followed the course of subacute thyroiditis (Woolf and Daly, 1976; Dorfman *et al.*, 1977; Gorman *et al.*, 1978). Initially they usually also show a transient thyrotoxicosis, and low 24 h^{131}I uptake. The condition has a transient, self-limited, but sometimes recurrent course, and usually needs only symptomatic treatment. The nature of this disorder is not fully understood and, consequently, its classification is uncertain. Volpé (1979) suggested that painless thyroiditis represents a variant of de Quervain's thyroiditis. However, in those few instances when thyroid biopsies have been performed, histological or cytological examination has revealed a lymphocytic infiltration and no granulomatous features (Gluck *et al.*, 1975). Furthermore, this condition differs clinically from typical de Quervain's thyroiditis, in the absence of fever, neck pain, tenderness or marked elevation of the sedimentation rate. In some patients, thyroid auto-antibodies against thyroglobulin have been reported to be raised. These findings indicate that at least some cases may represent an autoimmune process and, therefore, should rather be classified as variants of autoimmune thyroiditis (Dorfman *et al.*, 1977; Woolf, 1980).

4.3   Riedel's thyroiditis

This is an extremely rare condition, originally described by Riedel in 1896. Invasive fibrous thyroiditis, ligneous thyroiditis and struma fibromatosa are among its other names (Woolner *et al.*, 1957). The major feature is an inflammatory fibrosis involving the thyroid parenchyma as well as the adjacent soft tissues of the neck. Woolner *et al.* found this lesion in 0.05% of patients operated on. It is reported to occur mainly between 30 and 70 years of age (average 50 years), and is two to four times more common in women than in men. The presenting complaint is a hard lump in the neck, sometimes associated with a sense of pressure, dysphagia and dyspnoea. Clincially, the lesion is often thought to be carcinoma. Hypothyreosis may occur in patients with extensive involvement of the gland. Patients with only focal lesions tend to be euthyroid on admission, and remain so.

Grossly, the thyroid gland may be slightly enlarged or normal-sized. The gland, or part of it, its capsule, surrounding connective tissue and muscles are obliterated by a dense, extremely firm fibrous tissue. The affected part of the gland is stony hard and difficult to slice. Involvement may be asymmetrical in that only a lobe or a part thereof is affected. The spared parts of the gland have a normal appearance.

Histological examination reveals a fibrous tissue, often highly hyalinized and relatively acellular, that has replaced the involved area of the gland. In some areas, the fibrous tissue may contain infiltrates of inflammatory cells, mainly lymphocytes and plasma cells, and occasionally also neutrophils (Plate 4). In the affected parts of the gland there is an extensive destruction of the thyroid follicles. The few remaining follicles are atrophic without oxyphil cell changes. Parts not obliterated by the fibrous process appear histologically normal. The process typically involves the perithyroidal soft tissue including the adjacent striated muscles (Plate 5).

A perivasculitis and vasculitis, especially involving the veins, is often encountered in the extensive proliferation of fibrous tissue (Plate 5) (Meyer and Hausman, 1976; Harach and Williams, 1983).

Riedel's thyroiditis should not be confused with the more common fibrous variant of Hashimoto's thyroiditis (see p. 61 and Fig. 4.8). The major distinguishing histological features are that, in Riedel's thyroiditis, the fibrous tissue extends outside the gland, there is a fibro-occlusive phlebitis, and the spared thyroid tissue is relatively normal. In Hashimoto's thyroiditis, on the other hand, the fibrous tissue is confined to the thyroid gland, vasculitis is absent and the entire gland is involved by the process (Fig. 4.8).

The aetiology of this rare condition is unknown. It is sometimes associated with retroperitoneal fibrosis (Rao *et al.*, 1973), mediastinal fibrosis (Raphael *et al.*, 1966), orbital sclerosis (Arnott and Greaves, 1965) and sclerosing cholangitis (Bartholomew *et al.*, 1963). Currently, it is considered to be a manifestation of a spectrum of idiopathic connective tissue disorders, which are also associated with phlebitis, and referred to as inflammatory fibrosclerosis (Comings *et al.*, 1967).

4.4 Other forms of thyroiditis

4.4.1 Infective thyroiditis

Thyroiditis due to infection is very rare. Bacterial thyroiditis is seen in debilitated children and in the elderly, or in immunodeficient individuals. It may be caused by bacteria, such as staphylococci, streptococci, pneumococci, and *Escherichia coli* (Volpé 1978b). The thyroid involvement is generally a result of haematogenous spread from a distant focus. The disease may also complicate suppurative infections in the upper respiratory tract (Berger *et al.*, 1983; Hay, 1985) and there may be abscess formation which needs to be surgically drained.

Overwhelming systemic fungaemia in immunocompromised patients may, likewise, be complicated by fungal thyroiditis. Such cases are usually diagnosed at autopsy. They are most frequently caused by *Candida* or *Aspergillus*. Immunodeficiency may interfere with the appropriate inflammatory reaction in the thyroid and, therefore, the reaction is not always granulomatous. Instead, very little reaction or even an acute inflammation may be found.

Viral infection of the thyroid is thought to be very rare. In adults, however, it may occur in the presence of disorders of the immune system. Cytomegalo virus infection in the thyroid has been recorded in patients with the acquired immunodeficiency syndrome (AIDS) (Frank *et al.*, 1987).

Very rarely, tuberculosis may secondarily involve the thyroid gland. It may occur as scattered granulomas in a case of miliary tuberculosis. Tuberculosis of cervical lymph nodes or larynx may also spread by direct extension into the gland. Cases diagnosed in the past as tuberculous thyroiditis were probably examples of de Quervain's thyroiditis. The presence of well-formed granulomas with central caseation and haphazardly distributed in the thyroid parenchyma, and the demonstration of acid-fast bacilli are the features that distinguish the lesion from de Quervain's thyroiditis.

Tuberculosis must also be distinguished from sarcoidosis of the thyroid, which rarely may also involve the thyroid (Winnacker *et al.*, 1968). In this condition well-formed granulomas without caseation occur in the thyroid parenchyma. In contrast to de Quervain's thyroiditis, the granulomatous reaction in sarcoidosis is not specifically directed against spilled colloid, and the process leaves the intervening parts of the parenchyma unaffected.

4.4.2 Palpation thyroiditis

This thyroiditis is the result of repeated and zealous physical examination of the gland preoperatively. Histologically, the lesions are seen to have a focal distribution, involving scattered, isolated follicles which appear ruptured with accumulations of macrophages and foreign body giant cells around escaped colloid (Carney *et al.*, 1975). The lesion has most often been found in glands with adenomatous nodules or tumours.

References

Aichinger, G., Fill, H. and Wick, G. (1985) In situ immune complexes, lymphocyte subpopulations, and HLA-DR-positive epithelial cells in Hashimoto thyroiditis. *Lab. Invest.*, **52**, 132–40.

Albores-Saavedra, J., Monforte, H., Nadji, M. and Morales, A.R. (1988) C-cell hyperplasia in thyroid tissue adjacent to follicular cell tumors. *Hum. Pathol.*, **19**, 795–9.

Aozasa, K., Ueda, T., Katagiri, S. *et al.* (1987) Immunologic and immunohistologic analysis of 27 cases with thyroid lymphomas. *Cancer*, **60**, 969–73.

Arnott, E.J. and Greaves, D.P. (1965) Orbital involvement in Riedel's thyroiditis. *Br. J. Ophthalmol.*, **49**, 1–5.

Bartholomew, L.G., Cain, J.C., Woolner, L.B. *et al.* (1963) Sclerosing cholangitis: its possible association with Riedel's struma and fibrous retroperitonitis. *N. Engl. J. Med.*, **269**, 8–12.

Baschieri, L., Castagna, M., Fierabracci, A. *et al.* (1989) Distribution of calcitonin- and somatostatin-containing cells in thyroid lymphoma and in Hashimoto's thyroiditis. *Appl. Pathol.*, **7**, 99–104.

Bastenie, P.A. and Ermans, A.M. (1972) Thyroiditis and thyroid function. Clinical, morphological and physiological studies. In *International Series of Monographs in Pure and Applied Biology: Modern Trends in Physiological Sciences*, Vol. 36, Pergamon Press, Oxford.

Ben-Ezra, J., Wu, A. and Sheibani, K. (1988) Absence of clonal lymphoid gene rearrangements in Hashimoto's thyroiditis (abstract). *Lab. Invest.*, **68**, 9A.

Berger, S.A., Zonszein, J., Villamena, P. and Mittman, N. (1983) Infectious disease of the thyroid gland. *Rev. Infect. Dis.*, **5**, 108–22.

Berglund, J., Bondeson, L., Borup Christensen, S. and Tibblin, S. (1991) The influence of different degrees of chronic lymphocytic thyroiditis on thyroid function after surgery for benign, non-toxic goitre. *Eur. J. Surg.*, **157**, 257–60.

Biddinger, P.W., Brennen, M.F. and Rosen, P.P. (1991) Symptomatic C-cell hyperplasia associated with chronic lymphocytic thyroiditis. *Am. J. Surg. Pathol.*, **15**, 599–604.

Brown, J., Solomon, D.H., Beall, G.N. *et al.* (1978) Autoimmune thyroid disease – Graves' and Hashimoto's. *Ann. Intern. Med.*, **88**, 379–91.

Burek, C.L. and Rose, N.R. (1986) Cell-mediated immunity in autoimmune thyroid disease. *Hum. Pathol.*, **17**, 246–53.

Carney, J.A., Moore, S.B., Northcutt, R.C. *et al.* (1975) Palpation thyroiditis (multifocal granulomatous folliculitis). *Am. J. Clin. Pathol.*, **64**, 639–47.

Carpenter, C.C.J., Solomon, N., Silverberg, S.G. *et al.* (1964) Schmidt's syndrome (thyroid and adrenal insufficiency). A review of the literature and a report of fifteen new cases including ten instances of coexistent diabetes mellitus. *Medicine (Baltimore)*, **32**, 153–80.

Comings, D.E., Skubi, K.B., van Eyes, J. *et al.* (1967) Familial multifocal fibrosclerosis. *Ann. Intern. Med.*, **66**, 884–92.

Compagno, J. and Oertel, J.E. (1980) Malignant lymphoma and other lymphoproliferative disorders of the thyroid gland. A clinicopathologic study of 245 cases. *Am. J. Clin. Pathol.*, **74**, 1–11.

de Quervain, F. (1904) Die akute, nicht eitrige Thyreoiditis und die Beteiligung der Schilddrüse an akuten Intoxikationen und Infektionen überhaupt. *Mitt. Grenzgeb. Med. Chir.*, Suppl. 2, 1–165.

de Quervain, F. and Giordanengo, G. (1936) Die akute und subakute nicht eitrige Thyreoiditis. *Mitt. Grenzgeb. Med. Chir.*, **44**, 538–90.

Dhillon, A.P., Rode, J., Leathem, A. and Papadaki, L. (1982) Somatostatin: a paracrine contribution to hypothyroidism in Hashimoto's thyroiditis. *J. Clin. Pathol.*, **35**, 764–70.

Doniach, D. (1972) Autoimmunity in liver disease. In *Progress in Clinical Immunology*, vol. 1 (ed. R.S. Schwartz), Grune and Stratton, New York, pp. 45–70.

Doniach, D., Bottazzo, G.F. and Russell, R.C.G. (1979) Goitrous autoimmune thyroiditis. *Clin. Endocrinol. Metab.*, **8**, 63–80.

Dorfman, S.G., Cooperman, M.T., Nelson, R.L. *et al.* (1977) Painless thyroiditis and transient hyperthyroididism without goiter. *Ann. Intern. Med.*, **86**, 24–8.

Frank, T.S., LiVolsi, V.A. and Connor, A.M. (1987) Cytomegalovirus infection of the thyroid in immunocompromised adults. *Yale J. Biol. Med.*, **60**, 1–8.

Gavras, I. and Thomson, J.A. (1972) Late thyrotoxicosis complicating autoimmune thyroiditis. *Acta Endocrinol.*, **69**, 41–6.

Gluck, F.B., Nusynowitz, M.L. and Plymate, S. (1975) Chronic lymphocytic thyroiditis, thyrotoxicosis, and low radioactive iodine uptake. *N. Engl. J. Med.*, **293**, 624–8.

Gorman, C.A., Duick, D.S., Woolner, L.B. and Wahner, H.W. (1978) Transient hyperthyroidism in patients with lympohocytic thyroiditis. *Mayo Clin. Proc.*, **53**, 359–65.

Greene, J.N. (1971) Subacute thyroiditis. *Am. J. Med.*, **51**, 97–108.

Harach, H.R. and Williams, E.D. (1983) Fibrous thyroiditis – an immunopathological study. *Histopathology*, **7**, 739–51.

Hashimoto, H. (1912) Zur Kenntnis der lymphomatösen Veränderungen der Schilddrüse (Struma lymphomatosa). *Langenbecks Arch. Klin. Chir.*, **97**, 219–48.

Hay, I.D. (1985) Thyroiditis: a clinical update. *Mayo Clin. Proc.*, **60**, 836–43.

Holm, L.-E., Blomgren, H. and Löwhagen, T. (1985) Cancer risks in patients with chronic lymphocytic thyroiditis. *N. Engl. J. Med.*, **312**, 601–4.

Irvine, W.J. (1975) Autoimmunity in endocrine disease. *Clin. Endocrinol. Metab.*, **4**, 227–499.

Katz, S.M. and Vickery, A.L., Jr (1974) The fibrous variant of Hashimoto's thyroiditis. *Hum. Pathol.*, **5**, 161–70.

Kini, S.R. (1987) *Guides to Clinical Aspiration Biopsy. Thyroid.* Igaku-Shoin, New York and Tokyo, pp. 235–73.

Kontozoglou, T. and Mambo, N. (1983) The histopathologic features of lithium-associated thyroiditis. *Hum. Pathol.*, **14**, 737–9.

Lamberg, B-A. (1979) Aetiology of hypothyroidism. *Clin. Endocrinol. Metab.*, **8**, 3–19.

Libbey, N.P., Nowakowski, K.J. and Tucci, J.R. (1989) C-cell hyperplasia of the thyroid in a patient with goitrous hypothyroidism and Hashimoto's thyroiditis. *Am. J. Surg. Pathol.*, **13**, 71–7.

LiVolsi, V.A. (1988) The thyroid: new issues, problem areas, and recent advances. In *Pathology of the Head and Neck* (ed. D.R. Gnepp), Churchill Livingstone, New York, Edinburgh, London and Melbourne, pp. 517–45.

LiVolsi, V.A. and Merino, M.J. (1978) Squamous cells in the human thyroid. *Am. J. Surg. Pathol.*, **2**, 133–40.

LiVolsi, V.A. and Merino, M.J. (1981) Histopathologic differential diagnosis of the thyroid. *Pathol. Annu.*, **16** (2), 357–406.

Meachim, G. and Young, M.H. (1963) De Quervain's subacute granulomatous thyroiditis: histological identification and incidence. *J. Clin. Pathol.*, **16**, 189–99.

Meyer, S. and Hausman, R. (1976) Occlusive phlebitis in multifocal fibrosclerosis. *Am. J. Clin. Pathol.*, **65**, 274–83.

Rallison, M.L., Dobyns, B.M., Keating, F.R. *et al.* (1975) Occurrence and natural history of chronic lymphocytic thyroiditis in childhood. *J. Pediatr.*, **86**, 675–82.

Rao, C., Ferguson, G.C. and Kyle, V.N. (1973) Retroperitoneal fibrosis associated with Riedel's struma. *Can. Med. Assoc. J.*, **108**, 1019–21.

Raphael, H.A.L., Beahrs, O.H., Woolner, L.B. and Schulz, D.A. (1966) Riedel's struma associated with fibrous mediastinitis: report of a case. *Mayo Clin. Proc.*, **41**, 375–82.

Riedel, B. (1896) Die chronische, zur Bildung eisenharter Tumoren führende Entzündung dre Schilddrüse. *Verh. Dtsch. Ges. Chir.*, **25**, 101–5.

Rosai, J. (1989) Thyroid gland. In *Ackerman's Surgical Pathology*, Vol. 1 (ed. J. Rosai), C.V. Mosby, St Louis, Toronto and Washington, DC, pp. 391–447.

Segal, B.M. and Weintraub, M.I. (1976) Hashimoto's thyroiditis, myasthenia gravis, idiopathic thrombocytopenic purpura. *Ann. Intern. Med.*, **85**, 761–2.

Strakosch, C.R., Wenzel, B.E., Row, V.V. and Volpé, R. (1982) Immunology of autoimmune thyroid diseases. *N. Engl. J. Med.*, **307**, 1499–507.

Volpé, R. (1977) The role of autoimmunity in hypoendocrine and hyperendocrine function. *Ann. Intern. Med.*, **87**, 86–99.

Volpé, R. (1978a) The pathology of thyroiditis. *Hum. Pathol.*, **9**, 427–38.

Volpé, R. (1978b) Etiology, pathogenesis, and clinical aspects of thyroiditis. *Pathol. Annu.* **13** (2), 399–413.

Volpé, R. (1979) Subacute (de Quervain's) thyroiditis. *Clin. Endocrinol. Metab.*, **8**, 81–95.

Volpé, R., Farid, N.R., von Westarp, C. and Row, V.V. (1974) The pathogenesis of Graves' disease and Hashimoto's thyroiditis. *Clin. Endocrinol.*, **3**, 239–61.

Willems, J.-S. and Löwhagen, T. (1981) Fine-needle aspiration cytology in thyroid disease. *Clin. Endocrinol. Metab.*, **10**, 247–66.

Williams, E.D. and Doniach, I. (1962) The post-mortem incidence of focal thyroiditis. *J. Pathol. Bacteriol.*, **83**, 255–64.

Winnacker, J.L., Becker, K.L. and Katz, S. (1968) Endocrine aspects of sarcoidosis. *N. Engl. J. Med.*, **278**, 483–92.

Woolf, P.D. (1980) Transient painless thyroiditis with hyperthyroidism: a variant of lymphocytic thyroiditis? *Endocr. Rev.*, **1**, 411–20.

Woolf, P.D. and Daly, R. (1976) Thyrotoxicosis with painless thyroiditis. *Am. J. Med.*, **60**, 73–9.

Woolner, L.B., McConahay, W.M. and Beahrs, O.H. (1957) Invasive fibrous thyroiditis. (Riedel's struma). *J. Clin. Endocrinol. Metab.*, **17**, 201–20.

5 Non-toxic goitre

Non-toxic goitre is a descriptive rather than diagnostic term for a non-neoplastic enlargement of the thyroid gland which, traditionally, is considered to be the result of compensation or adaptation phenomena induced by an intermittent or persistent inadequate production of thyroid hormones. Other synonyms for this disorder include colloid goitre, nodular goitre, adenomatous goitre, and multinodular goitre.

The majority of cases are not associated with any clinical signs of thyroid malfunction. In some cases, however, the gland fails to produce sufficient functioning hormones, and hypothyreosis may ensue. Conversely, the disorder may rarely also be complicated by inappropriate, autonomous over-production of thyroid hormones leading to hyperthyroidism. Nodular goitre is the most common benign condition mistaken for a thyroid neoplasm (Meissner and Warren, 1969).

A deficiency of thyroid hormone may be due to dietary iodine deficiency, inborn errors of metabolism, dietary goitrogens, and goitrogenic drugs and chemicals, and various factors causing damage to the parenchyma, such as inflammation and radiation (Hennemann, 1979). Insufficient hormone production increases the secretion of pituitary thyroid-stimulating hormone (TSH), which results initially in increased activity of thyroid follicular cells. The thyroid gland will respond with epithelial hyperplasia with tall epithelial cells and small amounts of colloid (Studer and Greer, 1965; Studer et al., 1979). This is the so-called parenchymatous goitre, which later may develop follicular atrophy with extensive accumulation of colloid, causing further enlargement (so-called diffuse colloid goitre) and as time passes, progressive nodularity of the goitre (so-called nodular colloid goitre). Although TSH is thought to be the most important factor that initiates the structural changes in the gland, other mechanisms, either alone or as contributory factors, have also been suspected to be involved in the pathogenesis of certain forms of the disease.

5.1 Endemic goitre

By definition, endemic goitre is present in populations where its prevalence rate exceeds 10% (Ibbertson, 1979). This disorder occurs in parts of the world where water and soil have a low iodine content. In these areas, the disease is also common among domestic animals. In most developed countries iodized salt is used to supplement the iodine content of the diet and this measure can largely prevent the disease. Endemic goitre, however, is still present in many parts of the world, such as the Alpine countries of Europe, the Himalayan chain, South East Asia, certain areas of the Indonesian and Philippine archipelagos, the Andean Mountain chain of South America, and regions around the Great Lakes of North America (Kelly and Snedden, 1960; Schrimshaw, 1964). In these areas, chronic dietary deficiency of iodine still exists but iodine prophylaxis may not always eradicate the goitre. In such circumstances, subsidiary factors have been suspected in the aetiology of the disorder, including inborn errors of hormone synthesis, ingestion of natural goitrogens, polluted water supplies and infection (Ermans, 1979).

Endemic goitre occurs at all ages. Diffuse enlargement of the thyroid gland is often present in children of both sexes at puberty. This represents the early, compensatory stage of diffuse parenchymatous hyperplasia, characterized by depletion of colloid and follicular cell hyperplasia. From adolescence to old age there is a progressive development of nodularity. Nodular goitre is three to six times more common in women than men (Henneman, 1979). All types of thyroid enlargement may be seen, from a single nodule without any appreciable hyperplasia of the rest of the gland to enormous multinodular goitre, and the complications are principally the same as for the sporadic goitre.

As mentioned above, most patients are euthyroid since iodine is only relatively deficient. When there is an absolute lack of the iodine hypothyroidism occurs. This condition is especially prominent in children because the maternal supply of iodine to the fetus is insufficient, and endemic goitrous cretinism is the most severe complication of endemic goitre.

The early compensatory stage of diffuse parenchymatous goitre and diffuse colloid goitre are seldom seen by the pathologist as there is usually no need for surgical interference. The nodular goitre, however, may grow very large, cause much disfigurement or pressure symptoms, and even raise clinical suspicion of malignancy. Under these circumstances surgically removed thyroid specimens show the same gross and microscopic features as those described for the sporadic form of non-toxic goitre (see below). Himalayan endemic goitre is remarkable in that the hyperplasia tends to be very intense and dominate for decades without signs of involution (Ramalingaswami, 1969). Nodules may develop with age, but hyperplasia without colloid formation tends to persist in these nodules, showing very cellular patterns with marked cytological atypia which may cause confusion with carcinoma. Ramalingaswami (1969) commented on the similarity of these goitres to the dyshormonogenetic type of goitre and suggested that the difference between nodular endemic goitre in the Himalayan area and that of Europe and North America could be due to a more severe and continued dietary deficiency of iodine in the former.

5.2 Dyshormonogenetic goitre

There are several types of goitre caused by various enzyme defects in hormone synthesis (Kennedy, 1969). Defects in each major step in the synthesis of thyroid hormone have been described (Table 5.1), but in some instances the biochemical abnormality is not demonstrable (Barsano and DeGroot, 1979). The various forms of dyshormonogenetic goitre have a number of features in common. They are inherited, most often as a Mendelian recessive. The goitre is often associated with hypothyroidism and the disease is usually present in childhood. However, the spectrum of hypothyroidism may vary from severe mental and physical features of cretinism to clinical euthyroidism. Mildly hypothyroid or euthyroid patients with minimal goitre may sometimes be diagnosed only in adult life, often when screening family members of an obviously afflicted individual. Surgical resection of the goitre is invariably followed by recurrence, due to enlargement of the residual thyroid tissue, unless hormonal substitution is given. Dyshormonogenetic goitre is uncommon and comprises only a small proportion of all cases of non-toxic goitre. In addition to the six groups of enzymic defects listed in Table 5.1, a seventh category of dyshormonogenetic goitre may be included, which is secondary to end organ resistance to thyroid hormone (Refetoff et al., 1967). This syndrome is characterized by high serum levels of free thyroid hormones, although the patients are clinically euthyroid or show some signs of hypothyroidism. TSH secretion is normal or even elevated. In some cases a decreased affinity of nuclear T_3 receptor has been demonstrated (Ichikawa et al., 1987). In some families with this disorder, the goitre has been associated with deaf mutism, stippled bony epiphyses and a characteristic facies (Refetoff et al., 1967).

The dyshormonogenetic goitre is typically an asymmetrical, multinodular gland which may reach large size (weighing several hundred grams). The nodules tend to have a more solid and fleshy consistency than those in the common sporadic and endemic non-toxic goitre. Degenerative changes with cyst formation, focal fibrosis and calcification are also frequently seen.

Table 5.1 Description of thyroid hormone synthesis and metabolism, and the main biochemical defects causing dyshormonogenetic goitre

Normal hormone synthesis and metabolism	Enzymatic defects
Trapping and concentration of inorganic iodide.	Iodide trap defect. Impaired concentration of iodide from the circulation.
Oxidation of iodide to an active form of iodine. Iodination of tyrosyl residues in thyroglobulin (formation of mono- and diiodotyrosine, MIT and DIT). Oxidative condensation (coupling) of these to yield a variety of iodothyronines, e.g. T_4 T_3 and reverse T_3. Thyroid peroxidase plays a key catalytic role in all these reactions.	Organification group of defects (peroxidase abnormalities). A heterogeneous group with various subgroups, including Pendred's syndrome (congenital goitre and deafness). 'Coupling' defect. Results in lowered ratio of iodothyronines to iodotyrosines in the gland. The fundamental biochemical defect is unknown.
Hormone containing molecules of thyroglobulin, stored in the colloid, are engulfed by pinocytosis, and hydrolysed by lysosomal proteases to release thyroid hormones along with iodotyrosines that have not been coupled. Thyroid hormones are delivered to the circulation by diffusion. Superfluous idiotyrosines are reconverted into iodide and tyrosine by intrathyroidal dehalogenase and mainly used for subsequent hormone production. Thyroid hormones are deiodinated by peripheral tissues, a minor fraction is disposed of by biliary secretion of their glucuronide and sulphate conjugates.	Protease defect. Impairment of release of the thyroid hormones from thyroglobulin. Dehalogenase (deiodinase) defect. The thyroid and other tissues are deficient in breaking down iodotyrosines to release iodine. Iodotyrosines escape into the circulation and are excreted in the urine, resulting in iodine deficiency. Mental retardation, goitre and hypothyroidism are presenting features. Defects in iodoprotein synthesis. A heterogeneous group of inborn errors, the most severe forms of which cause goitrous hypothyroidism. Release of abnormal iodoproteins, such as iodinated albumin, from the thyroid. Structurally abnormal thyroglobulins have also been demonstrated.

Microscopically, tissue obtained at thyroidectomy often reveals an intense focal or diffuse hyperplasia. Nodules are often present, showing micro- or macrofollicular patterns, sometimes even very cellular, trabecular formations (Figs. 5.1 and 5.2). Cytological atypia may be marked, papillary structures may occur and even features mimicking capsular invasion. Such findings may prompt a diagnosis of follicular

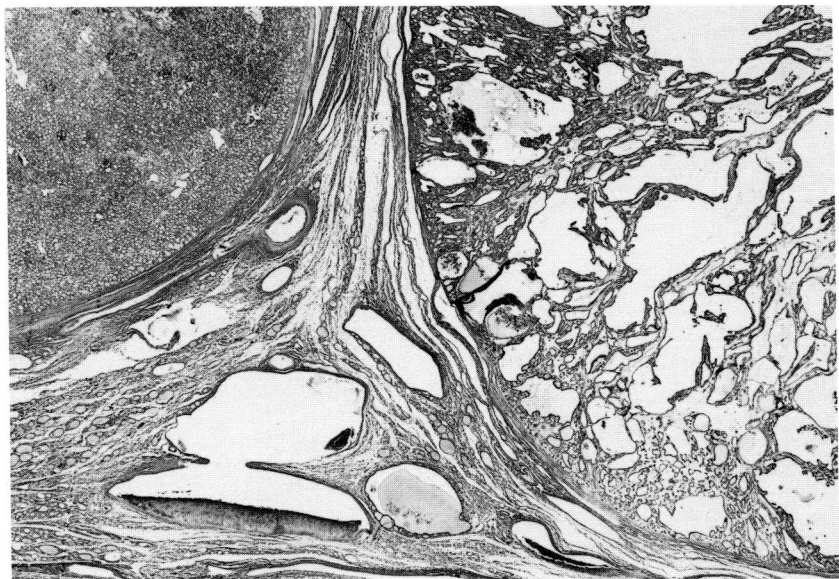

Figure 5.1 Dyshormonogenetic goitre (Pendred's syndrome). Nodules showing macro- and microfollicular patterns are embedded in atrophic thyroid parenchyma with some adjacent fibrosis (H&E, × 9).

or papillary carcinoma. No metastases have been recorded in such cases and the diagnosis of carcinoma must be avoided (Moore, 1962; Kennedy, 1969; Doniach, 1970). It is important to follow strictly the criteria used for the diagnosis of carcinoma including unequivocal capsular penetration and/or blood vessel invasion (Vickery, 1981). Although cases of carcinoma, usually of the follicular type, have been reported in patients with dyshormonogenetic goitre, this is a very rare observation and there is no convincing evidence of an increased cancer incidence among these patients.

5.3 Sporadic goitre

This term refers to a benign, diffuse or multinodular enlargement of the thyroid of unknown aetiology occurring sporadically in non-endemic areas where it constitutes the great majority of goitres seen in the general population. Other terms used to describe this condition are simple non-toxic goitre (Hennemann, 1979), simple goitre and colloid goitre.

The pathogenesis of this form of goitre is not known. However, it is often familial and genetic defects leading to an undisclosed,

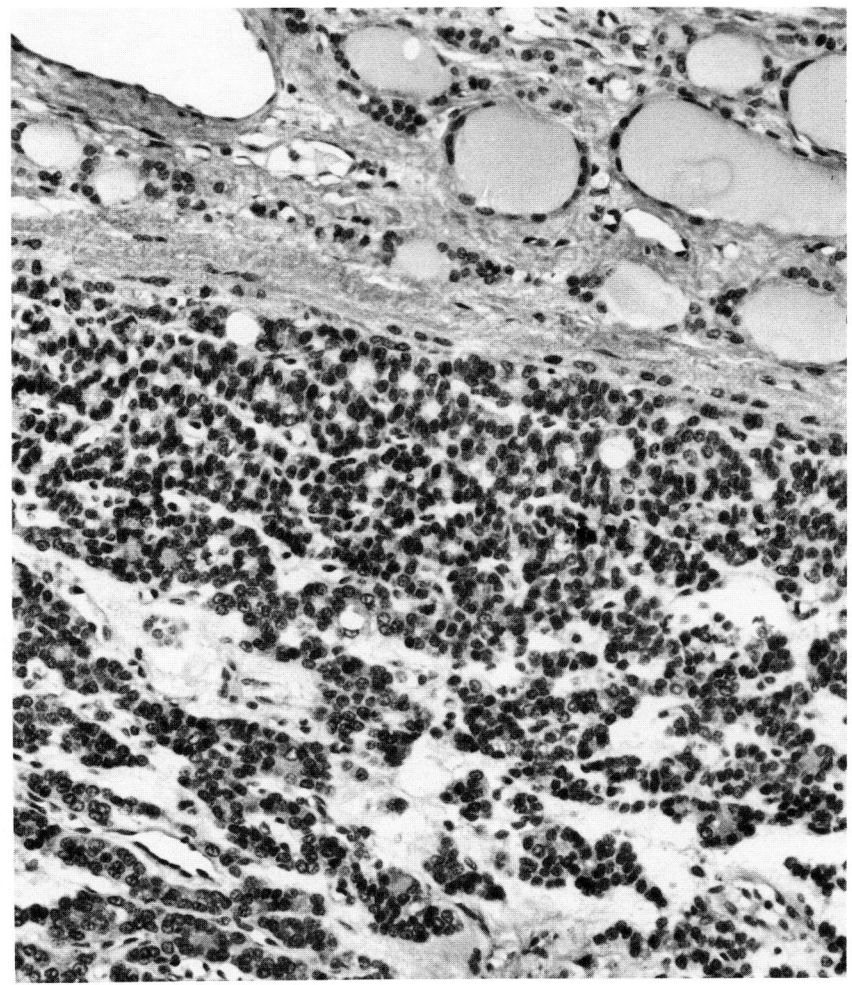

Figure 5.2 Same case as shown in Fig. 5.1. At the bottom is part of a cellular hyperplastic nodule with microfollicular and trabecular structures. At the top is part of a colloid nodule with distended follicles and atrophic epithelium (H&E, × 250).

slight impairment of thyroid hormone synthesis have been suspected to underlie the disorder. Other factors incriminated include mild dietary deficiency of iodine, increased iodine clearance by the kidneys, presence of weak goitrogens in the food, and autoimmune mechanisms with development of thyroid-stimulating immunoglobulins (Brown *et al.*, 1978; Hennemann, 1979; Studer and Ramelli, 1982). Sporadic nodular goitre is sometimes associated with lymphocytic thyroiditis or Hashimoto's thyroiditis. It has been suggested that thyroid enlargement results from a goitrogenic stimulus that must be a weak agent acting over a long period of time. This agent may be TSH, although in most patients, measurements of blood levels of TSH have shown normal values. Other stimulating factors may be growth-promoting immunoglobulins, or unknown, locally arising growth factors (Ramelli *et al.*, 1982; Thomas *et al.*, 1989). Marine (1924) and later Beckers (1979) proposed that, due to alternating periods of relative iodine deficiency and repletion, cyclic changes of hyperplasia and involution with colloid accumulation are induced in the gland, which finally result in multinodular goitre formation.

Multinodular goitre occurs clinically in 3–5% of the general adult population and there is a significant female preponderance. The incidence increases with age. In autopsy material, nodular thyroid glands have been found in 25–50% of the cases, but most would not have been clinically palpable (Mortensen *et al.*, 1955; Denham and Wills, 1980).

Clinically, the majority of patients are euthyroid and present with a multinodular goitre that may be very large and disfiguring, which is the most common reason for surgery. Less frequently, it may cause tracheal obstruction. Sometimes, a dominant nodule may present itself as a solitary tumour and its distinction from a true neoplasm may then be clinically difficult or impossible. Occasionally, the gland may grow very large and bulky, and even extend into the upper mediastinum and cause tracheal and venous compression. Acute haemorrhage within a nodule, often with cystic degeneration, may occur causing pain and sudden enlargement of the gland. Rarely, haemorrhage may be massive extending into the soft parts of the neck and causing death from blood loss and/or respiratory obstruction.

The nodules tend to be functioning autonomously (Beckers, 1979). This autonomy and hyperactivity of the nodule gradually disappears, but may then be taken over by other nodules in other parts of the goitrous gland. The iodine turnover rate within the gland tends to become increasingly heterogeneous. This is demonstrated in scintigraphic and autoradiographic studies usually showing a patchy distribution of the iodine isotope with alternating 'hot' and 'cold' areas (Beckers, 1979). This may also be reflected immnohistochemically by a difference in intensity of thyroglobulin immunoreactivity between nodules, in that cellular and hyperplastic nodules usually show an intense reactivity whereas macrofollicles in colloid-rich nodules mostly stain more weakly (Figs 5.3 and 5.4). In 'cold' nodules the turnover rate of iodine is low and such nodules are characterized

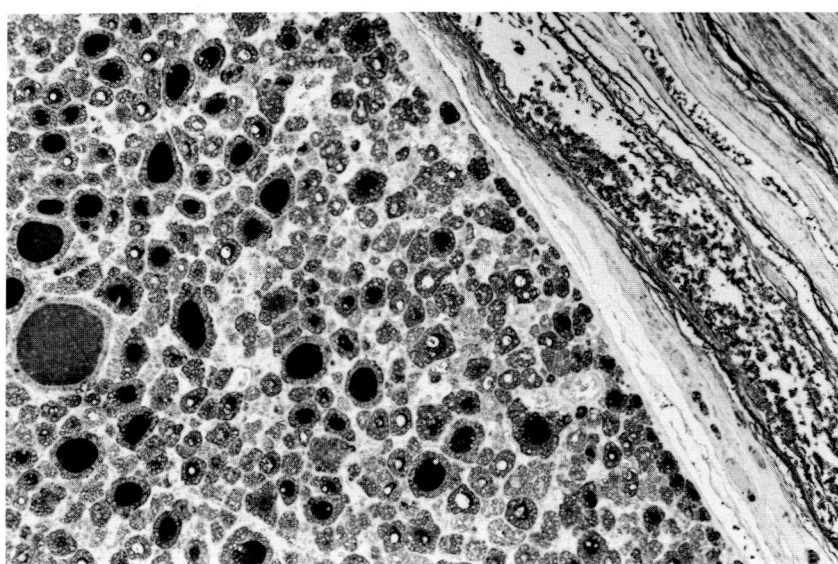

Figure 5.3 Sporadic, non-toxic nodular goitre. Hyperplastic nodule is mainly composed of small follicles with high follicular epithelium. Intense immunoreactivity for thyroglobulin in the colloid and follicular cells (ABC technique with antibodies against human thyroglobulin, × 90).

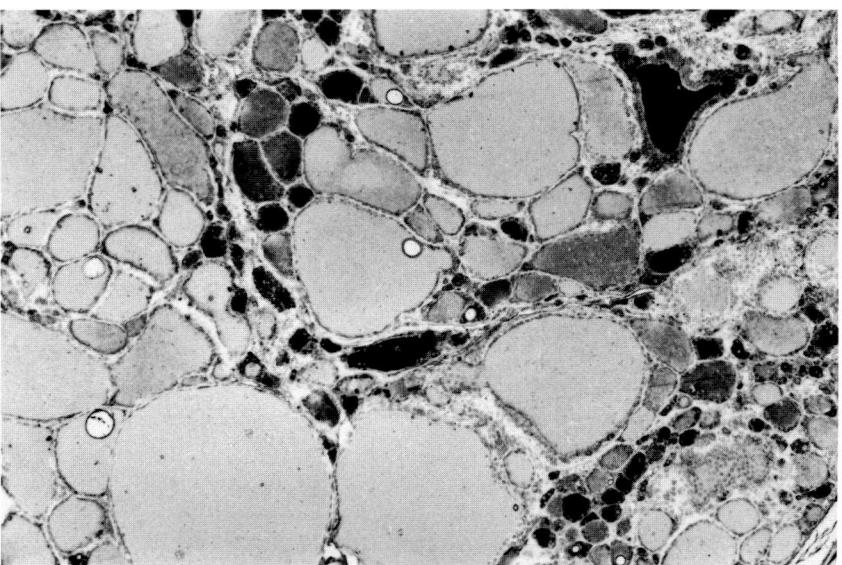

Figure 5.4 Same case as shown in Fig. 5.3. Part of an involutional nodule mainly composed of large follicles, which are distended with colloid and lined by flat, atrophic epithelium. The macrofollicles reveal only weak immunostaining for thyroglobulin (ABC technique with antibodies against human thyroglobulin, × 90).

by flat epithelium in their follicles distended with colloid. Large amounts of iodine may be sequestered in such nodules and become unavailable for hormone synthesis.

Although 'hot' nodules do not usually give rise to clinical hyperthyroidism, this may happen occasionally and cause a state of secondary hyperthyroidism which is known as toxic nodular goitre (see below).

5.4 Pathology of multinodular goitre

The patient with a euthyroid multinodular goitre will usually come to operation because of obstructive symptoms or for cosmetic reasons, more rarely because of clinical suspicion of neoplasia. The operation most commonly performed in these cases is a subtotal thyroidectomy. Grossly, the thyroid is much enlarged and distorted. Plate 6 shows the variable, irregular nodular appearance characteristic of the multinodular colloid goitre. The process is frequently asymmetrical and the size of the lobes may differ considerably. The surface of the gland shows a knobby configuration confined by a stretched but intact capsule. The cut surfaces disclose multiple various-sized nodules, some of which may be partially or completely encapsulated, others being

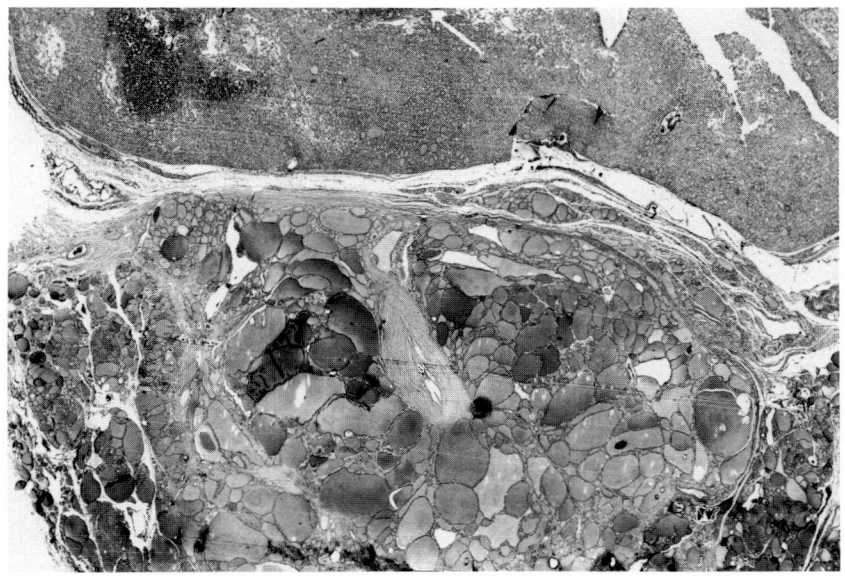

Figure 5.5 Multinodular goitre. An ill-defined colloid nodule is composed of dilated follicles with accumulation of colloid. At the top is part of another nodule which is hyperplastic and cellular, with scanty colloid (H&E, × 9).

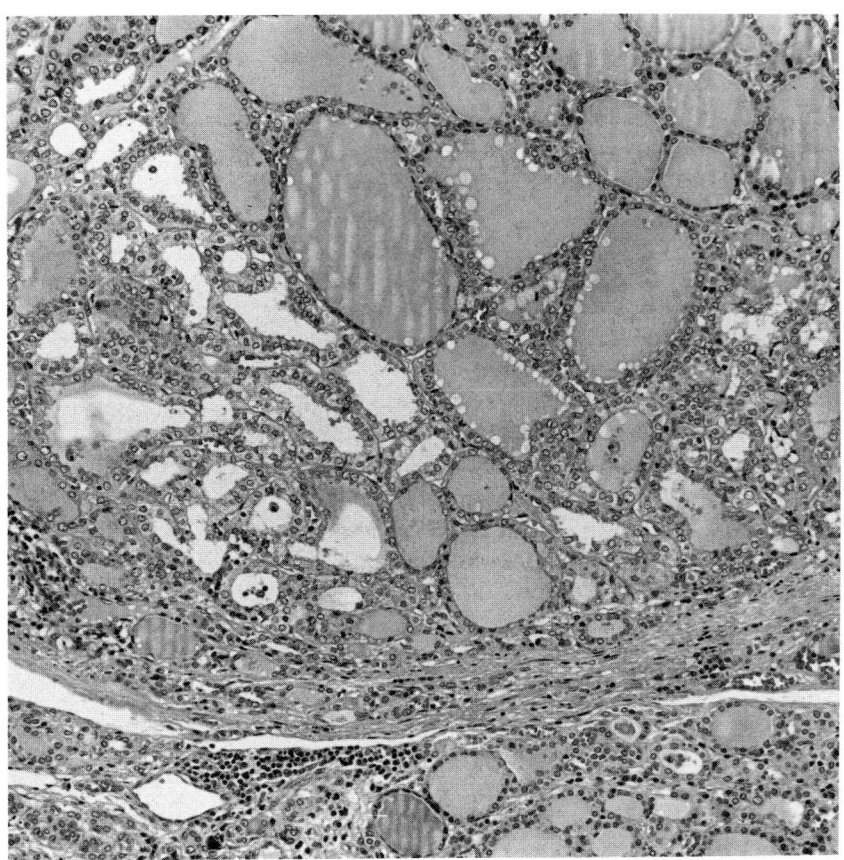

Figure 5.6 Multinodular goitre. This is a typical nodule showing follicles of different sizes, the larger lined by flat epithelium, the smaller usually by hyperplastic, high cuboidal or columnar follicular cells. There is some fibrosis around the nodule, but no true capsule, and little compression of adjacent thyroid parenchyma (H&E, × 125).

devoid of capsules and more or less well demarcated from the surrounding parenchyma. Nodules may be separated or transversed by whitish fibrous bands which may sometimes contain calcifications. Colloid-rich nodules are gelatinous with a yellowish, tan or red–brown, translucent appearance. More-cellular nodules appear fleshy or rubbery and resemble follicular tumours. Degenerative changes are common, such as signs of fresh or old haemorrhage, necrosis with fibrosis, cholesterol deposition, calcification, and cyst formation.

The microscopic picture shows great variation. On the one hand, there are nodules composed of very large follicles filled with large

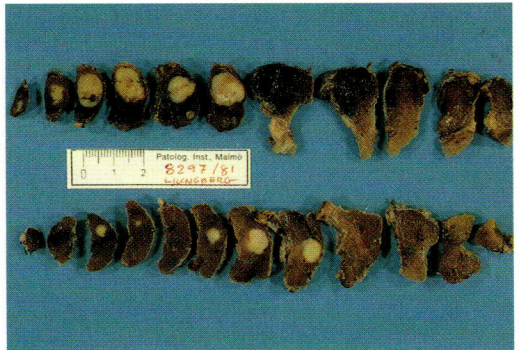

Plate 1 Transverse slicing of thyroid lobes in a case of familial medullary carcinoma in an early stage of development. Small, grey–white tumour nodules, seen on the cut surface are located in the upper halves of the lobes. The gland had been fixed in formalin.

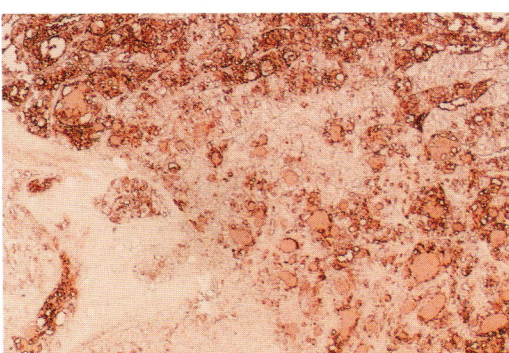

Plate 2 Vertebral metastasis of follicular thyroid carcinoma. Neoplastic follicles show positive immuno-staining for thyroglobulin (ABC technique with antibodies against human thyroglobulin).

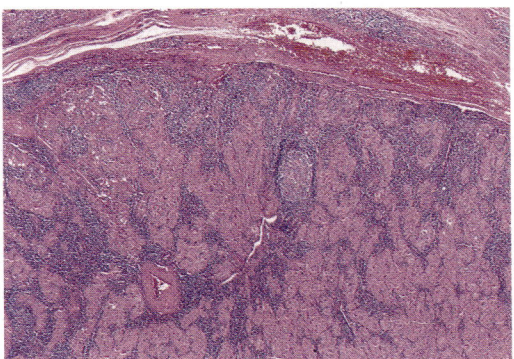

Plate 3 Hashimoto's thyroiditis with large oxyphil nodules.

Plate 4 Riedel's thyroiditis. There is pronounced destruction of thyroid parenchyma, which is replaced by a dense fibrous tissue with patchy infiltration of lymphocytes and plasma cells.

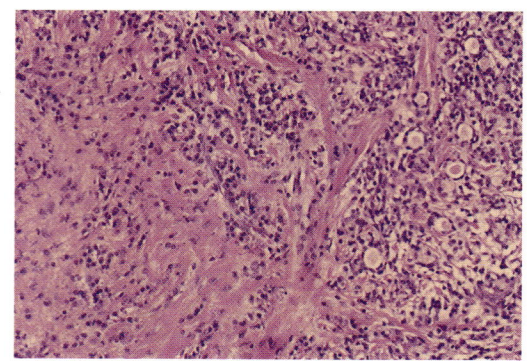

Plate 5 Riedel's thyroiditis, same case as shown in Plate 4. Fibrogenic chronic inflammation extends into the adjacent skeletal muscle, with marked destruction of muscle fibres, replaced by the inflammatory fibrous tissue. A degenerated muscle fibre is seen to the left. In the centre and to the right, there are two small veins with thickened walls and occluded lumina.

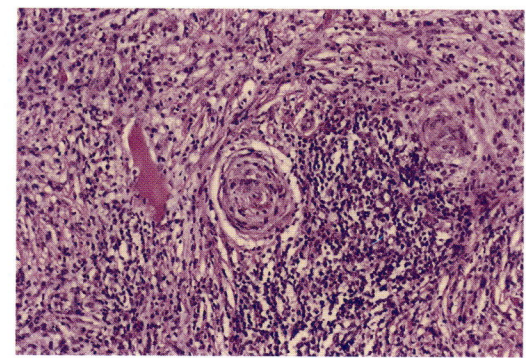

Plate 6 Multinodular goitre. The lobes are deformed by nodules of various sizes, some consisting of a friable greyish-red tissue, others showing cyst formation with accumulation of viscous, glistening colloid, discoloured by old haemorrhage. Occasional nodules are surrounded by fibrous capsules.

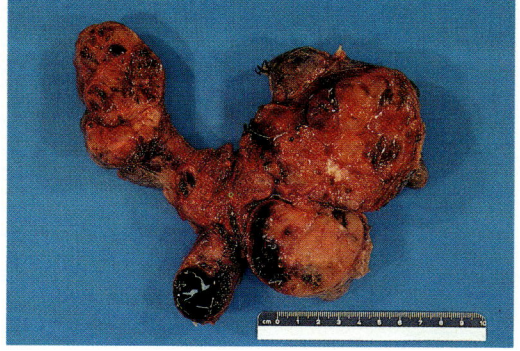

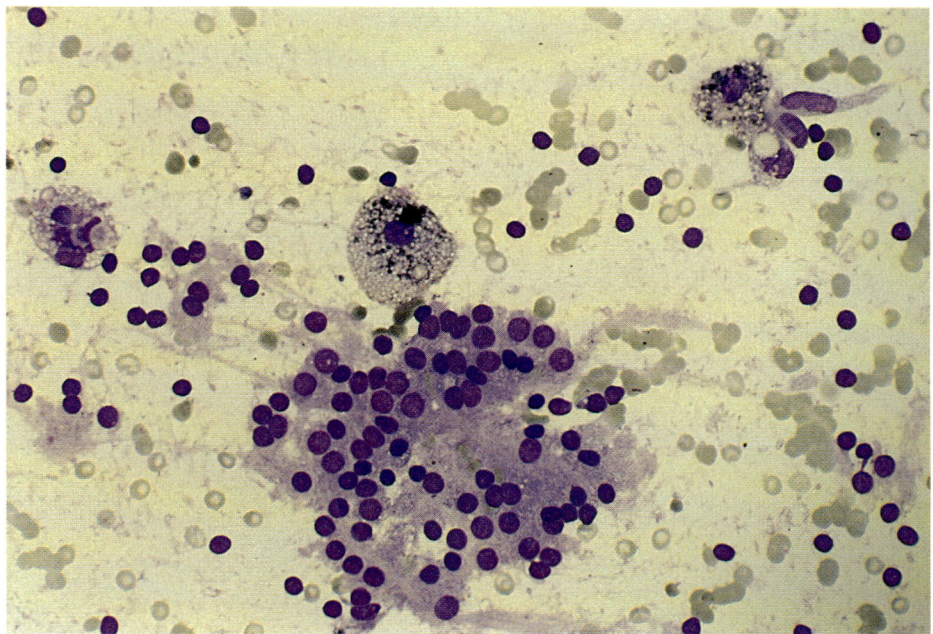

Plate 7 Fine needle aspirate from a nodular goitre. A fragment of benign-looking follicular cells with uniform nuclei and evenly distributed chromatin is seen. There are also isolated follicular cells and naked nuclei of similar appearance. The follicular cells have rather abundant, faintly greyish-blue, finely granular cytoplasm, indicating a tendency towards oxyphil cell change. Some haemosiderin-laden macrophages and an occasional spindle-shaped fibroblast (upper right corner) are also present, suggesting degenerative changes (May–Grünwald Giemsa preparation).

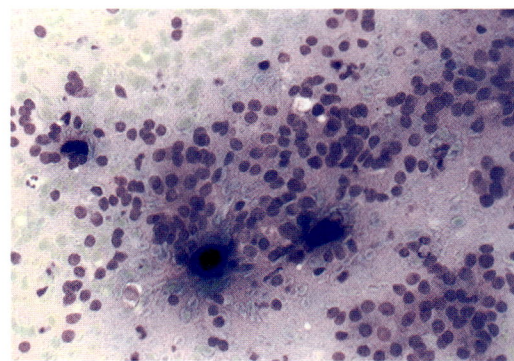

Plate 8 Fine needle aspirate from a solitary nodule. The smear is cellular with abundant tissue fragments and isolated cells. The follicular epithelium presents a microfollicular pattern showing some clumps of colloid. Small uniform nuclei with smooth nuclear membranes and rather even chromatin distribution, some also containing distinct nucleoli are seen. The high cellularity of this smear makes it impossible to exclude a follicular neoplasm and surgical extirpation for histological diagnosis is justified. Histologically, the lesion in this case was found to be a solitary, microfollicular adenoma (May–Grünwald Giemsa preparation).

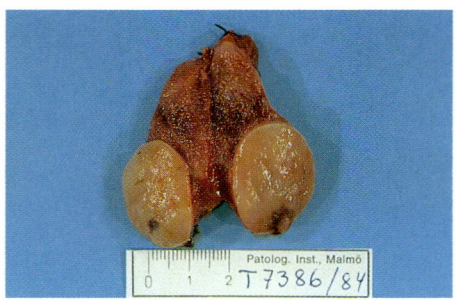

Plate 9 Follicular adenoma, clearly defined by a thin fibrous capsule.

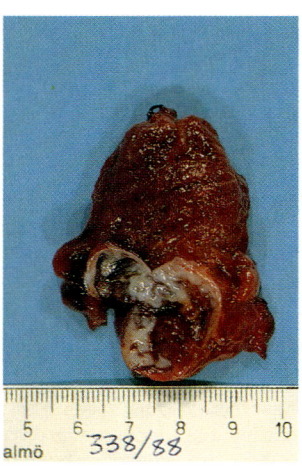

Plate 10 Follicular adenoma which has undergone degenerative changes leaving a cyst with a distinct grey–white fibrous wall and only small patches of red–brown adenomatous tissue covering its inner surface.

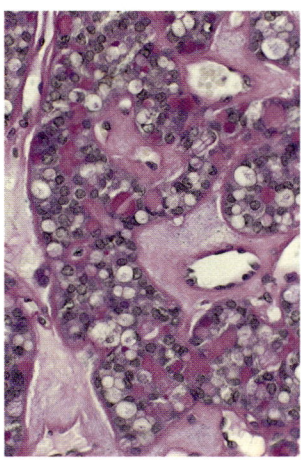

Plate 11 Follicular adenoma with signet-ring cells. Some intracytoplasmic vacuoles contain PAS-positive material.

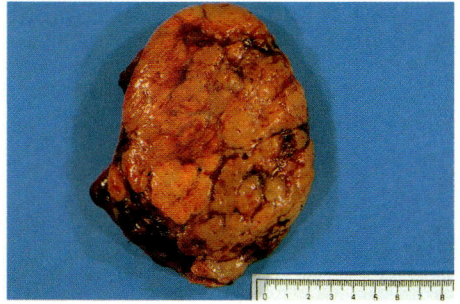

Plate 12 So-called atypical adenoma. Note the fleshy appearance of the tumour, having a bulging, yellow–brown and irregularly lobulated cut surface with areas of haemorrhage.

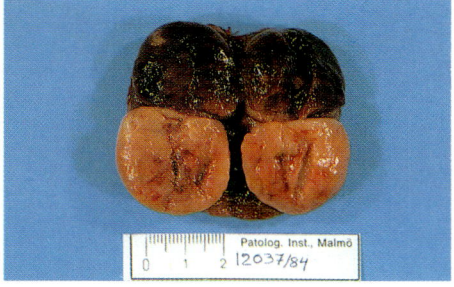

Plate 13 Thyroid lobe involved by two distinct tumours; the lower, with pale, yellowish-grey cut surface is a hyalinizing trabecular adenoma; the upper tumour, red-brown in colour, is a simple microfollicular adenoma.

amounts of colloid and lined by flat epithelium, and on the other there are nodules which are highly cellular with intensely hyperplastic proliferations of follicular cells sometimes forming solid accumulations with no or very scanty colloid (Fig. 5.5). The parenchyma between nodules may show the microscopic changes of colloid goitre with colloid-rich distended follicles often with flat epithelium; some areas, however, may appear normal. Individual nodules often show a variegated structure displaying areas of microfollicles interspersed among large distended follicles and cysts containing lakes of colloid (Fig. 5.7). Some nodules may still show hyperplastic changes where epithelium is tall and sometimes even grows in fronds or papillae or exhibits infoldings into large follicles. Microcalcifications and psammoma bodies may even be seen in these areas (Fig. 5.7). Such features may be mistaken for neoplastic papillae of papillary carcinoma (Fig. 5.8). However, the epithelial cells are devoid of the nuclear features, such as ground glass appearance and nuclear grooves, characteristics of papillary carcinoma. In addition, the stroma of the papillae in nodules frequently contains closely packed microfollicles, a feature seldom encountered in carcinomas.

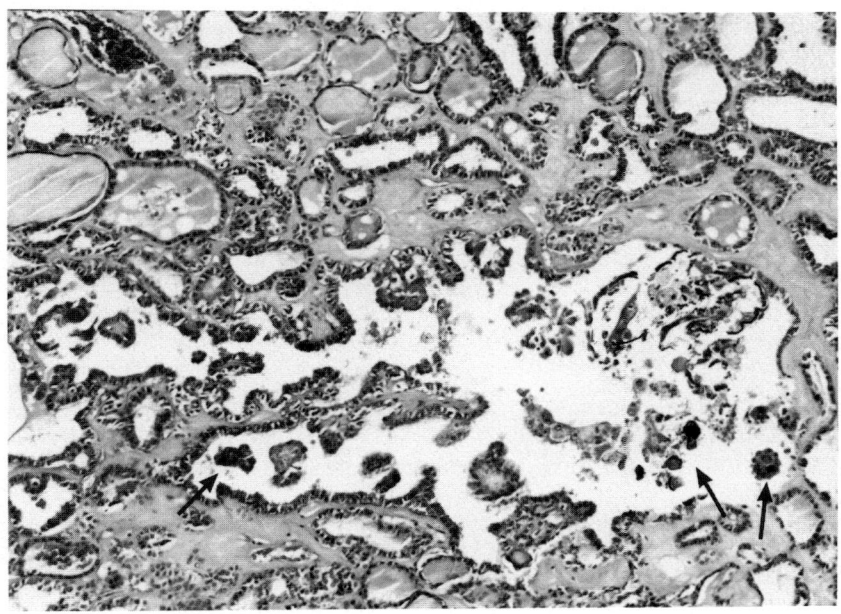

Figure 5.7 Multinodular goitre. Focal hyperplasia with papillary structures, covered by tall columnar cells and pointing into irregular lumina. Some papillary tips show degeneration with microcalcification (arrows) (H&E, × 90).

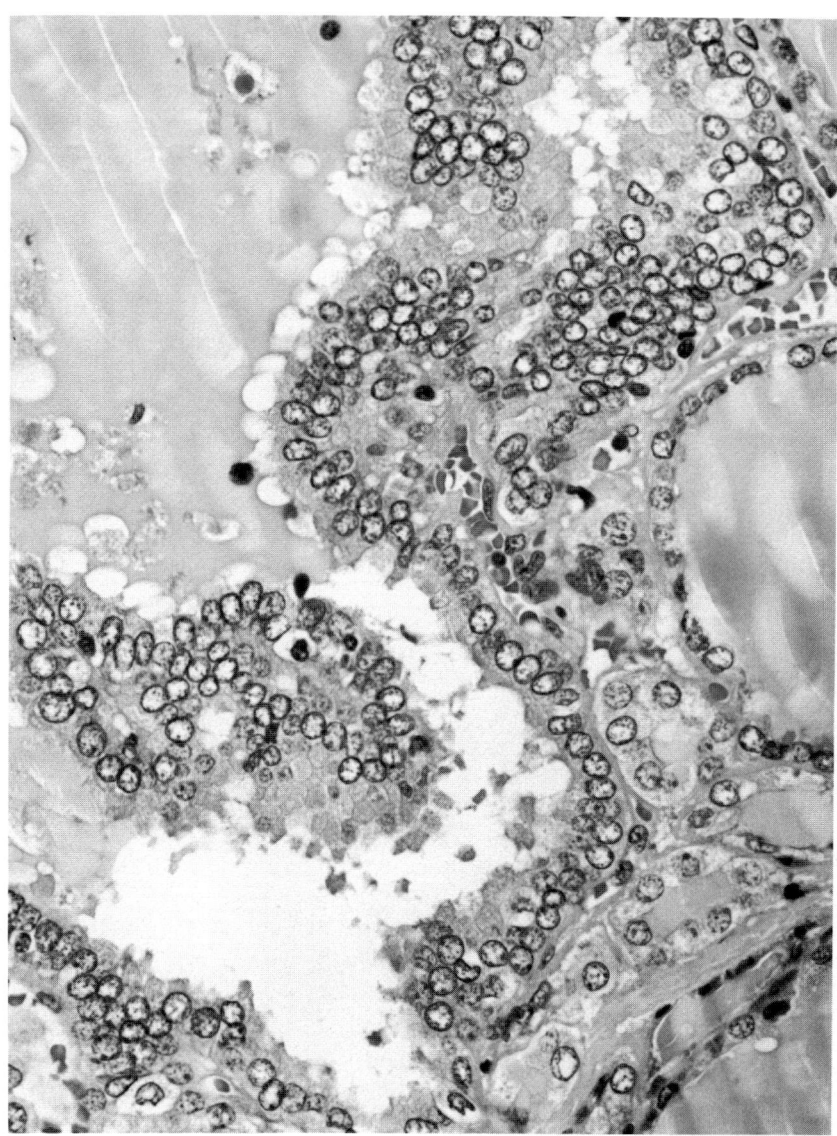

Figure 5.8 Multinodular goitre. Focus of hyperplastic epithelium with heaping up into papillary-like formations that may be mistaken for neoplastic papillae. The nuclei are vesicular, resembling ground glass nuclei of papillary carcinoma. However, in contrast to such nuclei, they are rounded with smooth nuclear membranes, lack nuclear grooves and have more irregularly distributed chromatin (H&E, × 500).

Although nodules may appear well circumscribed macroscopically, they are often less so at the microscopic level (Fig 5.6). In the marginal zone between the nodule and surrounding parenchyma there may be signs of compression with atrophy of follicles and varying degrees of fibrosis, sometimes even formation of a hyalinized fibrous zone that mimics a tumour capsule (Fig. 5.9). In contrast to the capsule around neoplasms, a fibrous zone around nodules is usually not sharply demarcated against the epithelial structures of the nodule, but there is a gradual vanishing of the fibrosis towards the underlying parts of the lesion (Fig. 5.9). This phenomenon indicates that the fibrous zone is not a true tumour capsule but the result of degeneration in the peripheral parts of the nodule and in the proximate zone of the surrounding parenchyma. This may be a helpful histological detail in the differential diagnosis against a follicular adenoma, and also against a follicular carcinoma. Follicles are sometimes secondarily included in the degenerative fibrosis around a nodule, a phenomenon which may be misinterpreted as an indication of active infiltration.

Nodules may also be encountered in which the epithelial cells have undergone varying degrees of transformation to oxyphil or clear

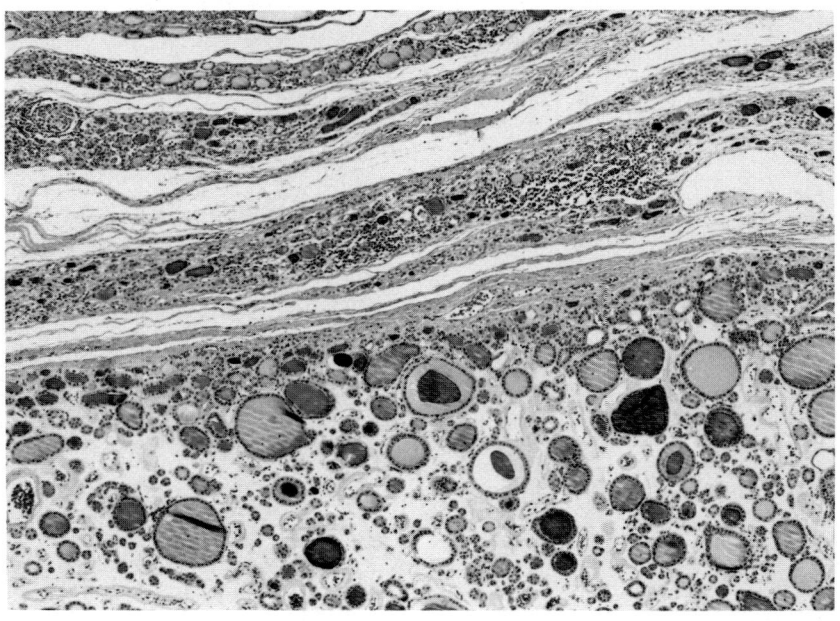

Figure 5.9 Multinodular goitre. Nodule surrounded by thin fibrous zone simulating a capsule, which, however, is diffusely demarcated against the under-lying epithelial structures of the nodule. There is some compression of adjacent parenchyma (H&E, × 50).

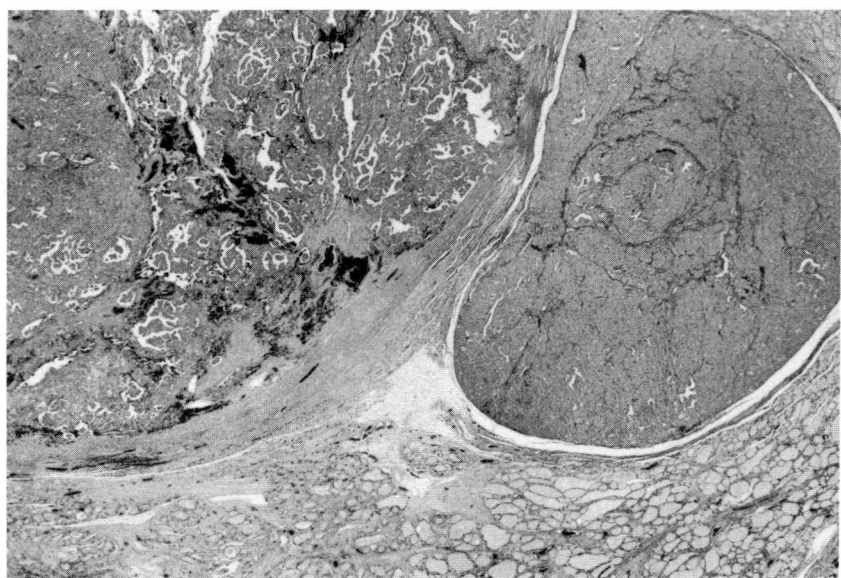

Figure 5.10 Multinodular goitre with oxyphil nodules (H&E, × 8).

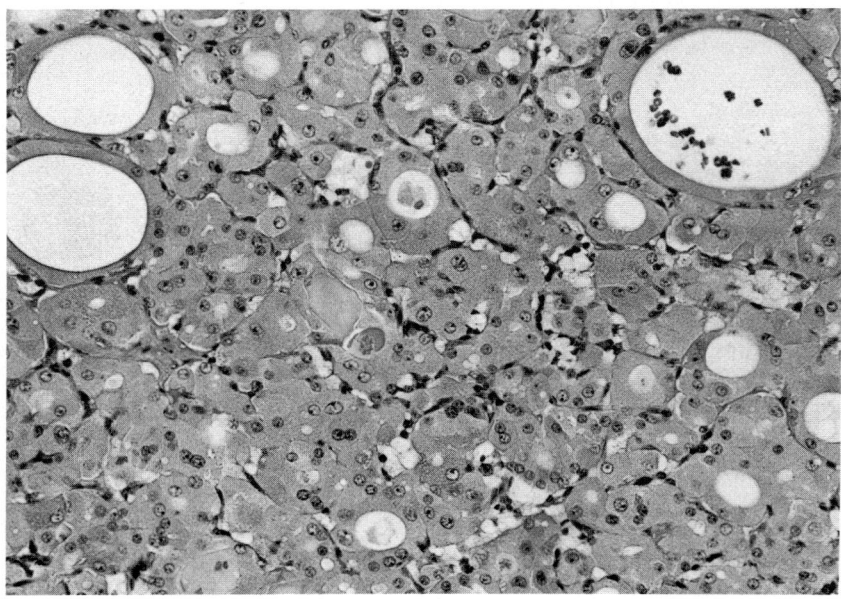

Figure 5.11 Multinodular goitre with oxyphil nodules. Same case as Fig. 5.10. The picture shows a cellular nodule exclusively composed of oxyphil cells (H&E, × 160).

cells, and occasionally nodules may consist exclusively of cells of these types. Such nodules may have a follicular and even solid structure (Figs 5.10 and 5.11). Since cellular and nuclear atypia may occur in such nodules, suspicion of an oxyphil or clear cell carcinoma may arise. Useful points, however, are the variegated histological appearance of the nodules and absence of infiltration and blood vessel invasion.

Degenerative changes include granulomatous reaction against colloid spilled through rupture of follicles, with accumulations of histiocytes and foreign body giant cells. Sometimes there are collections of cholesterol crystals and foam cells as well as histiocytes with haemosiderin, indicating previous haemorrhage. Nodules may be transversed by trabeculae of fibrous tissue, contain central irregular scars, or even be totally replaced by almost acellular hyalinized fibrous tissue. Calcification is common in areas of old scar tissue and metaplastic bone formation may occur (Fig. 5.12). As mentioned earlier, the perinodular stroma in some multinodular goitres may contain varying amounts of lymphoid cells indicating the presence of lymphoid thyroiditis (Fig. 5.13). This is especially prominent in some cases associated with hypothyroidism and circulating thyroid autoantibodies.

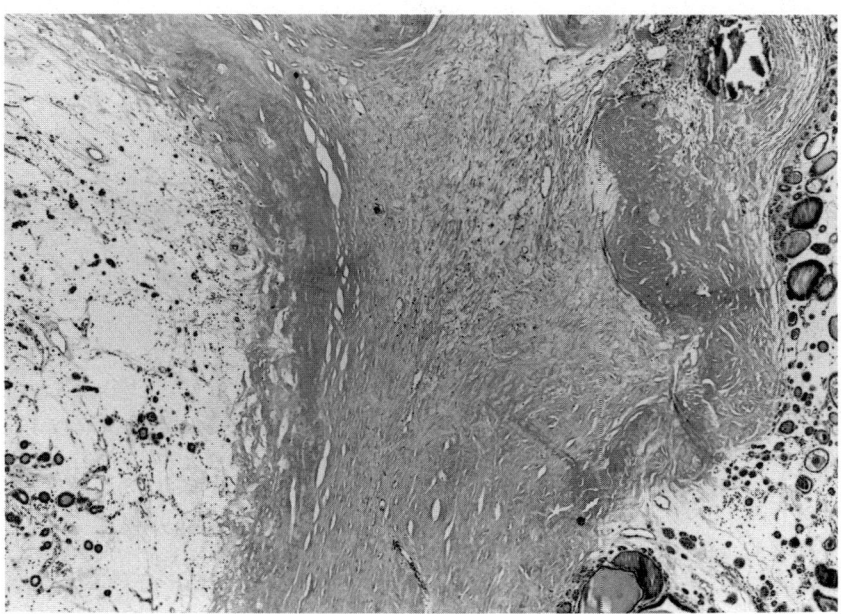

Figure 5.12 Multinodular goitre. Nodules show advanced degeneration together with loss of follicles, increased amounts of oedematous stroma and peripheral fibrosis with hyalinization and calcification (H&E, × 35).

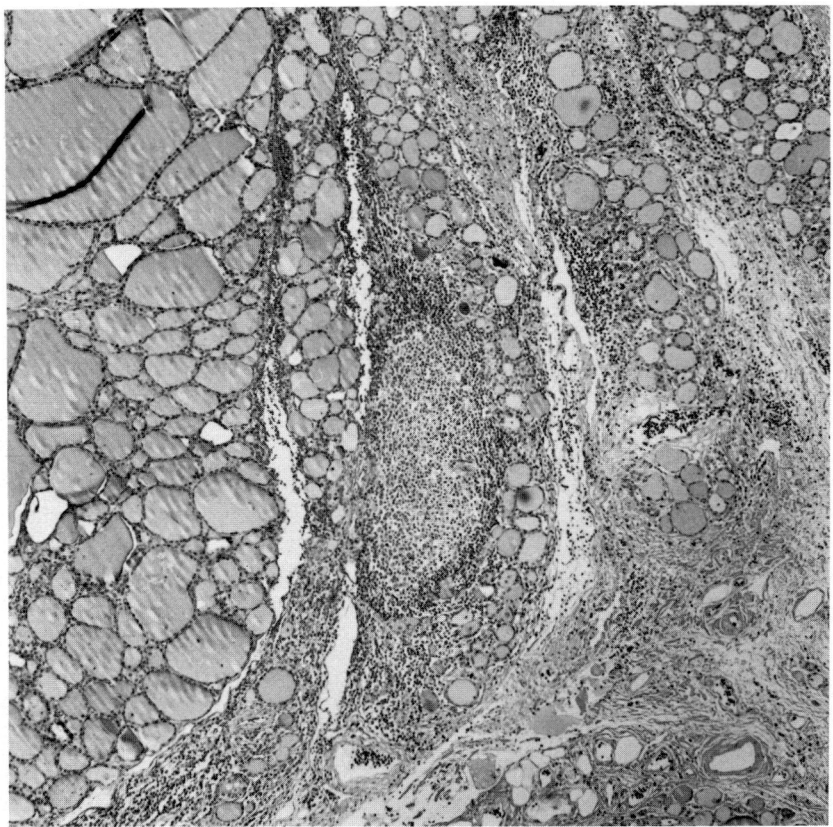

Figure 5.13 Multinodular goitre with chronic lymphocytic thyroiditis (H&E, × 65).

This histological differential diagnosis between a dominant nodule and a true follicular adenoma is based on a number of arbitrary criteria. The dominant nodule is always associated with several other nodules and incompletely encapsulated, sometimes with an indistinct border between fibrous capsular zone and the underlying epithelial structures. It usually shows a variable follicular pattern, some follicles being much larger than those in the surrounding thyroid parenchyma. The adenoma, on the other hand, is solitary and fully encapsulated, usually with a sharp demarcation between capsule and tumour tissue. In addition, adenomas tend to have a uniform follicular pattern composed of follicles that are smaller than those of the surrounding thyroid tissue. Table 5.2 lists some of the features that may help to discriminate between adenomas and hyperplastic nodules. In some instances the distinction may be very difficult or impossible to make, e.g. in cases

Table 5.2 Histological characteristics distinguishing follicular adenoma from a hyperplastic nodule

Follicular adenoma	Nodule
Solitary lesion	Dominant nodule in multinodular goitre
Complete fibrous encapsulation, usually with sharp demarcation between capsule and adenomatous tissue	Capsule incomplete or absent; when present indistinct border between fibrous capsular zone and epithelial structures of the nodule
Uniform structure	Variable structure
Histological pattern different from the pattern in surrounding thyroid parenchyma	Histological pattern resembles the structural variations of the adjacent glandular parenchyma
Degeneration infrequent	Degeneration frequent

where only a single microscopic section from the lesion is available, or in cases of multiple adenomas or when an adenoma develops in the background of a nodular goitre. Under such circumstances a descriptive diagnosis may be appropriate.

5.4.1 Nodular goitre and thyroid carcinoma

There are conflicting opinions about the association between nodular goitre and the risk of developing thyroid carcinoma, mainly of the follicular type. Considering the common occurrence of non-toxic goitre, however, an increased risk, if any, is insignificant from a practical point of view and does not warrant prophylactic surgery (Studer *et al.*, 1979). The carcinomas found in nodular goitres are mostly small well-differentiated papillary carcinomas, detected incidentally by the pathologist and with little clinical significance (so-called occult papillary carcinomas or papillary microcarcinomas, see Chapter 9).

5.5 The solitary thyroid nodule

Some nodular goitres contain a dominant nodule, which may clinically present as a solitary thyroid mass. In addition, such nodules tend to be scintigraphically hypo- or inactive. Solitary, 'cold' nodules, therefore, may raise suspicion of neoplasia. The evaluation of the patient with an apparently solitary thyroid nodule is in fact the most common thyroid problem facing the clinician and the pathologist. About 4% of the general adult population have one or more palpable thyroid nodules and most of these cases represent benign nodular goitre. However, it has been estimated that 10–20% of clinically solitary nodules, which are found to be 'cold' on scintiscan, turn out to be carcinomas on histopathological examination (Brown, 1981). The majority of these carcinomas are not associated with multinodular goitre. The main problem is to select as efficiently as possible those cases to be referred to surgery and at the same time include most of the carcinomas. Several factors should be considered when making this selection. These include a number of clinical features, such as age and sex of the patient, size and growth rate of the goitre, presence or absence of associated regional lymphadenopathy, functional status of the lesion(s), as well as fine-needle biopsy. Of these, the fine-needle biopsy, when performed by experienced cytopathologists, has proved to be the best predictor of malignancy (Willems and Löwhagen, 1981).

5.5.1 Fine-needle aspiration of the solitary nodule

It is true that the morbidity and mortality of thyroid carcinoma is a minor public health problem. Nevertheless, the large number of various diagnostic tests and lobectomies which have been performed in order to find the relatively small number of carcinomas hiding in the great bulk of lesions comprising thyroid nodular disease, makes thyroid carcinoma a disease of great economic importance. Against this background, fine-needle aspiration has become a very popular technique in the evaluation of the solitary thyroid nodule. It has proved to be the least expensive and most effective method for selecting patients with nodules for diagnostic lobectomy (Blum and Rothschild, 1980). The increased use of this diagnostic method in recent years has considerably lowered the number of operations prescribed, and at the same time has greatly increased the proportion of carcinomas found among the operated patients. Miller *et al.* (1980) for instance, report that the use of needle biopsy has halved the number of operations and doubled the proportion of malignant neoplasms identified among the operated cases.

Between 80 and 90% of all thyroid nodules represent some sort of highly differentiated follicular lesion. The great majority of these represent nodular goitre, in which the clinically apparent solitary nodule is the dominating nodule in a background of smaller nodules.

The cytological presentation of nodular goitre reflects the histopathological changes and corresponds to the stage of the disease and to secondary changes, when present, such as degeneration with necrosis, haemorrhage, chronic inflammation and fibrosis, and sometimes with calcification. Typically, aspirates from such lesions show colloid, which may be abundant in cases of colloid nodules and often show a characteristic cracking pattern. There is an admixture of varying numbers of benign-looking follicular cells, occurring singly and in clusters or large fragments with orderly arranged cells, mimicking a honeycomb pattern (Nguyen *et al.*, 1991). Some cells may show oxyphil cell change (Plate 7). Lesions with degenerative changes show many macrophages with haemosiderin, fibroblasts, some round cells and atrophic follicular cells. A characteristic feature of well-differentiated follicular neoplasms is the high cellularity of the smears.

Hyperplastic nodules, however, may also yield very cellular smears with sparse amounts of colloid and thus may be impossible to distinguish from a follicular neoplasm solely on a cytological basis (Willems and Löwhagen, 1981). A follicular adenoma and a highly differentiated follicular carcinoma may show similar cytological features and can usually not be distinguished (Plate 8). Smears from both types of lesion are typically cellular and often show a microfollicular pattern. It has been claimed that the regular arrangement of the follicular cells in the larger tissue fragment, forming a so-called orderly honeycomb pattern indicates a nodular goitre, whereas the occurrence of fragments with a syncytial-type of pattern, showing crowding and overlapping nuclei suggests a follicular neoplasm (Ravinsky and Safneck, 1990). If the cytological smears show features suggesting a follicular neoplasm, the case may be selected for diagnostic hemistrumectomy for a histological diagnosis, giving further guidance on whether or not an extended operation will be necessary. The same programme may apply to oxyphil cell tumours, in which the cytological distinction between a benign and malignant lesion is particularly difficult (Willems and Löwhagen, 1981).

The cytological diagnosis of poorly differentiated and undifferentiated carcinomas is usually less problematic and, in the hands of an experience cytopathologist, this is also the case with classic medullary carcinoma (see also p. 219). Typical papillary carcinoma, showing papillary structures and characteristic nuclear features, can also be identified with ease. The follicular variant of papillary carcinoma, on the other hand, may be dificult to recognize in cytological smears (Hugh *et al.*, 1988), as it is in histological sections.

5.6 Sequestered thyroid nodules

In cases of nodular goitre, superficially located nodules may be extruded from the surface of the gland. Usually, such nodules preserve their connection with the main gland by a stalk of varying length. But occasionally, the nodules may be disconnected from the gland and displaced into the lateral parts of the neck or, more often, caudally into the restrosternal area (Klinck, 1964). The mechanism of sequestration is unknown. Sisson *et al.* (1964) believe that mechanical separation may occur by the action of the neck muscles. When sequestered

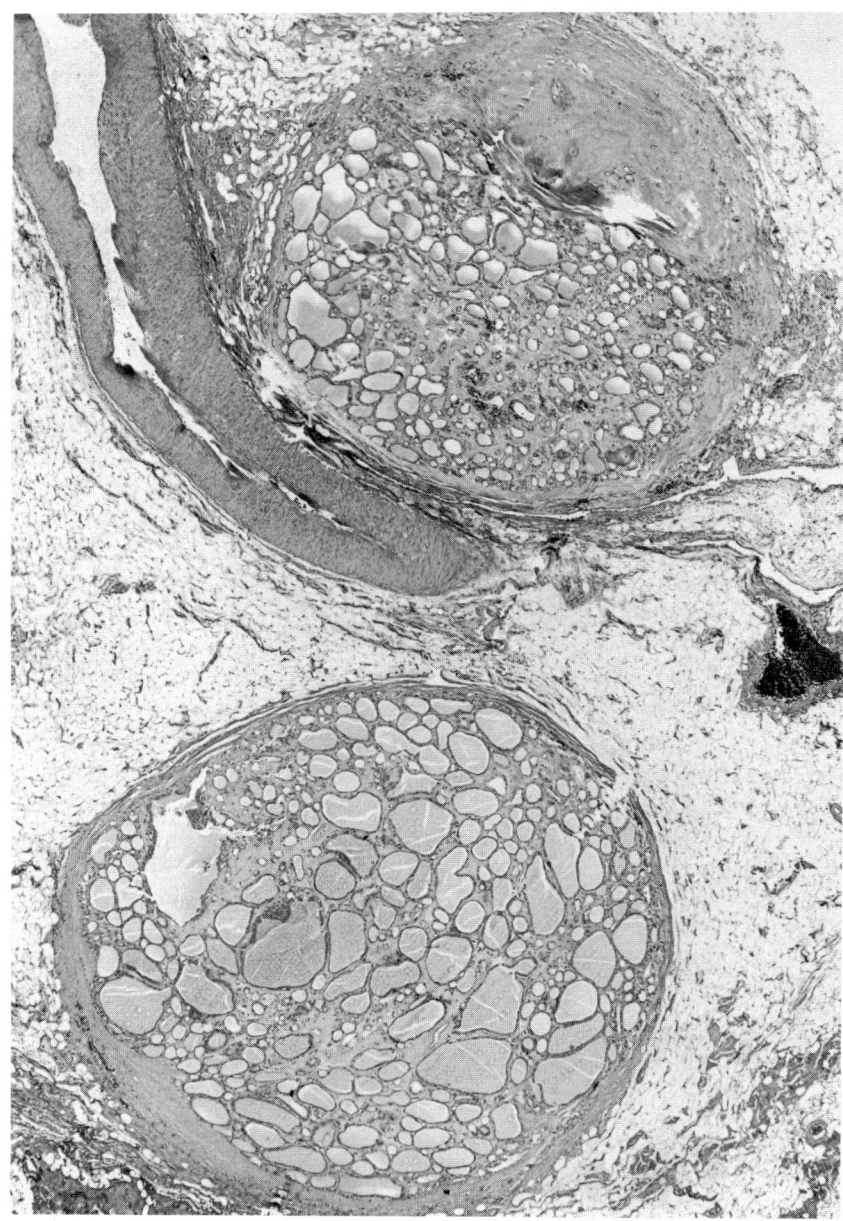

Figure 5.14 Isolated thyroid nodules embedded in adipose tissue close to the right sternocleidomastoid muscle. They had developed about 20 years after a right-sided hemithyroidectomy for benign multinodular goitre. The nodules are sharply demarcated by hyalinized fibrous capsules and are composed of benign-looking follicles, with some interstitial fibrosis. There is no lymph node tissue present and no histological evidence of malignancy (H&E, × 25).

laterally, such nodules are often found in the region of the jugular lymph nodes. Extruded thyroid nodules may be subject to the same lesions as the gland itself. For example, if the gland shows evidence of hyperplasia, as in Graves' disease, or Hashimoto's thyroiditis, the sequestered nodules will reveal the same histological changes. In cases of lymphocytic thyroiditis such nodules, therefore, may closely resemble lymph nodes containing thyroid tissue and may be confused with lymph node metastases of thyroid carcinoma. Ironically, the nodules tend to lie in close association with true lymph nodes, and care must be taken to avoid misinterpretation. However, a close study of the lymphoid tissue reveals that the nodules are devoid of a marginal sinus and other features characteristic of lymph nodes. In addition, knowledge of the nature of the lesion in the gland proper will aid a correct diagnosis.

Exceptionally, operation for nodular goitre may apparently later give rise to the development of extrathyroidal nodules in the neck. Figure 5.14 illustrates such a case, in which multiple nodules developed chiefly in the lateral neck regions, about 20 years after extirpation of the right thyroid lobe. In this case, foreign body granulomas with residual suture material were found adjacent to some of the nodules suggesting that these had developed from thyroid tissue which had been implanted in the extrathyroidal tissues during surgery. Review of histological sections from the original surgical specimen revealed a benign multinodular goitre with degenerative changes. The remaining left lobe was later removed and showed similar histopathological features, with no evidence of malignancy.

5.6.1 So-called benign ovarian strumatosis

A similar sequestration of nodules composed of benign-looking thyroid tissue may occasionally occur from a struma ovarii, especially when this has undergone multinodular colloid changes. Under such circumstances, nodules may be extruded from the ovarian lesion and seeded on the peritoneum and omentum (so-called benign strumatosis) where they may easily be mistaken for tumour metastases.

The author has recently had the opportunity of studying such a case (kindly provided by Dr Poul Boiesen, Jönköping). The patient was a 43-year-old woman who underwent cholecystectomy for gallstones. At the operation, multiple spherical nodules, up to 2.5 cm in diameter, were found scattered over the large omentum and tumour metastases were suspected. This suspicion was further strengthened by the finding of a large polycyclic tumour (9 cm in

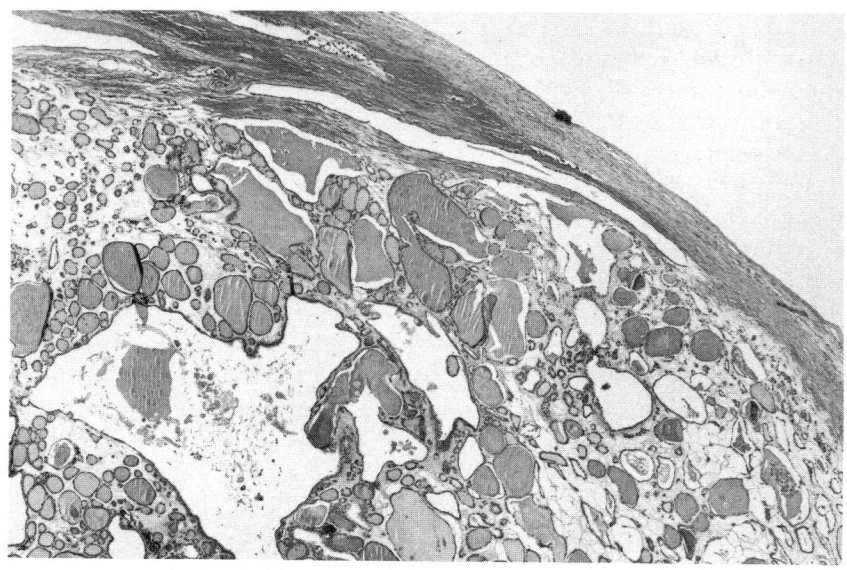

Figure 5.15 Benign struma ovarii. Part of a colloid nodule with signs of involution and degeneration (H&E, × 30).

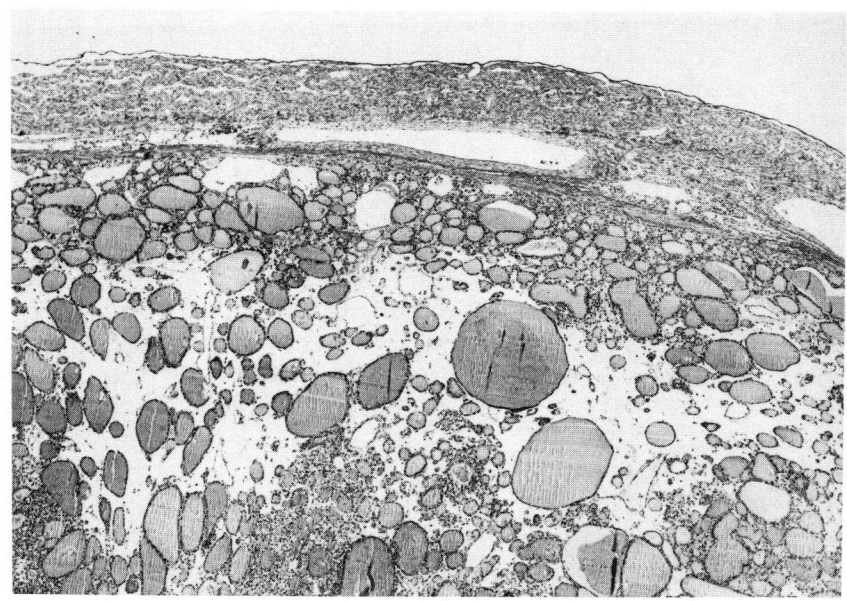

Figure 5.16 Same case as illustrated in Fig. 5.15. Benign ovarian strumatosis. Implantation to the large omentum of a thyroid nodule, which has the same histological appearance as the ovarian struma, shown in Fig. 5.15 (H&E, × 30).

largest diameter) in the right ovary. However, on microscopy, the ovarian lesion was found to be a large struma ovarii, composed of multiple various-sized nodules, mostly comprising large follicles distended with colloid, some, however, showing a microfollicular, hyperplastic pattern but without any evidence of malignant change (Fig. 5.15). The omental nodules showed the microscopic picture of benign colloid-rich thyroid tissue with degenerative changes (Fig. 5.16).

Hastleton *et al*. (1978) pointed out that, although malignant tumours do occasionally develop in struma ovarii, many of the previously reported cases of malignant struma ovarii, in which the diagnosis of malignancy had been based on the presence of peritoneal implants of thyroid tissue, were really misinterpreted cases of benign strumatosis. Typically these cases have a benign clinical course (Scully, 1970). Early reports of similar findings include those by Werth (1928), Shapiro (1930) and Emge (1940). A recent case has been reported by Willemse *et al*. (1987).

5.7 Toxic nodular goitre

As previously stated, functionally autonomous nodules in non-toxic multinodular goitre may become hyperactive and cause hyper-thyroidism. This condition has also been referred to as Plummer's disease. It occurs predominantly in the middle-aged and elderly and the great majority of patients are women. The condition usually develops insidiously over many years and, therefore, may be very difficult to recognize, especially in older patients who may have other diseases (Davis and Davis, 1974). Scintiscan will show hyperfunc-tioning areas in the gland. The histological findings are similar to those of non-toxic nodular goitre, although in some nodules hyperplasia may be marked, with tall columnar epithelium and peripherally scalloped colloid (Figs 5.17 and 5.18).

In a number of patients with toxic multinodular goitre the excess of thyroid hormones is due to hyperfunction of the internodular thyroid tissue rather than the nodules themselves. These patients are likely to have Graves' disease superimposed on a pre-existing non-toxic nodular goitre. The thyroid tissue between nodules may show hyperplastic changes, and, not infrequently, lymphoid infiltration is present in the stroma. Nodules may show oxyphil or even clear cell changes. In some cases both nodules and internodular parenchyma show intense epithelial hyperplasia (Figs 5.19 and 5.20). The patients may have thyroid-stimulating antibodies in the blood and other signs, such as eye symptoms, characteristic of Graves' disease.

It should be mentioned here that individuals with multinodular non-toxic goitre are liable to develop hyperthyroidism following iodine administration (Vidor *et al.*, 1973). This rare condition is known as Jod–Basedow disease.

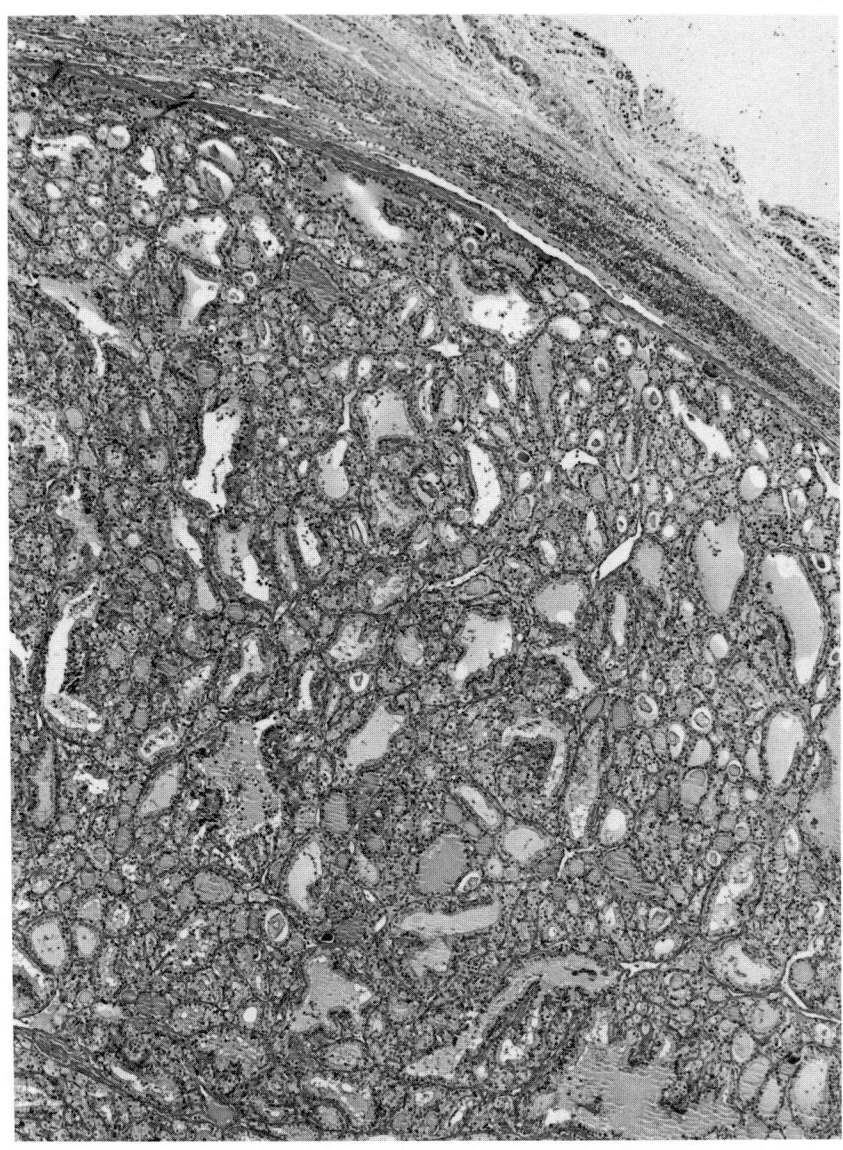

Figure 5.17 Multinodular goitre showing a toxic nodule with highly hyperplastic epithelium. The surrounding thyroid parenchyma is atrophic (H&E, × 50).

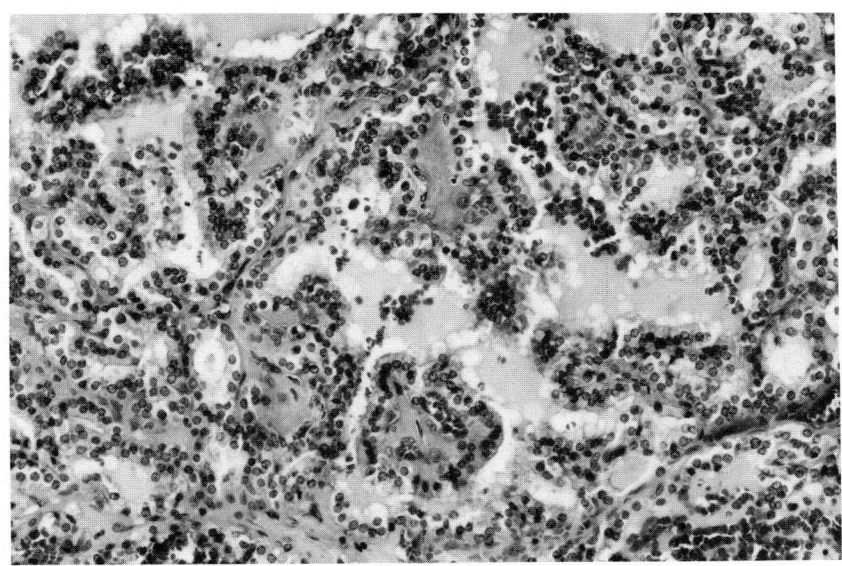

Figure 5.18 Toxic nodular goitre. Nodule with marked hyperplasia showing irregular follicles and papillary-like structures, lined by high, columnar epithelium and peripherally scalloped colloid (H&E, × 160).

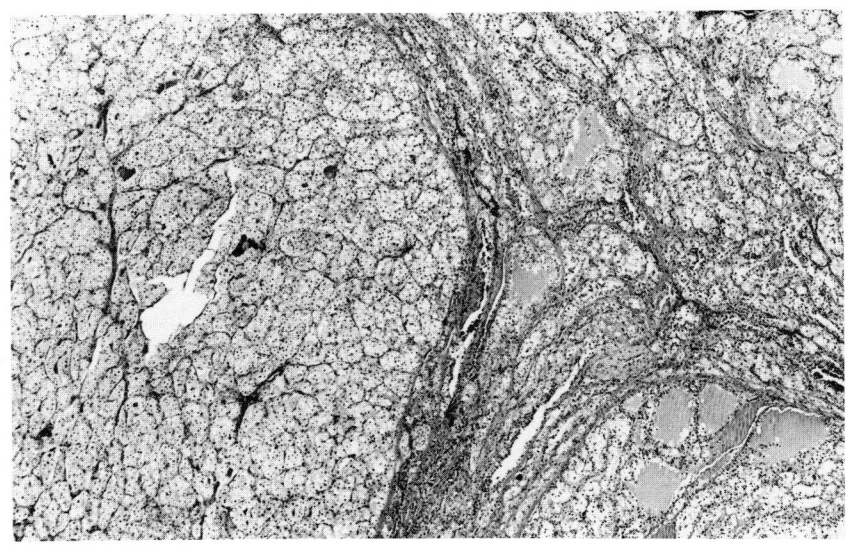

Figure 5.19 Toxic nodular goitre. Both nodules and intervening thyroid parenchyma show prominent hyperplasia (H&E, × 30).

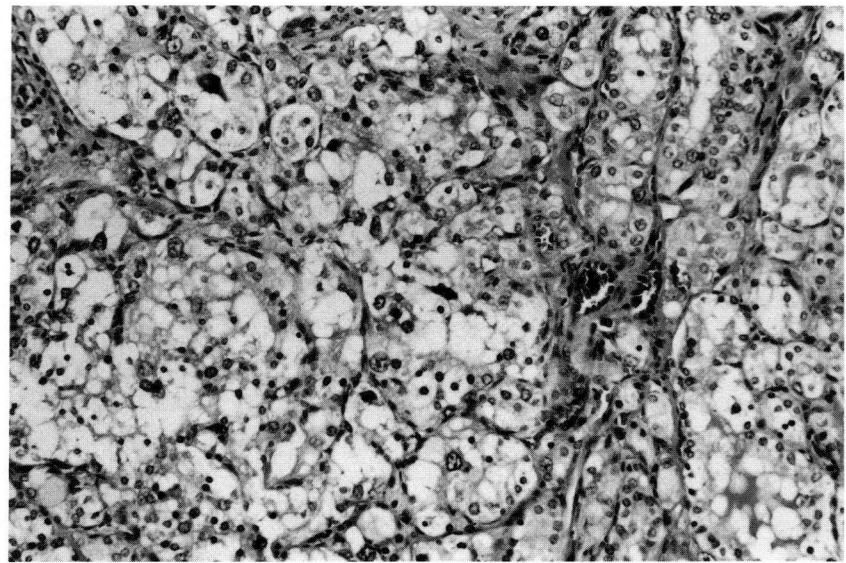

Figure 5.20 Same case as illustrated in Fig. 5.19. A cellular nodule is composed of small follicles with scanty colloid and solid clusters of follicular cells with pronounced clear cell change. There is a marked cytological atypia with pleomorphic nuclei, which should not be interpreted as evidence of malignancy (H&E, 145).

References

Barsano, C.P. and DeGroot, L.J. (1979) Dyshormonogenetic goitre. *Clin. Endocrinol. Metab.*, **8**, 145–65.

Beckers, C. (1979) Thyroid nodules. *Clin. Endocrinol. Metab.*, **8**, 181–92.

Blum, M. and Rothschild, M. (1980) Improved nonoperative diagnosis of the solitary 'cold' thyroid nodule. *JAMA*, **243**, 242–5.

Brown, C.L. (1981) Pathology of the cold nodule. *Clin. Endocrinol. Metab.*, **10**, 235–45.

Brown, R.S., Pohl, S.L., Jackson, I.M.D. and Reichlin, S. (1978) Do thyroid-stimulating immunoglobulins cause non-toxic and toxic multinodular goitre? *Lancet*, **i**, 904–6.

Davis, P.J. and Davis, F.B. (1974) Hyperthyroidism in patients over the age of 60 years. Clinical features in 85 patients. *Medicine*, **53**, 161–81.

Denham, M.J. and Wills, E.J. (1980) A clinico-pathological survey of thyroid glands in old age. *Gerontology*, **26**, 160–6.

Doniach, I. (1970) Aetiological consideration of thyroid carcinoma. In *Tumours of The Thyroid Gland* (ed. D. Smithers), E. & S. Livingstone, Edinburgh and London, pp. 55–72.

Emge, L.A. (1940) Functional and growth characteristics of struma ovarii. *Am. J. Obstet. Gynecol.*, **40**, 738–50.

Ermans, A.M. (1979) Endemic goiter and endemic cretinism. In *Endocrinology*,

Vol. 1, (eds L.J. DeGroot *et al.*), Grune and Stratton, New York, San Francisco and London, pp. 501–8.

Hasleton, P.S., Kelehan, P., Whittaker, J.S. *et al.* (1978) Benign and malignant struma ovarii. *Arch. Pathol. Lab. Med.*, **102**, 180–4.

Hennemann, G. (1979) Non-toxic goitre. *Clin. Endocrinol. Metab.*, **8**, 167–79.

Hugh, J.C., Duggan, M.A. and Chan-Poon, V. (1988) The fine needle-aspiration appearance of the follicular variant of thyroid papillary carcinoma: a report of three cases. *Diagn. Cytopathol.*, **4**, 196–201.

Ibbertson, H.K. (1979) Endemic goitre and cretinism. *Clin. Endocrinol. Metab.*, **8**, 97–128.

Ichikawa, K., Hughes, I.A., Horwitz, A.L. and DeGroot, L.J. (1987) Characterization of nuclear thyroid hormone receptors of cultured skin fibroblasts from patients with resistence to thyroid hormone. *Metabolism*, **36**, 392–9.

Kelly, F.C. and Snedden, W.W. (1960) Prevalence and geographical distribution of endemic goiter. In *Endemic Goiter*. World Health Organization Monograph Series no. 44, pp. 27–233.

Kennedy, J.S. (1969) The pathology of dyshormonogenetic goitre. *J. Pathol.*, **99**, 251–64.

Klinck, G.H. (1964) Structure of the thyroid. In *The Thyroid* (eds J.B. Hazard and D.E. Smith), Williams and Wilkins, International Academy of Pathology Monographs in Pathology, No. 5, Baltimore, pp. 1–31.

Marine, D. (1924) Etiology and prevention of simple goiter. *Medicine*, **3**, 453–79.

Meissner, W.A. and Warren, S. (1969) Tumors of the thyroid gland. In *Atlas of Tumor Pathology*, Fascicle 4, Second Series, Armed Forces Institute of Pathology, Washington, DC.

Miller, J.M,. Hamburger, J.I. and Kini, S.R. (1980) The impact of needle biopsy on the preoperative diagnosis of thyroid nodules. *Henry Ford Hosp. Med. J.*, **28**, 145.

Moore, G.H. (1962) The thyroid in sporadic goitrous cretinism. *Arch. Pathol.*, **74**, 35–46.

Mortensen, J.D., Woolner, L.B. and Bennett, W.T. (1955) Gross and microscopic findings in clinically normal thyroid glands. *J. Clin. Endocrinol. Metab.*, **15**, 1270–80.

Nguyen, G.-K., Ginsber, J. and Crockford, P.M. (1991) Fine-needle aspiration biopsy cytology of the thyroid. Its value and limitations in the diagnosis and management of solitary thyroid nodules. *Pathol. Annu.*, **26** (1), 63–91.

Ramalingaswami, V. (1969) Iodine and thyroid cancer in man. In *Thyroid Cancer* (ed. Chr. E. Hedinger), UICC Monograph Series, Volume 12, Springer-Verlag, Berlin, Heidelberg and New York, pp. 111–23.

Ramelli, F., Studer, H. and Bruggisser, D. (1982) Pathogenesis of thyroid nodules in multinodular goiter. *Am. J. Pathol.*, **109**, 215–23.

Ravinsky, E. and Safneck, J.R. (1990) Fine needle aspirates of follicular lesions of the thyroid gland. The intermediate-type smear. *Acta Cytol.*, **34**, 813–20.

Refetoff, S., DeWind, L.T. and DeGroot, L.J. (1967) Familial syndrome combining deaf-mutism, stippled epiphyses, goiter, and abnormally high PBI: possible target organ refractoriness to thyroid hormone. *J. Clin. Endocrinol. Metab.*, **27**, 279–94.

Schrimshaw, N.S. (1964) The geographic pathology of thyroid disease. In *The Thyroid* (eds J.B. Hazard and D.E. Smith), International Academy of Pathology Monographs in Pathology, No. 5, Williams and Wilkins,

Baltimore, pp. 100–22.

Scully, R.E. (1970) Recent progress in ovarian cancer. *Hum. Pathol.*, **1**, 73–98.

Shapiro, P.F. (1930) Metastasis of thyroid tissue to abdominal organs. With special case report of a struma ovarii metastasizing to the omentum. *Ann. Surg.*, **92**, 1031–42.

Sisson, J.C., Schmidt, R.W. and Beierwaltes, W.H. (1964) Sequestered nodular goiter. *N. Engl. J. Med.*, **270**, 927–32.

Studer, H. and Greer, M.A. (1965) A study of the mechanisms involved in the production of iodine deficient goitre. *Acta Endocrinol.*, **49**, 610–28.

Studer, H. and Ramelli, F. (1982) Simple goiter and its variants. Euthyroid and hyperthyroid multinodular goiters. *Endocr. Rev.*, **3**, 40–61.

Studer, H., Riek, M.M. and Greer, M.A. (1979) Multinodular goiter. In *Endocrinology*, Vol. 1 (eds L.J. DeGroot *et al.*), Grune and Stratton, New York, San Francisco and London, pp. 489–99.

Thomas, G.A., Williams, D. and Williams, E.D. (1989) The clonal origin of thyroid nodules and adenomas. *Am. J. Pathol.*, **134**, 141–7.

Vickery, A.L., Jr (1981) The diagnosis of malignancy in dyshormonogenetic goitre. *Clin. Endocrinol. Metab.*, **10**, 317–35.

Vidor, G.I., Stewart, J.C., Wall, J.R. *et al.* (1973) Pathogenesis of iodine-induced thyrotoxicosis: studies in Northern Tasmania. *J. Clin. Endocrinol. Metab.*, **37**, 901–9.

Werth, G. (1928) Beitrag zur Pathologie und Klinik der Struma ovarii. *Zentralbl. Gynäkol.*, **52**, 2944–57.

Willems, J.-S. and Löwhagen, T. (1981) The role of fine-needle aspiration cytology in the management of thyroid disease. *Clin. Endocrinol. Metab.*, **10**, 267–73.

Willemse, P.M.B., Oosterhuis, J.W. , Aalders, J.G. *et al.* (1987) Malignant struma ovarii treated by ovariectomy, thyroidectomy and ^{131}I administration. *Cancer*, **60**, 178–82.

6 Graves' disease

Graves' disease (Basedow's disease, diffuse toxic hyperplasia, exophthalmic goitre) is a syndrome comprising diffuse, hyperplastic goitre, hyperthyroidism, and eye signs. Less frequent features are localized (pretibial) myxoedema (5% of cases) and very rarely, so-called thyroid acropachy, i.e. clubbing of fingers and toes and swelling of extremities due to periosteal new bone formation. The disease is relatively common, occurring predominantly in the third to fifth decades, and females are affected about ten times more often than males. The patient with Graves' disease typically presents with diffuse goitre, exophthalmus, heat intolerance, weight loss despite increased appetite, muscle weakness, tachycardia, sometimes with atrial fibrillation, and irritability.

6.1 Immunological aspects

Graves' disease is closely linked to autoimmune thyroiditis (see p. 52). There is an increased prevalence of Hashimoto's thyroiditis, Graves' disease, and hypothyroidism in relatives of patients with Graves' disease, and not infrequently, several generations may be affected. An association with the histocompatibility type HLA-DR3 and other HLA antigens has been demonstrated (Strakosch et al., 1982; Hanfusa et al., 1983; Schleusner et al., 1983; Jansson et al., 1984; Aichinger et al., 1985). It has been shown that common immunogenetic factors are involved in the pathogenesis of both Hashimoto's and Graves' disease. Some workers, therefore, believe that both disorders are variants of the same autoimmune disease (Tamai et al., 1985).

There may also be a clinical overlap and both diseases are associated with high serum titres of autoantibodies against various thyroid antigens (see also p. 58). As in Hashimoto's disease, the basic immunological defect appears to be a deficiency of antigen-specific T-suppressor lymphocytes, which is thought to lead to excessive production of thyroid antibodies (Volpé, 1977). This view has recently received further support by the demonstration of local aberrant expression of HLA-DR antigens on the follicular cells in both diseases. These

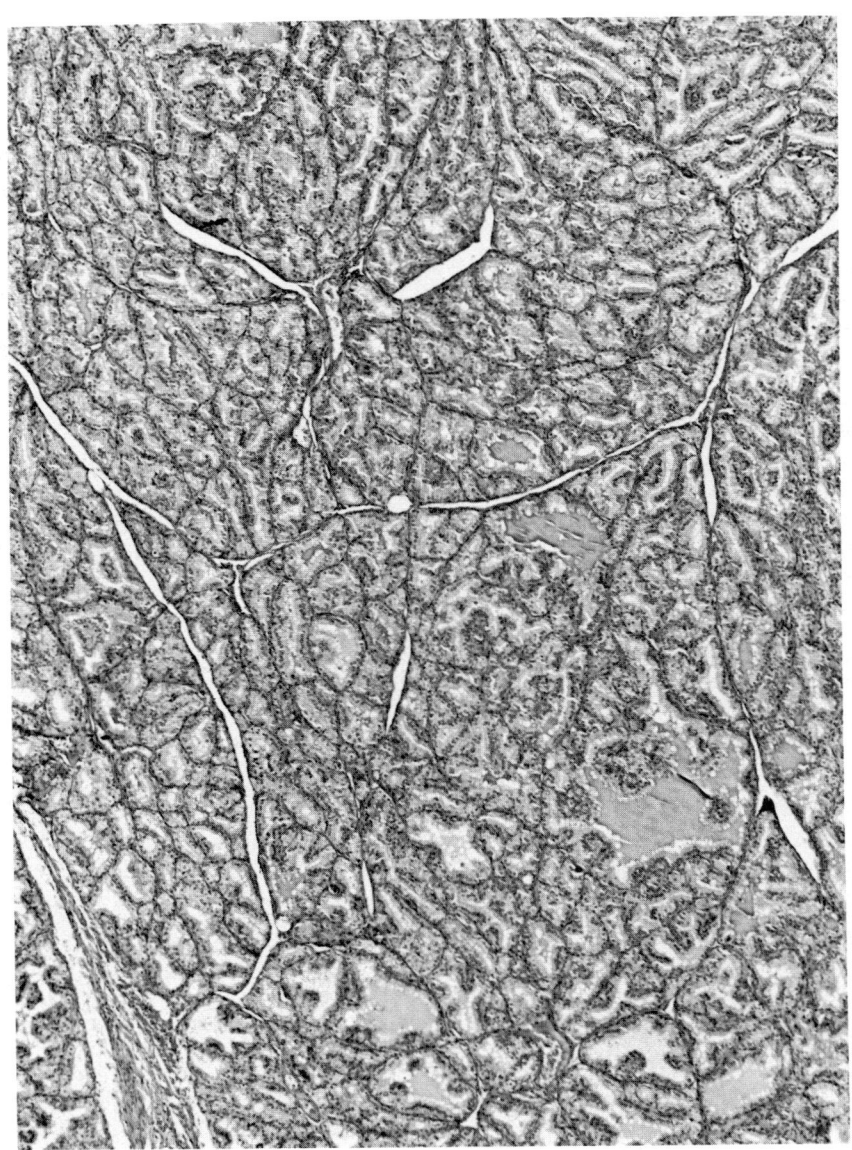

Figure 6.1 Graves' disease. There is a pronounced diffuse hyperplasia of the thyroid parenchyma. Preoperative treatment with thiamazole (H&E, × 50).

DR-positive follicular cells are thought to initiate the immune responses by presenting thyroidal cell surface autoantigens to the immuno-competent lymphoid cells. The presence of a concomitant impairment of suppressor cell activity in these patients may lead to an excessive production of autoantibodies in the thyroid gland (Thielemans *et al.*, 1981; Bottazzo *et al.*, 1983; Ueki *et al.*, 1987).

It is now generally agreed that the hyperthyroidism and goitre of Graves' disease result from the action of a group of autoantibodies usually referred to as thyroid-stimulating antibodies (TSAb). These antibodies, which belong to the IgG class of immunoglobulins, are directed against the TSH receptor antigens of the follicular cells. They mimic the effect of TSH and act via the adenyl cyclase mechanism. Recently, two distinct classes of TSAb have been identified, i.e. growth-stimulating antibodies, responsible for the cellular hyperplasia of the thyroid gland, and hormone-stimulating antibodies, inducing hyper-function of the gland (Drexhage *et al.*, 1980; Smyth *et al.*, 1982). Antibodies that can block these thyroid autoantibodies have also been described. Drexhage *et al.* (1981) have suggested that the clinical and pathological differences between Graves' disease and Hashimoto's thyroiditis may be due to differences in the expression of these blocking immunoglobulins. Long-acting thyroid stimulator (LATS, Adams, 1962) which is also an IgG immunoglobulin, was initially believed to be the causal factor in Graves' disease. However, LATS does not occur in all patients and has been found to be more closely correlated with exophthalmus and pretibial myxoedema than with hyperthyroidism (Munro, 1970).

The pathogenesis of the eye changes remain uncertain. Ophthalmogenic immunoglobulins (OIg), which have been detected in some patients with active Graves' ophthalmopathy, have been found to react with soluble eye-muscle antigens (Kodama *et al.*, 1982). However, their role in the development of the eye lesions is unknown.

6.2 Pathology

The thyroid gland is usually moderately enlarged. In men, the goitre tends to be small. The thyroid enlargement is mostly symmetrical and diffuse, although there are many exceptions. The gland may be asymmetrical, with one or more gross nodules, and it may show the features of a multinodular goitre, especially in elderly patients (toxic nodular goitre, p. 97). The cut surface of the diffusely enlarged gland is glistening, reddish, or tan coloured, depending on the content of blood, and often clearly lobulated, like pancreatic tissue.

The microscopic appearance (Fig. 6.1) is characterized by a diffuse follicular epithelial hyperplasia and decreased amounts of colloid. The follicular cells are crowded, lining the follicular walls in several layers, sometimes heaping up into papillary-like proliferations into the follicular lumina. Follicular lumina diminish and become irregularly outlined, slit-like or branched. Some follicles may be entirely filled with epithelial cells, with no colloid left. The epithelial cells undergo hypertrophy and become moderately or notably columnar, with pale, sometimes vacuolated cytoplasm and enlarged, basally located nuclei. The amount of colloid is reduced and stains weakly eosinophilic. Marginal vacuoles are often seen in the colloid at the luminal surface of the follicular epithelial cells. Foci of oxyphil cells may also occur, as well as areas where the follicular cells have undergone a foamy change of their cytoplasm, features that are similar to those seen in Hashimoto's thyroiditis (see also p. 56). Papillary projections of columnar epithelium, growing on a fibrous stalk may protrude into some follicles and simulate neoplastic papillae (Fig. 6.2). In Graves' disease, however, the nuclei of epithelial cells are usually uniform, round or oval with smooth outlines and have fine chromatin granules which, although occasionally concentrated at the periphery of the nucleus, tend to be rather evenly dispersed over the entire profile of the nucleus. These features are quite similar to those occurring in hyperplastic nodules (see also p. 85 and Figs 5.8 and 5.17). In papillary carcinoma, on the other hand, the nuclei are crowded and overlapping, their profiles tend to be irregularly outlined, often square with rounded angles and sometimes with invaginations (so-called nuclear grooves), and have the typical ground glass appearance (see also p.147). In the diffusely affected gland the microscopic changes are uniform. The lobulation may be exaggerated to some degree, but the general histological architecture of the normal parenchyma is usually preserved. This is a helpful, additional diagnostic point in the differential diagnosis against a malignant process.

The stroma is sparse with a rich network of capillary vessels. It frequently contains focal lymphoid aggregates, sometimes with germinal centres. Immunohistochemically, the great majority of lymphocytes have been found to express HLA-DR and most cells are of T cell type, helper/inducer cells being in the majority and present in the interfollicular infiltrates, and suppressor/cytotoxic cells being rare and scattered throughout the follicles (Jansson et al., 1984). In long-standing cases, there may be some interstitial fibrosis.

As pointed out in Chapter 3 (p. 48), hyperplastic thyroid tissue may be encountered outside the thyroid gland, often in surrounding skeletal muscles. This is especially prominent in florid cases of Graves' disease and should not be misinterpreted as evidence of infiltrating thyroid carcinoma.

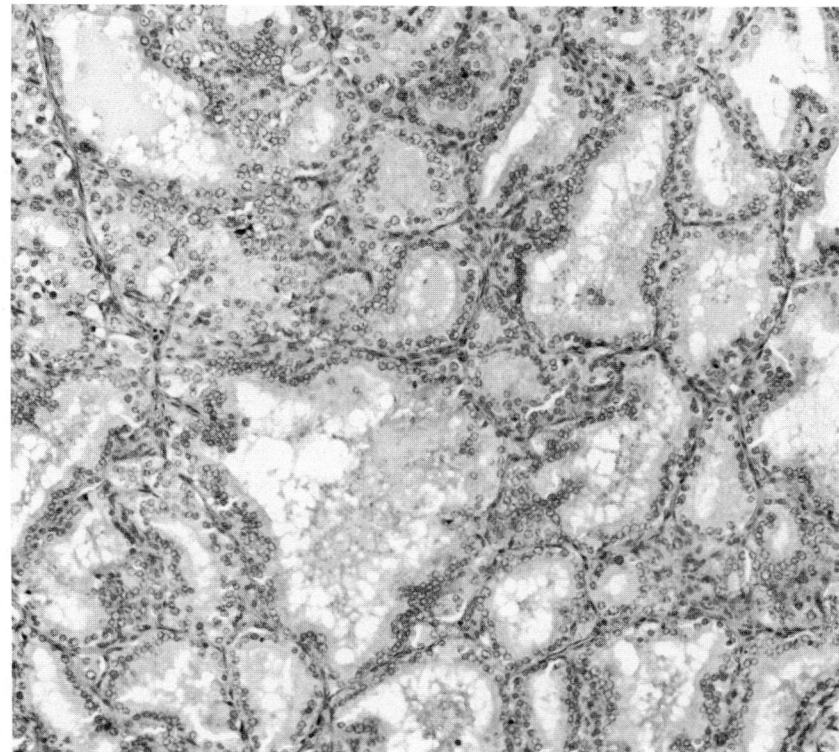

Figure 6.2 Graves' disease. There is marked hyperplasia showing irregular follicles with foamy colloid and tall columnar epithelium, at some places heaped up into papillary projections.These features may be confused with those of papillary carcinoma (H&E, × 125).

6.3 Treated Graves' disease

The microscopic features described above do not represent the original state of the thyroid, since the patient has usually received various types of preoperative medical treatment which produce other histological changes (Eggen and Seljelid, 1973).

Iodine treatment, seldom used today, diminishes the vascularity of the parenchyma and produces various degrees of involutionary change, often unevenly distributed, and characterized by accumulation of colloid, dilation of follicles, and flattening of the epithelium.

Treatment with sulphur-containing antithyroid drugs, such as propylthiouracil, methimazole, thianazole and carbimazole are usually combined with administration of thyroid hormone. This type of treatment, either alone or followed by subtotal thyroidectomy, is the one most commonly used today, especially in younger patients. When a combined treatment is given, the hyperplastic changes in the gland are mainly preserved, although there may be areas where the intensity of the hyperplastic changes appears to be reduced, with some increase of colloid and dilation of follicles. It should be pointed out that in some cases treated with a combination of antithyroid drugs and thyroid hormone, the degree of cellular and nuclear atypia may be very pronounced (McGowan and Sandler, 1967). In these thyroid specimens, areas may be found which show highly polymorphic follicular cells with hyperchromatic, sometimes quite bizarre nuclear forms (Fig. 6.3). Such features may easily be misinterpreted as signs of malignancy, as may be the fine needle aspirates which show highly polymorphic epithelial cells mimicking carcinoma cells. However, it should be remembered that the basal architecture of the parenchyma is generally preserved. In addition, the few mitoses among these cells, and the frequent occurrence of signs of cellular disintegration indicate that the cytological features are the result of degenerative alterations rather than neoplastic

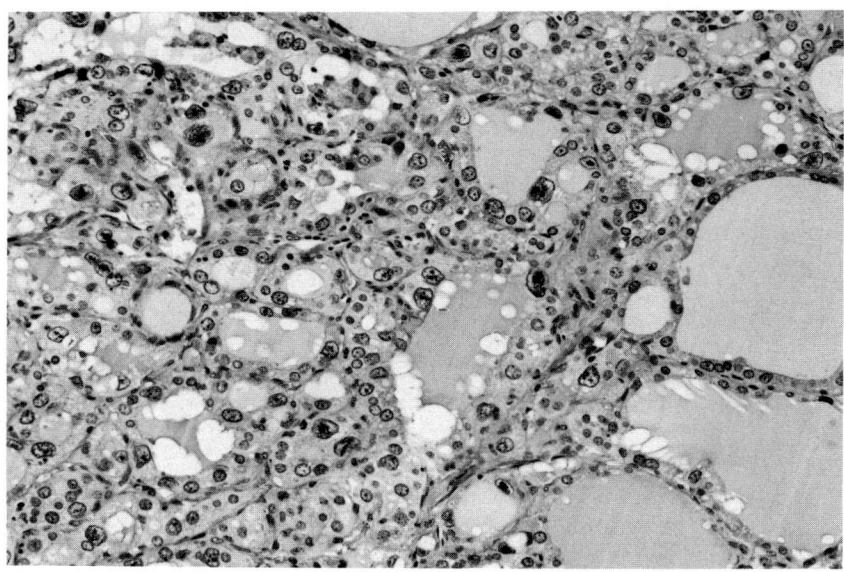

Figure 6.3 Graves' disease. Area showing pronounced cytological atypia: highly polymorphic nuclei with irregular chromatin pattern, but no mitoses. Cytological changes of this type may occur after combined treatment with sulphur containing antithyroid drugs and thyroid hormone, and should not be misinterpreted as evidence of neoplastic change (H&E, × 165).

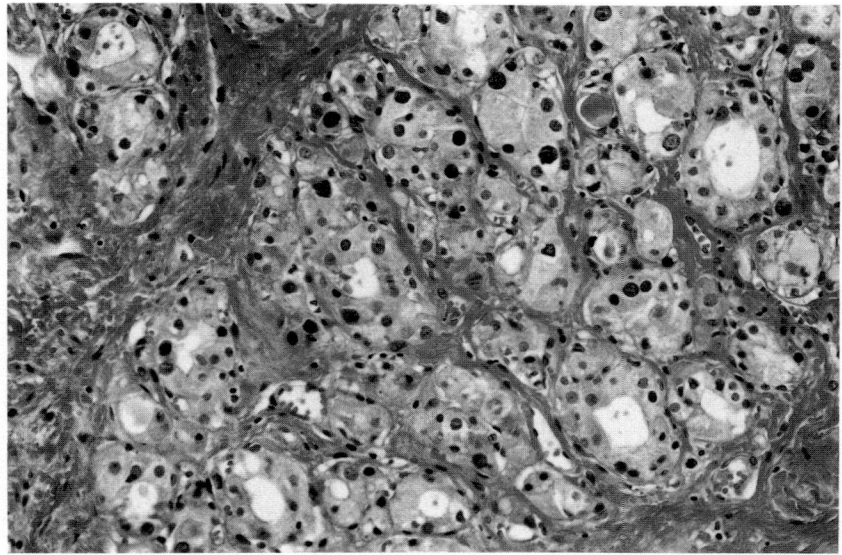

Figure 6.4 Graves' disease treated with radioiodine. Failure to remedy the condition with this treatment, necessitated subtotal thyroidectomy. The thyroid parenchyma shows atrophy with degeneration and disintegration of follicles and interstitial fibrosis. Residual follicles are small with atypical epithelial cells having pleomorphic, often hyperchromatic nuclei (H&E, × 160).

change. Finally, it should be borne in mind that thyroid cancer associated with Graves' disease is rare. Such carcinomas are usually incidental findings and mostly small papillary carcinomas without clinical significance (Olen and Klinck, 1966; Farbota et al., 1985).

Treatment of Graves' disease with radioactive iodine is mainly restricted to older patients. Such a treatment induces nuclear changes and disintegration of epithelial cells. There may be marked cytological atypia with highly polymorphic nuclei (Fig. 6.4), features that may resemble those seen in glands treated with antithyroid drugs and thyroid hormone. Later, however, progressive atrophy occurs, with disappearance of follicles and increasing fibrosis. There is a high incidence of hypothyroidism as a long-term complication of this therapy. Within 10–20 years the cumulative incidence of hypothyroidism may reach levels up to 70% (Glanzmann et al., 1975). However, such treatment does not seem to be associated with any increase in the incidence of malignant thyroid tumours (Holm et al., 1980).

References

Adams, D.D. (1961) Bioassay of long-acting thyroid stimulator (L.A.T.S.); the dose–response relationship. *J. Clin. Endocrinol.*, **21**, 799–805.

Aichinger, G., Fill, H. and Wick, G. (1985) *In situ* immune complexes, lymphocyte populations, and HLA-DR-positive epithelial cells in Hashimoto thyroiditis. Autoantibodies. *Lab. Invest.*, **52**, 132–40.

Bottazzo, G.F., Pujol-Borrell, R. and Hanafusa, T. (1983) Role of aberrant HLA-DR expression and antigen presentation in induction of endocrine autoimmunity. *Lancet.*, **ii**, 1115–9.

Drexhage, H.A. Bottazzo, G. F. Bitensky, L. *et al.* (1981) Thyroid growth-blocking antibodies in primary myxoedema. *Nature*, **289**, 594–6.

Drexhage, H.A., Bottazzo, G.F., Doniach, D. *et al.* (1980) Evidence for thyroid-growth-stimulating immunoglobulins in some goitrous thyroid diseases. *Lancet*, **ii**, 287–92.

Eggan, P.C. and Seljelid, R. (1973) The histological appearance of hyperfunctioning thyroids following various pre-operative treatments. *Acta Pathol. Microbiol. Scand.* [A], **81**, 16–20.

Farbota, L.M., Calandra, D.B., Lawrence, A.M. and Paloyan, E. (1985) Thyroid carcinoma in Graves' disease. *Surgery*, **98**, 1148–52.

Glanzmann, Ch., Kaestner, F. and Horst, W. (1975) Therapie der Hyperthyreose mit Radio-Isotopen des Jods: Erfahrungen bei über 2000 Patienten. *Klin. Wochenschr.*, **53**, 669–78.

Hanafusa, T., Pujol-Borrell, R., Chiovato, L. *et al.* (1983) Aberrant expression of HLA-DR antigen of thyrocytes in Graves' disease: relevance for auto-immunity. *Lancet*, **ii**, 1111–5.

Holm, L.-E., Dahlqvist, I., Israelsson, A. and Lundell, G. (1980) Malignant thyroid tumours after iodine-131 therapy: a retrospective cohort study. *N. Engl. J. Med.*, **303**, 188–91.

Jansson, R., Karlsson, A. and Forsum, U. (1984) Intrathyroidal HLA-DR expression and T lymphocyte phenotypes in Graves' thyrotoxicosis, Hashimoto's thyroiditis and nodular colloid goitre. *Clin. Exp. Immunol.*, **58**, 264–72.

Kodama, K., Sikorska, H., Bandy-Dafoe, P. *et al.* (1982) Demonstration of a circulating autoantibody against a soluble eye-muscle antigen in Graves' ophthalmopathy. *Lancet*, **ii**, 1353–6.

McGowan, G.K. and Sandler, M. (1967) Symposium on the thyroid gland. *J. Clin. Pathol.*, **20**, 309–413.

Munro, D.S. (1970) The long-acting thyroid stimulator. *Br. J. Hosp. Med.*, **3**, 421–4.

Olen, E. and Klinck, G.H. (1966) Hyperthyroidism and thyroid cancer. *Arch. Pathol.*, **81**, 531–5.

Schleusner, H., Schernthaner, G., Mayr, W.R. *et al.* (1983) HLA-DR3 and HLA-DR5 associated thyrotoxicosis in two different types of toxic diffuse goiter. *J. Clin. Endocrinol. Metab.*, **56**, 781–5.

Smyth, P.P.A., Neylan, D., O'Donovan, D.K. (1982) The prevalence of thyroid-stimulating antibodies in goitrous diseases assessed by cytochemical section bioassay. *J. Clin. Endocrinol. Metab.*, **54**, 357–67.

Strakosch, C.R., Wenzel, B., Row, V.V. and Volpé, R. (1982) Immunology of autoimmune thyroid disease. *N. Engl. J. Med.*, **307**, 1499–507.

Tamai, H., Uno, H., Hirota, Y., Matsubayashi, S. *et al.* (1985) Immunogenetics of Hashimoto's and Graves' diseases. *J. Clin. Endocrinol. Metab.*, **60**, 62–6.

Thielemans, C., Vanhaelst, L., De Waele, M. *et al.* (1981) Autoimmune thyroiditis: a condition related to a decrease in T-suppressor cells. *Clin. Endocrinol.*, **15**, 259–63.

Ueki, Y., Eguchi, K., Fukuda, T. *et al.* (1987) Dysfunction of suppressor T cells in thyroid glands from patients with Graves' disease. *J. Clin. Endocrinol. Metab.*, **65**, 922–8.

Volpé, R. (1977) The role of autoimmunity in hypoendocrine and hyperendocrine function. With special emphasis on autoimmune thyroid diseases. *Ann. Intern. Med.*, **87**, 86–99.

7 Classification of thyroid neoplasms

7.1 General aspects

It is generally agreed that the classification of thyroid tumours into various histopathological types is of great importance. This is because of the strong link that exists between the morphological type of tumour and its epidemiology, natural history, function and prognosis – features that, in turn, influence the management and choice of therapy. During the 1960s and 1970s, much progress was made to achieve a standard and consistent histological nomenclature and classification of thyroid tumours. Important contributions in this regard include, for example, the extensive clinicopathological studies by Woolner *et al.* (1961), Russell *et al.* (1963), Hazard (1964), and others, but also the classifications proposed by the Armed Forces Institute of Pathology fascicles on Tumors of the Thyroid Gland (Warren and Meissner, 1953; Meissner and Warren, 1969), and by the American Thyroid Association (Werner *et al.* 1969). More recently, the World Health Organization published its first edition (Hedinger and Sobin, 1974), on histological classification of thyroid tumours, which was mainly based on the previous classifications and which has since been adopted throughout the world. The WHO classification has been found to work well in daily diagnostic practice. It has been of great value both for providing clinicopathological correlations of prognostic and therapeutic importance to the clinician, and in scientific work, forming the basis for many clinical, pathological and epidemiological investigations, and allowing comparative epidemiological studies between different geographic areas.

However, no tumour classification can be static, but must be subject to change according to new knowledge acquired about the nature and behaviour of the tumours and the introduction of new diagnostic and therapeutic measures. Since 1974, there have been considerable advances in several areas of thyroid pathology, partly due to the introduction of new investigative techniques for diagnosis and prognosis. Immunohistochemistry represents a major new development in this respect, and the value of nuclear DNA analysis as an adjuvant for prognosis estimation is currently being investigated. It has become increasingly evident that the distinction between certain main groups of tumours may not be as clear as previously believed and that, on the other hand, certain tumour types may have to be abandoned or included in other main groups. Attempts have been made in recent years to delineate further various morphological tumour entities in order to meet increasing demands from our clinical colleagues for sharper prognostic and therapeutic distinctions. As a result of these developments, WHO has recently published the second edition of its classification of thyroid tumours (Hedinger *et al.*, 1988). This revised WHO classification has basically preserved the common, main groups, but with further simplification to facilitate its practical use.

The most important changes include the elimination of certain rare tumours as major types (e.g. squamous carcinoma) or as subtypes (e.g. certain variants of undifferentiated carcinoma), the introduction of malignant lymphoma as a major group, and a revision of the non-epithelial and miscellaneous groups of tumours. A commentary on the revised classification has been published by Hedinger *et al.* (1989).

7.2 Present histological classification

The following presentation will adhere to the main principles of the WHO classification, though on a few points different views will be discussed. The classification adopted is given in Table 7.1 and is a modification of that proposed by the World Health Organization in 1988 (Hedinger *et al.* 1988).

The large majority of clinically evident thyroid tumours are primary and epithelial. They are divided into two major categories, adenomas and carcinomas. From a histogenetic point of view the majority of the epithelial neoplasms may be regarded as tumours showing follicular or parafollicular (C) cell differentiation. Recently, a third, rare category has been identified, which exhibits both follicular and parafollicular cell differentiation and may even develop ultimobranchial features. This tumour has been referred to as the intermediate type of thyroid carcinoma and regarded as a distinct category (Chapter 9, p. 227). Following the suggestion of Carcangiu *et al.* (1984), the poorly differentiated carcinoma has also been introduced as a major group because of its distinct clinicopathological features, which are clearly separable from those of the differentiated and anaplastic tumours.

Non-epithelial thyroid tumours are very rare. Increased attention has been paid to malignant lymphomas, which to a great extent have formerly been diagnosed erroneously as small cell undifferentiated carcinomas. In accordance with the WHO classification, these tumours have been given a major separate heading.

Table 7.1 Classification of thyroid tumours

I.	*Epithelial tumours*
	A. Benign
	1. Follicular adenoma
	2. Others
	B. Malignant
	1. Follicular carcinoma
	2. Papillary carcinoma
	3. Medullary carcinoma
	4. Intermediate type carcinoma
	5. Poorly differentiated carcinoma
	6. Undifferentiated (anaplastic) carcinoma
	7. Others
II.	*Non-epithelial tumours*
	A. Benign
	B. Malignant
III.	*Malignant lymphomas*
IV.	*Miscellaneous tumours*
V.	*Secondary tumours*
VI.	*Unclassified tumours*
VII.	*Tumour-like lesions*

References

Carcangiu, M.L., Zampi, G. and Rosai, J. (1984) Poorly differentiated (insular) thyroid carcinoma. A reinterpretation of Langhans' 'wuchernde struma'. *Am. J. Surg. Pathol.*, **8.**, 655–68.

Hazard, J.B. (1964) Neoplasia. In *The Thyroid*. International Academy of Pathology Monograph No. 5 (eds J.B. Hazard and D.E. Smith), Williams and Wilkins, Baltimore, pp. 236–55.

Hedinger, C. and Sobin, L.H. (1974) Histological typing of thyroid tumours. In *International Histological Classification of Tumours*, no. 11, World Health Organization, Geneva.

Hedinger, C., Williams, E.D. And Sobin, L.H. (1988) Histological typing of thyroid tumours, 2nd edn. In *International Histological Classification of Tumours*, no. 11, World Health Organization, Springer-Verlag, Berlin, Heidelberg, New York, London, Paris, Tokyo, Hong Kong.

Hedinger, C., Williams, E.D. and Sobin, L.H. (1989) The WHO histological classification of thyroid tumours: a commentary on the second edition. *Cancer*, **63**, 908–9.

Meissner, W.A. and Warren, S. (1969) Tumors of the thyroid gland, 2nd series, Fascicle 4. In *Atlas of Tumor Pathology*, Armed Forces Institute of Pathology, Washington, DC.

Russell, W.O. Ibanez, M.L., Clark, R.L. and White, E.C. (1963) Thyroid carcinoma. Classification, intraglandular dissemination and clinicopathological study based upon whole organ sections of 80 glands. *Cancer,* **16**, 1425–60.

Warren, S. and Meissner, W.A. (1953) Tumors of the thyroid gland. 1st series, Fascicle 4. In *Atlas of Tumor Pathology,* Armed Forces Institute of Pathology, Washington, DC.

Werner, S.C., Beierwaltes, W., DeGroot, L.J. *et al.* (1969) Classification of thyroid disease. Report of the committee on nomenclature of the American Thyroid Association. *J. Clin. Endocrinol. Metab.,* **29**, 860–2.

Woolner, L.B., Beahrs, O.H., Black, B.M. *et al.* (1961) Classification and prognosis of thyroid carcinoma. A study of 885 cases observed in a thirty year period. *Am. J. Surg.,* **102**, 354–87.

8 Thyroid adenoma

Adenomas are benign, usually solitary and encapsulated lesions composed of follicular cells. Thyroid adenomas occur most frequently between 30 and 50 years of age and 4–6 times more often in females than males. It should be noted that the clinically solitary nodule in the great majority of cases is in fact the dominant nodule in a multinodular goitre, and rarely represents a true neoplasm (Brown, 1981). The proportion of neoplasms diagnosed in cases of clinically solitary nodules selected for surgery has, however, increased in recent years, mainly due to the greater use of more accurate diagnostic means, the most important of which is the fine-needle aspiration technique (Willems and Löwhagen, 1981; see also chapter 5, p. 92).

8.1 Follicular adenomas

Follicular adenomas are tumours with morphological and histochemical evidence of follicular cell differentiation.

These tumours are the most common of all thyroid neoplasms. Adenomas grow slowly, may remain dormant for years and typically are asymptomatic. They must reach the size of 0.5–1 cm before they can be palpated and they are typically discovered accidentally by the patient or physician. Occasionally, haemorrhage into the tumour may cause local pain and tenderness, and even transient thyrotoxicosis due to leakage of hormones into the circulation. Bleeding may be followed by spontaneous regression of the lump or result in cyst formation (DeGroot, 1979). Approximately 70% of thyroid nodules are hypo-functional as defined by the degree of accumulation of radioiodine. About 10% are considered hyperfunctional, producing T_3 and reverse T_3, and thus causing thyrotoxicosis and suppressing the functional activity of the remainder of the gland (Malamos *et al.*, 1968; Reinwein *et al.*, 1977; Panke *et al.*, 1978).

Grossly, the follicular adenoma typically ranges from 1 to 3 cm, is well demarcated and has a fibrous capsule that is intact (Plate 9). The cut surface is bulging, and has a soft, greyish, tan or brownish appearance. Tumours with little colloid appear dry and brittle,

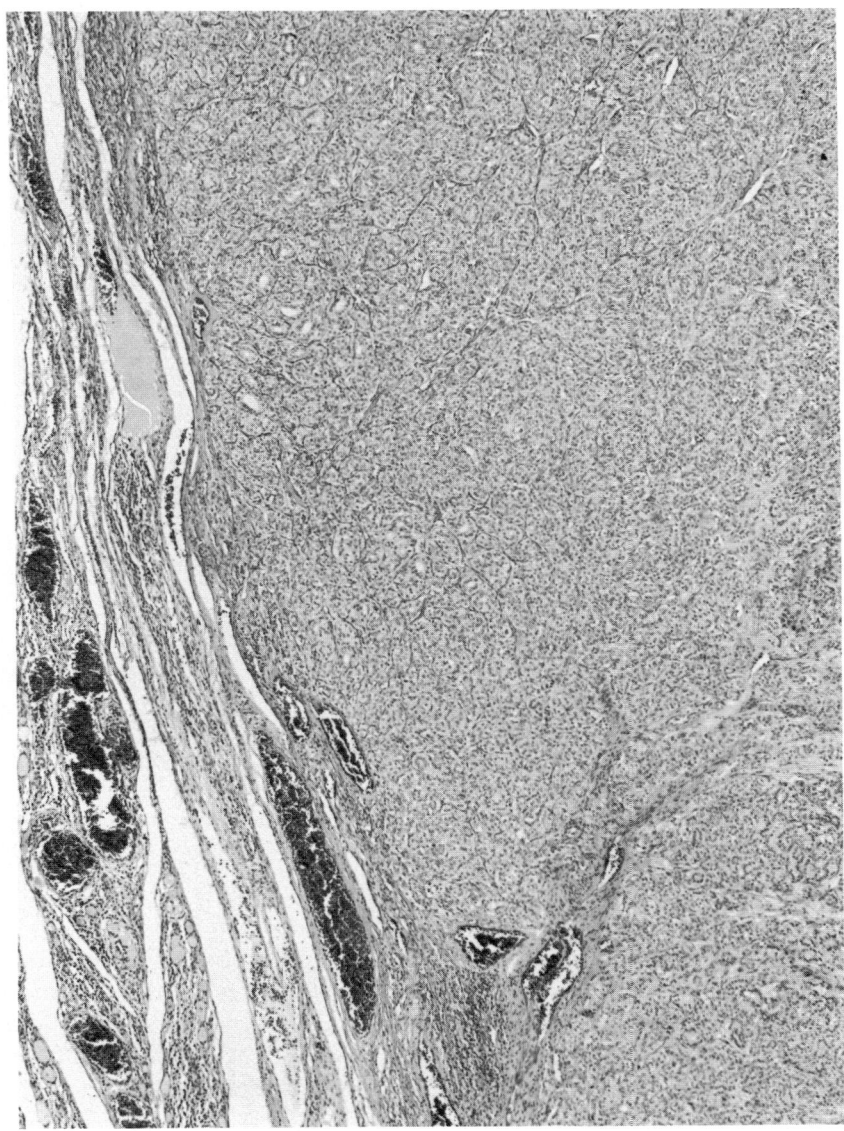

Figure 8.1 Follicular adenoma, clearly separated from the surrounding atrophic parenchyma by a thin fibrous capsule. Slit-like and thin-walled blood vessels are often seen in the capsular zone (H&E, × 50).

whereas colloid-rich lesions are shining, sometimes gelatinous in character. The tumour tends to compress the neighbouring thyroid parenchyma. Degenerative changes may occur, especially in larger tumours, as well as grey–white, firm areas of fibrosis, sometimes with

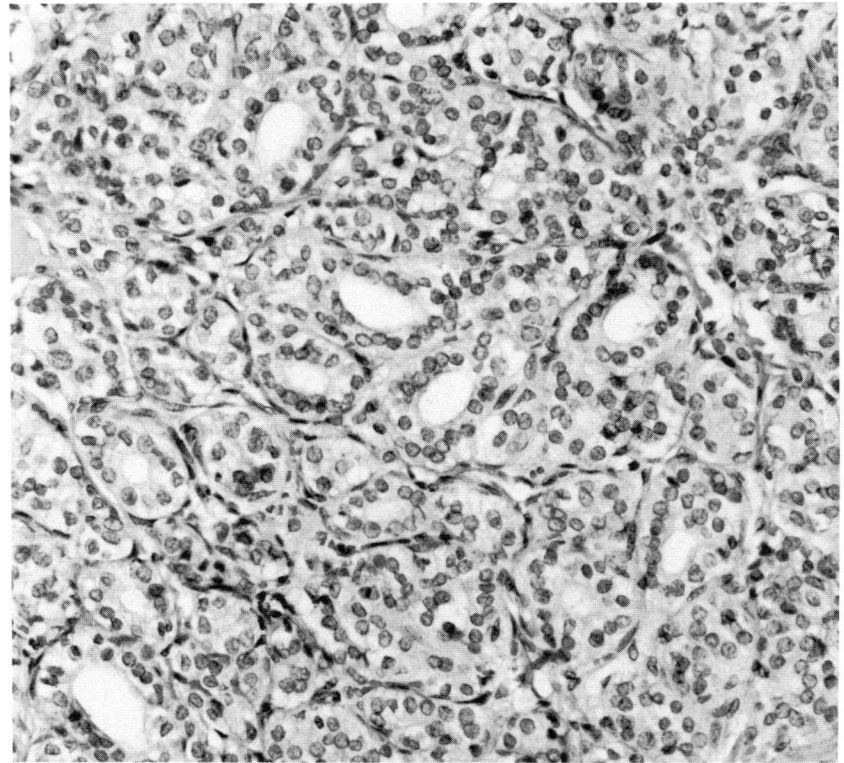

Figure 8.2 Detail of the adenoma depicted in Fig. 8.1 showing a regular micro-follicular and solid architectural pattern with little cytological atypia (H&E, × 250).

calcification and bone formation. Necrosis and haemorrhage may cause secondary cyst formation. Sometimes virtually all tumour tissue may disintegrate, leaving only a cyst with fibrous tissue in its wall and containing fluid with liquefied necrotic material and residuum from blood decomposition (Plate 10). However, cyst formation is more frequently seen in colloid nodules.

Follicular adenomas have histological patterns which are usually different from that of the remaining normal parenchyma. The criteria used to differentiate between follicular adenoma and nodules in multinodular goitre have been discussed in Chapter 5 (p. 90). The characteristic features used to differentiate between adenoma and nodule may be distinct in most instances; but some nodules may share one or more of the features of an adenoma, making the distinction uncertain or impossible.

Follicular adenomas have traditionally been subclassified according to the architectural differentiation of the follicles within the neoplasm, e.g. colloid or macrofollicular, simple or normofollicular, fetal or microfollicular, and embryonal or trabecular/solid (Figs 8.1 and 8.2). These various patterns may occur singly or in combinations. It is generally agreed that differences in structural pattern have no apparent clinical significance. They may, however, be of help for the cytologist in cytohistological correlation and in understanding overlapping cytopathological patterns (Kini, 1987).

Adenomas are histologically well demarcated and surrounded by a fibrous capsule, which may be thin and indistinct. As will be discussed in the section on follicular carcinoma (section 9.3), it is crucial to examine the capsular area of the tumour with great care, in order to establish whether or not capsular penetration and/or invasion of blood vessels is present. Such signs indicate that the tumour is malignant and should be referred to the solitary encapsulated, minimally invasive variant of follicular carcinoma. An adequate number of tissue blocks must be taken from the capsular region since signs of malignancy may be focal and easily missed (Lang *et al.*, 1980; Franssila *et al.*, 1985; and Chapter 2, p. 26). If the initial examination shows signs of equivocal significance, such as the presence of tumour aggregates within, but without full penetration of, the capsule, or incipient small invaginations of tumour tissue into blood vessels but no larger polypoid projections of tumour masses into vessel lumina or obvious intravasal tumour thrombi (Fig. 8.3 and see Fig. 8.9), additional blocks should be taken to search for definite indicators of malignancy. If such signs are not found, the lesion should be considered benign. Encapsulated follicular carcinomas, in contrast to adenomas, also tend to have thick capsules (Evans, 1984). Thyroglobulin is not a useful marker for making the differential diagnosis between follicular adenomas and their malignant counterparts, since its distribution pattern is very similar in both neoplasms. Recently, other glycoproteins, unrelated to thyroid function, such as lactoferrin and ceruloplasmin have been suggested as additional new immunohistochemical markers claimed to be of value in making this separation (Tuccari and Barresi, 1985, 1987; see also p. 33).

The architectural pattern or the cytological features of the tumour tissue are not helpful in the distinction between adenoma and follicular carcinoma. Mitoses are usually rare in adenomas; their presence indicates an increased growth rate of the tumour. When present in appreciable numbers the possibility of malignancy should be carefully considered and the margins of the lesion examined using many sections. DNA aneuploidy is reported to be common in histologically benign follicular adenomas, but does not appear to be associated with

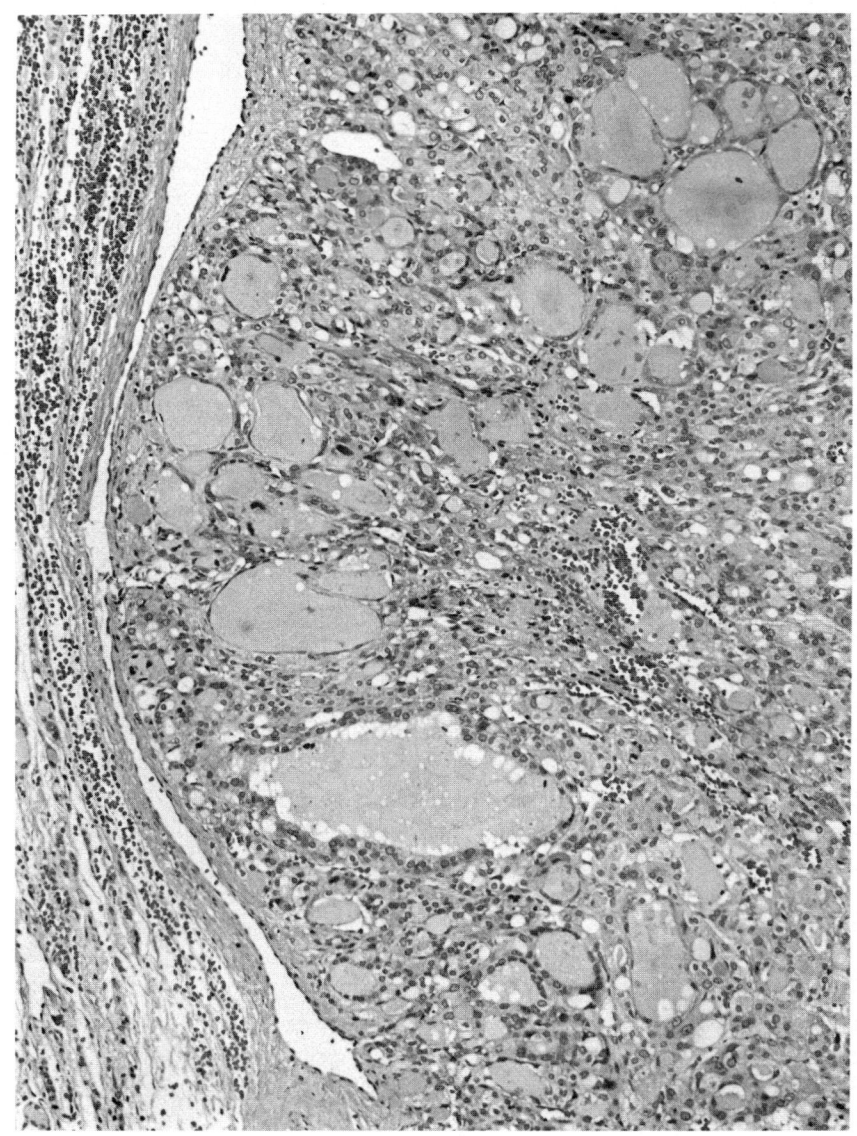

Figure 8.3 Follicular thyroid adenoma with solid and follicular structures and pronounced cellular proliferation. A tongue of tumour tissue is seen to push the wall of a large capsular blood vessel, but without penetrating into its lumen. This feature is insufficient evidence of blood vessel invasion and, thus, does not justify a malignant diagnosis. However, such a finding calls for extended histological examination of the tumour capsule (H&E, × 125).

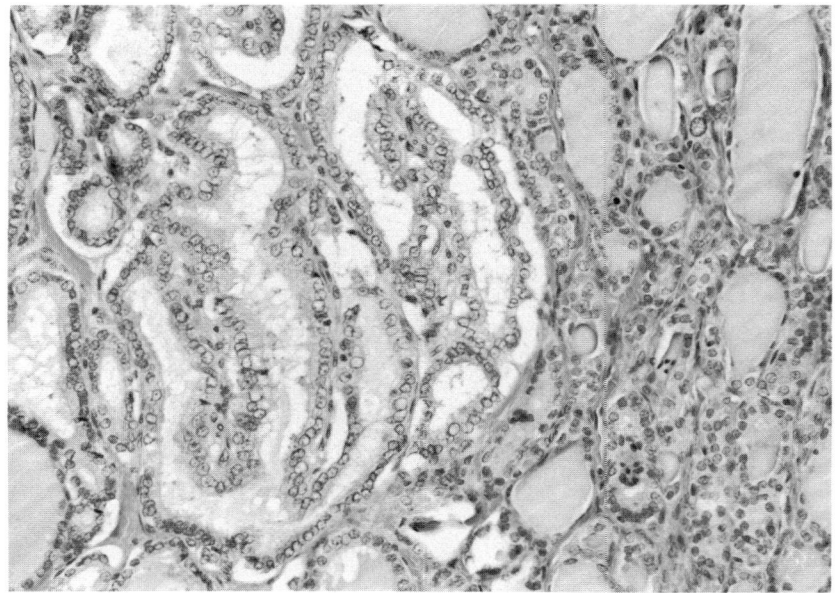

Figure 8.4 A minute focus of papillary structures and irregular follicles, with cytological and nuclear features compatible with papillary carcinoma, is located within an otherwise benign-looking follicular adenoma (H&E, × 175).

an adverse prognosis (Joensuu *et al.*, 1986). The discrimination of follicular adenomas and follicular carcinomas does not seem to be possible by means of DNA measurements (Sprenger *et al.*, 1977; Johannessen *et al.*, 1982; Auer *et al.*, 1985).

Follicular adenoma should not be confused with the rare, encapsulated variant of papillary carcinoma which may be mainly or (almost) entirely follicular in structure (Rosai *et al.*, 1983; see also Fig. 9.24). The main differential diagnostic point is the appearance of the tumour cells and, more importantly, their nuclei. Neoplastic cells in the follicular variant of papillary carcinoma are rather large, cuboidal or cylindrical, and have overlapping, large and irregularly outlined, clear nuclei of the ground-glass type (Chen and Rosai, 1977). In addition, the encapsulated papillary carcinoma, although appearing sharply delineated, on close examination frequently shows evidence of infiltrating aggregates of tumour cells outside the fibrous capsule, growing in between the normal follicles of the adjacent parenchyma. It is especially important to search for signs of infiltrative growth as evidence of papillary carcinoma, where typical cytological or nuclear ground-glass features appear less distinctive, or are only focally present

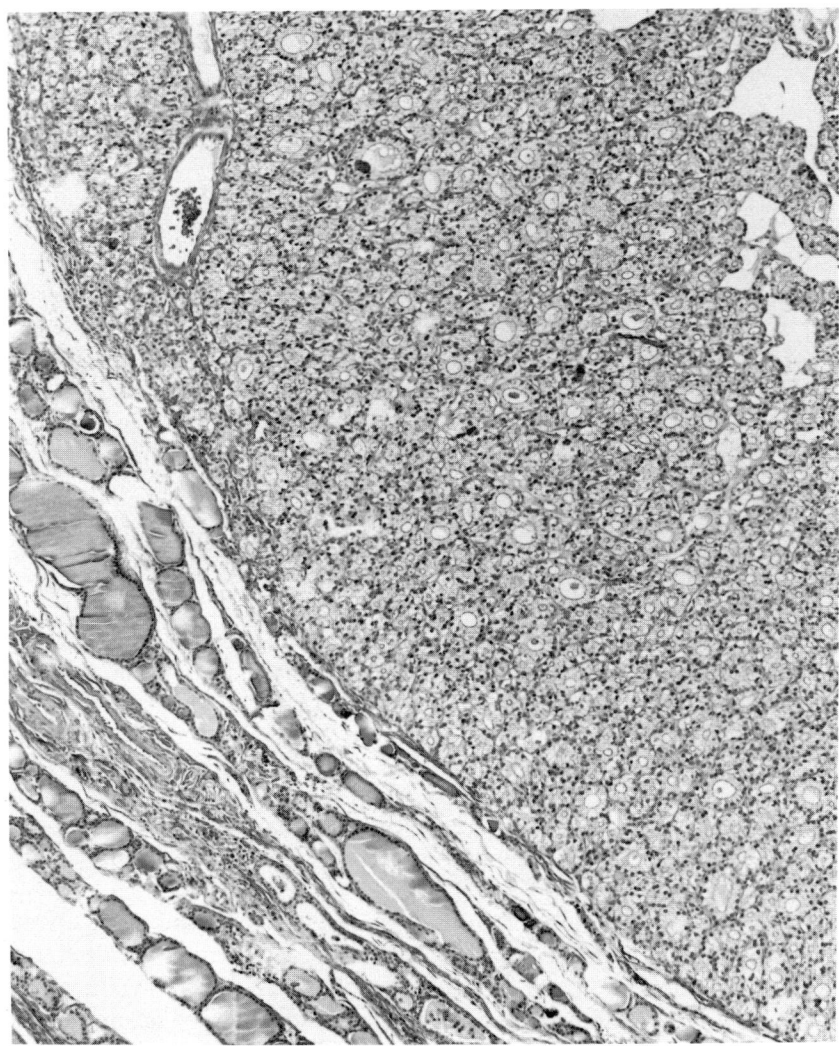

Figure 8.5 Clear cell adenoma with microfollicular histological pattern (H&E, × 65).

(Rosai and Carcangiu, 1987). Vickery (1983) and Vickery *et al.* (1985) reported a few cases of focal papillary carcinoma occurring within an otherwise benign-looking adenomatous lesion and suggested that some encapsulated papillary carcinomas may arise from pre-existing follicular adenomas. A similar tumour chiefly composed of follicles lined by cuboidal or flat, benign-looking cells, but with an area showing a gradual transition into irregular follicles and papillary structures with nuclear features characteristic of papillary carcinoma has also been encountered (Fig. 8.4).

8.2   Other adenomas

Several forms of follicular adenoma have been described which are, to various extents, composed of cells considered to be morphological and functional variants of the follicular cells. Other adenomas contain an admixture of cellular elements distinct from the follicular cells, e.g. fat cells. This group also includes adenomas of uncertain nature, which differ from ordinary follicular adenomas by having atypical structure or by showing marked cytological and/or nuclear atypia.

8.2.1   Oxyphil adenoma

This tumour is also referred to as Askanazy cell, Hürthle cell or large-cell eosinophil adenoma. It is defined as a neoplasm which largely or entirely is made up of mitochondrion-rich, oxyphil cells. This variant, also having a malignant counterpart, is discussed in the section on carcinomas, on p. 190.

8.2.2   Clear cell adenoma

Tumours of this category are composed of cells with vacuolated cytoplasm and distinct cell boundaries. Some adenomas are made up exclusively of clear cells; others show only focal clear cell change. Although this change may be seen in follicular adenomas of all the aforementioned microscopic types, the pure clear cell adenoma usually has a trabecular/microfollicular structure (Figs 8.5–8.7). The clear cell change is not considered to be of any clinical significance and from this point of view, adenomas with clear cell change should be judged as follicular adenomas in general, i.e. diagnosing as malignant only those in which unequivocal evidence of blood vessel invasion and/or capsular penetration can be demonstrated (Rosai and Carcangiu, 1984; Carcangiu et al., 1985). It may also occur in various forms of thyroid carcinoma (Chapter 9) and occasionally, in non-neoplastic disorders, such as Hashimoto's thyroiditis (see p. 56). Oxyphil tumours are most liable to undergo secondary clear cell changes, suggesting that oxyphil cells and clear cells occurring in these neoplasms may be related. Ultrastructurally, the clearing of the cytoplasm in these tumours appears to be the result of accumulation of vesicles, which may represent a vesicular swelling of degenerating mitochondria or derivatives of granular endoplasmic reticulum (Saleiro et al., 1981;

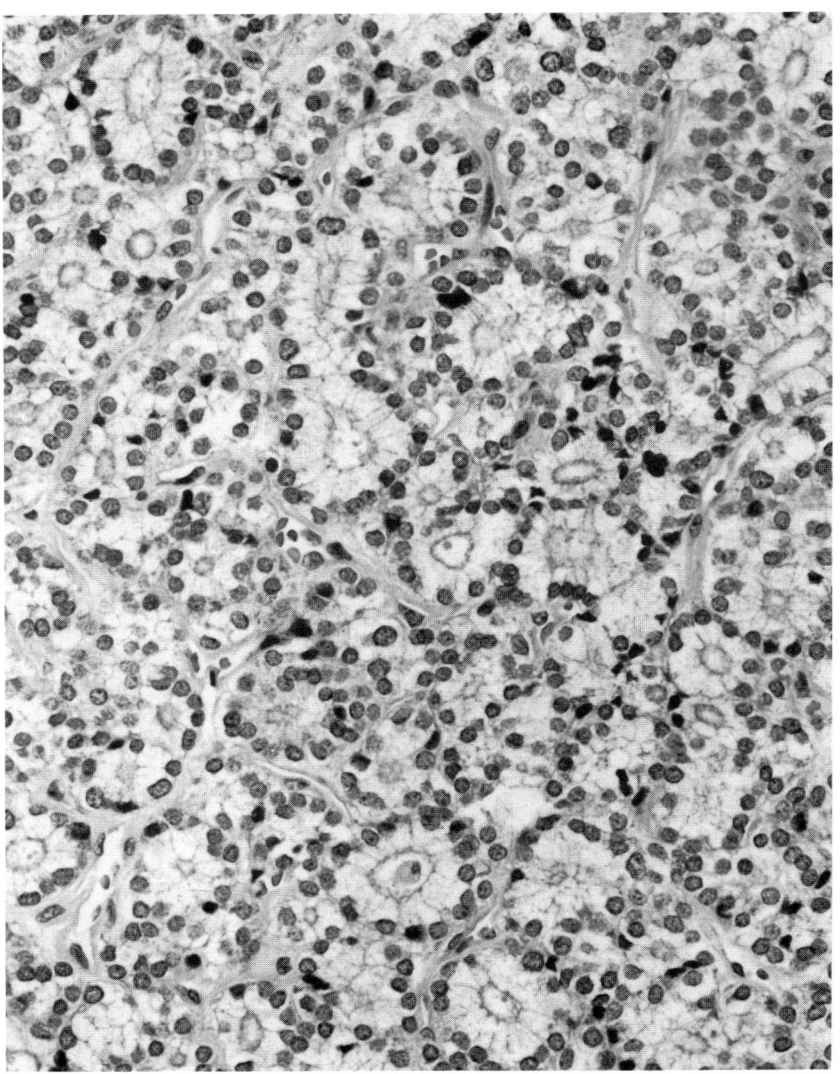

Figure 8.6 Detail of clear cell adenoma shown in Fig. 8.5. Follicles are lined by cuboidal or columnar epithelial cells with basally located, uniform nuclei and clear, finely granulated cytoplasm (H&E, × 330).

Carcangiu *et al.*, 1985). Clear cell adenomas are very rare. Stoll and Lietz (1973), in a retrospective study of 4271 surgical thyroid specimens, found only two morphologically benign and six obviously malignant clear cell tumours. The most important differential diagnosis for thyroid clear cell adenoma is with a metastasis from renal clear cell carcinoma,

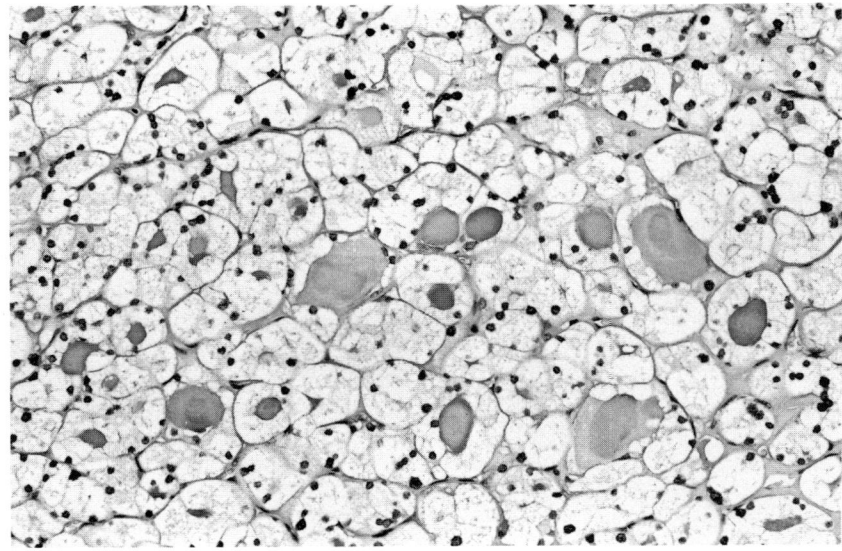

Figure 8.7 Another case of clear cell adenoma with microfollicular structure and pronounced clearing of the follicular cells giving their cytoplasm a foamy appearance. The nuclei are more irregularly distributed, smaller and more hyperchromatic than those seen in Fig. 8.6. Some follicles contain colloid (H&E, × 170).

which is discussed in Chapter 12, p. 284). A parathyroid adenoma composed of vacuolated chief cells or water-clear cells may be confused with a thyroid clear cell adenoma, especially in the rare event of an intrathyroidal location. Immunohistochemical demonstration of thyroglobulin is helpful in this situation (Böcker *et al.*, 1978). According to the author's experience this is a specific and sensitive method for identification of primary clear cell tumours of the thyroid.

8.2.3 Signet-ring cell adenoma

This very rare variant of follicular adenoma is mainly composed of cells with a signet-ring configuration (Mendelsohn, 1984; Rigaud *et al.*, 1985; Gherardi, 1987). These tumours are encapsulated and consist of nests and columns of closely packed, pale-staining cells, each showing eccentric nuclei and a large central clear vacuole (Plate 11). Areas with a microfollicular structure may also be present. The stroma tends to be hyalinized and may contain microcalcifications and haemosiderin collections. The intracytoplasmic vacuoles contain both immunoreactive thyroglobulin and material which stains as mucin with various histochemical methods. It is believed that the signet-ring

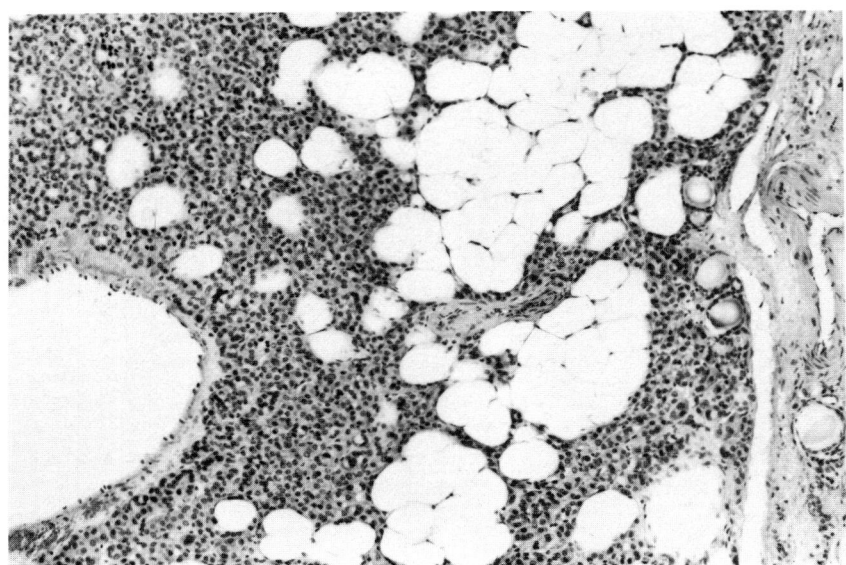

Figure 8.8 Thyroid adenolipoma. The tumour consists of a mixture of adipose tissue and follicular cells arranged in microfollicular and solid patterns (H&E, × 80).

change is of a degenerative nature, associated with the accumulation of thyroglobulin, and of protein–polysaccharide complexes derived from the degradation of thyroglobulin, which may account for the positive mucin staining (Gherardi, 1987).

8.2.4 Adenolipoma

This tumour, also called thyrolipoma, is an extremely rare form of follicular adenoma showing adipose metaplasia of the stroma (Fig. 8.8) (Schröder *et al.*, 1984; Hjorth *et al.*, 1986; De Rienzo and Truong, 1989). Adipose tissue may also occur in the stroma of papillary and follicular thyroid carcinomas (DeRienzo and Truong, 1989; Gnepp *et al.*, 1989; see also p. 147).

8.2.5 Atypical adenoma

The term 'atypical adenoma' was initially most often used by American authors for non-invasive encapsulated thyroid tumours with atypical features. Grossly, this tumour lacks the translucency of the typical

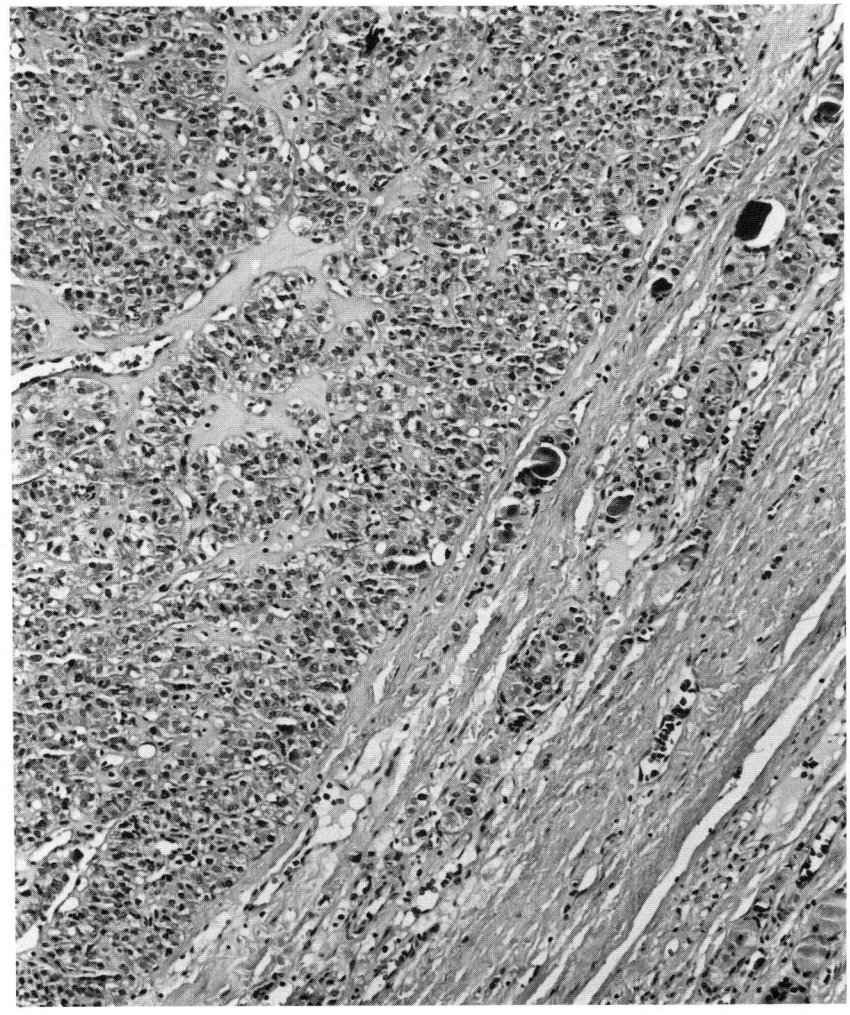

Figure 8.9 Atypical adenoma, showing a compact cell arrangement with irregular distribution of tumour cells. Small islands of tumour cells extend into, but do not penetrate, the fibrous capsule (H&E, × 125).

follicular adenoma, being solid and fleshy in appearance (Plate 12). Microscopically, it is characterized by a compact cell arrangement with pronounced cellular proliferation and irregular distribution of epithelial and stromal elements, but without evidence of capsular or blood vessel invasion (Fig. 8.9) (Hazard and Kenyon, 1954). According to Hazard and Kenyon, about 2% of thyroid adenomas are atypical. The degree

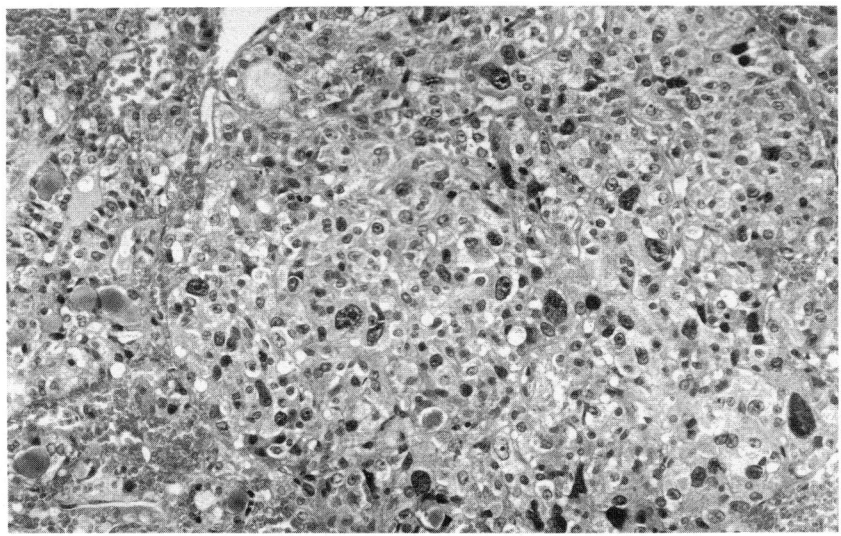

Figure 8.10 Atypical adenoma with solid growth pattern and focally pronounced cellular and nuclear pleomorphism (H&E, × 160).

of cytological atypia is variable (Fig. 8.10), some tumours showing areas with highly atypical cells with bizarre nuclei (Hazard, 1968). The term 'atypical adenoma' has also been used by some workers for encapsulated follicular tumours with equivocal capsular or vascular invasion, but in which a diagnosis of carcinoma cannot be established with certainty (Lang *et al.*, 1980). It is essential to examine the capsular portion of all atypical adenomas with many sections in order to exclude encapsulated (minimally invasive) follicular carcinoma. The encapsulated follicular carcinoma carries a small but definite risk for a malignant clinical behaviour, whereas atypical adenomas are invariably benign (Evans, 1984; Schröder *et al.*, 1984; Lang *et al.*, 1986).

8.2.6 Hyalinizing trabecular adenoma

Hyalinizing trabecular adenoma is a term proposed by Carney *et al.* (1987) to describe a rare form of atypical thyroid adenoma with distinctive histological and cytological features. It is important because microscopically it may be confused with medullary as well as papillary carcinoma. This tumour may occur in any part of the thyroid gland. It is typically 1–3 cm in diameter, well circumscribed, sometimes with faintly lobulated outer contours and usually, surrounded by a fibrous capsule of variable thickness. It has a slightly bulging,

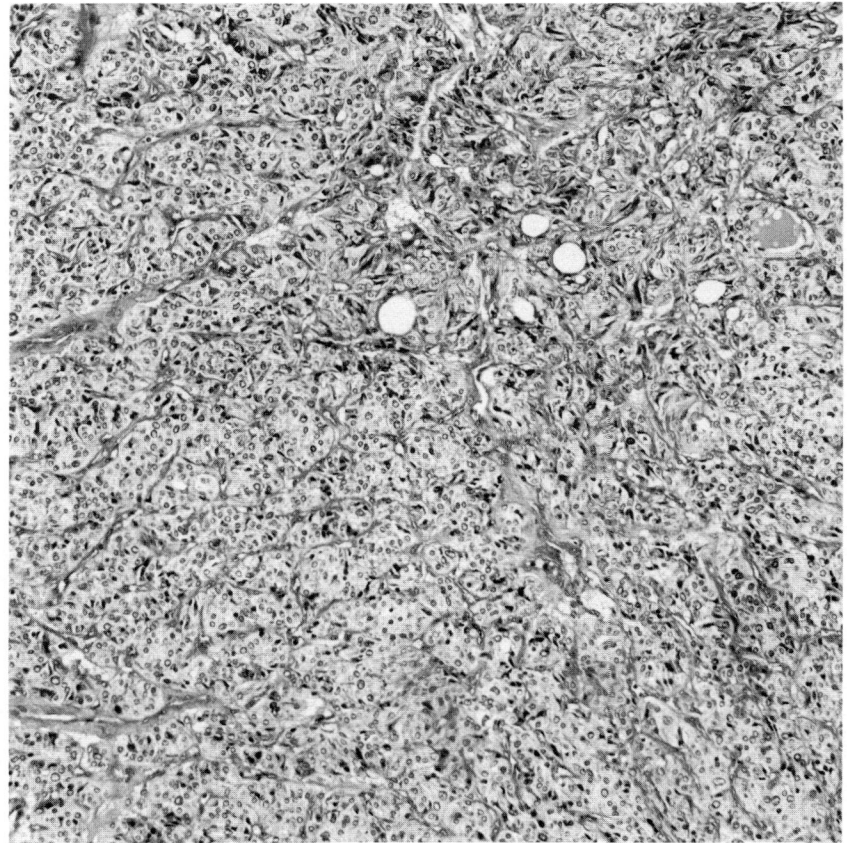

Figure 8.11 Hyalinizing trabecular adenoma. The cellular area has closely arranged nests and trabecular formations in a sparse amount of stroma. Note the typical 'empty', rounded spaces of various sizes, scattered among the cells (H&E, × 125).

tan, yellow–tan or yellowish–grey, homogeneous cut surface, sometimes with white streaks and flecks (Plate 13).

Histologically, the tumour has very cellular parts, composed of closely arranged, compact round or trabecular formations of polyhedric tumour cells in a sparse amount of stroma (Fig. 8.11). There may be different types of neoplastic cells. Some areas contain smaller cells with vacuolated cytoplasm and chromatin-dense nuclei, other parts are composed of large polyhedric cells with abundant, weakly eosinophil and fibrillary cytoplasm and light nuclei (Fig. 8.12). Despite marked cellular atypia sometimes, mitoses are seldom found. Less cellular areas may also be found with scattered small nests or thin

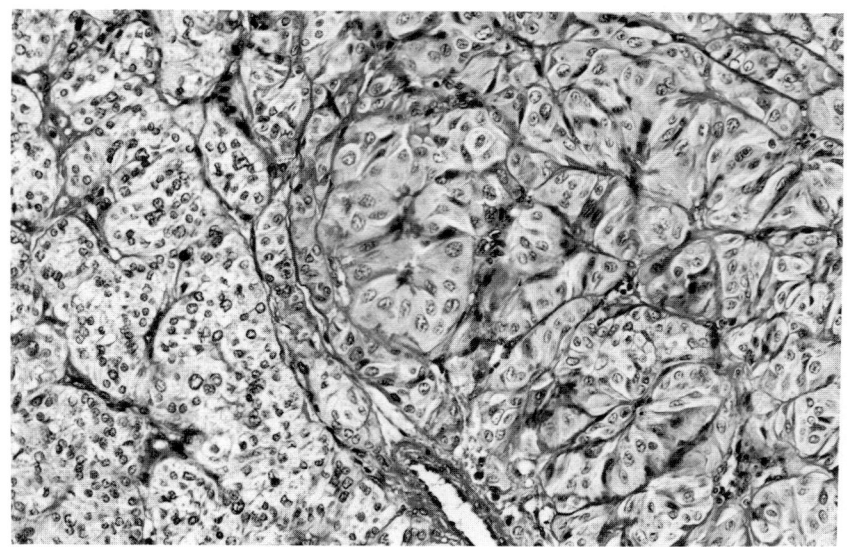

Figure 8.12 Hyalinizing trabecular adenoma. Cellular area with 'Zellballen'-like pattern and two different types of cells. On the right, alveoli of large cylindrical or triangular cells with weakly eosinophil and fibrillary cytoplasm whose apices point to the centres of the alveoli in rosette-like arrangements. On the left, more irregular nests of smaller, polygonal cells with vacuolated cytoplasm and small dark nuclei (H&E, × 140)

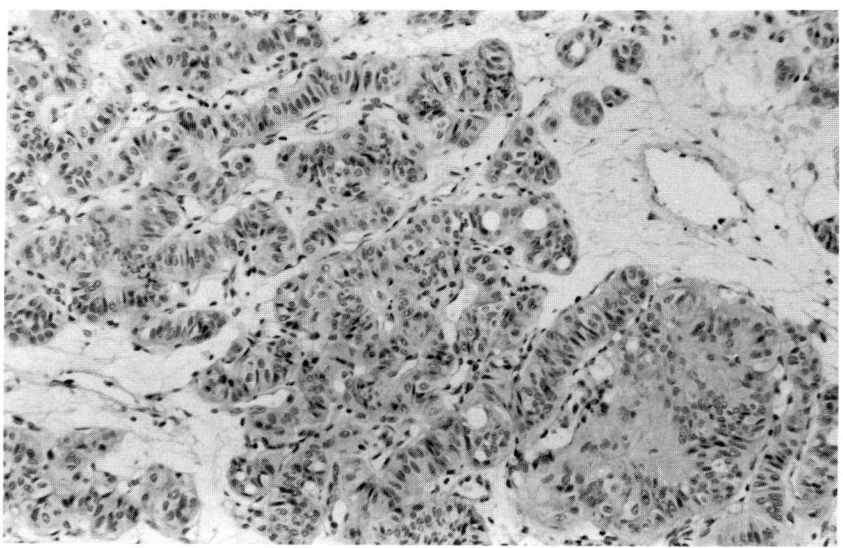

Figure 8.13 Hyalinizing trabecular adenoma. Area with loosely arranged trabeculae and columns consisting of palisading, transversely positioned cells. Abundant, oedematous and hyalinized stroma with some thin-walled blood vessels (H&E, × 140).

trabeculae composed of a single or a few rows of tumour cells embedded in a rich amount of oedematous or hyalinized stroma with dilated sinusoidal vessels (Fig. 8.13). The tumour often shows 'Zellballen'-like patterns (Fig. 8.12) and frequently, columns, which often are peculiarly curved, and composed of palisading, transversely positioned, elongated tumour cells (Fig. 8.13), both features that closely mimic medullary carcinoma. Tumour cells may also be arranged in rosette-like structures, around small hyaline clumps that may resemble amyloid but are not stained by the Congo red technique. Other nests may contain variable-sized empty-looking spaces that appear to be lined by flattened cells, smaller spaces possibly being located within the cytoplasm of tumour cells (Figs 8.11 and 8.13). In addition, there may be areas containing neoplastic follicles with colloid. In the central parts of the tumour, peculiar irregular spaces or tubular structures are sometimes found, lined by flattened cells, and appearing empty or containing eosinophilic proteinaceous material (Fig. 8.14). Still another peculiarity is the frequent occurrence of nuclear cytoplasmic invaginations, which may appear as so-called pseudonucleoli (Fig. 8.15). These pseudonucleoli are also seen in fine-needle aspirates from the tumour, which together with the frequent presence of papillary-like cell aggregates, may result in the erroneous cytological diagnosis of papillary carcinoma (Fig. 8.16) (Carney et al., 1987; Goellner and Carney, 1989). On the other hand, some histological features, such as whirling and palisading of tumour cells, presence of spindle cell features and of amorphous hyaline material in the stroma, which suggest amyloid may falsely direct the pathologist to the erroneous diagnosis of medullary carcinoma. However, Congo red staining invariably gives negative results and immunostains for calcitonin are also negative.

The nature of hyalinizing trabecular adenomas is enigmatic. The tumour cells show no immunoreactivity for calcitonin, neurotensin or somatostatin as far as can be judged from the seven cases that this author has studied. Grimelius' silver stain is negative as is the immunostaining for chromogranin A. Ultrastructural studies have failed to demonstrate cells with secretory granules of neuroendocrine type (Carney et al., 1987; Sambade et al., 1989). On the other hand, at least some areas of the tumour show strong immunopositivity for thyroglobulin (Fig. 8.17) (Bronner et al., 1988), indicating a follicular cell differentiation in these areas of the tumour. From a histogenetic point of view, it may be significant that ultrastructurally, tumour cells are frequently found which contain bundles or whorls of filaments (Fig. 8.18), similar to those described in certain cells present in normal human ultimobranchial remnants (Yamaoka, 1973, see also p. 16), in ultimobranchial structures of the rat thyroid (Calvert and Isler, 1970),

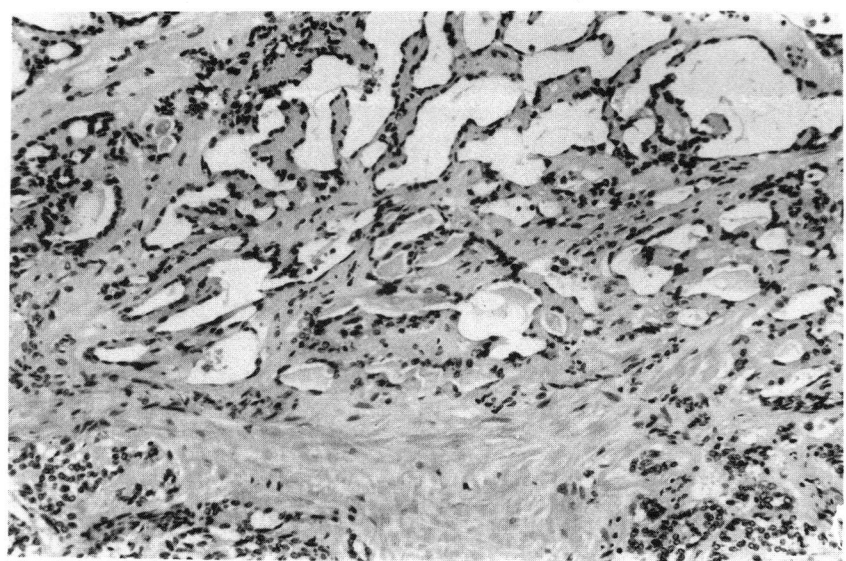

Figure 8.14 Hyalinizing trabecular adenoma. Sclerotic part with abundant hyalinized stroma containing irregularly shaped tubules and cysts, lined by flattened tumour cells and looking empty or with some eosinophilic, proteinaceous material (H&E, × 140).

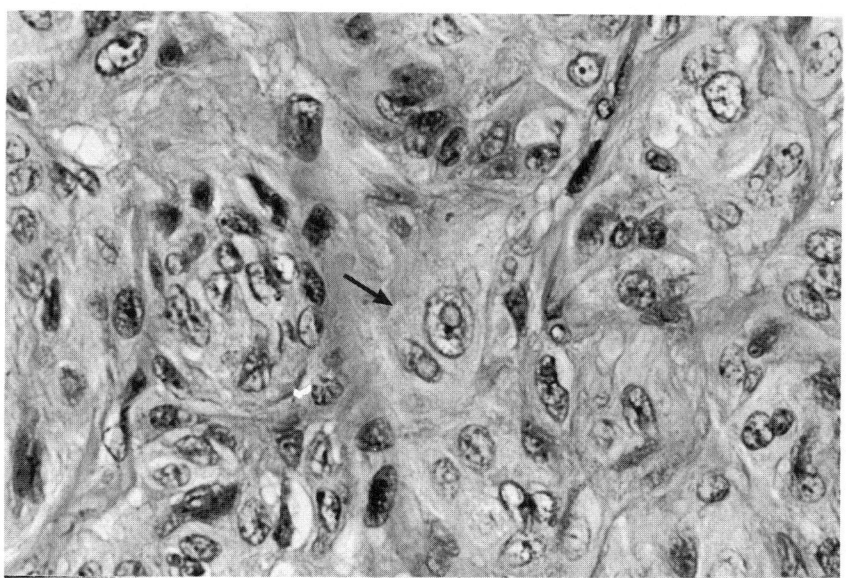

Figure 8.15 Hyalinizing trabecular adenoma. High-power field showing the indistinct cell borders of the tumour cells and the light, fibrillary appearance of their cytoplasm. In the centre are two nuclei with cytoplasmic invaginations (arrow) (H&E, × 525).

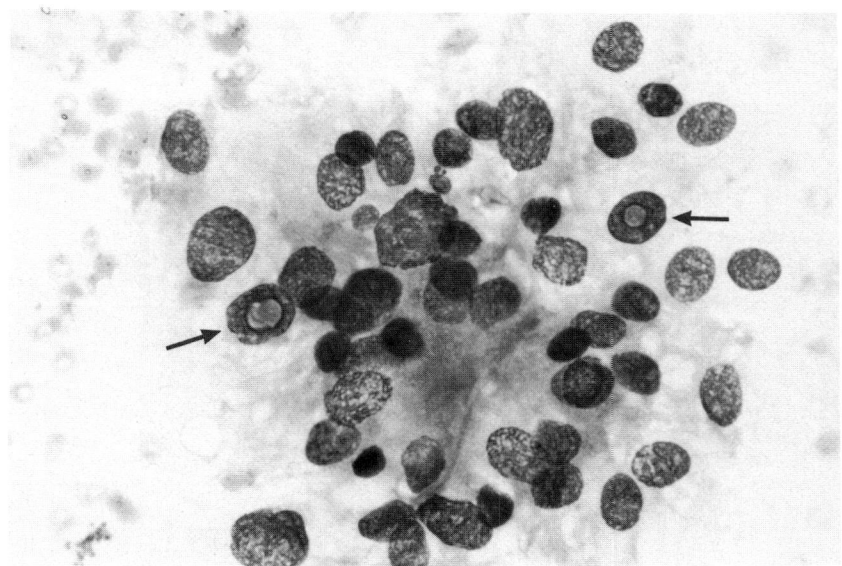

Figure 8.16 Fine needle aspirate from a hyalinizing trabecular adenoma showing a tissue fragment of tumour cells and naked nuclei with a few pseudonucleoli (arrows). The arrangement of the cells has a rosette-like appearance that may simulate a papilla. This feature, in conjunction with the presence of pseudonucleoli, may lead to an erroneous diagnosis of papillary carcinoma (May Grünwald-Giemsa stain, × 425).

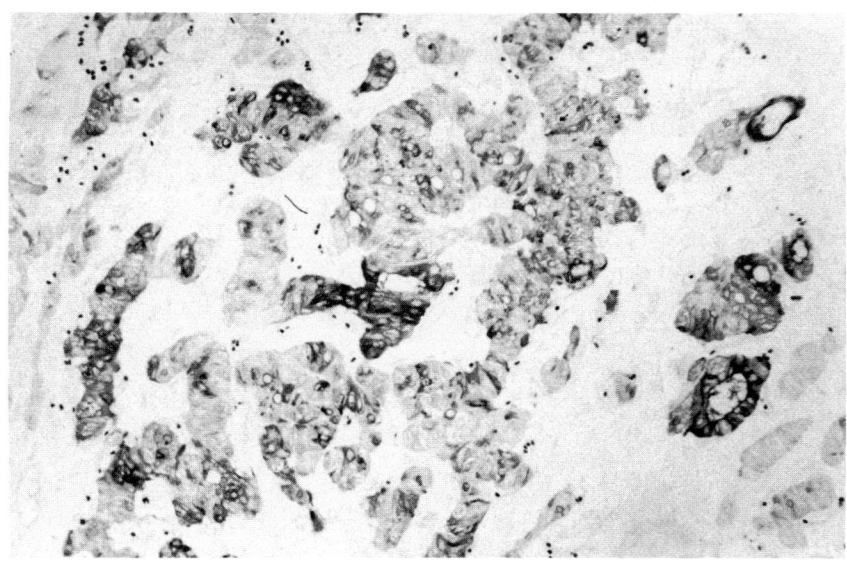

Figure 8.17 Hyalinizing trabecular adenoma. Tumour cells show immunoreactivity for thyroglobulin of varying intensity (PAP technique with antibodies against human thyroglobulin, × 145).

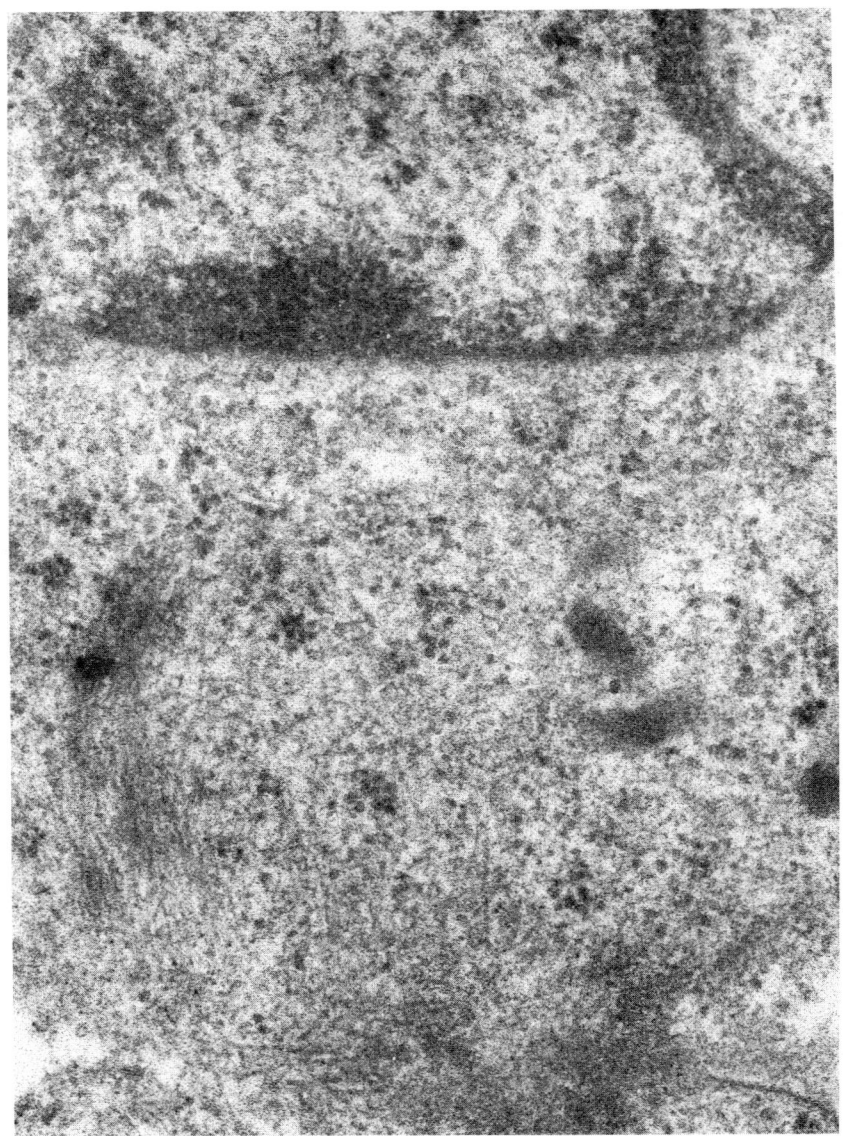

Figure 8.18 Electron micrograph of hyalinizing trabecular adenoma. Part of a tumour cell with bundles of cytoplasmic intermediate filaments (× 64 000).

as well as in ultimobranchial tumours of bulls (Black *et al.*, 1973). The irregularly formed tubules as well as empty spaces found in the hyalinizing trabecular adenoma resemble similar structures in bull ultimobranchial tumours (Ljungberg and Nilsson, 1985; see also p. 240). These bull tumours may also develop follicles with immuno-reactive thyroglobulin, as well as differentiated neuroendocrine cells with immunoreactive calcitonin and/or somatostatin. Although, as yet, there has been no report that the human hyalinizing trabecular adenoma is capable of parafollicular cell differentiation, other struc-tural features of the two tumours show such a degree of resemblance, that it can be speculated that the hyalinizing trabecular adenoma is of ultimobranchial nature, and, if so, possibly represents the benign counterpart of the intermediate type of thyroid carcinoma.

References

Auer, G.U., Bäckdahl, M., Forsslund, G.M. and Askensten, U.G. (1985) Ploidy levels in nonneoplastic and neoplastic thyroid cells. *Anal. Quant. Cytol. Histol.*, **7**, 97–106.

Black, H.E., Capen, C.C. and Young, D.M. (1973) Ultimobranchial thyroid neoplasms in bulls. *Cancer*, **32**, 865–78.

Böcker, W., Dralle, H., Koch, G. *et al.* (1978) Immunohistochemical and electron microscope analysis of adenomas of the thyroid gland. *Virchows Arch. [A]*, **380**, 205–20.

Bronner, M.P., LiVolsi, V.A. and Jennings, T.A. (1988) Plat: paraganglioma-like adenomas of the thyroid. *Surg. Pathol.*, **1**, 383–9.

Brown, C.L. (1981) Pathology of the cold nodule. *Clin. Endocriol. Metab.*, **10**, 235–45.

Calvert, R. and Isler, H. (1970) Fine structure of a third epithelial component of the thyroid gland in the rat. *Anat. Rec.*, **168**, 23–42.

Carcangiu, M.L., Sibley, R.K. and Rosai, J. (1985) Clear cell changes in primary thyroid tumours. A study of 38 cases. *Am. J. Surg. Pathol.*, **9**, 705–22.

Carney, J.A., Ryan, J. and Goellner, J.R. (1987) Hyalinizing trabecular adenoma of the thyroid gland. *Am. J. Surg. Pathol.*, **11**, 583–91.

Chen, K.T.K. and Rosai, J. (1977) Follicular variant of thyroid papillary carcinoma: a clinicopathologic study of six cases. *Am. J. Surg. Pathol.*, **1**, 124–30.

DeGroot, L. J. (1979) Thyroid neoplasia. In *Endocrinology*, Vol. 1 (eds L. J. DeGroot, G. Cahill, Jr, W.D. Odell *et al.*), Grune and Stratton, New York, San Francisco and London, pp. 509–21.

DeRienzo, D. and Truong, L. (1989) Thyroid neoplasms containing mature fat: a report of two cases and review of the literature. *Mod. Pathol.*, **2**, 506–10.

Evans, H.L. (1984) Follicular neoplasms of the thyroid. A study of 44 cases followed for a minimum of 10 years, with emphasis on differential diagnosis. *Cancer*, **54**, 535–40.

Franssila, K., Ackerman, L., Brown, C.L. and Hedinger, C.E. (1985) Session II: Follicular carcinoma. *Semin. Diagn. Pathol.*, **2**, 101–22.

Gherardi, G. (1987) Signet ring cell 'mucinous' thyroid adenoma: a follicle cell tumour with abnormal accumulation of thyroglobulin and a peculiar histochemical profile. *Histopathology*, **11**, 317–26.

Gnepp, D.R., Ogorzalek, J.M. and Heffess, C.S. (1989) Fat-containing lesions of the thyroid gland. *Am. J. Surg. Pathol.*, **13**, 605–12.

Goellner, J.R. and Carney, J.A. (1989) Cytologic features of fine-needle aspirates of hyalinizing trabecular adenoma of the thyroid. *Am. J. Clin. Pathol.*, **91**, 115–9.

Hazard, J.B. (1968) Nomenclature of thyroid tumours. In *Thyroid Neoplasia* (eds S. Young and D.R. Inman), Academic Press, London, New York, pp. 3–37.

Hazard, J.B. and Kenyon, R. (1954), Atypical adenoma of the thyroid. *Arch. Pathol.*, **58**, 554–63.

Hjorth, L., Thomsen, L.B. and Nielsen, V.T. (1986) Adenolipoma of the thyroid gland. *Histopathology*, **10**, 91–6.

Johannessen, J.V., Sobrinho-Simões,, M., Lindmo, T. and Tangen, K.O. (1982) The diagnostic value of flow cytometric DNA measurements in selected disorders of the human thyroid. *Am. J. Clin. Pathol.*, **77**, 20–5.

Joensuu, H., Klemi,P. and Eerola, E. (1986) DNA aneuploidy in follicular adenomas of the thyroid gland. *Am. J. Pathol.*, **124**, 373–6.

Kini, S.R. (1987) *Guides to Clinical Aspiration Biopsy. Thyroid.* Igaku-Shoin, New York, Tokyo, p. 57.

Lang, W., Choritz, H. and Hundeshagen, H. (1986) Risk factors in follicular thyroid carcinomas: a retrospective follow-up study covering a 14-year period with emphasis on morphological findings. *Am. J. Surg. Pathol.*, **10**, 246–55.

Lang, W., Georgii, A., Stauch, G. and Kienzle, E. (1980) The differentiation of atypical adenomas and encapsulated follicular carcinomas in the thyroid gland. *Virchows Arch.* [A] **385**, 125–41.

Ljungberg, O. and Nilsson, P-O. (1985) Hyperplastic and neoplastic changes in ultimobranchial remnants and in parafollicular (C) cells in bulls: a histologic and immunohistochemical study. *Vet. Pathol.*, **22**, 95–103.

Malamos, B. Koutras, D.A., Fringeli, D. and Tassopoulos, C.N. (1968) Toxic adenoma of the thyroid. *Horm. Metab. Res.*, **1**, 19–25.

Mendelsohn, G. (1984) Signet-cell-simulating microfollicular adenoma of the thyroid. *Am. J. Surg. Pathol.*, **8**, 705–8.

Panke, T.W., Croxson, M.S., Parker, J.W. *et al.* (1978) Triiodothyronine-secreting (toxic) adenoma of the thyroid gland. Light and electron microscopic characteristics. *Cancer*, **41**, 528–37.

Reinwein, D., Durrer, H., Emrich, D. *et al.* (1977) The thyroidal production of reverse triiodothyronine in autonomous adenoma. *Clin. Endocrinol.*, **7**, 171–3.

Rigaud, C., Peltier, F. and Bogomoletz, W.V. (1985) Mucin producing microfollicular adenoma of the thyroid. *J. Clin. Pathol.*, **38**, 277–80.

Rosai, J. and Carcangiu, M.L. (1984) Pathology of thyroid tumours: some recent and old questions. *Hum. Pathol.*, **15**, 1008–12.

Rosai, J. and Carcangiu, M.L. (1987) Pitfalls in the diagnosis of thyroid neoplasms. *Pathol. Res. Pract.*, **182**, 169–79.

Rosai, J., Zampi, G. and Carcangiu, M.L. (1983) Papillary carcinoma of the thyroid: a discussion of its several morphological expressions, with particular emphasis on the follicular variant. *Am. J. Surg. Pathol.*, **7**, 809–17.

Saleiro, J.V., Faria, V. and Oliveira, M.C. (1981) Clear cell tumour of the thyroid gland. *J. Submicrosc. Cytol.*, **13**, 75–7.

Sambade, C., Sarabando, F., Nesland, J.M. and Sobrinho-Simões, M. (1989) Intriguing case. Hyalinizing trabecular adenoma of the thyroid (Case of the Ullensvang Course). *Ultrastruct. Pathol.*, **13**, 275–80.

Schröder, S., Böcker, W., Hüsselmann, H. and Dralle, H. (1984) Adenolipoma (thyrolipoma) of the thyroid gland. Report of two cases and review of literature. *Virchows Arch.* [A], **404**, 99–103.

Schröder S., Pfannenschmidt, M., Dralle, H. *et al.* (1984) The encapsulated follicular carcinoma of the thyroid. A clinicopathologic study of 35 cases. *Virchows Arch.* [A], **402**, 259–73.

Sprenger, E., Löwenhagen, T. and Vogt-Schaden, M. (1977) Differential diagnosis between follicular adenoma and follicular carcinoma of the thyroid by nuclear DNA determination. *Acta Cytol.*, **21**, 528–30.

Stoll, W. and Lietz, H. (1973) Zur Kenntnis und Problematik des hellzelligen Adenomes in der Schilddrüse. *Virchows Arch.* [A], **361**, 163–73.

Tuccari, G. and Barresi, G.B. (1985) Immunohistochemical demonstration of lactoferrin in follicular adenomas and thyroid carcinomas. *Virchows Arch.* [A], **406**, 67–74.

Tuccari, G. and Barresi, G.B. (1987) Immunohistochemical demonstration of ceruloplasmin in follicular adenomas and thyroid carcinomas. *Histopathology*, **11**, 723–31.

Vickery, A.L. (1983) Thyroid papillary carcinoma. Pathological and philosophical controversies. *Am. J. Surg. Pathol.*, **7**, 797–807.

Vickery, A.L., Carcangiu, M.L., Johannessen, J.V. and Sobrinho-Simões, M. (1985) Session I: Papillary carcinoma. *Semin. Diagn. Pathol.*, **2**, 90–100.

Willems, J.-S and Löwhagen, T. (1981) Fine-needle aspiration cytology in the management of thyroid disease. *Clin. Endocrinol. Metab.*, **10**, 267–73.

Yamaoka, Y. (1973) Solid cell nest (SCN) of the human thyroid gland. *Acta Pathol. Jpn*, **23**, 493–506.

9 Thyroid carcinoma

9.1 Introduction

Clinically apparent thyroid carcinoma is relatively uncommon. The reported annual incidence ranges from 0.3 to 10 per 100 000 persons (Sancho-Garnier, 1977; Schottenfeld and Gershman, 1977). There is a great variation in the incidence figures throughout the world (Waterhouse *et al.*, 1976); Iceland, Israel and Hawaii have the highest reported rates of clinically detected thyroid cancer, whereas Great Britain, Rumania and Hungary show a low frequency. In Hawaii, thyroid cancer accounted for 2.3% of all non-skin cancers between 1973 and 1977 compared with a national estimate of only 1.3% of all cancers for the United States (Harber and Lipkovic, 1970; Goodman *et al.*, 1988). Among the Nordic countries, Iceland has the highest and Finland and Denmark the lowest incidence (Franssila *et al.*, 1981; Hakulinen *et al.*, 1986). During the years 1960–1977, the annual incidence of clinically diagnosed thyroid carcinoma in Malmö was, on average, 2.4 per 100 000 (Borup Christensen *et al.*, 1984), which was somewhat lower than the average incidence (3.6 per 100 000 in the whole of Sweden during the same period.

Thorough pathological post-mortem studies together with extensive population surveys have shown that the prevalence of thyroid carcinoma is considerably higher than would appear from the annual incidence figures (Maruchi *et al.*, 1971; Fukunaga and Yatani, 1975). This is mainly due to the frequent occurrence of so-called occult carcinomas, i.e. very small lesions, which were clinically unsuspected, but revealed as incidental findings at autopsy or by a thorough histopathological examination of surgical thyroid specimens removed for other reasons. The prevalence rate of occult thyroid carcinoma (i.e. 1.5 cm or less in great diameter and clinically unsuspected; chiefly of papillary type) has been reported to vary from about 6% to 35% of cases in population-based autopsy series (Fukunaga and Yatani, 1975; Harach *et al.*, 1985). In the Malmö region, a non-endemic area with a littoral urban population, the prevalence rate of thyroid carcinomas, 1.5 cm or less and found by thorough examination of complete glands from autopsies (histological sampling of all lesions grossly detected by cutting the tissue at 1–2 mm intervals) was found to be 7–8%, the vast majority being of papillary type (Bondeson and Ljungberg, 1981, 1984). A similar study from a Swedish goitre area showed a comparable prevalence (Nielsen and Zetterlund, 1985). These figures suggest that the number of cases of clinically diagnosed thyroid carcinoma is strongly dependent on the frequency and quality of clinical examination and health screening procedures, the extent of thyroidectomy, and on the accuracy of the histopathological examination of surgically excised thyroid tissue.

Clinically apparent thyroid carcinoma is more common among females. On average, one male is affected for every three females. Although different sex ratios have been reported from different geographical areas, varying from 2/3 to 1/4, the preponderance of females is an invariable finding. The sex difference is particularly marked in papillary and follicular carcinoma, but is much less so in medullary and undifferentiated carcinoma (Woolner et al., 1961).

Interestingly, the marked female preponderance seen in clinical thyroid carcinoma is not present in studies of the prevalence of occult thyroid carcinoma (Sampson, 1977; Bondeson and Ljungberg, 1981). This preponderance completely disappears, and even reverses, if the examination of the thyroid provides an adequate histopathological representation of the smallest tumours (i.e. when serial sectioning of the thyroid gland is performed). It appears that tumours in females tend to be larger and the greater ease with which they can be identified is responsible for the difference in frequency observed between the two sexes in clinical material. The reason for larger thyroid carcinomas in females than in males is not known; the presence of a promoting (hormonal?) factor in females, which is lacking in males has been suggested as a possible explanation (Sampson et al., 1971). Conversely, the existence of an inhibiting mechanism in the male, which is absent in females may also be considered.

9.2 Papillary carcinoma

Papillary carcinoma is the most common form of differentiated thyroid carcinoma, accounting for 60–70% of all thyroid carcinomas in most series. Clinically it is diagnosed two to four times more often among females than in males, and it may occur at any age, the mean age at the time of the initial diagnosis being about 40 years. It is the type most often encountered in children, accounting for at least 90% of thyroid malignancies in the prepubertal age group. There is an association with previous external radiation to the head and neck region, especially when given to children and young adults (Winship and

Rosvoll, 1970; DeGroot and Paloyan, 1973; Hempelmann, 1977; Roudebush and DeGroot, 1977; Fjälling *et al.*, 1986). Papillary thyroid carcinoma may occasionally be inherited (Lote *et al.*, 1980; Phade *et al.*, 1981; Borup Christensen and Ljungberg, 1983). It has also been reported in families with stigmata of Gardner's syndrome or familial adenomatosis of the colon (Camiel *et al.*, 1968; Lee and MacKinnon, 1981; Thompson *et al.*, 1983).

When considering the incidence of papillary carcinoma, it is important to make a distinction between clinically apparent tumours and the small and usually harmless lesions found coincidentally at microscopy in carefully examined resected thyroid glands. As pointed out above, the thoroughness with which the thyroid gland is examined is the main factor determining prevalence figures. This is especially true for papillary carcinoma, because this is the dominant histological type among the so-called occult thyroid carcinomas.

The most common presenting symptom is a lump in the thyroid region or enlarged cervical lymph nodes, or both. A special variant, the diffuse sclerosing form of papillary carcinoma (see below), typically causes a diffuse, sometimes painful enlargement of both lobes which may be associated with local pain and lead to a suspicion of thyroiditis (Carcangiu and Bianchi, 1989). Very rarely, the patient lacking clinical evidence of disease in the neck may present with lung or bone metastases (Carcangiu *et al.*, 1985b).

Some 5–10% of patients report a previous history of external head and neck irradiation during infancy or adolescence (see Figs 9.1–9.3 and 13.1 and 13.3) (Mazzaferri *et al.*, 1977; Carcangiu *et al.*, 1985b). The non-neoplastic parts of glands from such cases frequently show other abnormalities, such as follicular adenomas and hyperplastic nodules which may show marked cytological and nuclear atypia, sometimes with oxyphil cell changes, chronic lymphocytic thyroiditis and evidence of previous parenchymatous destruction with loss of epithelial elements and lobular fibrosis (Straus and Spitalnik, 1977; Hanson *et al.*, 1983). The thyroid gland in children and adolescents is considered to be more sensitive (with regard to proliferative activity) to ionizing radiation than that of adults (Hempelmann *et al.*, 1967). This may explain why previous administration of radioactive iodine, chiefly used in adult individuals, and regardless of whether given in diagnostic or therapeutic doses, does not seem to be associated with the same risk of later development of clinically evident thyroid carcinoma (Holm *et al.* 1980a, b).

As pointed out in Chapter 6 (see page 109), it is questionable whether there is an increased frequency of papillary carcinoma in Graves' disease. Papillary carcinomas found in glands with Graves' disease are rare. The majority are small, clinically latent tumours found

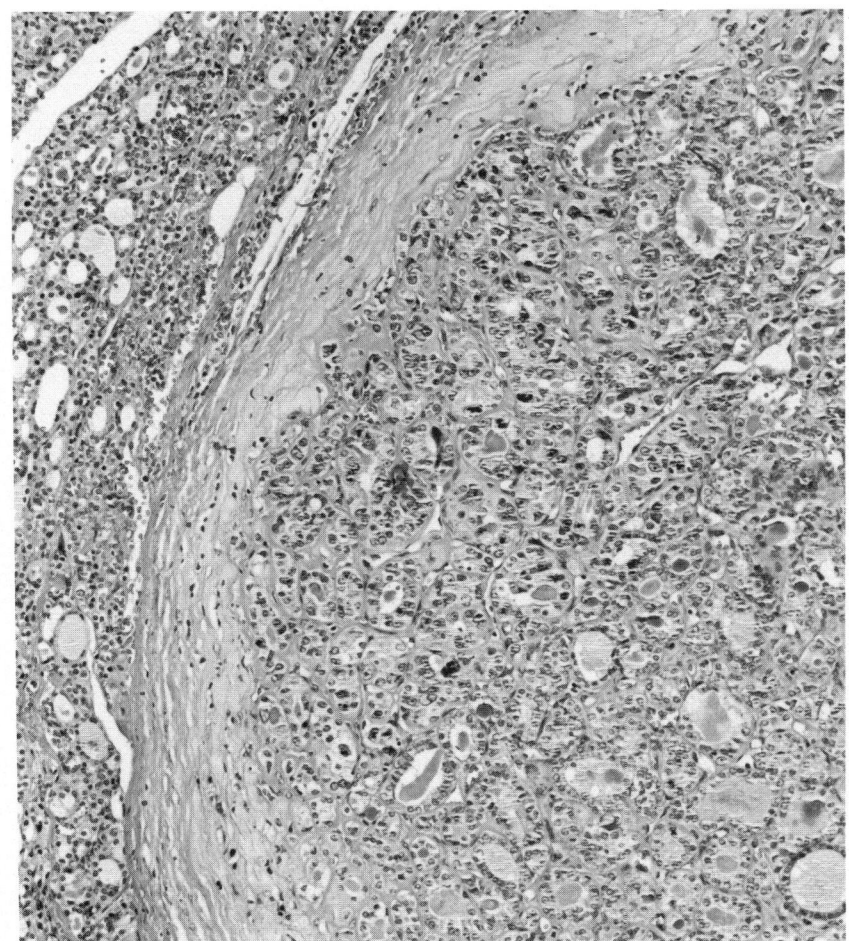

Figure 9.1 Nodular goitre in a 62-year-old woman treated with external irradiation for cervical lymph node tuberculosis at the age of 22. The gland was involved by multiple nodules of hyperplastic and involutionary type. This picture shows part of a cellular, hyperplastic nodule which is well defined by a thick hyalinized capsule. Note the pronounced cytological pleomorphism of the lesion (H&E, × 125).

incidentally by the pathologist, and usually without clinical significance. As is now recognized, small papillary carcinomas are also common in the general population, and their frequency is highly dependent on the extent of histopathological examination. It may be assumed, therefore, that any such an association is difficult to demonstrate in clinical material, and is of little significance. Similar

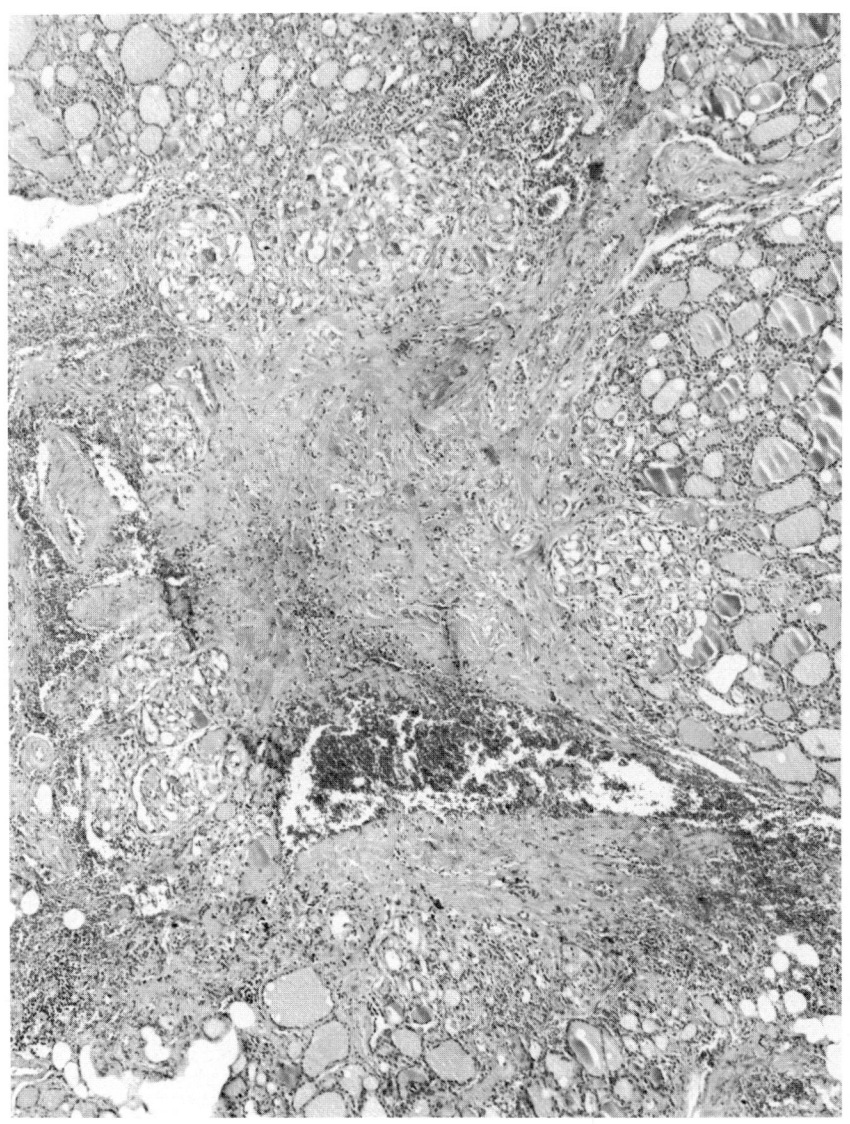

Figure 9.2 Same case as illustrated in Fig. 9.1. The internodular parenchyma contained a minute sclerotic and ill-defined papillary microcarcinoma, less than 2 mm in size (H&E, × 66).

considerations apply to the positive association sometimes reported to exist between Hashimoto's thyroiditis and nodular goitre respectively, and papillary thyroid carcinoma (Ott *et al.*, 1985) (see pages 66 and 91).

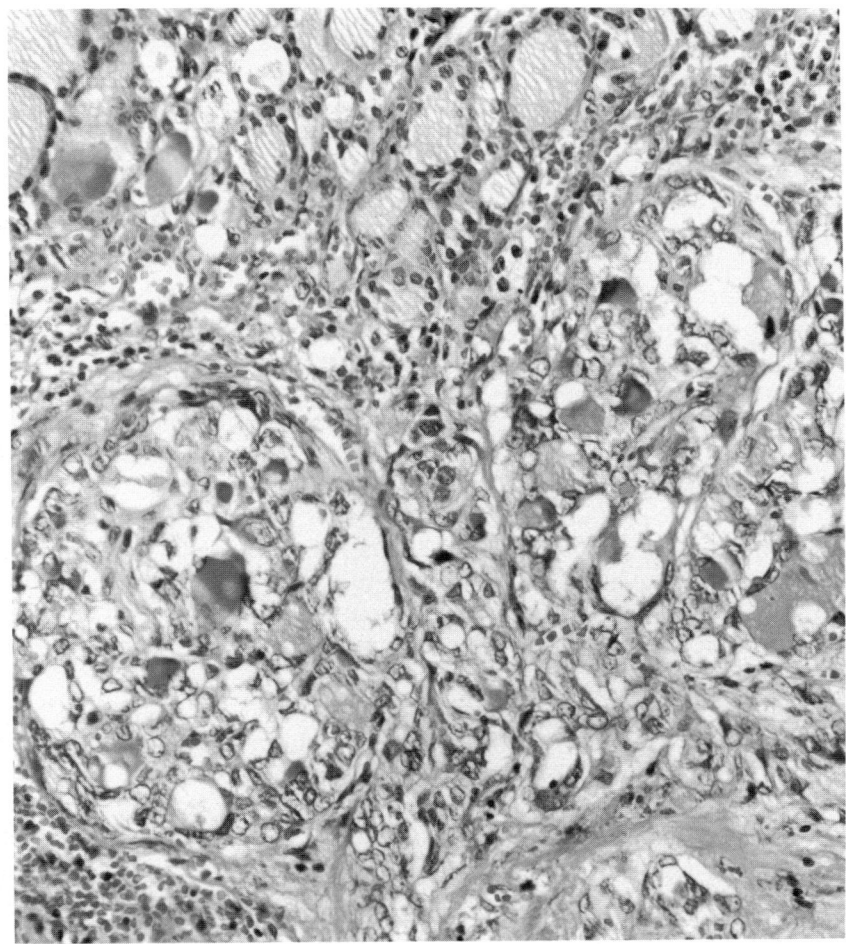

Figure 9.3 Detail of the papillary microcarcinoma depicted in Fig. 9.2. Irregular clusters, cribriform and follicular structures of highly atypical large cells with ground glass nuclei are seen to infiltrate into the adjacent parenchyma (H&E, × 250).

9.2.1 General pathological features

Papillary thyroid carcinoma carries a great propensity to invade lymph vessels and regional lymph nodes. This tendency is particularly marked in the younger age groups. Microscopic blood vessel invasion can be seen in 10–15% of neoplasms; this is far from the frequency found with follicular carcinoma and does not appear to influence prognosis (Hazard, 1968). Microscopically, confirmed positive cervical lymph nodes have been found, at some time during the evolution of the disease, in up to 50% of cases (Carcangiu *et al.*, 1985b). Woolner *et al.* (1961) in their study, including 509 cases of papillary carcinoma, reported cervical node metastases in 39.1% of these patients. The frequency of lymph node metastases seems to bear little relation to the size of the primary tumour. Woolner *et al.* (1961) thus found positive nodes in 43.2% of patients with primary lesions smaller than 1.5 cm, in 32.4% of those with lesions that were larger than 1.5 cm but present within the confines of the gland, and in 56.6% of patients with massively invasive tumours extending beyond the thyroid capsule. Tscholl-Ducommun and Hedinger (1982) found the following figures for the corresponding tumour groups: 25%, 43% and 32% respectively.

As noted in Chapter 3 (p. 45), the term 'lateral aberrant thyroid' is inappropriate, since thyroid tissue in a cervical lymph node is practically always a metastasis from a papillary thyroid carcinoma, which may well be very small and sometimes not revealed until the ipsilateral lobe has been serially sectioned. This also applies to cases where the thyroid tissue found within the lymph node is composed of normal-appearing follicles. It is advisable in such cases to section the tissue block at several levels. Very often, this procedure will disclose at least a minute focus of typical papillary structures with cytological features characteristic of papillary carcinoma, revealing the metastatic nature of the lesion (Figs 3.8 and 3.9).

Papillary carcinoma commonly occurs as multiple lesions in both thyroid lobes. The tumour may present as multiple gross lesions. Most frequently however, there is one dominating, clinically apparent tumour which, on subsequent histological examination, is found to be associated with multiple microscopic lesions in the ipsilateral and/or the contralateral lobe. Most studies have reported multifocality in about 30–40% of cases. On whole gland serial sectioning, multiple minute tumour nests have been detected in more than 80% (Russell *et al.*, 1963). Lymph node metastases are more frequently found in patients with multiple thyroid carcinomas than in patients with a single lesion. This suggests that the multifocality of the papillary carcinoma represents an intraglandular lymphatic permeation rather than multicentric tumours (Iida *et al.*, 1969).

9.2.2 Macroscopic appearance

The size of papillary carcinomas ranges from microscopic to massively invasive lesions that may be huge. The smallest tumours identifiable by the naked eye usually present as tiny whitish scars, often well circumscribed but without encapsulation, and some may contain calcifications (Plate 14). These tumours tend to lie superficially in the gland, often adjoining its capsule but rarely extending beyond it

(Hazard, 1960). Otherwise, they may be located anywhere in either lobe, and tumours may also occur in the isthmus and may even be multifocal. Such lesions are usually chance findings in thyroid glands removed for other reasons, or have been revealed by meticulous microscopic examination as the origin of clinically evident lymph node metastasis in a patient with no palpable thyroid mass. According to the definition originally proposed by Woolner *et al.* (1961), the term 'occult' has been used for tumours measuring 1.5 cm or less in greatest diameter. However, the large majority of these tumours are much less than 1 cm. Therefore, an upper size limit of 1.0 cm instead of 1.5 cm has been suggested. In addition, they are not always occult from a clinical point of view: some of the larger lesions may be clinically evident and some may reveal themselves clinically by cervical lymph node metastases. The term 'papillary microcarcinoma' has been suggested as more appropriate to designate these small tumours and this new definition has been adopted by the World Health Organization (Hedinger *et al.*, 1988).

Larger tumours are solid, grey–white or tan with lobulated or radiating outer contours and often with a firm whitish, sclerotic centre (Plate 15). They tend to have roughened cut surfaces due to the presence of papillae. Most tumours lack a capsule (Fig 9.4), about

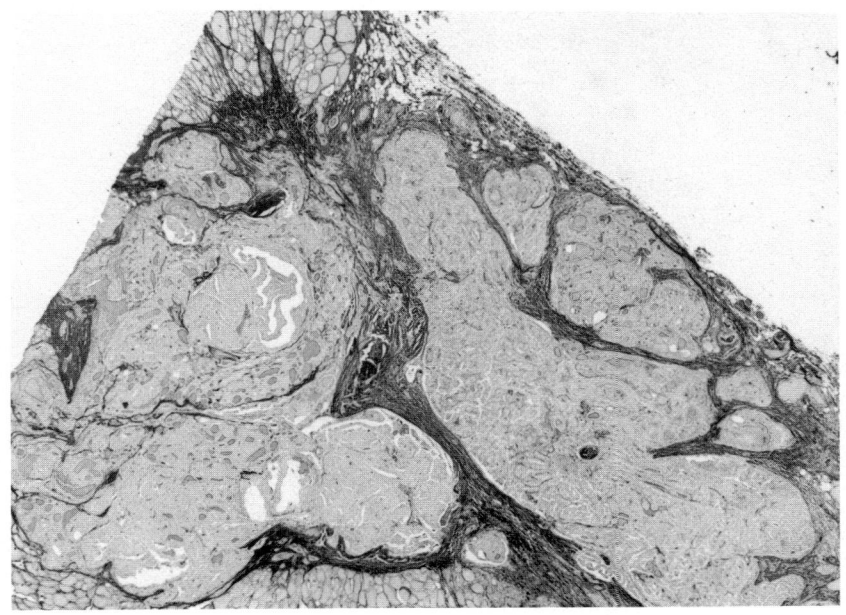

Figure 9.4 Papillary carcinoma with characteristic, lobulated growth pattern and areas of sclerosis (Van Gieson–Hansen stain, × 11).

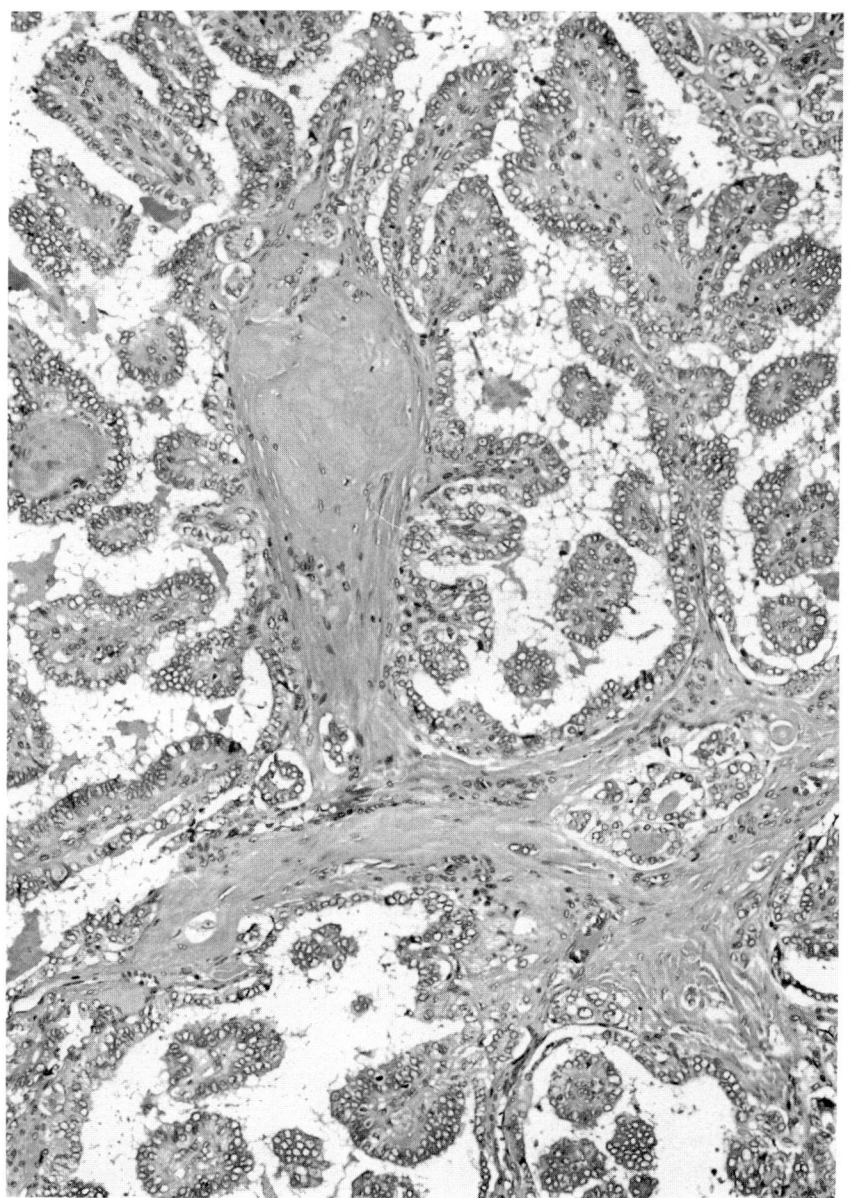

Figure 9.5 Papillary carcinoma showing unusually well developed papillae (H&E, × 125).

10% are mainly or wholly encapsulated. Variable-sized cysts are common in the larger tumours, and may reach several centimetres in diameter and usually contain a watery brownish fluid. Papillary formations project from the walls of the cysts. Calcification and even bone formation may occur in fibrotic areas; this may be so extensive that the tumour has to be decalcified before it can be cut.

9.2.3 Microscopic appearance

Microscopically, the diagnosis of papillary thyroid carcinoma is based on certain characteristic architectural and cytological features. None is pathognomonic, but when combined they are highly useful in identifying the tumour (Carcangiu et al., 1985c).

The presence of papillae, formed by a central fibrovascular stalk and covered by neoplastic epithelial cells remains a central hallmark of most cases of papillary carcinoma (Fig. 9.5). The well-developed papillae are long and slender, forming complex arborescent structures, often projecting into cystic spaces. The stroma is represented by loose connective tissue with variable-sized thin-walled blood vessels. It may be swollen by oedema or hyalinized and acellular, and psammoma bodies and other microcalcifications may be present. Sometimes, however, there are accumulations of foamy macrophages or haemosiderin (Fig. 9.6). In some cases, there is a heavy infiltration of lymphocytes variably admixed with plasma cells. Rarely, adipose tissue may also be present (Fig. 9.7). The papillae may be rudimentary, without discernible fibrovascular stalks. Only few papillary carcinomas are composed exclusively of papillary structures. The majority contain some follicular elements, although in varying proportions, and some may even appear purely follicular in structure. It should be noted here that a primary tumour, that is mainly composed of papillary structures, may give rise to cervical lymph node metastases that are chiefly or almost exclusively follicular in structure. Since follicular carcinoma very rarely metastasizes to lymph nodes, the presence of follicular structures in a cervical lymph node, even when looking quite benign, almost invariably represents metastasis of a papillary carcinoma. The admixture of follicular structures does not alter the biological behaviour or prognosis of the tumour. It is now generally agreed that tumours with a mixed papillary–follicular pattern, as well as tumours appearing exclusively follicular, providing they fulfil the cytological criteria of papillary thyroid (see below), should all be included within the same main tumour group, and thus classified as papillary carcinoma (Hedinger et al., 1988).

The cytological characteristics of the tumour cells refer more to the nucleus than to the cytoplasm. In most cases, the cells are larger than normal, cuboidal or polygonal, sometimes tall cylindrical in shape, and

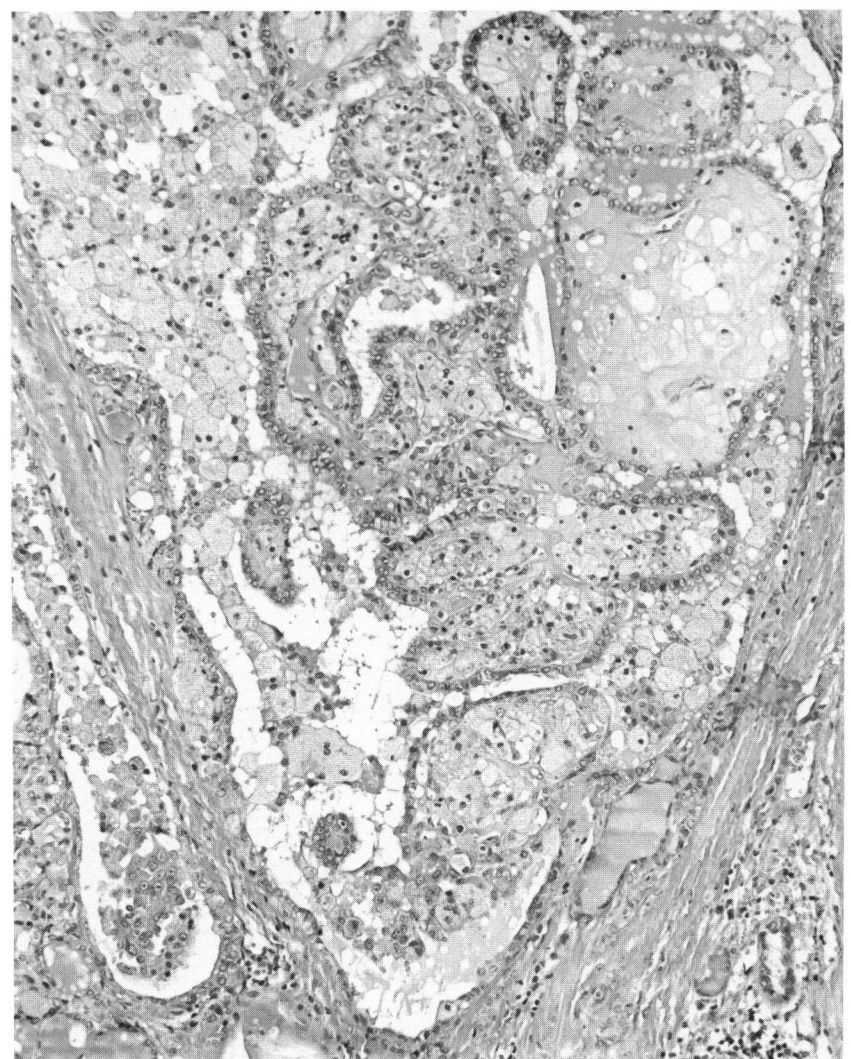

Figure 9.6 Papillary carcinoma with oedematous stroma and accumulation of foam cells (H&E, × 125).

have weakly eosinophilic or amphophilic cytoplasm. The diagnostically more important nuclear changes include a relatively large size, pale-staining with ground-glass appearance, irregular outlines with deep grooves, small nucleoli frequently lying eccentrically close to the nuclear membrane, and so-called pseudoinclusions or pseudo-nucleoli due to cytoplasmic invaginations. Nuclei typically over-lap, giving the impression of marked nuclear crowding (Fig. 9.8).

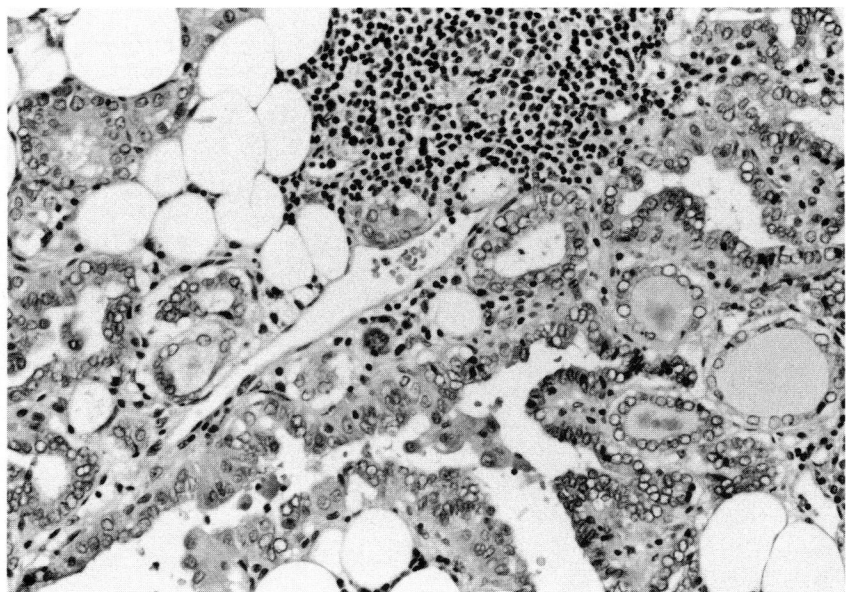

Figure 9.7 Papillary carcinoma with admixture of adipose tissue. Lymphocytic stromal infiltration (H&E, × 165).

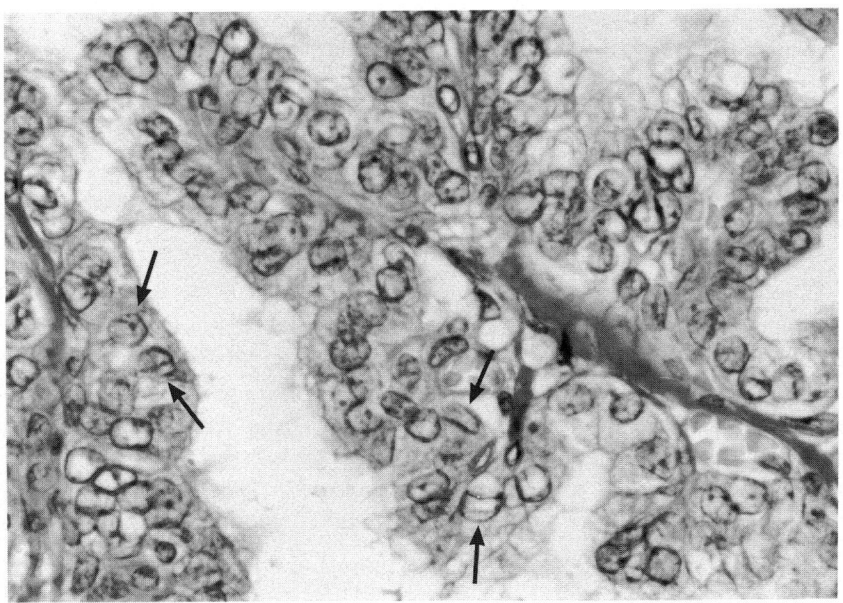

Figure 9.8 Cytological features of papillary carcinoma. Large, columnar tumour cells with irregularly distributed, overlapping nuclei. Nuclei show an irregular contour, with weakly angular shape; they appear empty, with marginal distribution of chromatin and nucleoli, and some have nuclear grooves (arrows) (H&E, × 540).

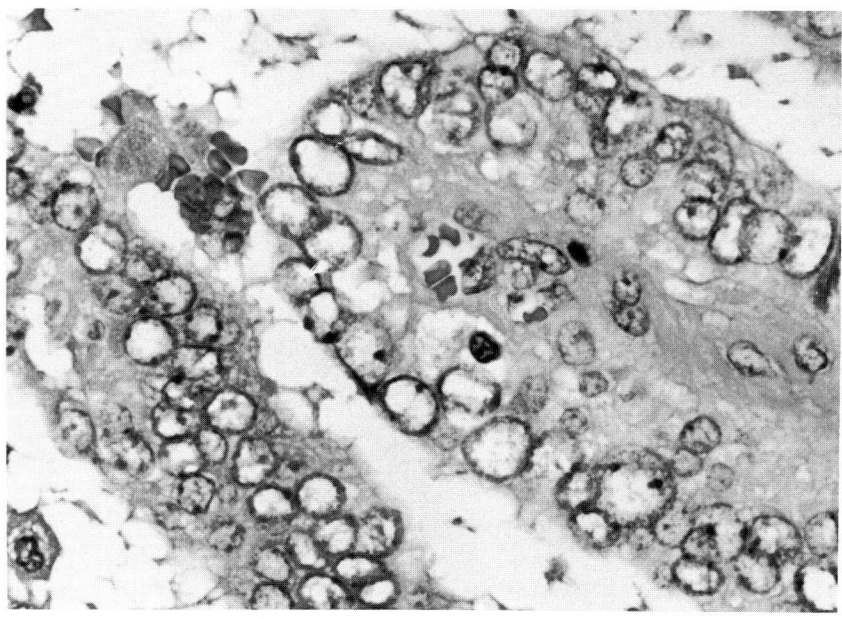

Figure 9.9 Nuclear features of papillary carcinoma. Chromatin strands are distributed peripherally, lying close to the nuclear membrane which looks thickened, with an irregular inner contour (H&E, × 660).

However, true stratification of the cells lining the papillae is infrequent.

Mitoses are typically rare in papillary carcinoma. This agrees with the low incidence of cells in S-phase demonstrated in these tumours by flow cytometric analysis (Johannessen *et al.*, 1981).

The 'empty' appearance of the nucleoplasm is due to the peripheral displacement of the chromatin strands and the nucleolus. The chromatin material appears to be apposed to the inner surface of the nuclear membrane which looks thickened with an irregular inner contour (Fig.9.9). Nuclei with this appearance are described under various, more or less imaginative terms, such as ground-glass, pale, clear, empty, and 'Little Orphan Annie' appearance. It is likely, that these features are mainly due to artifacts induced by fixation and/or embedding procedures. Hapke and Dehner (1979) showed that they are inconspicuous or absent in frozen sections and cytological smears or touch preparations from cases which exhibit them in sections from the formalin or Bouin-fixed, paraffin-embedded material.

Another common characteristic of the nuclei is the presence of deep cytoplasmic invaginations. When cut transversely, such a pseudo-

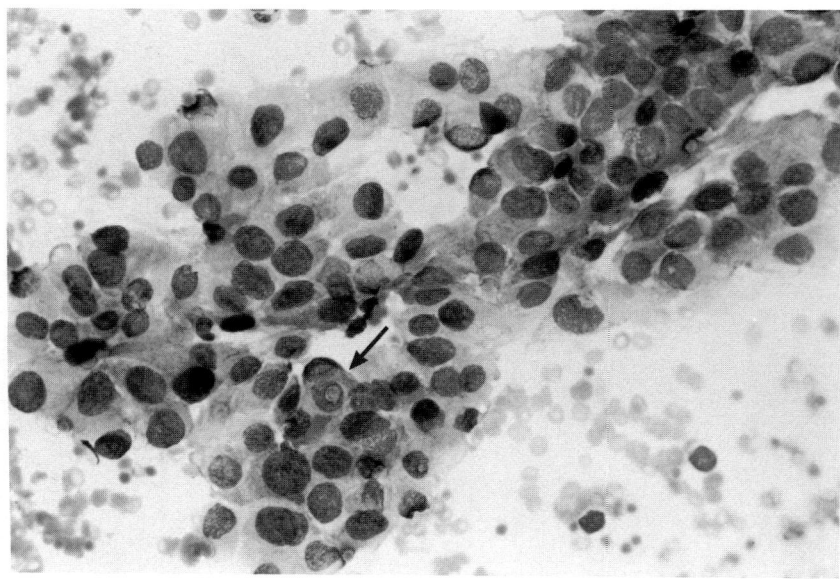

Figure 9.10 Fine needle aspirate from papillary carcinoma. Papillary tissue fragments with crowding nuclei with occasional pseundonucleoli (arrow) are seen. Note that in this type of preparation the chromatin is finely granular and evenly distributed over the entire nuclear profile. Ground glass features and nuclear grooves are better visualized in alcohol-fixed preparations stained with Papanicolaou or haematoxylin–eosin (air-dried smear, May-Grünwald Giemsa stain, × 340)

inclusion will appear as a small, membrane-limited island of cytoplasm with accompanying organelles, lying within the nucleus. Intranuclear cytoplasmic pseudoinclusions ('pseudonucleoli') have been reported to occur in histological sections in about 50% of cases (Chan and Saw, 1986), but are much more frequent in smears of cytological aspirates, being identified in more than 90% of cases (Fig. 9.10) (Christ and Haja, 1979). Pseudoinclusions are used as an important criterion for the identification of papillary carcinoma in cytology smears. But it should be emphasized that they are diagnostically important only when present in a proper setting. Such inclusions may also be present in other thyroid neoplasms, such as medullary carcinomas, oxyphil tumours and undifferentiated carcinomas. Likewise, they are common in the so-called hyalinizing trabecular adenoma (Fig. 8.16). Since this tumour may also show psammoma bodies and even papillary structures it may easily be mistaken cytologically for a papillary carcinoma (Goellner and Carney, 1989; see also Chapter 8, p. 131). Frequently, the invaginations take the form of deep linear infoldings which give rise to the so-called grooving of the nuclear membrane (Fig. 9.8). There may

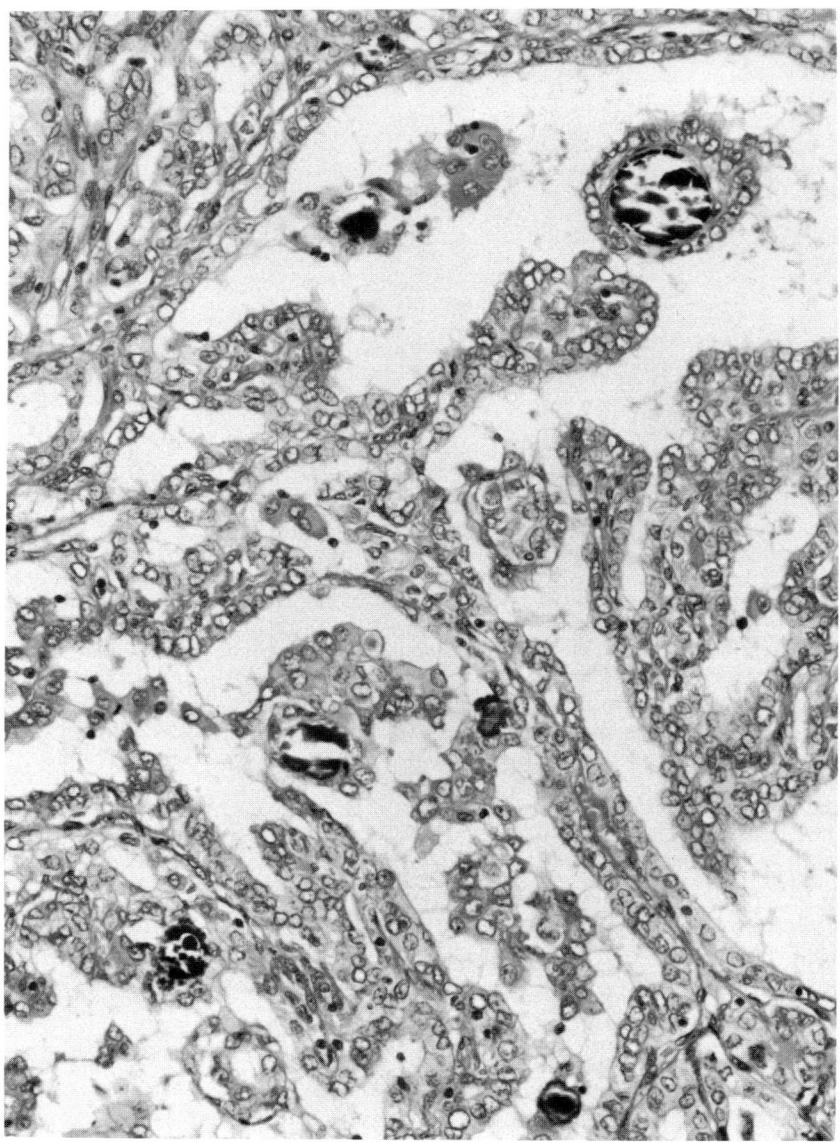

Figure 9.11 Papillary carcinoma with psammoma bodies mainly located at the tips of the papillae (H&E, × 250).

be a single linear infolding along the long axis of the oval nucleus, or more complex multiple infoldings and superficial notches. These phenomena are well known from many electron microscopic studies (Albores-Saavedra *et al.*, 1971; Valenta and Bechet, 1977; Johannessen *et al.*, 1978; Beaumont *et al.*, 1981). More recently, these features have been emphasized as very useful diagnostic criteria of papillary carcinoma at the light microscopic level (Chan and Saw, 1986). To distinguish the nuclear groove it is often necessary to focus at various levels on the nucleus to make sure it is not confused with the nuclear membrane of an adjacent, overlapping nucleus. The identification can be facilitated by examining thinner (2 μm) sections from paraffin-embedded tissue. Nuclear grooves are easily discernible in aspiration cytology smears, preferably after alcohol fixation and staining with the Papanicolaou stain or haematoxylin–eosin. However, like nuclear pseudoinclusions, these grooves may also occur in smears of aspirates from other lesions, and should be interpreted with caution.

Psammoma bodies, i.e. round, calcified concretions showing concentric lamination (Fig. 9.11), are found in roughly 40–60% of papillary carcinomas (Klinck and Winship, 1959; Batsakis *et al.*, 1960; Meissner and Warren, 1969; Franssila, 1973). They appear to be more common in individuals under 20 years of age. Psammoma bodies measure up to 70 μm or so in diameter. They seem to arise among tumour cells and are probably a result of cell necrosis associated with insufficient blood supply to the tips of the papillae (Johannessen and Sobrinho-Simões, 1980). They should be distinguished from amorphous, irregular calcifications frequently seen in nodular goitre, and in any tumour in which haemorrhage or necrosis undergoes organization. They are particularly common in areas where the tumour has a papillary growth pattern and are often seen within the hyalinized stroma at the tips of the papillae (Fig. 9.11), sometimes occurring in aggregates. They are sometimes seen within cribriform or plexiform epithelial structures, and they frequently also occur within clusters of tumour cells that are formed from the lining of cystic spaces of the growth. Psammoma bodies are highly diagnostic of papillary carcinoma, since they are very rare in other lesions of the thyroid. It should be borne in mind however, that laminated calcified bodies closely resembling psammoma bodies and probably representing inspissated and calcified colloid, may occur in the trabecular hyalinizing adenoma (Carney *et al.*, 1987) and may, exceptionally, also be encountered in toxic goitres (Fig. 9.12) (Patchefsky and Hoch, 1972).

'Naked' psammoma bodies may occur in normal thyroid tissue adjacent to or even distant from the carcinoma (Fig. 9.13), and have also been observed within the sinuses of regional lymph nodes in cases of thyroid papillary carcinoma. These probably represent lymphatic spread of tumour cells which have subsequently degenerated, leaving the psammoma bodies as a residue (Klinck and Winship, 1959). Thus the discovery of a psammoma body within otherwise normal thyroid tissue or regional lymph nodes indicates a high likelihood of the presence of papillary carcinoma in its vicinity, and consequently further histological sectioning

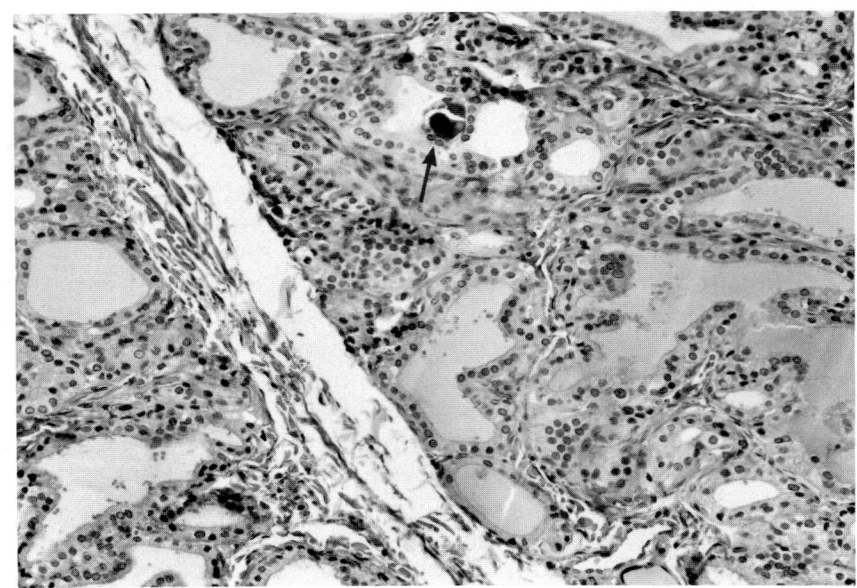

Figure 9.12 Psammoma body in Graves' disease (arrow) (H&E, × 185).

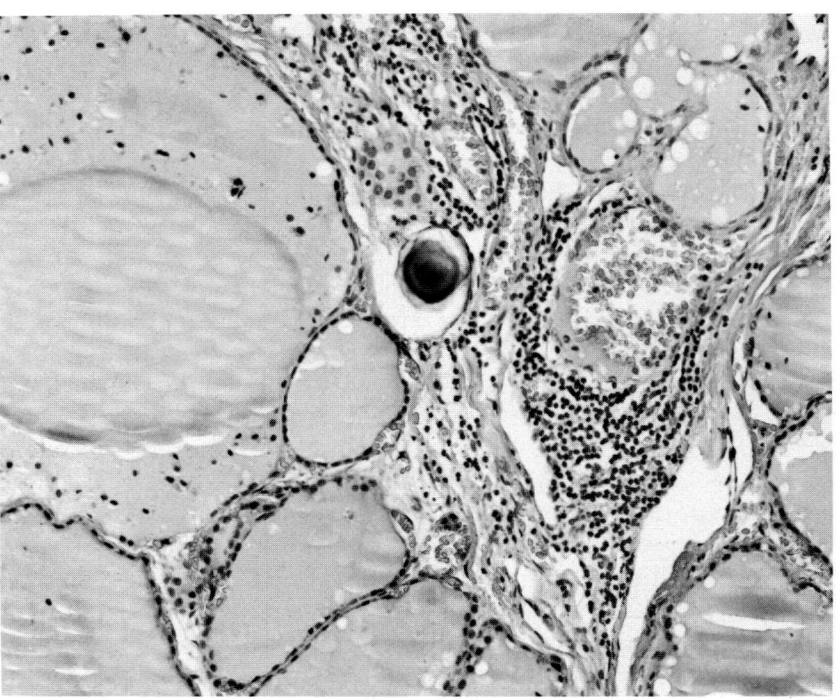

Figure 9.13 'Naked' psammoma body, lying in the interfollicular stroma, possibly in a lymph vessel. A papillary carcinoma was present in another part of the lobe (H&E, × 165).

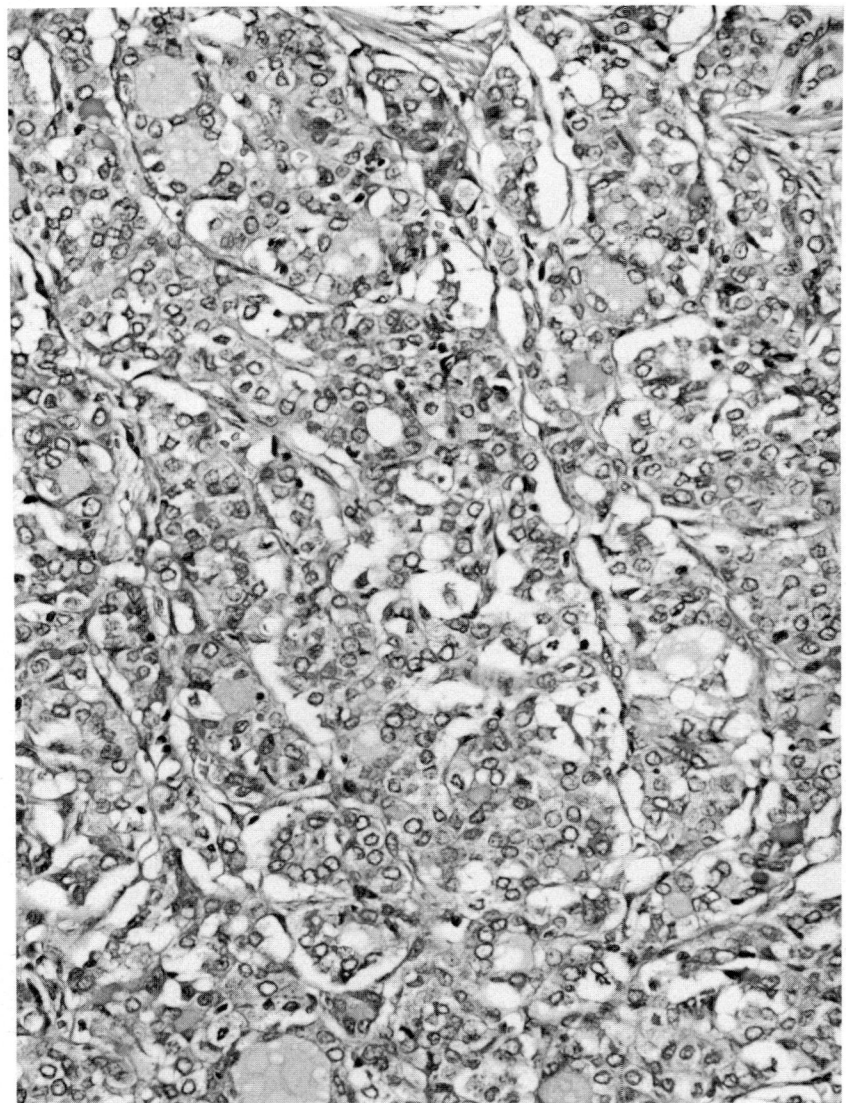

Figure 9.14 Papillary carcinoma with irregular histological architecture, forming cribriform structures with occasional follicles and papillae (H&E, × 250).

sectioning should be instituted. Its presence is also a strong predictor of lymph node metastases.

Papillary carcinomas are seldom purely papillary, but contain an admixture of follicular structures in varying proportions. In some it may be hard to find any papillary structures at all (so-called follicular variant). Occasionally, the tumour cells may be arranged in cribriform patterns of less well-defined plexiform structures (Fig. 9.14). Not infrequently, transitions into solid masses of cells may be seen, sometimes also squamous metaplasia, but usually without frank keratinization (Fig. 9.15).

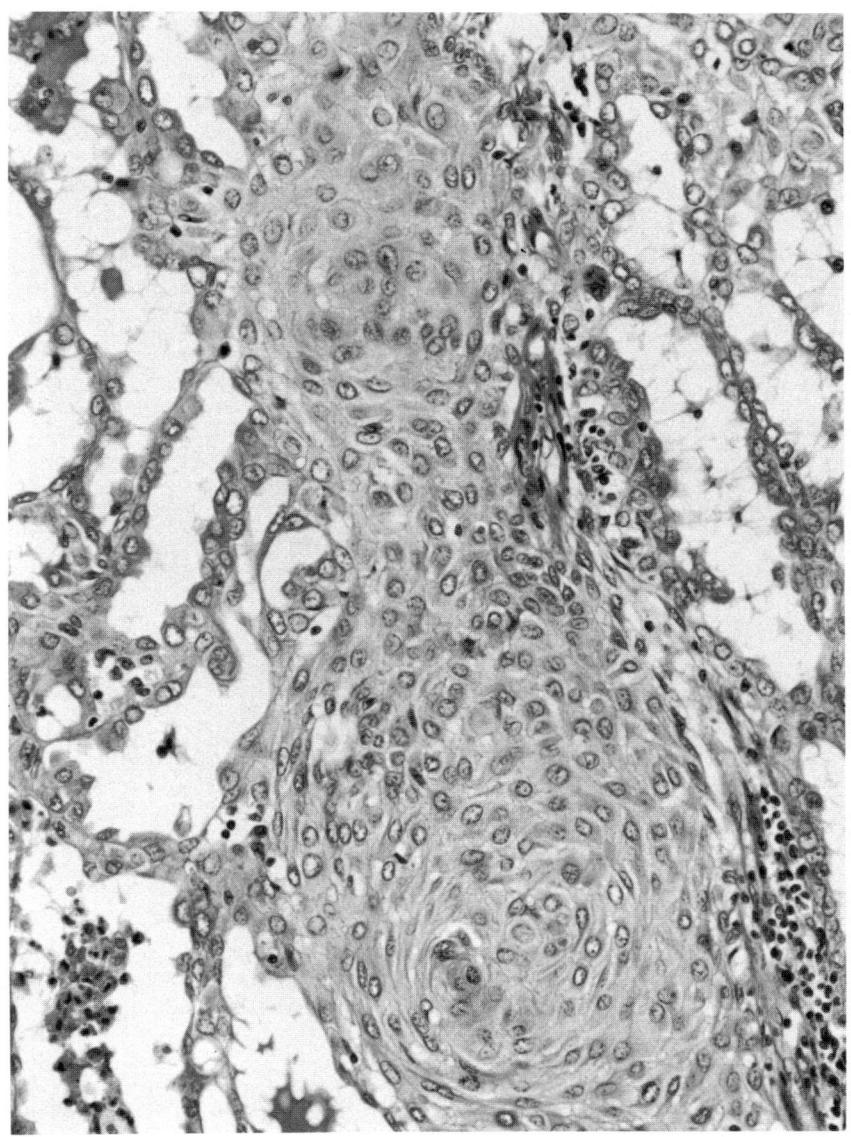

Figure 9.15 Papillary carcinoma with marked focal squamous metaplasia but without frank keratinization (H&E, × 250).

Nevertheless, tumours with such non-papillary structures reveal, at least in some areas (Fig. 9.16), the characteristic cytological features of papillary carcinoma and the nuclear ground-glass features are

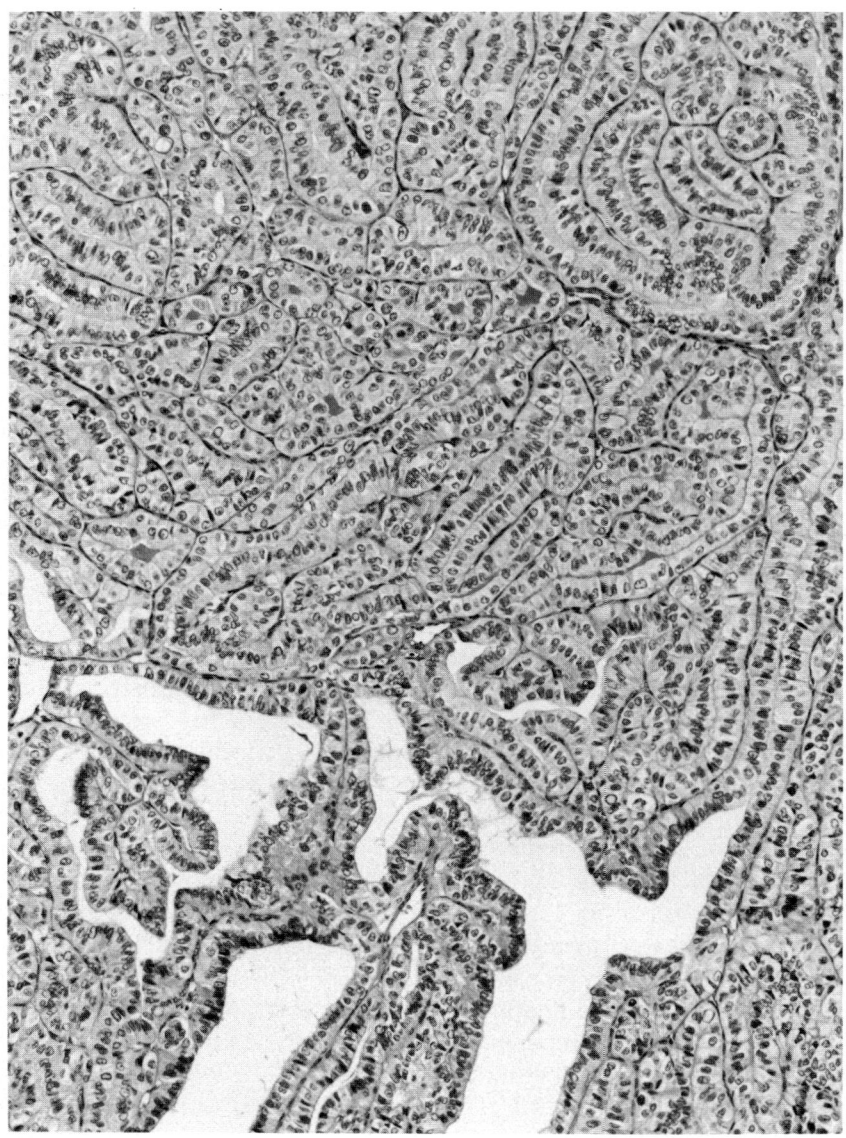

Figure 9.16 Unusual papillary carcinoma with solid growth of closely packed columns of cylindrical cells arranged transversely in single rows. Unequivocal papillary structures are only focally present (lower left) (H&E, × 125).

frequently preserved, even in squamous areas. They show the same biological behaviour as classic papillary carcinoma, and should be classified as such. Lymphocytic and plasma cell stromal infiltration is seen in about 25% of cases (Woolner *et al.*, 1961; Carcangiu *et al.*, 1985b). This infiltration may be confined to the tumour stroma, but quite often the non-neoplastic thyroid tissue is also affected by lymphocytic or Hashimoto's thyroiditis (Fig. 9.24). It is unknown whether this represents a reaction to the neoplasm, or whether the tumour has merely been involved by the same events that led to the lymphocytic inflammation in the remainder of the gland. Schröder *et al.* (1988a) have pointed out that stromal infiltration of dendritic/Langerhans' cells, identified by a positive immunostaining for S-100 protein, is a frequent finding in papillary carcinomas, and that a high density of such cells appears to be associated with a benign disease course.

9.2.4 *Immunohistochemical findings*

The tumour cells of papillary carcinoma regularly exhibit immunoreactivity for thyroglobulin, but usually less intensely than in follicular carcinomas. The staining may be confined to the apical parts of the cells, or to the apical surface of the cells. Other cells exhibit diffuse cytoplasmic staining. Follicular structures, when present, usually show a staining pattern similar to that of follicular tumours, whereas areas of solid cell proliferation or squamous metaplasia react weakly or not at all (Albores-Saavedra *et al.*, 1983; Ryff-de Léche *et al.*, 1986).

Immunohistochemical staining with antibodies against various cytokeratins has been used in an attempt to distinguish papillary carcinoma from other types of thyroid neoplasms. Investigators have mostly used broad-spectrum polyclonal high-molecular-weight keratin antisera, largely with disappointing results. Schelfhout *et al.* (1989), however, reported that immunohistochemical staining of frozen sections using monoclonal antibodies against low-molecular-weight keratin (Moll no. 19) might be of value in discriminating papillary carcinoma from other thyroid lesions. Keratin 19 immunoreactivity was demonstrated in all tumour cells of papillary carcinomas, but was absent or only focally present in follicular adenomas and carcinomas. We have recently studied a new monoclonal antibody, keratin NCL-5D3 (purchased from Milab AB, Malmö), reacting with low-molecular-weight cytokeratins, principally Moll no. 8, but also with nos. 18 and 19, and claimed to be useful on formalin-fixed, paraffin-embedded material (Angus *et al.*, 1987). All papillary carcinomas tested were reactive with this antibody. However, the intensity and distribution of positive staining varied; it was usually confined to the marginal parts of the lesions (Figs 9.17 and 9.18;

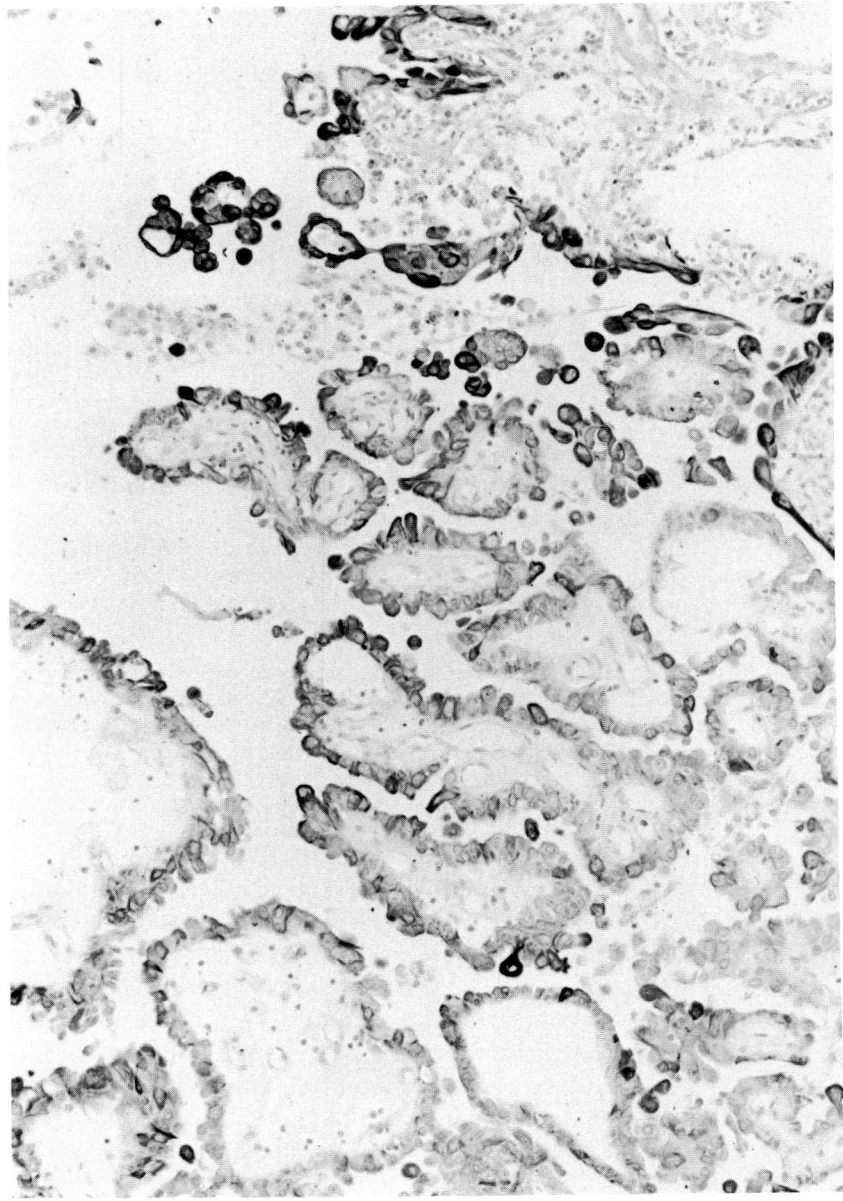

Figure 9.17 Papillary carcinoma. Immunostaining with monoclonal antibodies against low-molecular-weight cytokeratins (keratin NCL-5D3, Milab AB, Malmö), showing positive reaction in tumour cells, most intensely in peripheral parts of the papillae (ABC technique, × 125).

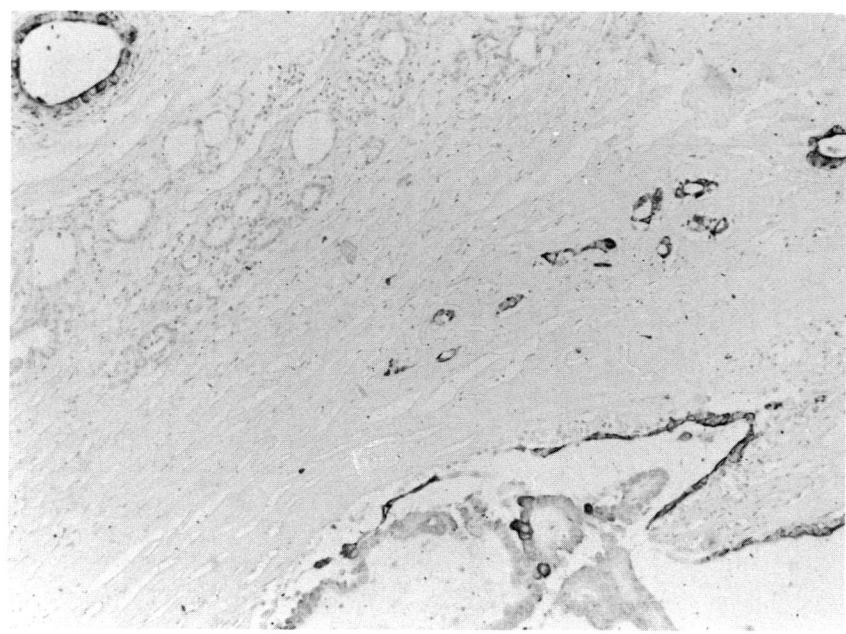

Figure 9.18 Cystic papillary carcinoma. Immunostaining for low-molecular-weight cytokeratins. A positive immunoreaction is seen in the epithelium lining the cyst as well as in papillary structures of the tumour, and in small tumour elements growing in the thick capsule. A positively stained, neoplastic follicle is seen outside the capsule, separated from it by a cluster of non-reactive normal follicles (upper left) (ABC technique, × 80).

Plate 16). On the other hand, follicular adenomas, follicular carcinomas, oxyphil neoplasms, classic medullary carcinomas, colloid nodules, and normal thyroid parenchyma were non-reactive or showed positive staining only focally. When present in normal thyroid tissue or thyroid lesions other than papillary carcinoma, staining was seen in single or small groups of epithelial cells, and sometimes in occasional follicles. Vimentin is also expressed by the tumour cells of papillary carcinoma, often together with cytokeratin in the same cell (Schröder *et al.*, 1986). However, vimentin occurs in other thyroid lesions as well and is, therefore, of little use in the differential diagnosis of thyroid neoplasms (Viale *et al.*, 1989).

9.2.5 Morphological variants of papillary carcinoma

(a) Papillary microcarcinoma
The papillary microcarcinoma (so-called occult papillary carcinoma)

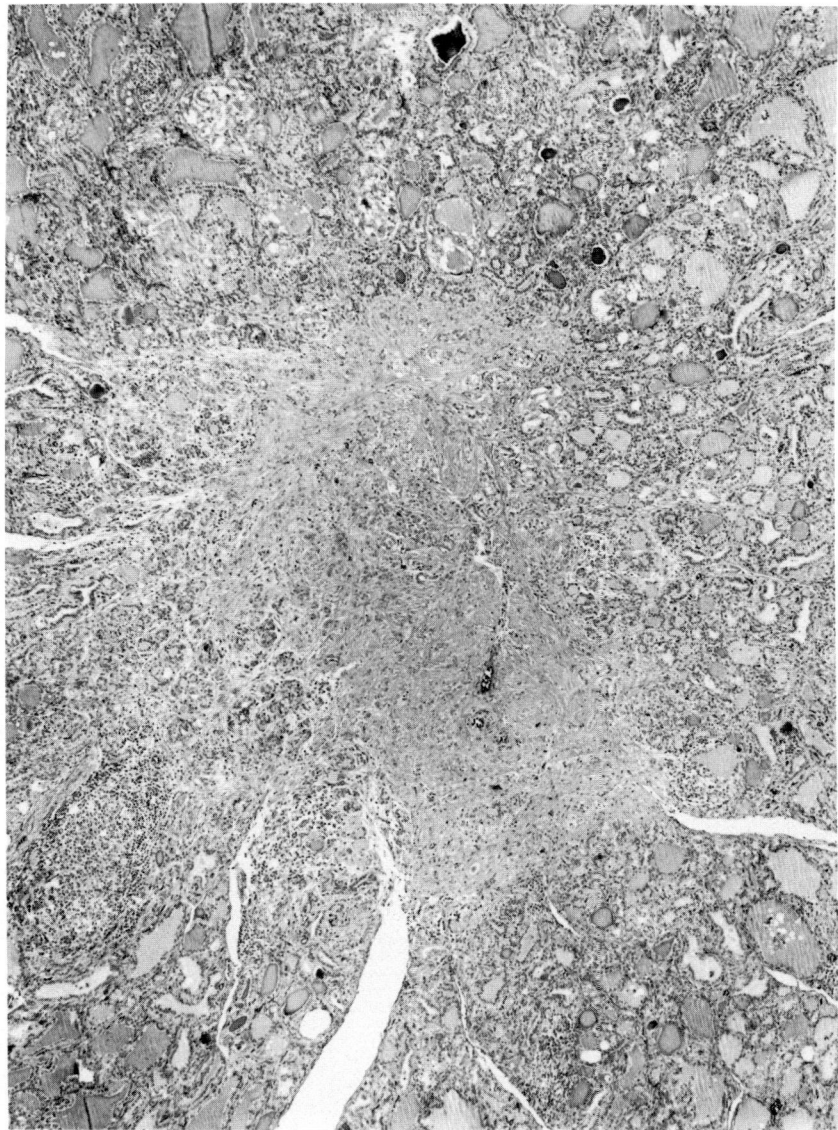

Figure 9.19 Papillary microcarcinoma. The lesion is ill-defined, having a sclerotic cell-poor centre with fibrous extensions radiating into the surrounding thyroid parenchyma (H&E, × 50).

is now generally defined as a papillary carcinoma 1.0 cm or less in diameter (see also p. 145). This lesion was originally described by Hazard *et al.* (1949), by the term 'non-encapsulated sclerosing tumour', which stressed the characteristic and frequent sclerotic reaction of the tumour stroma (Fig. 9.19). It is discovered following biopsy of an enlarged metastatic lymph node or as an incidental finding in a thyroid surgical specimen. As a group, these tumours show a benign clinical behaviour. Deaths from disseminated disease are extremely rare (Strate *et al.*, 1984) and life expectancy of patients with papillary micro-carcinomas is no shorter than that of the population as a whole,

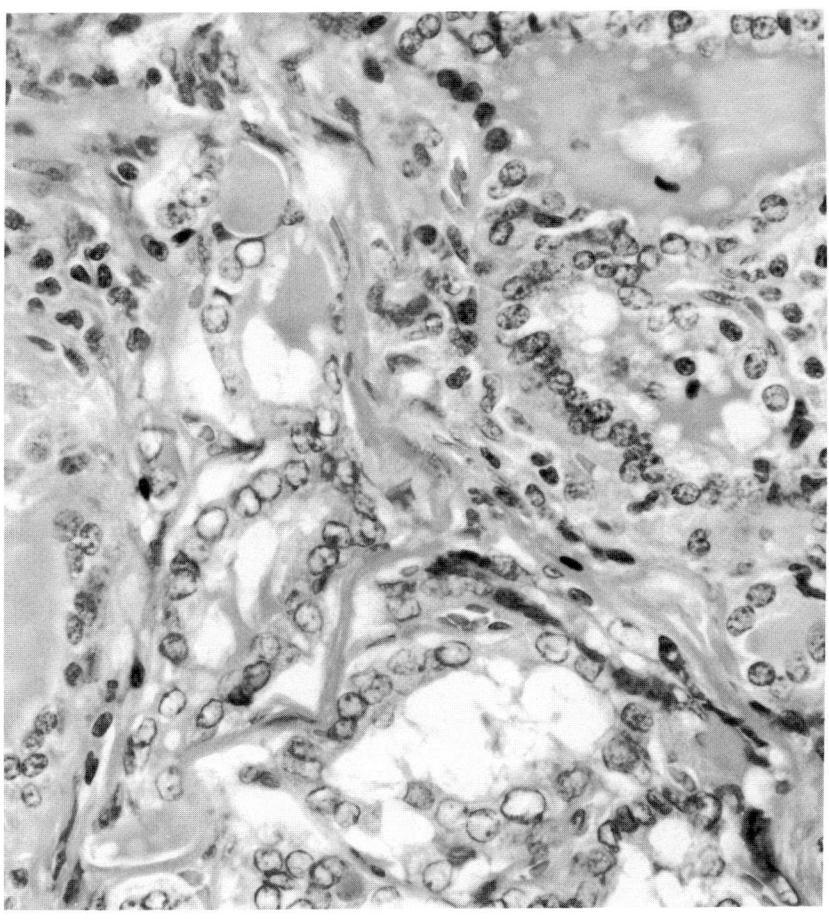

Figure 9.20 High-power field from the peripheral part of the lesion shown in Fig. 9.19. Neoplastic follicles (bottom) are seen to infiltrate between adjacent normal follicles. Note the large, irregular ground glass nuclei of the former (H&E, × 500).

regardless of whether cervical lymph nodes are present (Woolner *et al.*, 1961). A conservative attitude is now generally advocated in their management and treatment (Hubert *et al.*, 1980; Cady, 1981).

Microscopically, papillary microcarcinoma most often presents as an ill-defined radiate lesion having a sclerotic centre with abundant elastoid deposits. Interestingly, the general architecture and characteristic stromal reaction of this form show a close resemblance to the so-called radial scar of the breast, and its neoplastic counterpart, the tubular carcinoma (Linell *et al.*, 1980). In the peripheral parts of the lesion there are finger-like projections of the stroma, radiating into the surrounding parenchyma. These often harbour neoplastic follicles, plexiform structures or small groups of neoplastic cells which may also be seen to infiltrate into the surrounding normal parenchyma and blend with the normal follicles. The infiltrative component may be inconspicuous and easily overlooked, unless diligently sought. The greater size of the tumour cells, as well as their nuclear features, characteristic of papillary carcinoma, are usually clearly recognizable, making it possible to discriminate them from the normal follicular cells (Fig. 9.20). Rarely, the tumour is more cellular, with less stromal sclerosis. Some tumours may be partially or totally encapsulated (Fig. 9.21), although evidence

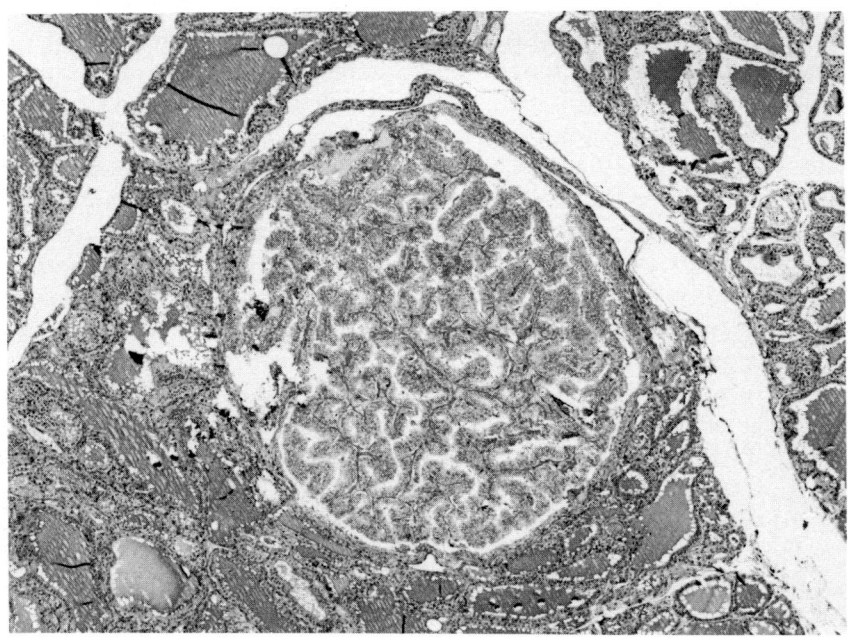

Figure 9.21 Well-defined, encapsulated papillary microcarcinoma, composed entirely of papillary structures (H&E, × 50).

of capsular invasion is usually present. Rarely, the lesion may be circumscribed but lacks a capsule as well as signs of infiltration. Fully developed papillary structures may be inconspicuous and difficult to find. Lymph node metastases not infrequently show only sparse stromal fibrosis. In addition, they may contain a large and even dominant component of follicular elements which may look benign. Careful examination, however, will usually reveal focal papillary structures as well as cytological features typical of papillary carcinoma (Figs 3.8 and 3.9). As pointed out in Chapter 3 (p. 45), it should be remembered that the finding of thyroid tissue in a cervical lymph node is a strong indication of the presence of a papillary carcinoma in the gland proper.

(b) Encapsulated variant
The encapsulated variant of papillary carcinoma is completely surrounded by a capsule (Figs 9.22 and 9.23) and comprises 3–14% of cases of papillary carcinoma (Hazard, 1968; Vickery, 1983; Schröder *et al.*, 1984a; Schröder, 1988). The separation of this subtype is justified from clinical and prognostic points of view, because it has been found to have an exceptionally favourable prognosis. Its propensity for lymph node metastasis is less than for other forms of papillary carcinoma. Local recurrences are extremely rare, and distant metastases and tumour deaths have not been reported (Hazard, 1964; Schröder *et al.*, 1984a; Carcangiu *et al.*, 1985b). The proportion of papillary structures may vary over a broad spectrum, from purely papillary tumours to lesions which are exclusively follicular with the typical nuclear hallmarks of papillary carcinoma. The encapsulated form usually shows histological signs of invasion of the tumour capsule, although this may be focal and difficult to detect. Rarely, histological evidence of invasive growth is absent. Under such circumstances the diagnosis of malignancy may be brought into question, especially in cases where the histological structure of the tumour appears mainly or entirely follicular (Evans, 1987). Schröder *et al.* (1984a) found regional lymph node metastases in some of their cases which lacked clear histological evidence of invasion of the tumour capsule. It appears justifiable, therefore, to regard such lesions as papillary carcinomas, provided they show the cytological hallmarks of papillary carcinoma. Vickery (1983) described occasional examples of encapsulated follicular tumours being mainly made up of benign-looking follicular structures with a small focus of unequivocal papillary carcinoma, suggesting that some encapsulated papillary carcinomas may arise from pre-existing follicular adenomas (see also Fig. 8.4).

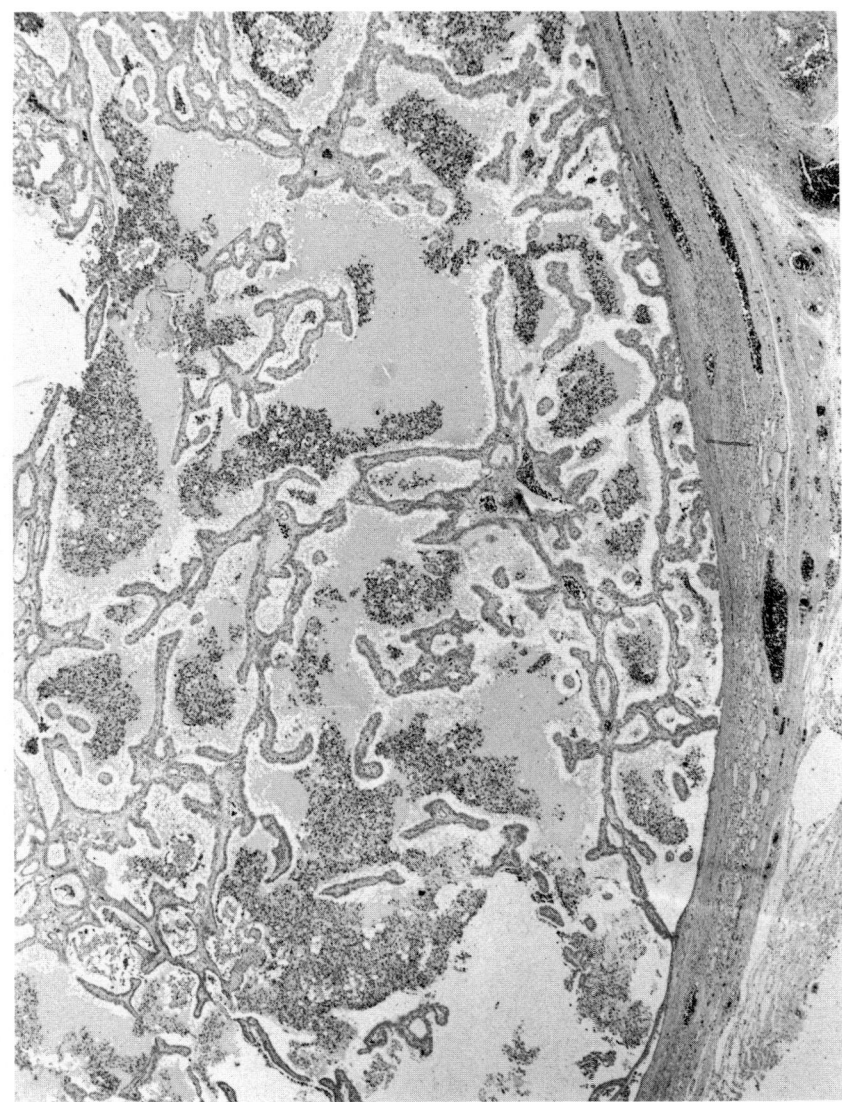

Figure 9.22 Encapsulated variant of papillary carcinoma. This example shows an unusual architectural pattern, formed by thin, richly branched papillary fronds and a thin septa separating large, irregular, colloid-filled spaces (H&E, × 25).

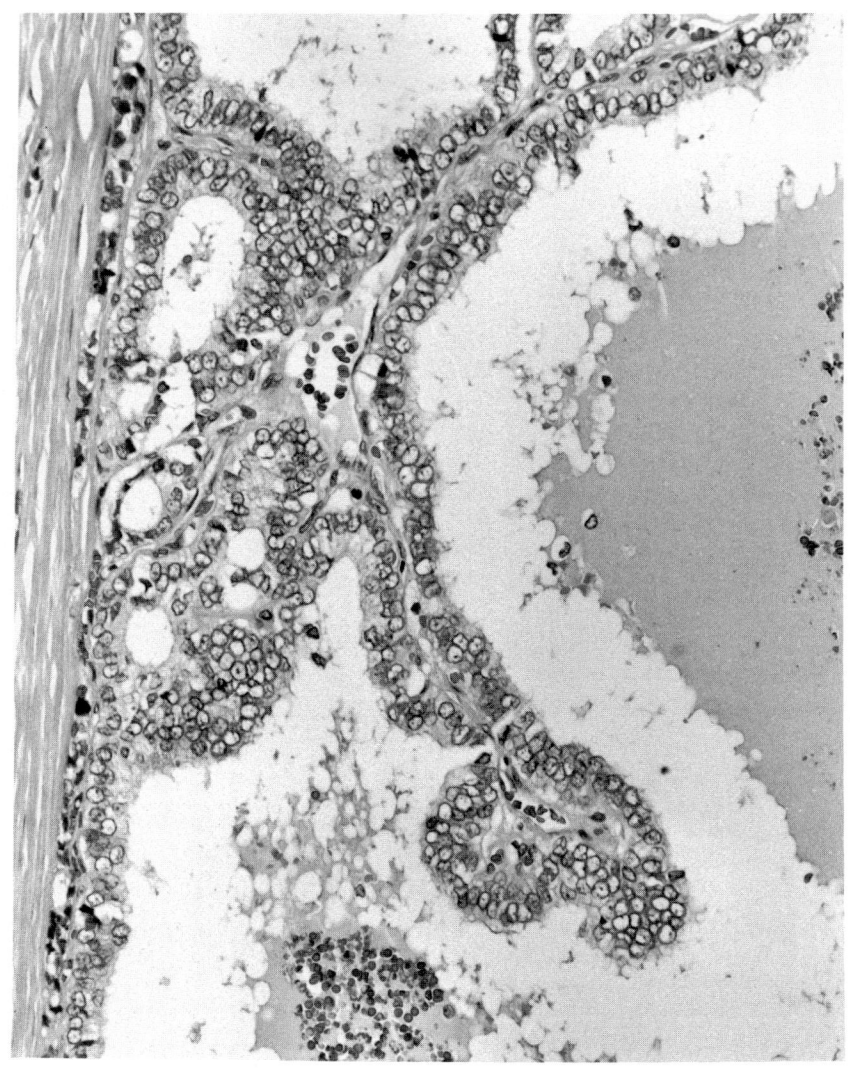

Figure 9.23 Detail of the lesion illustrated in Fig. 9.22 showing the cytological features, typical of papillary carcinoma (H&E, × 250).

(c) Follicular variant

The follicular variant of papillary carcinoma was described by Crile and Hazard (1953). Lindsay (1960) originally thought it was a sub-type of follicular carcinoma, although he later classified it among the papillary carcinomas. Chen and Rosai (1977) gave a detailed clinocopathological description of this tumour and confirmed it as a

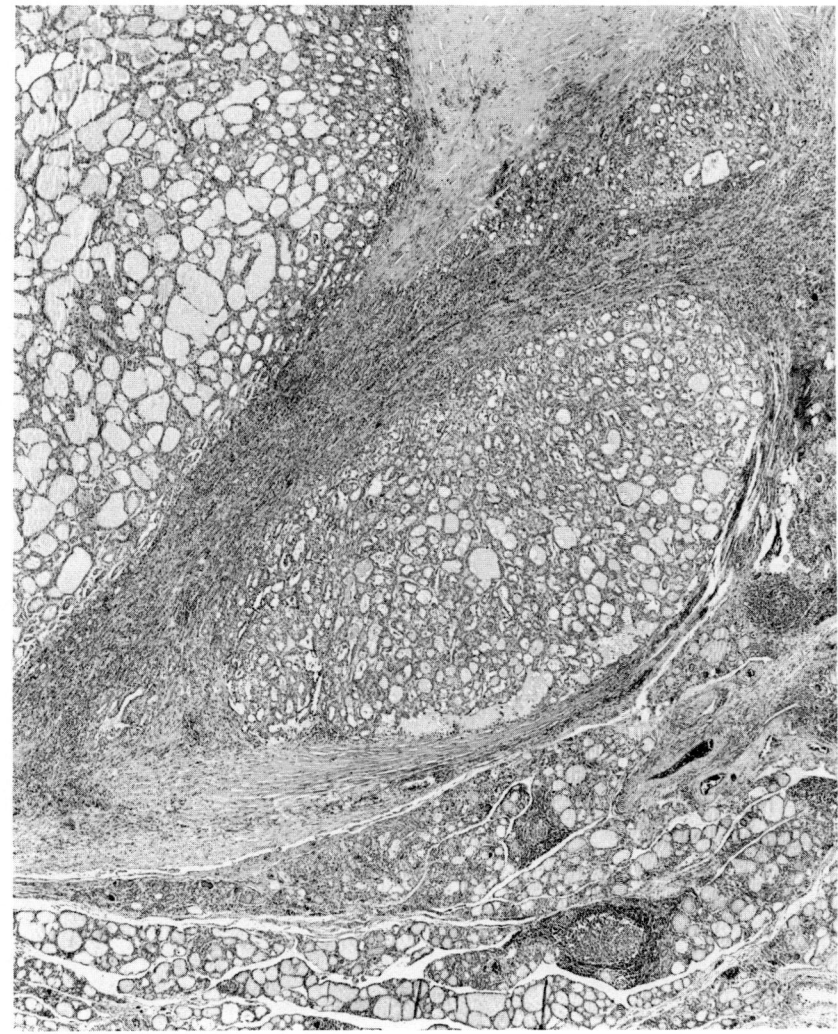

Figure 9.24 Follicular variant of papillary carcinoma, well defined and partly encapsulated. A chronic lymphocytic thyroiditis is seen in the surrounding thyroid parenchyma (H&E, × 50).

variant of papillary carcinoma with similar biological behaviour. These tumours may or may not be encapsulated. When present, the capsule is usually incomplete (Fig. 9.24). The tumour tissue is almost entirely composed of follicles (Fig. 9.25). Occasional irregular folds, buds or other intrafollicular protrusions which may look like abortive papillae

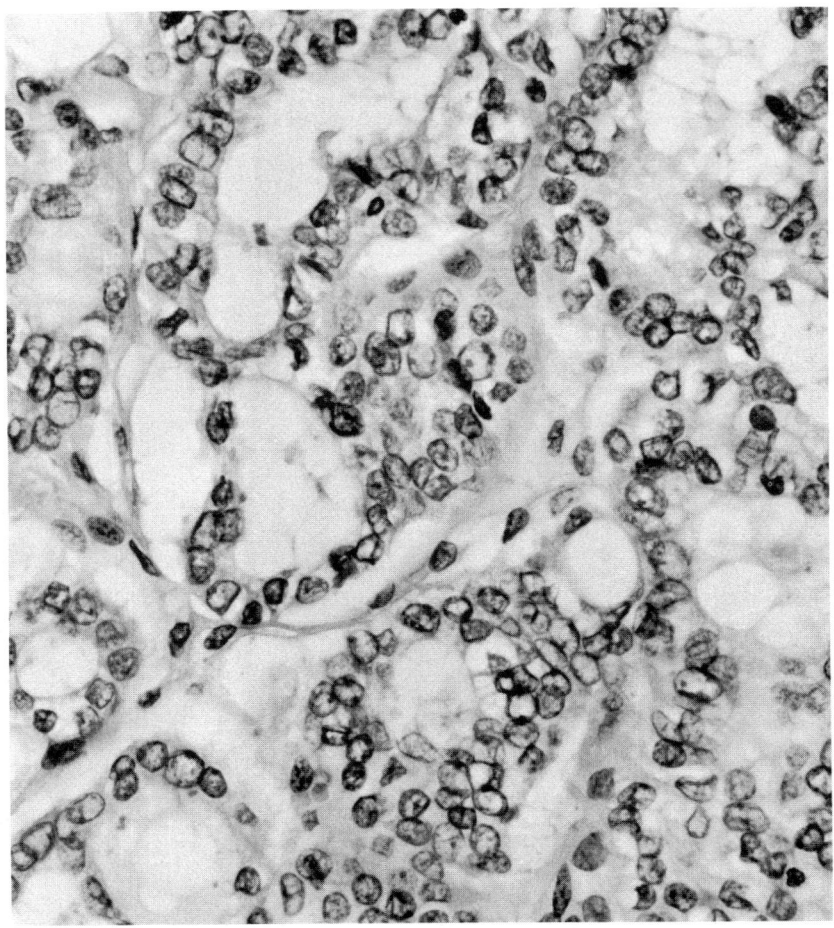

Figure 9.25 Same case as shown in Fig. 9.24. The neoplastic follicles are lined by tall atypical follicular cells with crowding, irregular nuclei showing ground-glass features and nuclear grooves (H&E, × 500).

or forerunners thereof can often be found (Fig. 9.26). Sometimes the tumour cells may form irregular plexiform or glomeruloid formations and there may be foci of solid growth. The most important diagnostic criterion is the presence of ground-glass nuclei (Fig. 9.25). Features supporting the diagnosis include invasive growth pattern, foci of fibrosis which may be extensive, presence of psammoma bodies, and marginal vacuoles in the colloid (Fig. 9.26) (Rosai *et al.*, 1983). Although fully developed papillae may be absent in the primary tumour, areas with a well-formed papillary growth pattern may be

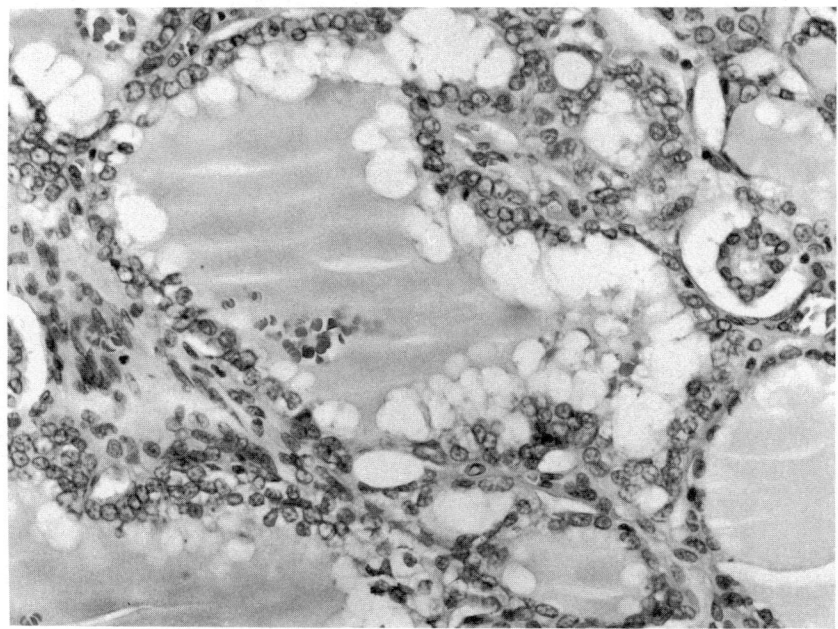

Figure 9.26 Same case as in Figs 9.24 and 9.25. A few abortive papillae were found in the lesion. Note the marginal vacuoles in the colloid (H&E, × 500).

seen in its regional lymph node metastases. Immunohistochemically, a positive reaction for low-molecular-weight cytokeratins (p. 158), may be helpful in differentiating this tumour type from follicular adenoma and follicular carcinoma (Plate 16).

(d) Diffuse sclerosing variant

The diffuse sclerosing variant is an unusual form characterized by diffuse involvement of one or both thyroid lobes. It shows a widespread dense fibrous stroma with scattered islands of papillary carcinoma, often growing in endothelial-lined spaces consistent with lymph vessels (Fig. 9.27). Although papillary structures with cytological features typical of papillary carcinoma do occur, tumour cells often form solid nests with foci of squamous metaplasia. The tumour also shows abundant psammoma bodies, and a marked lymphoplasmacytic stromal infiltration. Although the number of cases reported so far is small, current information suggests that this tumour differs from conventional papillary carcinoma in that it occurs among younger individuals and behaves aggressively, with a higher incidence of lymph node and lung metastases and a lower disease-free survival rate

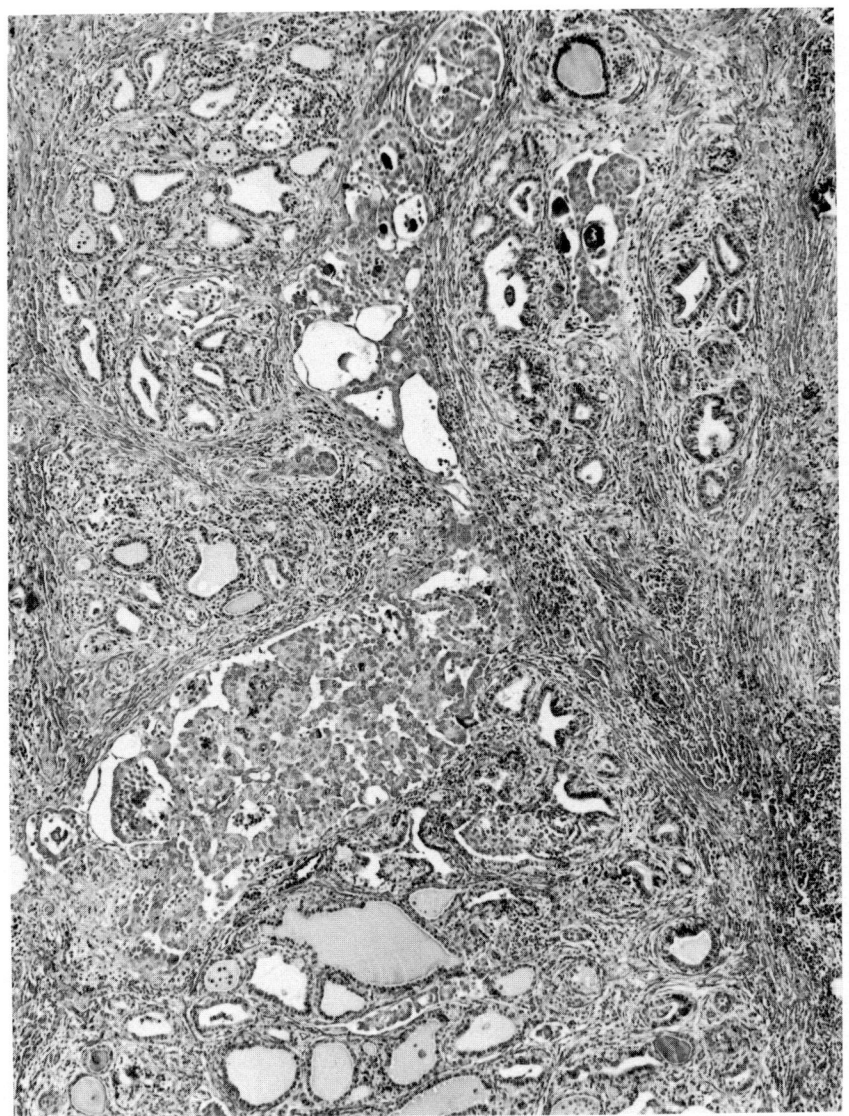

Figure 9.27 Diffuse sclerosing variant of papillary carcinoma. The tumour is growing diffusely in the stroma and in lymph vessels. There is marked destruction of thyroid parenchyma with extensive fibrosis and a chronic lymphocytic infiltrate. Many psammoma bodies are also seen (H&E, × 50).

(Vickery *et al.*, 1985; Chan *et al.*, 1987; Carcangiu and Bianchi, 1989). As stated previously (p. 140), this variant may present with a diffuse, often bilateral and sometimes painful enlargement of the thyroid gland which may erroneously lead to a clinical diagnosis of thyroiditis, and consequently, to a delay in the institution of appropriate treatment. Sobrinho-Simões *et al.* (1990) have recently described a diffuse follicular variant of papillary carcinoma, also occurring in young patients and showing aggressive clinical behaviour. This form, however, lacks stromal sclerosis and usually presents with a diffuse or multinodular gland enlargement, simulating a benign goitre.

(e) Oxyphil variant and clear cell papillary carcinoma
The oxyphil variant of papillary carcinoma and clear cell papillary carcinoma are discussed on pp. 197 and 199.

(f) Tall cell variant
Tall cell variant is characterized by tall columnar tumour cells, their height being twice their breadth (Hawk and Hazard, 1976). It often shows a highly papilliferous histological pattern, although it may sometimes also form follicular structures. The tumour cells may be strikingly oxyphil. This tumour variant tends to produce large bulky lesions and shows a more aggressive clinical course than other subtypes (Tscholl-Ducommun and Hedinger, 1982; Johnson *et al.*, 1988). Because of its poor prognosis, it has been suggested that tall cell carcinoma should perhaps be separated from the papillary carcinoma group (Evans, 1986; Sobrinho-Simões *et al.*, 1988).

(g) Papillary carcinoma with poorly differentiated, undifferentiated or epidermoid components
It is generally agreed that the presence of solid areas with or without squamous metaplasia does not have any relationship to prognosis. However, there appears to be a rare variant of papillary carcinoma with an aggressive behaviour, showing a less differentiated component composed of small, uniform cells with polymorphic nuclei. The tumour shows a high mitotic rate and grows in solid clusters or trabeculae as well as in microfollicular or cribriform patterns, often with foci of necrosis. This tumour type corresponds to the poorly differentiated variant of papillary carcinoma, described by Tscholl-Ducommun and Hedinger (1982). A similar tumour was designated 'poorly differentiated' ('insular') thyroid carcinoma' by Carcangiu *et al.* (1984), who believe it to be a distinct clinicopathological entity, probably related to the 'wuchernde Struma' of Langhans (1907). The tumour shows a high incidence of lymphatic spread, but in addition, and in contrast to classic papillary carcinoma, a considerable propensity

for blood vessel invasion. Carcangiu *et al.* (1984) feel uncertain whether all cases of this tumour type are related to papillary carcinoma. Some fail to develop papillary structures and may be related to follicular carcinoma. They conclude that although it is not always clear whether this tumour should be included in the papillary or follicular group, the important issue is that it is poorly differentiated and biologically occupies an intermediate zone between the well differentiated tumours and the totally undifferentiated group. It has a great propensity for local recurrence and metastases to the lymph nodes, lungs and bone, characterisics which require aggressive therapy and suggest considera-tion of alternative or supplementary forms of therapy (Carcangiu *et al.*, 1984).

Highly differentiated papillary carcinoma may occasionally show areas of undifferentiated carcinoma (Hutter *et al.*, 1965). This is in accordance with the generally accepted view that many cases of undifferentiated thyroid carcinoma develop from a well-differentiated papillary or follicular carcinoma by a process of dedifferentiation or transformation (Harada *et al.*, 1977). The undifferentiated component may be composed of highly atypical spindle cells or giant cells. Occasionally there may be focal development into epidermoid carcinoma (Ross, 1947; Cooke and Carrera, 1964; Motoyama and Watanabe, 1983). The presence of foci of undifferentiated or squamous carcinoma in otherwise highly differentiated papillary carcinoma has serious prognostic consequences, and such a neoplasm should be categorized as an undifferentiated carcinoma or squamous carcinoma respectively, as recommended by the WHO classification (Hedinger *et al.*, 1988). Therefore, it is of great importance to distinguish between squamous carcinoma and foci of squamous metaplasia, the latter being a common feature of highly differentiated papillary carcinomas but not having any influence on prognosis. The most helpful differential diagnostic points are cellular atypia and mitoses. In addition, intercellular bridges and foci of keratinization are frequent findings in squamous carcinoma, but are seldom seen in squamous metaplasia.

(h) Papillary carcinoma with signet-ring cells
Rarely papillary carcinoma may show tumour cells resembling signet-ring cells. These cells have eccentrically placed nuclei and a large cytoplasmic vacuole giving a 'targetoid' or 'bull's eye' appearance similar to that seen in lobular carcinoma of the breast. The vacuoles usually contain material composed of a central PAS-positive core, surrounded by a peripheral Alcian blue-positive rim, which is usual-ly stained for thyroglobulin. The material is probably not a true mucosubstance but rather carbohydrate components related to thyroglobulin or breakdown products of thyroglobulin or colloid (Chan

and Tse, 1988). It may be similar in nature to the material found in follicular signet-ring cell adenomas described in Chapter 8 (see p. 125) and occasionally in follicular carcinoma (see below) (Gherardi, 1987; Rigaud and Bogomoletz, 1987).

9.2.6 Differential diagnostic problems

The most important differential diagnostic problem concerns lesions showing papillary epithelial formations and/or ground-glass nuclei (see also p. 85). Hyperplastic nodules may show areas where the epithelium is high and even grows in papillary-like structures, often as infoldings into large follicles. However, the stroma of such infoldings is frequently filled with small follicles, which are seldom seen in papillary carcinoma. Although the nuclei may be vesicular and mimic ground-glass nuclei, their size is smaller, their outline is usually smooth, and nuclear grooves do not occur. In Graves' disease papillary projections may also be encountered, protruding into irregularly formed follicles and giving the impression of a focus of papillary carcinoma. However, the cells lack the nuclear features of the carcinoma (see also p. 106). The problem of distinguishing a follicular adenoma from the follicular, encapsulated variant of papillary carcinoma is discussed on p. 186. Metastatic papillary carcinoma may occasionally present as a primary thyroid neoplasm. Papillary carcinomas, especially of the lung, but also of the breast and ovary may occasionally metastasize to the thyroid. Some cases may even show numerous psammoma bodies. In the thyroid the metastatic tumour tends to grow extensively in the lymph vessels; a dominating interstitial tumour infiltrate is often missing (Fig. 12.4). This feature may be helpful in distinguishing it from the majority of primary papillary carcinomas (the diffuse sclerosing variant excluded). Close microscopic analysis of the nuclear characteristics, however, seems to be the most reliable histological clue to a correct diagnosis (Fig. 12.5).

9.2.7 Prognostic factors

Generally, the prognosis of papillary carcinoma is excellent (Mazzaferri et al., 1977; McConahey et al., 1986). However, the outcome depends on various pathological and clinical factors, some of which have already been dealt with. From a pathological point of view the most important prognostic predictor is the size of the primary tumour. Tumours small enough to be included within the confines of the gland have the best prognosis. The outcome is significantly worse for patients with large tumours that have extended beyond the capsule of the gland. From a prognostic standpoint, it is relevant to separate classic papillary carcinoma by size into three groups: (1) papillary micro-carcinoma, being 1 cm or less in greatest diameter; (2) intrathyroidal papillary carcinoma, exceeding 1 cm in diameter but confined within the gland; and (3) extrathyroidal papillary carcinoma, which is grossly infiltrating papillary carcinoma extending beyond the thyroid capsule. Tumours which are encapsulated or have smooth, pushing margins have a better prognosis than lesions with marked invasiveness. Cases with multiple tumours show a greater tendency for metastases and a decreased disease-free survival rate compared to those with solitary lesions (Carcangiu et al., 1985b). The presence of distant metastases, which occurs most frequently in the older age groups, is another important predictor of prognosis (Høie et al., 1988). The specific site of metastasis has a distinct bearing on prognosis, in that patients with lung metastases have a much better outcome than patients with involvement of other organs, such as the skeletal system (Schlumberger et al., 1986).

From a clinical point of view, age is the most important prognostic predictor. Children and young adults have the best prognosis. Nearly all deaths occur among patients in whom the carcinoma has been diagnosed after 40 years of age (Woolner et al., 1961; Borup Christensen et al., 1984; Carcangiu et al., 1985b; DeGroot et al., 1990). Although it has been claimed that females have a better outcome than males (Byar et al., 1979; Frauenhoffer et al., 1979; Ito et al., 1980; Cady et al., 1985), other studies have not found any significant difference between sexes (Borup Christensen et al., 1984; Carcangiu et al., 1985b; Tennvall et al., 1986).

The value of DNA ploidy as a prognostic index for papillary carcinoma is disputed. Some studies indicate a good correlation between aneuploidy and lethal course of the disease, and between diploid DNA values and survival (Cohn et al., 1984; Auer et al., 1985; McIntosh et al., 1986). Other studies report no such correlation and consider DNA measurements to have no prognostic predictive value (Johannessen et al., 1981; Schröder, 1988).

9.3   Follicular carcinoma

Follicular thyroid carcinoma can be defined as a malignant thyroid epithelial tumour showing evidence of follicular cell differentiation, but lacking the diagnostic features of papillary carcinoma (Hedinger et al., 1988).

Follicular carcinoma is the second most common type of thyroid carcinoma, accounting for 10–30% of thyroid carcinomas in most series (Woolner et al., 1961; Franssila, 1971; Heitz et al., 1976; Borup Christensen et al., 1984; Schröder 1988). The marked variation in incidence reported in different series may be due partly to geographic variation and environmental factors. There is some evidence, for instance, that the relative frequency of follicular carcinoma is higher in endemic goitre areas than in non-endemic areas (Cuello *et al.*, 1969; Williams *et al.*, 1977; Hedinger, 1981). However, it should be noted that the differences found may also, at least in part, be dependent on differences in surgical policy, and in methodology and classification criteria used by the pathologists diagnosing the lesions (Fransilla *et al.*, 1981). The histological diagnosis of follicular carcinoma is often more problematic than the diagnosis of other types of thyroid carcinoma (Saxén *et al.*, 1978). Follicular carcinoma may be difficult to separate from some follicular adenomas, especially atypical variants, and from the follicular variant of papillary carcinoma. Oxyphil carcinomas are included within the follicular category by some workers but not by others. In addition, there is uncertainty as to where to include rare intermediate forms of thyroid carcinoma with traits of both follicular and medullary carcinoma (see p. 227), as well as poorly differentiated variants of thyroid carcinoma (see p. 187).

Follicular carcinoma has a similar predilection for females as papillary carcinoma. The average age at diagnosis ranges between 48 and 58 years (Woolner *et al.*, 1961; Franssila, 1971; Lang *et al.*, 1986, Schröder, 1988).

In contrast to papillary carcinoma, follicular carcinoma tends to invade blood vessels and to give rise to haematogenous metastases. Secondaries are most frequently seen in the skeleton and the lungs, and less often at other sites, including visceral organs and lymph nodes (Heitz *et al.*, 1976; Borup Christensen and Ljungberg, 1984). The skeletal metastases are often located in the shoulder girdle, sternum, skull and iliac bone (Meissner and Warren, 1969; Massin *et al.*, 1984). Occasionally, solitary metastases are found which are amenable to resection (Woolner, 1971). Not infrequently, a metastatic lesion occurs as the presenting complaint (Meissner and Warren, 1969). Such a solitary metastasis, often in bone, may show histologically well-differentiated follicular structures imitating normal thyroid tissue (so-called metastasizing colloid goitre, Fig. 9.28). The primary tumour is often quite small and clinically inconspicuous, but histologically, it usually shows less well differentiated structures as well as signs of minimal invasion, revealing its malignant nature (Figs 9.29 and 9.30). Metastases may not be evident for many years and when detected, may remain dormant for long periods. Deaths due to haematogenous metastases may not occur until 10–20 years after initial surgery (Hirabayashi and Lindsay, 1961; Woolner, 1971; Beaugie *et al.*, 1976).

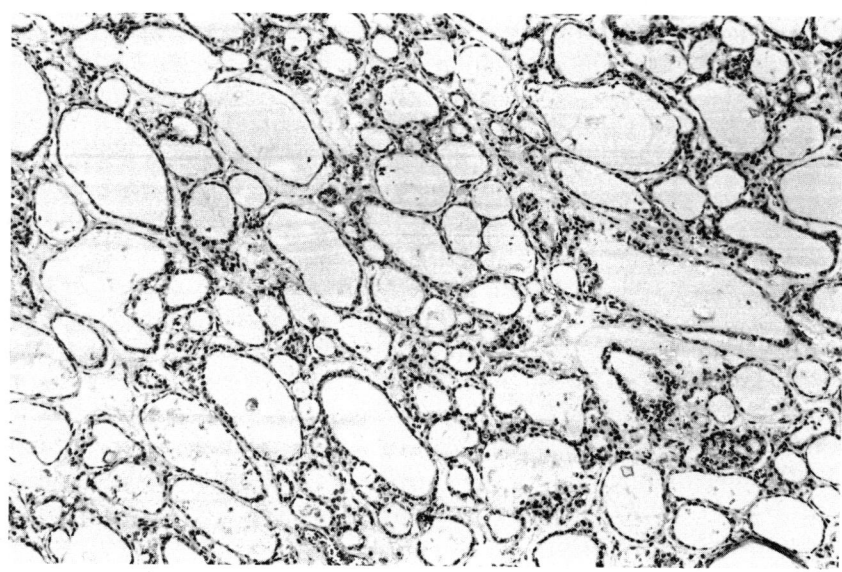

Figure 9.28 Autopsy case, 70-year-old woman. Solitary metastasis in frontal bone of a follicular thyroid carcinoma. Were it not for its abnormal location, the tissue would histologically have been accepted as benign colloid-rich thyroid tissue (H&E, × 85).

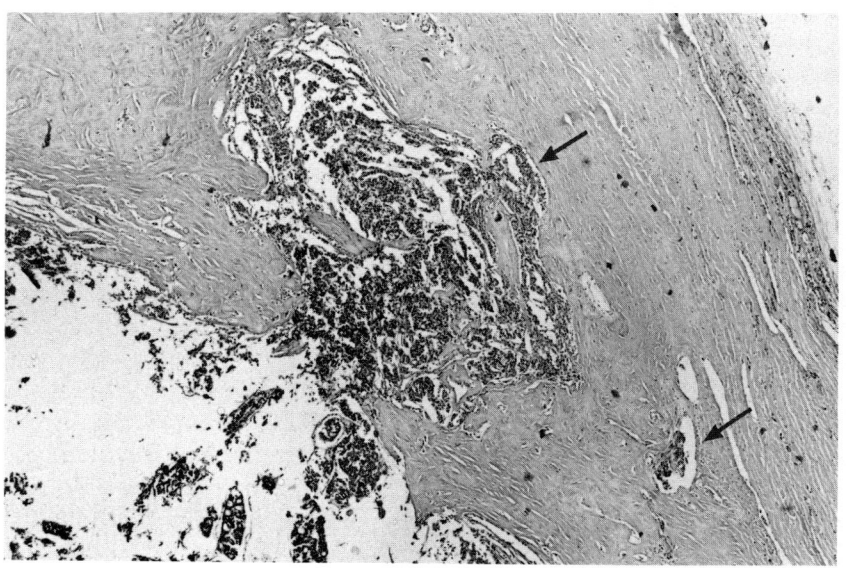

Figure 9.29 The primary of the metastasis shown in Fig. 9.28. The thyroid tumour, the size of a walnut, showed extensive degenerative changes with cyst formation and had developed a thick, hyalinized capsule with calcification. Tumour tissue is infiltrating the capsular area and invades blood vessels (arrows) (H&E, × 35).

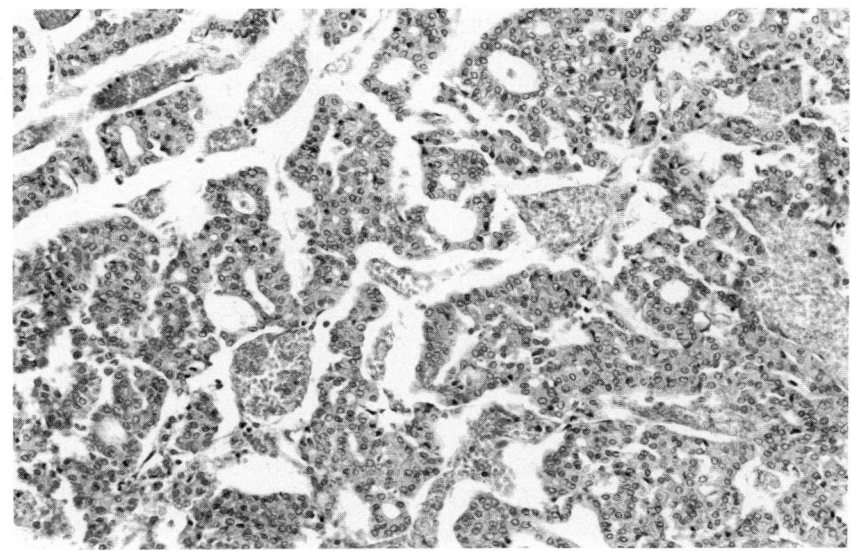

Figure 9.30 Detail of the primary follicular carcinoma shown in Fig. 9.29. It is less differentiated than the metastasis, showing follicular, cribriform and trabecular structures (H&E, × 165).

9.3.1 Macroscopic features

Grossly, a distinction should be made between the minimally invasive and the widely invasive forms of follicular carcinoma. On gross examination, minimally invasive carcinoma appears totally encapsulated and may be indistinguishable from a follicular adenoma. However, it shows microscopic evidence of capsular and/or blood vessel invasion. As a rule, it is a fleshy tumour with a tan or greyish colour (Plate 17). As pointed out earlier (p. 119), its capsule, in contrast to that of adenomas, tends to be thick (Evans, 1984). Larger tumours show degenerative changes including central necrosis, haemorrhage and scarring, sometimes with calcification. Widely invasive carcinoma is only partly encapsulated or totally devoid of a capsule and shows marked invasiveness into surrounding tissues and/or blood vessels (Plate 18), features which are usually apparent macroscopically. In contrast to the minimally invasive variant, the widely invasive carcinoma is usually a large lesion occupying a whole lobe, or the entire gland and, not infrequently, extends into the extrathyroidal tissues. The importance of separating these two variants lies in the marked difference in prognosis. The 10-year survival rate for the minimally invasive, well-encapsulated form is 70–95%, compared to only

30–45% for the widely invasive variant (Woolner *et al.*, 1961; Franssila, 1975). The rate of distant metastases and deaths is highly influenced by the degree of invasiveness of the primary tumour and invasiveness appears to have more weight regarding prognosis than the microscopic growth pattern (with the exception of poorly differentiated forms) (Lang *et al.*, 1986).

The average age of patients with minimally invasive follicular carcinoma is about 10 years younger than that of those affected by the widely invasive form (Franssila, 1971; Woolner, 1971; Beaugie *et al.*, 1976; Schröder *et al.*, 1984c). Traces of a capsule are identified in some widely invasive carcinomas suggesting that the encapsulated carcinoma may represent a precursor of the widely invasive form.

9.3.2 *Microscopic features*

The microscopic appearance of follicular carcinoma is as variable as is that of follicular adenoma. Some tumours may be highly differentiated with large normal-looking follicles lined by regular follicular cells (Fig. 9.31). Others may be more cellular, mimicking hyperplastic thyroid tissue, or show solid patterns formed by nests or anastomosing trabeculae separated by a delicate connective tissue septa. Solid parts often contain areas with some degree of follicular differentiation, developing cribriform or microfollicular patterns with small lumina containing clumps of colloid (Figs 9.32 and 9.33). Rarely, a peculiar pattern is seen with clusters of tumour cells which resemble signet-ring cells. These cells contain cytoplasmic vacuoles, which are usually PAS and Alcian blue-positive and show thyroglobulin immunoreactivity. As noted above, similar cytological patterns may be seen in follicular adenomas and papillary carcinomas (Plate 11).

The architectural and cytological features and mitotic activity have no bearing on prognosis and are of little help in the differential diagnosis between follicular adenoma and carcinoma. An exception in this respect, however, is a highly malignant group of poorly differentiated carcinomas, sometimes referred to as insular carcinomas, which are composed of solid clusters or trabeculae of rather small, basophilic tumour cells with high mitotic activity and a markedly infiltrative growth pattern (Sakamoto *et al.*, 1983; Carcangiu *et al.*, 1984; and p. 187). Some tumours in this category are probably related to follicular carcinoma, others to papillary carcinoma, although a large number may be of uncertain nature (Rosai *et al.*, 1985).

The crucial histological features which determine the biological behaviour of the tumour are found at the peripheral, capsular area of the lesion. They concern signs of penetration of the capsule and/or invasion of blood vessels. Either or both of these features place the lesion in the malignant group, no matter how highly differentiated it is. Features of invasion may only be microscopic and focal. As has been pointed out several times (pp. 26, 119 and 128), extensive histological sampling from the capsular region of the tumour is mandatory in order to detect or exclude evidence of such an invasion.

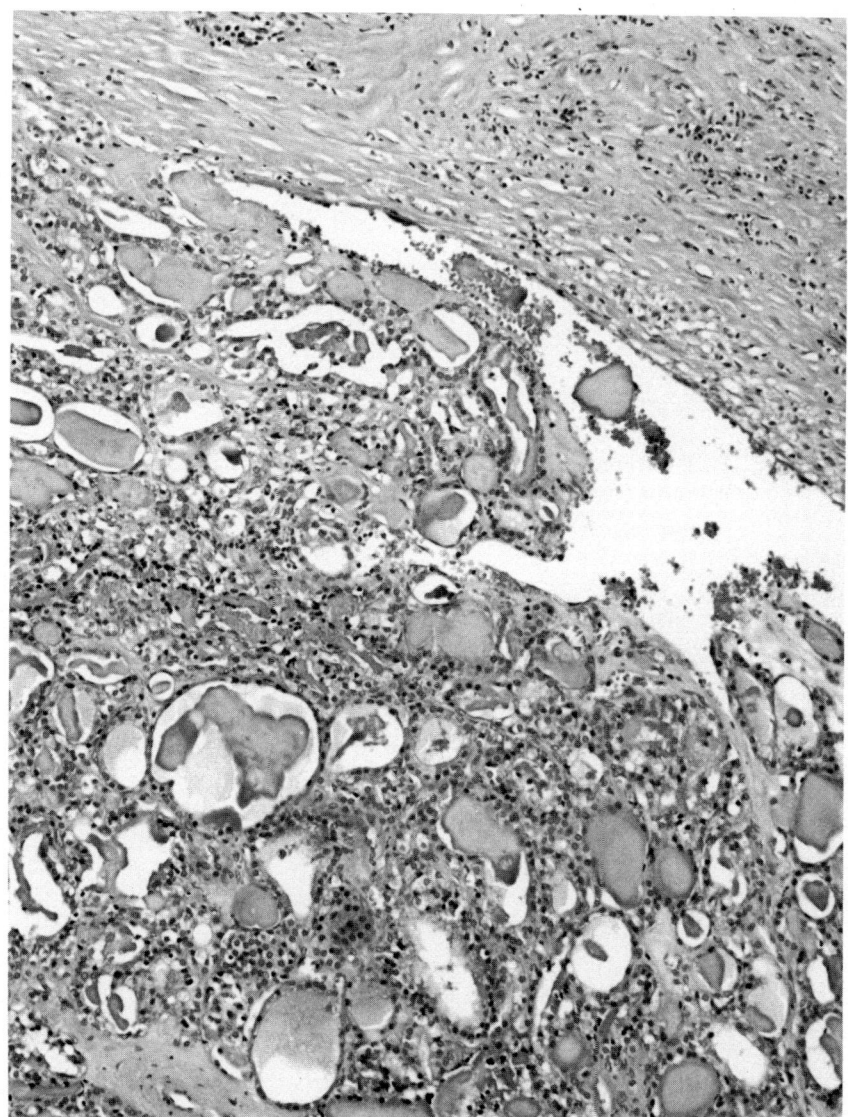

Figure 9.31 Highly differentiated follicular carcinoma composed of variable-sized benign-looking follicles. However, the tumour is seen to invade a blood vessel in the capsule, indicating its malignant nature (H&E, × 125).

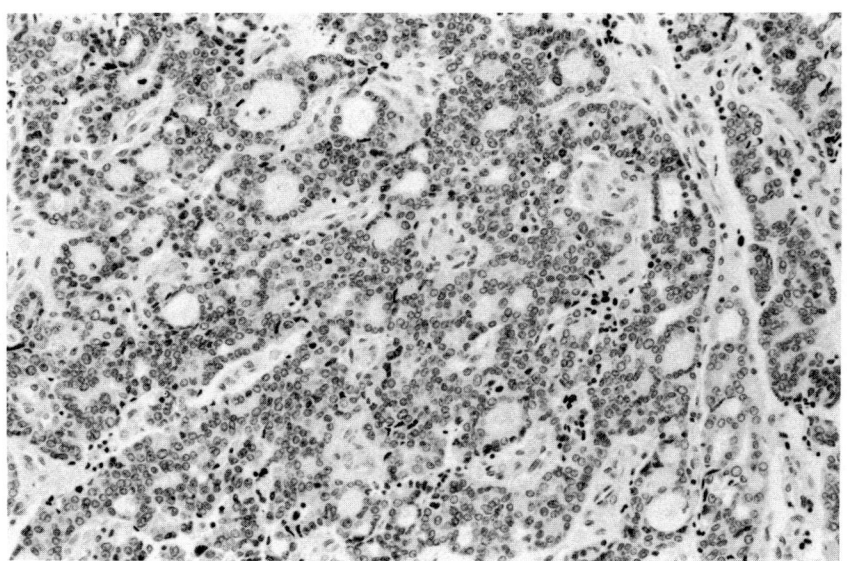

Figure 9.32 Well-differentiated follicular carcinoma forming small follicles and cribriform structures (H&E, × 160).

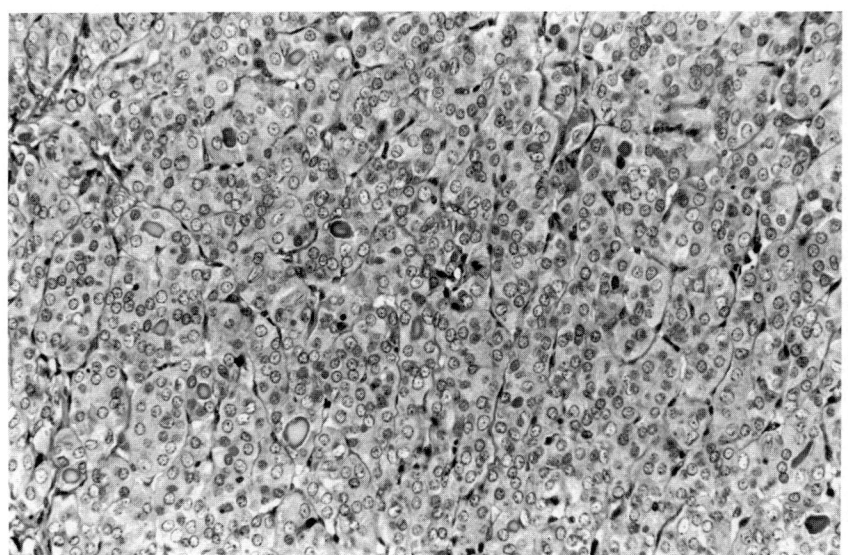

Figure 9.33 Well-differentiated follicular carcinoma with a mainly solid structure, consisting of small solid nests separated by a thin connective tissue septa. Occasional small follicles with dense colloid are also seen (H&E, × 165).

Penetration of the capsule and/or unequivocal invasion of blood vessels within or immediately outside the tumour capsule are the features generally regarded as the prerequisites for a malignant diagnosis (Franssila *et al.*, 1985). Capsular invasion should be accepted as a criterion of malignancy only if total penetration of the capsule can be demonstrated (Fig.9.34) (Kahn and Perzin, 1983; Lang *et al.*,

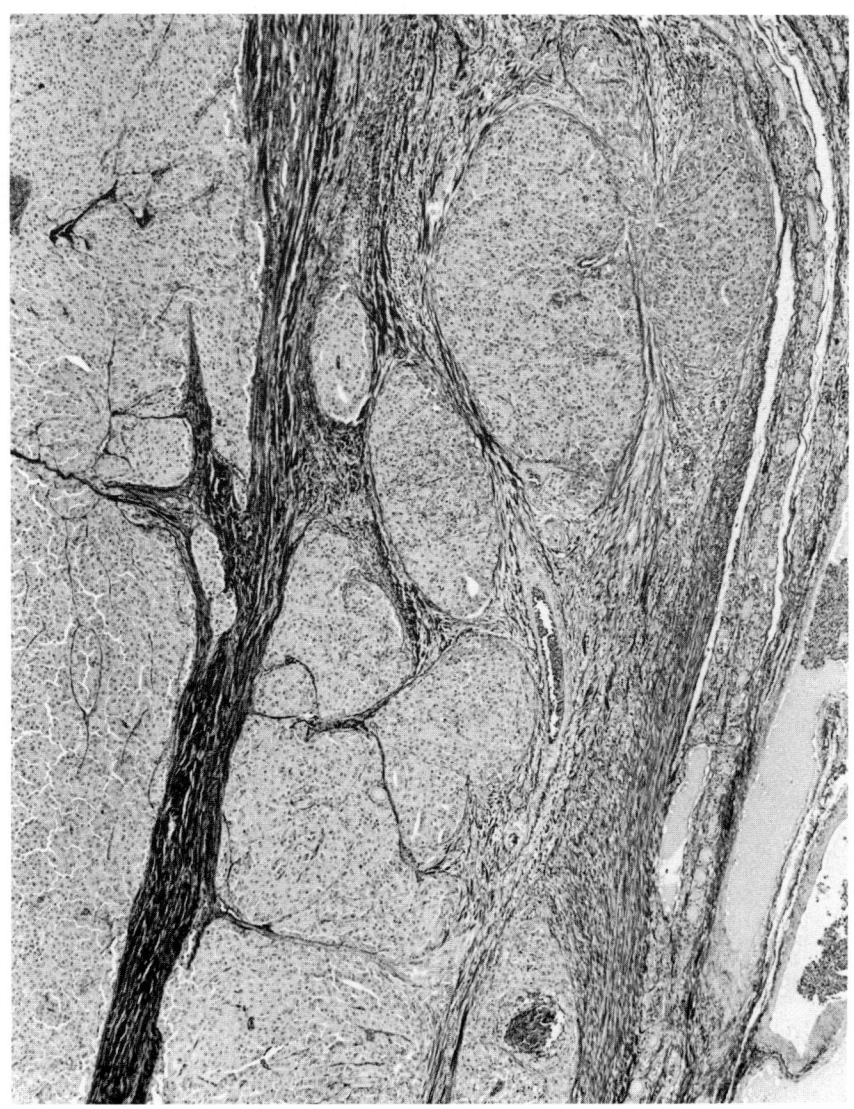

Figure 9.34 Follicular carcinoma of solid structure, marginally splitting up into small nodules which penetrate the tumour capsule (Van Gieson-Hansen stain, × 50).

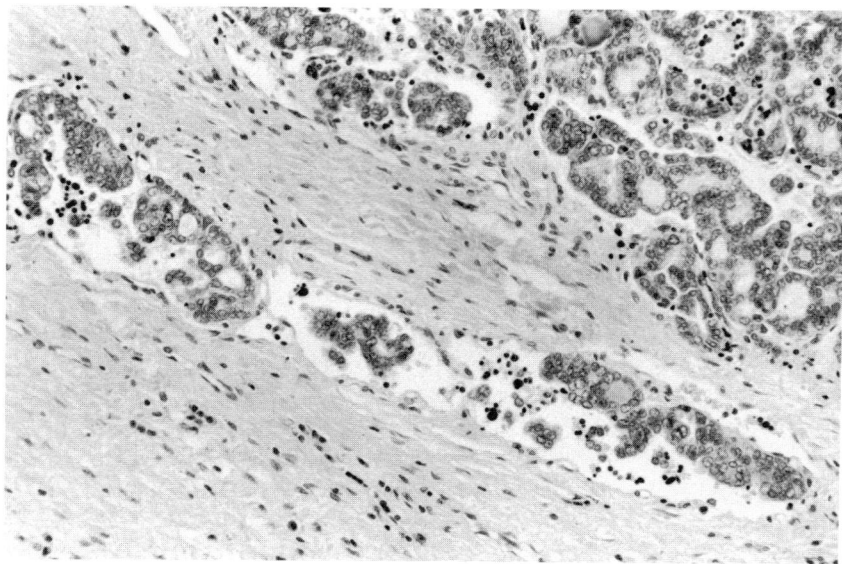

Figure 9.35 Follicular carcinoma growing in small blood vessels in the tumour capsule. Although some clusters of tumour cells are lying freely in the lumen, others are clearly attached to the inner surface of the vessel wall (H&E, × 250).

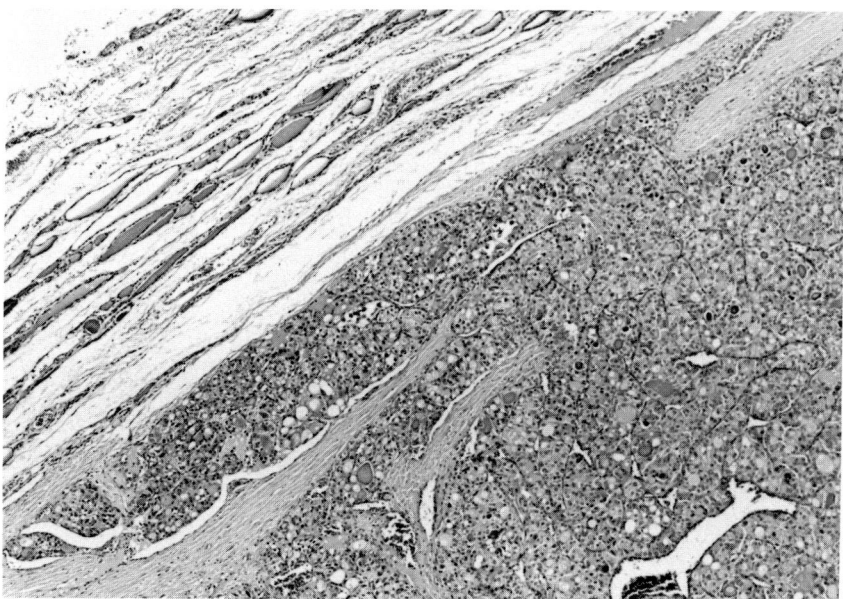

Figure 9.36 Follicular carcinoma showing unequivocal invasion of a large blood vessel in the tumour capsule. There is direct continuity between the main tumour mass and its extension which is bulging into, and almost occluding, the vessel lumen (H&E, × 45).

1980). Smoothly contoured, discrete, often almond-shaped islands of tumour within the capsule, tongues of tumour extending into the capsule but without penetrating it, or pressing the wall of a blood vessel causing incipient invagination into the lumen but without penetrating the wall (Figs 8.3 and 8.9) should be regarded as an observandum, indicating that additional blocks should be taken from the capsular region for histological examination (Hazard and Kenyon, 1954).

Blood vessel invasion appears to be a more important criterion of malignancy than infiltration of the capsule (Hazard and Kenyon, 1954; Lang *et al.*, 1980). Only invasion into blood vessels within or immediately outside the capsule should be regarded as diagnostic of carcinoma (Franssila *et al.*, 1985). Invasion of blood vessels is considered to be present when a mass or cluster of tumour cells is present within the lumen, attached to the inner surface of the wall and protruding into the lumen (Figs 9.35 and 9.36). It should preferably be covered by endothelium in a manner similar to that of an ordinary thrombus. But tumour masses without endothelial cover may also be accepted as evidence of true invasion, provided such clusters are clearly attached to the wall (Fig. 9.37). Clusters of tumour cells lying freely in the lumen of the vessel can only be accepted when covered by endothelial cells; this feature excludes that they are artefactually displaced tumour fragments.

Elastin stains have been found less valuable in identifying blood vessels involved by tumour. Immunohistochemical stains for factor VIII as a marker of endothelial cells, likewise, appear to be of little help. More recently, *Ulex europaeus* I agglutinin has been recommended as a more reliable marker of endothelial cells (González-Cámpora *et al.*, 1986a; Stephenson *et al.*, 1986). Immunohistochemical analysis of this marker may be helpful in determining the true nature of structures surrounding solid nests of tumour, and appearing equivocal in routine-stained sections.

9.3.3 *Immunohistochemical and electron microscopic examination and DNA analysis*

Thyroglobulin immunoreactivity is helpful to confirm the thyroid origin of metastatic follicular carcinoma (Plate 2) (Franklin *et al.*, 1982). Positive staining for thyroglobulin and negative staining for calcitonin may be useful to distinguish solid forms of follicular carcinoma and medullary carcinoma, especially when the latter shows atypical features and lacks amyloid. In fact, it may be worthwhile analysing those tumours having a solid component with antibodies against both follicular cell markers (such as thyroglobulin) and neuroendocrine cell markers (such as chromogranin A, calcitonin, somatostatin and neurotensin). Such

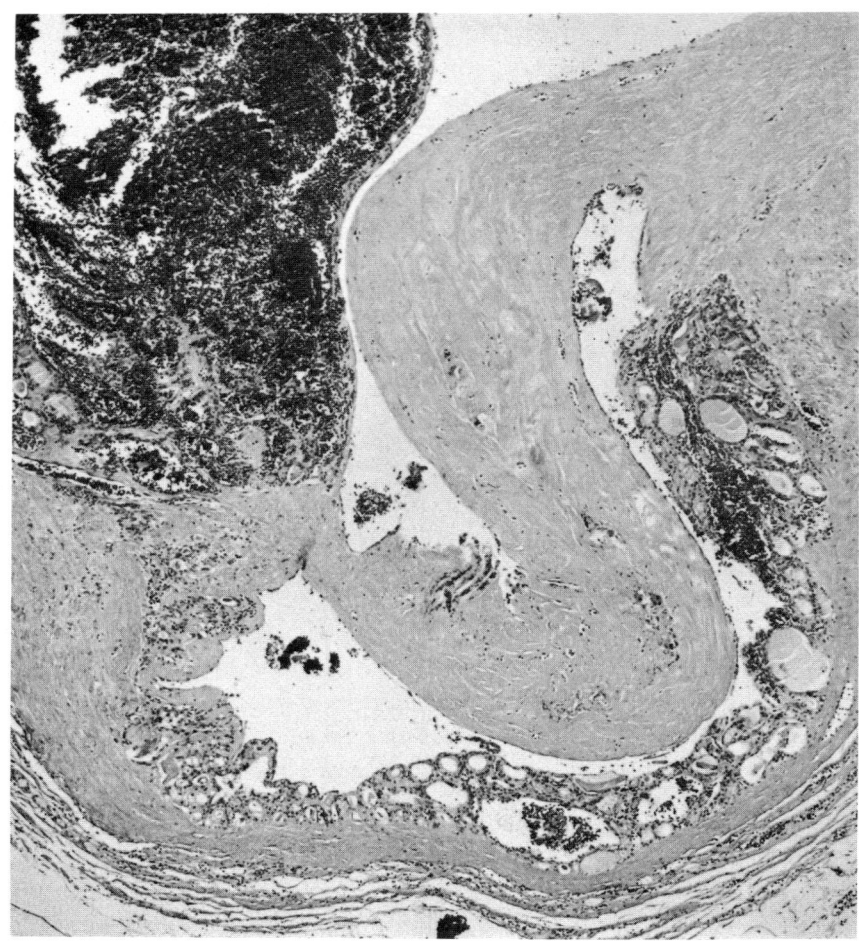

Figure 9.37 Highly differentiated follicular carcinoma with cystic degeneration, from a 22-year-old woman. There is unequivocal invasion of a large capsular blood vessel, whose inner surface is partly covered by a border of neoplastic follicles (H&E, × 50).

tumours may, occasionally, turn out to be intermediate forms with traits of both follicular and medullary carcinoma (p. 232).

As pointed out in the section on papillary carcinoma (p. 158), immunostaining for low-molecular-weight cytokeratins may be helpful in the distinction between follicular carcinoma and the follicular variant of papillary carcinoma. The usefulness of lactoferrin and ceruloplasmin immunoreactivity in the differential diagnosis between benign and malignant follicular neoplasms remains to be investigated further (p. 33).

The electron microscopic features of follicular carcinoma are of limited diagnostic value and do not allow the distinction between follicular adenoma and carcinoma (Sobrinho-Simões *et al.*, 1985).

DNA aneuploidy appears to be common in histologically and clinicaly benign follicular tumours (Joensuu *et al.*, 1986). The prognosis of aneuploid follicular adenomas does not seem to be inferior to that of diploid adenomas. Histologically and clinically malignant follicular neoplasms may be either diploid or aneuploid. Thus, a diploid modal DNA content in a follicular tumour does not exclude malignancy, and aneuploidy is not an unequivocal sign of malignancy (Johannessen *et al.*, 1982).

9.3.4 *Fine-needle aspiration cytology*

Since the diagnosis of malignancy in an encapsulated follicular tumour depends on vascular invasion and/or capsular penetration and since these features will not be demonstrable in most needle biopsies, the differential diagnosis between a benign and malignant follicular lesion is not possible with this technique. About 80–90% of thyroid nodules represent a follicular lesion of some sort. The great majority of these are nodular goitres which do not need surgical intervention. When the cellularity is higher than that in the common hyperplastic nodule and when the nuclear features of papillary carcinoma are not present, the cytological diagnosis of a follicular neoplasm may be justified. This is an indication for surgical removal of the lesion. Thus, with respect to follicular lesions, fine-needle aspiration may be regarded as a valuable screening method for the selection of cases for surgical treatment, rather than a diagnostic technique (see also p. 92). Core needle biposy may be helpful in confirming the diagnosis in an advanced carcinoma (Vickery, 1981), but like fine-needle biopsy, it is of little help in distinguishing follicular adenoma from minimally invasive follicular carcinoma.

9.3.5 *Frozen section diagnosis*

Widely invasive follicular carcinoma can usually be diagnosed without difficulty by frozen section. Minimally invasive, encapsulated carcinomas however, generally cannot be differentiated from follicular adenomas, since vascular and/or capsular invasion, when present, is often not detected. It may not be practically feasible to prepare multiple bocks to cover the entire capsule for frozen section and hence, it is usually not possible to diagnose or exclude invasion. However, this diagnostic deficiency need not cause any serious clinical problems, since both lesions are effectively treated by unilateral lobectomy, or, at most, subtotal thyroidectomy. As a consequence, the frozen section procedure in cases of this category, has been abandoned at many centres.

9.3.6 Differential diagnosis

The most common differential diagnostic problem, i.e. versus follicular adenoma, has already been discussed.

Most often, papillary carcinoma and follicular carcinoma can be distinguished without difficulty. Follicular carcinoma is usually well demarcated and when markedly invasive, tends to grow in cohesive lobular or nodular masses. Papillary carcinoma is more diffusely invasive, with marked fibrosis and, in contrast to its follicular counterpart, tends to grow in the lymph vessels. However, well encapsulated papillary carcinomas do occur. Nevertheless, they usually exhibit scattered small groups of infiltrating tumour cells within or outside the capsule, features that are generally not seen in follicular carcinomas. Follicular carcinomas, like adenomas and hyperplastic nodules, may show papillary infoldings into the follicles which may erroneously be interpreted as evidence of papillary carcinoma. However, the characteristic nuclear features of papillary carcinoma are absent.

Hashimoto's thyroiditis may show marked cytological atypia, especially when hyperplastic foci are present. Similar cytological features may be seen in Graves' disease treated with antithyroid drugs or radioiodine (see also pp. 107). However, in contrast to cases of follicular carcinoma, the basic architecture of the glandular parenchyma is generally rather well preserved. In addition, nuclear and cytological polymorphism as pronounced as in these disorders is uncommon in follicular carcinoma, as is lymphoplasmacytic stromal infiltration.

Medullary carcinoma, when histologically typical, can easily be separated from follicular carcinoma. But when medullary carcinoma is devoid of amyloid it may be difficult to distinguish from solid and trabecular variants of follicular carcinoma. As mentioned previously, such lesions should be analysed immunohistochemically, with antibodies against thyroglobulin and neuroendocrine markers, which will usually solve the differential diagnostic problem. The relationship between follicular and medullary carcinomas will be discussed in section 9.7.

9.4   Poorly differentiated carcinoma

There is growing evidence of the existence of a group of thyroid carcinomas that, from clinical prognostic and morphological points of view, may be positioned between the less aggressive group of highly

differentiated papillary and follicular carcinomas and the highly
malignant group of undifferentiated or anaplastic carcinomas.
Carcangiu et al. (1984) have identified such a group under the name
of insular carcinoma. Sakamoto et al. (1983) also stress the existence
of a highly aggressive form of poorly differentiated thyroid carcinoma,
that is more aggressive than well-differentiated carcinoma, but rather
mild compared with undifferentiated carcinoma.

These tumours tend to occur in middle-aged adults with a
female/male ratio of 2:1. They behave aggressively. Of the 24 patients
reported by Carcangiu et al. (1984), only four were alive on follow-up
after 1–8 years, and 21 had developed recurrent or metastatic disease.
Metastases occur mainly in lymph nodes, lungs and bone.

These tumours have characteristic histological features (Figs 9.38
and 9.39), showing a predominantly solid growth pattern with
formation of solid nests or 'insulae' of tumour cells, and sometimes
also trabecular structures. Due to extensive necrosis, a
'peritheliomatous' appearance may be seen, with preservation of the
tumour cells as a rim around the blood vessels. Sometimes the 'insulae'
consist of closely packed small follicles or show cribriform patterns.
Carcangiu et al. (1984) pointed out that thin clefts are often created
between the tumour cell nests and the stroma due to artificial retraction from the stroma. Tumour cells are typically small, somewhat
basophilic and rather monotonous in appearance, and may show many
mitoses. The tumour has marked infiltrative properties and tends to
invade both blood and lymph vessels (Fig. 9.40). The overall
microscopic picture may simulate that of medullary carcinoma, but
Carcangiu et al. reported negative immunostaining for calcitonin (see
also p. 212). Conversely, thyroglobulin is invariably present (Fig. 9.41)
(Rosai et al., 1985).

The exact position of this tumour in the classification of thyroid
neoplasia has been debated. Many cases show areas of follicular
differentiation and may histogenetically be regarded as examples of
a poorly differentiated type of follicular carcinoma. But some cases
show gradual transitions towards areas of typical papillary carcinoma,
and rather fit within the papillary group. Still others, however, may
not reveal their histogenetic relationship. Interestingly, some examples
of the intermediate type of thyroid carcinoma with
immunohistochemical features of both follicular and medullary
carcinoma, show morphological similarities to tumours of the poorly
differentiated group, as defined here. This group is possibly rather
heterogeneous, comprising less diffentiated forms of several tumour
types. From a clinical stand-point, however, the important thing
is to report all these cases as poorly differentiated carcinomas,
implying the need for aggressive therapy. This may include radical

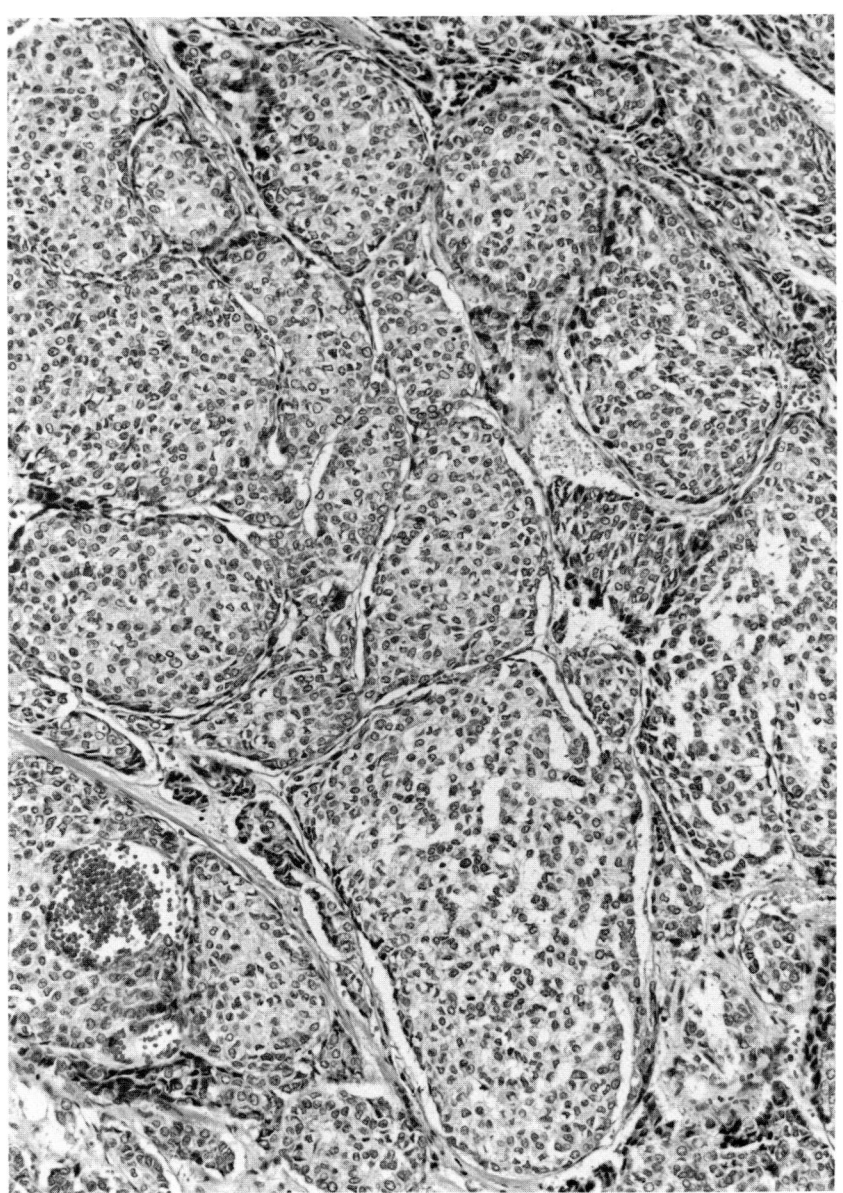

Figure 9.38 Poorly differentiated thyroid carcinoma ('insular' type). A solid growth pattern is formed by compact nests of highly atypical tumour cells. Note the thin clefts between the tumour alveoli and the stroma (H&E, × 125).

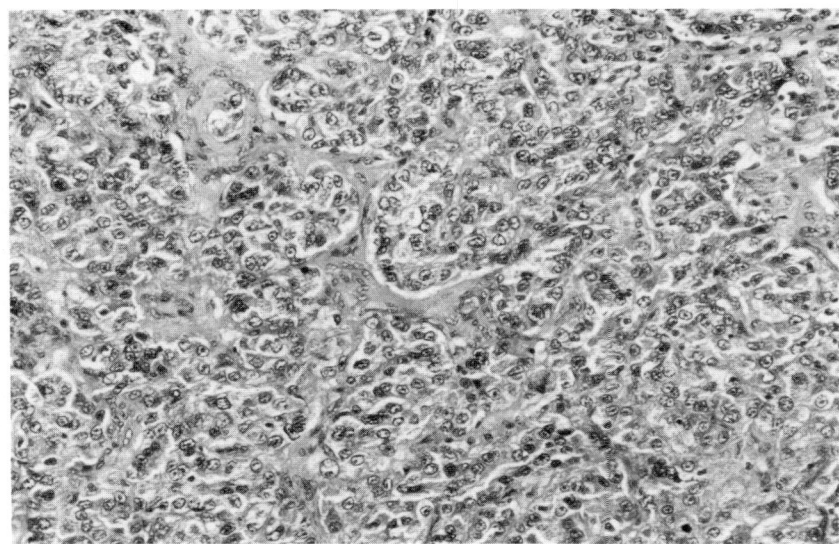

Figure 9.39 Poorly differentiated thyroid carcinoma. Less distinct alveolar pattern; small, basophil and highly atypical tumour cells are lying in irregular clusters and trabeculae (H&E, × 165).

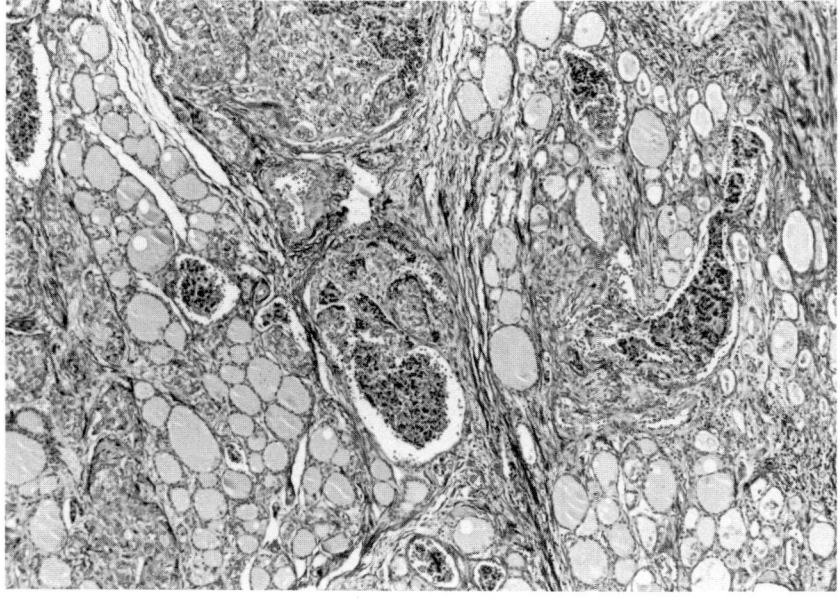

Figure 9.40 Poorly differentiated thyroid carcinoma. The picture shows extensive blood and lymph vessel invasion in the thyroid parenchyma surrounding the primary lesion (H&E, × 45).

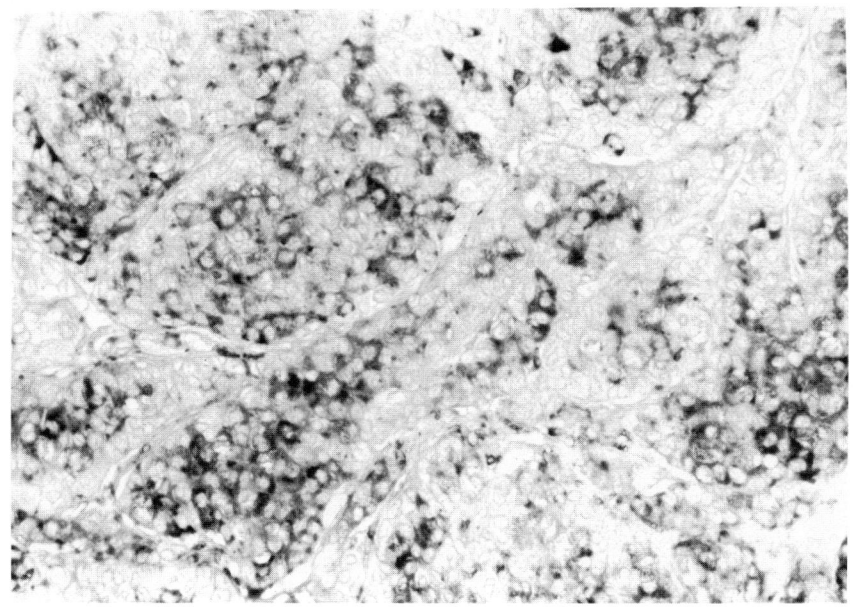

Figure 9.41 Poorly differentiated thyroid carcinoma. The tumour tissue shows a patchy thyroglobulin immunoreactivity of varying intensity (ABC technique, × 165).

surgery, radical neck dissection and administration of radioiodine, and perhaps even supplementary measures such as external radiation and systemic chemotherapy, as has been suggested by Carcangiu *et al.* (1984).

9.5   Oxyphil cell tumours

Thyroid oxyphil cell tumours still remain a controversial subject regarding pathological diagnosis and classification, prognosis and treatment (Flint and Lloyd, 1990). These tumours are composed predominantly or exclusively of mitochondrion-rich, oxyphil cells (Fig. 9.42) closely related to the oxyphil cells found in certain non-neoplastic thyroid disorders e.g. Hashimoto's thyroiditis (p. 55), nodular goitre (p. 87) and Graves' disease (p. 106). They are often also called Hürthle cell tumours. The term 'Hürthle cell' however, is a misnomer, since the cells that Hürthle described in 1894 were in fact canine parafollicular cells and not the oxyphil cells under discussion, which were instead correctly described by Askanazy in 1898 (see also p. 8). Consequently, 'Askanazy cell tumours' should be the appropriate term, a fact

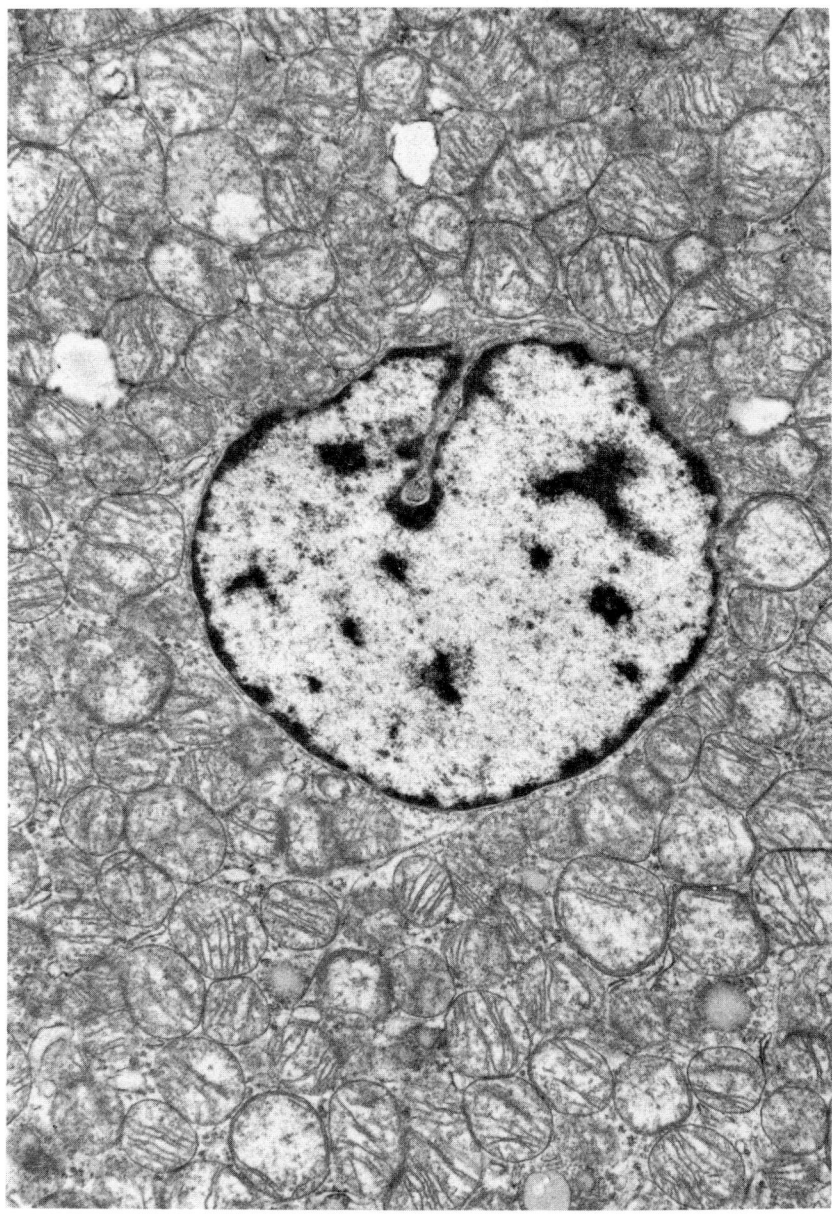

Figure 9.42 Electron micrograph of oxyphil thyroid carcinoma. Portion of an oxyphil tumour cell whose cytoplasm is packed with enlarged mitochondria (× 11 300).

that unfortunately has been given little attention in the literature. The term 'eosinophil large cell tumour' has recently also been suggested for these tumours. This is because they should be separated from lesions formed by 'ergastoplasm-rich cells', which are also eosinophilic but have been shown to contain abnormal amounts of rough endoplasmic reticulum and few mitochondria. 'Ergastoplasm-rich cells' are smaller than the typical Askanazy cells (Böcker *et al.*, 1978; Krayen-bühl and Hedinger, 1985). In Europe, 'oxyphil cell tumour' is the term most often used, and is the one also adopted by the World Health Organization.

There is still much debate about the nature of oxyphil tumours. Oxyphil cell transformation has been interpreted by some workers as a secondary inconsequential phenomenon that can occur in any pre-existing, well differentiated follicular cell tumour and does not affect the behaviour of the tumour (Woolner *et al.*, 1961; Hazard, 1968; LiVolsi and Merino, 1981; Hedinger *et al.*, 1988). Hence, oxyphil cell tumours are regarded as morphological variants of follicular adenoma, follicular carcinoma and papillary carcinoma, depending on their architectural growth pattern. The other view, to which this author adheres, is that thyroid tumours which are composed predominantly or exclusively of oxyphil, mitochondrion-rich cells, should be regarded as a separate entity having distinct features which are more dependent on specific characters of the oxyphil cell itself, rather than on its architectural arrangement (Tollefsen *et al.*, 1975; Rosai and Carcangiu, 1984).

Oxyphil cell tumours are rare, representing approximately 5–9% of all thyroid neoplasms (González-Cámpora *et al.*, 1986b; Flint and Lloyd, 1990). However, reported frequency figures vary considerably. This may be due to difficulties in making a precise distinction between a neoplasm and a hyperplastic nodule of oxyphil cells. Oxyphil tumours are usually encapsulated solitary growths. Non-neoplastic, oxyphil cell nodules, on the other hand, are most often unencapsulated and multiple. They are usually part of a multinodular goitre, or occur in a background of multifocal aggregates of oxyphil cells, associated with Hashimoto's lymphocytic thyroiditis. One should be reluctant to classify a tumour-like nodule of oxyphil cells as a neoplasm in this setting, provided the lesion has not a distinct capsule.

Oxyphil cell tumours mainly occur in middle-age and predominantly in females (González-Cámpora *et al.*, 1986b). Carcinomas tend to present about a decade later than their benign counterparts (González-Cámpora *et al.*, 1986b; Johnson *et al.*, 1987) and show a moderate degree of malignancy; the 5-year survival rate being 20– 40% in major reported series (Tollefsen *et al.*, 1975; Gundry *et al.*, 1983; Har-El *et al.*, 1986).

Metastases occur most often in the lungs and bones and more rarely in regional lymph nodes (Watson *et al.*, 1984; Har-El *et al.*, 1986).

9.5.1 Macroscopic features

The tumours are usually encapsulated growths, typically friable with homogeneous, tan cut surfaces. They are usually solitary, although multifocal lesions may occur (Gosain and Clark, 1984). Benign tumours tend to separate easily from the surrounding parenchyma. Malignant lesions may show signs of invasive growth macroscopically, and tend to have lobulated, less well-defined margins. Oxyphil tumours subjected to recent fine-needle aspiration frequently show extensive necrosis, sparing only the marginal parts of the tumour (Bauman and Strawbridge, 1983; Keyhani-Rofagha *et al.*, 1990; Layfield and Lones, 1991). The causal mechanism of this phenomenon is unknown. However, since oxyphil tumours are highly vascularized and enveloped by a rigid capsule, it is conceivable that fine-needle aspiration may easily cause haemorrhage and possibly a rise in the pressure within the tumour, compromising the circulation and resulting in necrosis.

9.5.2 Microscopic features

Oxyphil thyroid tumors should, by definition, exclusively or almost exclusively be composed of oxyphil, mitochondrion-rich epithelial cells. They tend to be cellular, with sparse amounts of stroma which tends to be highly vascularized. They may show solid areas with trabeculae or alveoli of tumour cells, but follicular and papilliferous patterns are also found (Figs 9.43–9.45).

Papillary-like structures may occur as a result of degeneration of cells in central parts of solid alveoli and persistence of cells surrounding the stromal trabeculae. Likewise, papillary patterns may form as a result of infoldings of tumour cells into large follicles. However, true papillae are rare as a dominant feature of the tumour. When present, such papillae usually lack the nuclear features typical of conventional papillary carcinoma (Figs 9.46 and 9.47). Inspissated colloid with calcification resembling psammoma bodies may be seen, but usually do not occur outside the follicles. As noted above, haemorrhage and necrosis are common phenomena, especially in malignant tumours.

Tumour cells are polygonal or oval, with abundant eosinophilic, granular cytoplasm (Fig. 9.44). With phosphotungstic acid haematoxylin (PTAH) stain, the cytoplasmic granules are clearly visualized

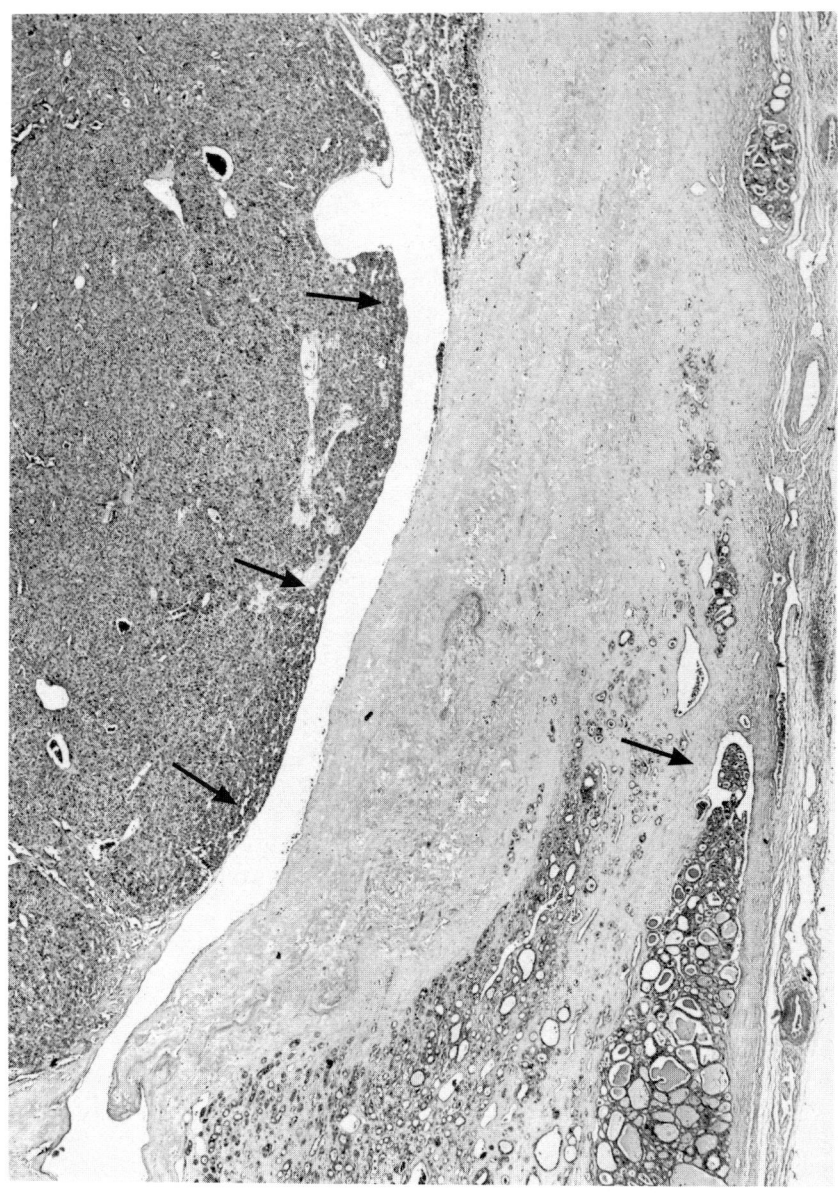

Figure 9.43 Oxyphil thyroid carcinoma with solid (left) and follicular (right) structures showing blood vessel invasion (arrows). The solid component is seen to break, on a broad front, into a large blood vessel in the inner capsular area. Follicular tumour structures, in addition, are seen to invade and partially fill the lumen of a smaller vessel in the outer part of the thickened capsule (H&E, × 25).

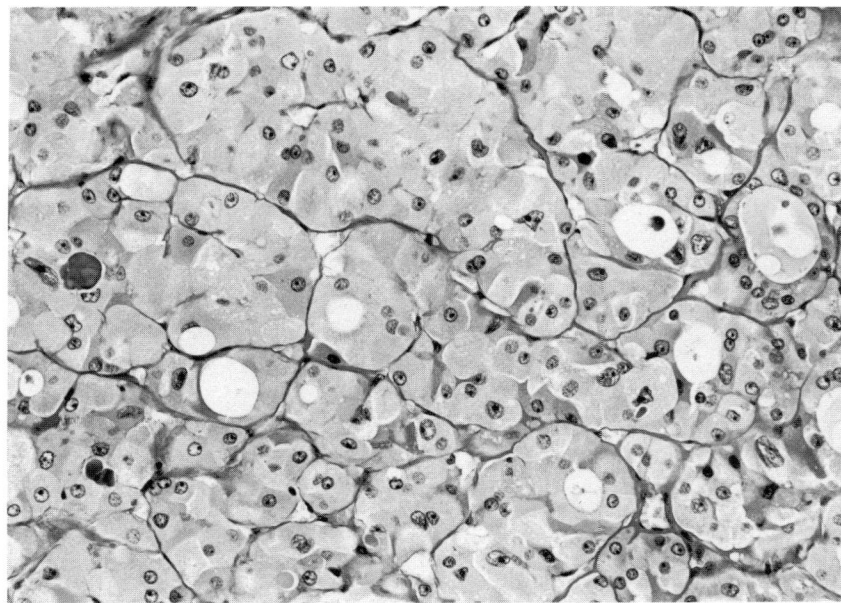

Figure 9.44 Same tumour as shown in Fig. 9.43. Detail of the tumour showing solid nests and occasional follicles, exclusively composed of typical oxyphil cells (H&E, × 170).

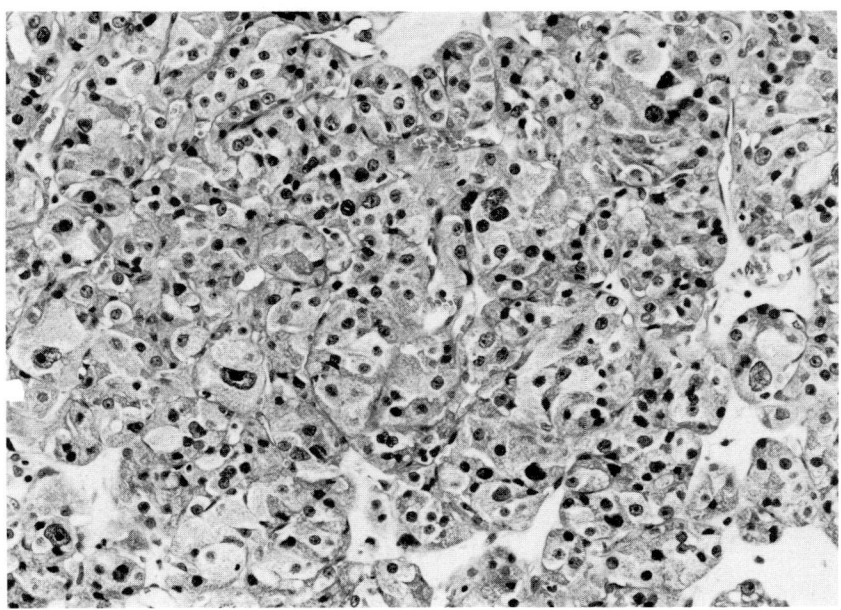

Figure 9.45 Same case as in Figs 9.43 and 9.44. Area with pronounced cellular and nuclear pleomorphism (H&E, × 170).

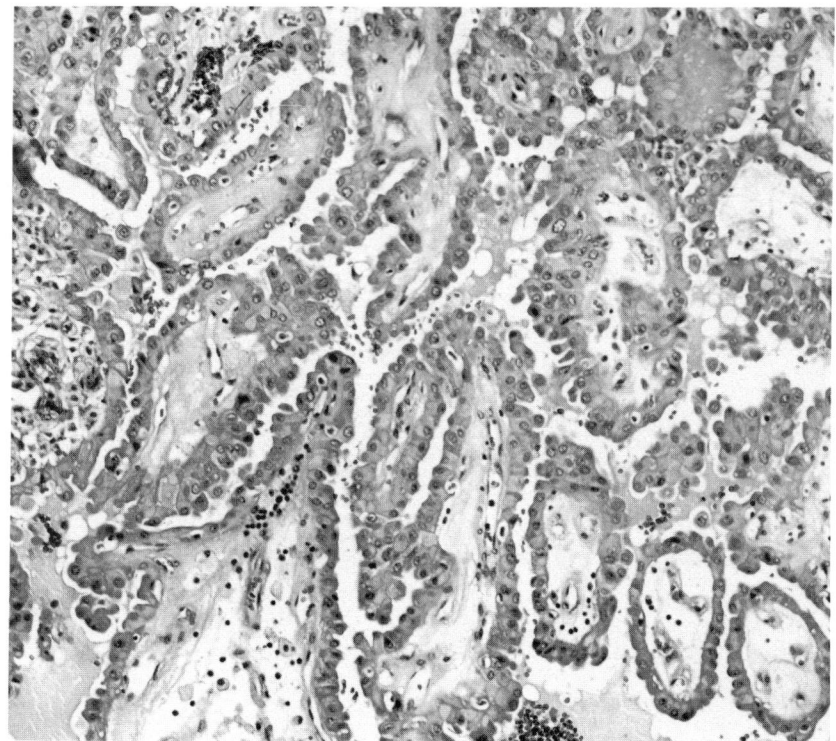

Figure 9.46 Oxyphil thyroid carcinoma. This example had a combined solid, follicular and papillary growth pattern. The picture shows an area of well-developed papillae with hyalinized stromal stalks, covered by oxyphil cells (H&E, × 125).

by their dark-blue colour. By electron microscopy, these granules are seen as closely packed mitochondria which appear abnormal in size and shape (Fig. 9.42) (Feldman *et al.*, 1972; Nesland *et al.*, 1985). In some oxyphil tumours, so-called clear cells may be found. The tumour cells show progressive clearing of their cytoplasm, which is associated with a dispersion and rarefaction of the eosinophilic granules. Ultrastructurally, this change is considered to be due to dilatation of the mitochondria and loss of their internal matrix (Dickersin *et al.*, 1980; Civantos *et al.*, 1984).

Cytological and nuclear pleomorphism may be marked in oxyphil tumours, sometimes with occurrence of bizarre cell forms (Fig. 9.45). These cytological features do not have any bearing on the behaviour of the tumour and should not be regarded as signs of malignancy (Bondeson *et al.*, 1983).

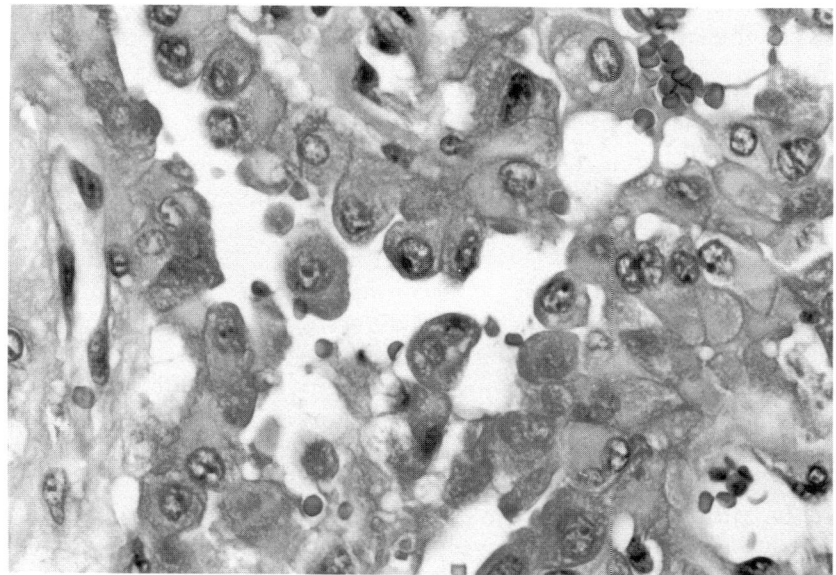

Figure 9.47 Detail of papillary portion of the tumour shown in Fig. 9.46. Pleomorphic oxyphil tumour cells whose nuclei lack the features characteristic of ordinary papillary carcinoma (H&E, × 500).

Some workers, though not the present author, believe that all oxyphil thyroid tumours should be considered malignant or potentially malignant (Thompson *et al.*, 1974; Miller *et al.*, 1983). According to our experience, these tumours are not especially prone to assume a malignant course and we have found no reason to treat them differently from other well-differentiated epithelial neoplasms (Bondeson *et al.*, 1981). Tumours with solid and follicular patterns should be evaluated according to the same histopathological criteria as those valid for differentiated follicular tumours of the ordinary type. Thus, unequivocal capsular penetration and/or blood vessel invasion, or metastases are the prerequisites for a malignant diagnosis (Fig. 9.43). Due to their rarity, little is known about the behaviour of oxyphil tumours with a predominantly papillary pattern. The present author is inclined to regard as benign those papillary lesions that are encapsulated, without signs of invasion and that lack the nuclear features of conventional papillary carcinoma. Occasional examples of papillary oxyphil tumours have been reported with ground-glass nuclei, nuclear grooves and other nuclear features typical of classic papillary carcinoma, at least focally (González-Cámpora *et al.*, 1986b). However, such neoplasms have been non-encapsulated and clearly invasive, and have shown a clinical behaviour similar to that of the ordinary papillary carcinoma.

9.5.3 *Immunohistochemical findings and DNA studies*

Oxyphil thyroid tumours may show focal immunoreactivity for thyroglobulin, but the reaction is generally weaker than that seen in conventional follicular cell tumours (Albores-Saavedra *et al.*, 1983; González-Cámpora *et al.*, 1986b; Johnson *et al.*, 1987). Some oxyphil tumours may be entirely negative (Ljungberg *et al.*, 1984; Logman and Jobsis, 1984). There appears to be no difference in the immunoreactivity pattern between benign and malignant tumour forms. Carcino-embryonic antigen (CEA) immunoreactivity in oxyphil tumours has been variable (Nesland *et al.*, 1985; Johnson *et al.*, 1987), probably due to differences in the affinities of the antibodies used.

DNA studies on oxyphil tumours indicate that analysis of DNA ploidy patterns appears to have limited diagnostic value, being unable to exclude malignancy. However, aneuploidy has been associated with a high probability of invasive growth (Bondeson *et al.*, 1986; Flint *et al.*, 1988).

9.6   Clear cell tumours

Some thyroid epithelial neoplasms consist partly or exclusively of cells with optically clear cytoplasm that is usually sharply outlined by distinct cell membranes. It has been commonly believed that such tumours may represent a specific entity. Increasing evidence in recent years, however, indicates that a clear cell change may result from a variety of mechanisms, that usually do not change the biology of the lesion (Schröder and Böcker, 1986). Clear cell change has been seen in follicular adenomas (p. 123), follicular, papillary and anaplastic carcinomas, oxyphil tumours (p. 196) and, rarely, even in medullary carcinomas (Carcangiu *et al.*, 1985d; Landon and Ordoñez, 1985). In addition, as has been mentioned in previous chapters, clear cells may also occur in Hashimoto's thyroiditis (p. 56), Graves' disease (pp. 106) and nodular goitre (p. 97).

Oxyphil cell tumours most frequently undergo clear cell change, as pointed out in section 9.5. Here the cytoplasmic clearing is due to swelling of the mitochondria and loss of their matrix (Dickersin *et al.*, 1980; Saleiro *et al.*, 1981; Flint and Lloyd, 1990). The process may take place evenly, throughout the cytoplasm of the cell, but it may also involve only a part of the cell. This has been described in some oxyphil papillary carcinomas where the clearing was present in the apical parts, but the basal portions of the cells had retained their oxyphil appearance (Carcangiu *et al.*, 1985d). Increased TSH stimulation has been suggested as a cause of the clear cell change (Civantos *et al.*, 1984).

Clear cell change in non-oxyphil follicular tumours may also be caused by swelling of mitochondria. Dilation of the rough endoplasmic reticulum may be another mechanism, resulting in cytoplasmic vacuolation seen by light microscopy (Ishimaru *et al.*, 1988). In some follicular tumours, malignant as well as benign, the clear cells show distinctive signet-ring or lipoblast-like appearances. In these instances, clearing of the cytoplasm is usually caused by abnormal accumulation of thyroglobulin or its degradation products (Gherardi, 1987). Ultrastructurally, intracytoplasmic vacuoles, lined by microvilli and suggesting intracytoplasmic follicular lumina, have been described in these cases. The signet-cell-simulating microfollicular adenoma, described by Mendelsohn (1984) and Rigaud *et al.* (1985), belongs to this category of clear cell tumours. Rarely, signet-ring appearance may be due to cytoplasmic accumulation of fat (Schröder *et al.*, 1984b).

Clear cell change in papillary carcinomas seems most often to be due to glycogen accumulation, although mitochondrial cystic dilation may also be a feature (Variakojis *et al.*, 1975).

Glycogen accumulation as a cause of cytoplasmic clearing has also been encountered in undifferentiated thyroid carcinomas, often in association with squamous differentiation (Fisher and Kim, 1977).

The main practical concern regarding clear cell thyroid carcinomas lies in their differential diagnosis from parathyroid tumours and, most important, a metastasis of clear cell renal carcinoma. Immunostaining for thyroglobulin is the most helpful technique to separate primary clear cell carcinoma of the thyroid from the other two. When interpreting the staining pattern, one should be aware of the fact that metastatic tumours in the thyroid may show focal immunoreactivity for thyroglobulin, due to non-specific diffusion of the stain or secondary uptake of thyroglobulin from residual normal follicular cells present within the tumour (see also p. 31). Only when the staining is widespread and diffuse should it be accepted as specific evidence of the presence of a primary thyroid clear cell tumour. In routinely stained sections, it may be impossible to make a distinction between a renal carcinoma metastasis and primary thyroid tumour. Features favouring the diagnosis of renal carcinoma metastasis include multifocality of the lesion, pronounced vascularization with many large thin-walled blood vessels, presence of tubular structures lacking colloid, but often containing blood, and absence of the fine cytoplasmic granularity, often seen in primary thyroid lesions.

9.7 Medullary carcinoma

Medullary carcinoma may be defined as a malignant epithelial neoplasm predominantly or exclusively composed of cells showing evidence of parafollicular (C) cell differentiation. It was defined as a distinct clinicopathological entity in 1959 by Hazard *et al.*, who stressed the presence of amyloid in the tumour as its most striking histopathological feature. In the early 1960s, reports appeared on the familial occurrence of thyroid carcinoma, and its frequent association with phaeochromocytoma (Sipple, 1961; Cushman, 1962; Friedell *et al.*, 1962; Nourok, 1964). In 1965, Schimke and Hartmann, and Williams emphasized that when thyroid carcinoma was familial, it was most often of the medullary type and then frequently combined with phaeochromocytoma. In 1965 a family with the syndrome was also discovered in South Sweden (Ljungberg, 1966; Ljungberg *et al.*, 1967).

Medullary carcinoma accounts for 4–12% of all cases of thyroid carcinoma (Ljungberg, 1972b). The sporadic form is the most common variant. It is nearly always a solitary lesion and the parafollicular (C) cell system in the residual parts of the gland appears unaffected. The familial form, accounting for approximately a quarter of the total number of cases (Bergholm *et al.*, 1989), is principally a multifocal process, where multiple tumours develop on the basis of a preceding general hyperplasia of the entire parafollicular cell system. In familial cases, the tumours therefore are accompanied by co-existing hyperplastic changes in the remaining parts of this system. Since the parafollicular cells are normally concentrated in the upper half of the lobes (see also p. 12), the tumours are most often found there, and typically, both lobes tend to be involved.

Sporadic medullary carcinoma usually presents in middle age (45–55 years) whereas symptomatic cases of the familial form most often become apparent a decade earlier. The group of familial cases detected by calcitonin screening are about another ten years younger (Bergholm *et al.*, 1989). In most series the sex distribution has shown a slight preponderance for females (Ljungberg, 1972b). Occasionally, medullary carcinoma may present clinically as a metastasis, usually in a cervical lymph node, the primary being an occult lesion, only revealing itself by increased serum calcitonin levels (White *et al.*, 1981; Aldabagh *et al.*, 1986). Figure 9.48 shows a tiny medullary carcinoma, only a few millimetres in size, that was discovered after repeated subserial sectioning of the thyroid gland removed from a patient who presented with a large cervical lymph node metastasis and was shown to have elevated serum calcitonin levels.

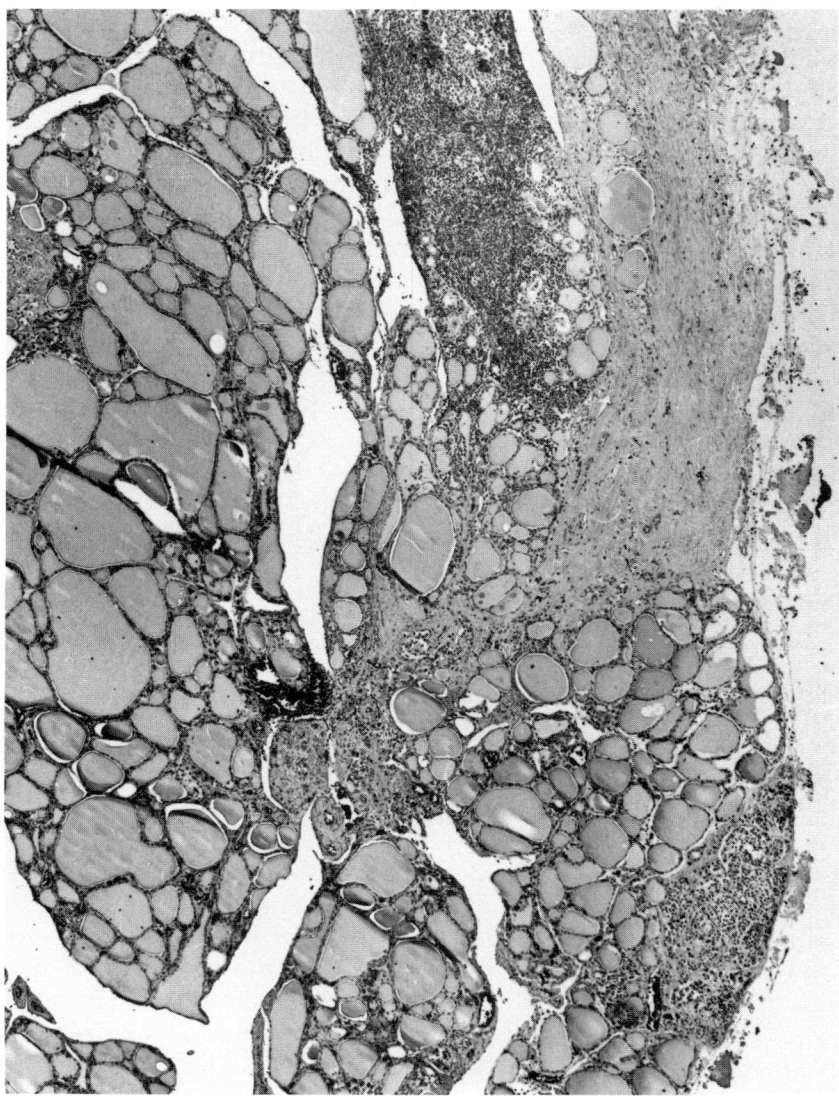

Figure 9.48 Microscopic medullary thyroid carcinoma in a 64-year-old man. The patient presented with a large cervical lymph node metastasis, histologically compatible with metastasis of medullary thyroid carcinoma. Very high serum calcitonin levels were found. Total thyroidectomy was performed. The thyroid tumour (only about 2 mm in size) was not found until the gland had been subserially sectioned. The tumour is seen to be ill-defined and located close to the surface of the gland. There is tumour infiltration within a sclerotic thickened portion of the glandular capsule (H&E, × 50).

9.7.1 Macroscopic appearance

Medullary carcinoma usually lacks a capsule, although it tends to be well demarcated, often with a slightly lobulated outer contour and shows bulging, greyish-white or tan, somewhat granular cut surfaces (Plate 19). Some tumours, however, may be ill-defined, hard and sclerotic, with radiating contours. Old tumours may show extensive calcification, and even bone formation (Ljungberg, 1972b). Today, as a result of screening procedures with calcitonin stimulation tests, the familial form is often detected at a very early stage. In such cases, the tumours may be less than a few millimetres in size and some may not be detectable macroscopically (Plate 20). Naturally, in this instance, it is essential to cut the entire lobes into thin (1 mm or so) slices and inspect the cut surfaces closely for minute lesions. All slices should be processed for later histological examination. Since the parafollicular cells are concentrated in the upper halves of the lobes, paraffin blocks from these parts should be sectioned at several levels, in order to get an adequate microscopic picture of this cell system (see also p. 27).

9.7.2 Microscopic features

Partial encapsulation of the tumour may be seen, but usually there is a clear microscopic evidence of infiltration into the surrounding thyroid parenchyma. Blood vessel invasion is common, and involvement of lymph vessels and regional lymph nodes are also frequent findings. Amyloid deposition is a typical but not invariable feature. In cases of multiple tumours, amyloid may be found in some, but not in others. Whereas the primary tumour may show such deposits, metastases may be completely devoid of the substance. When present, it is a highly diagnostic feature, hardly ever found in any of the other main types of thyroid carcinoma. The amyloid deposits are eosinophilic, amorphous, and are stained by the common amyloid staining methods, such as the alkaline Congo red technique (Puchtler et al., 1962). So stained, it also exhibits yellow and apple-green birefringence in polarized light. In the author's experience, minute amounts of amyloid are more easily detected when sections stained with alkaline Congo red are examined in the fluorescence microscope equipped with filters for rhodamine (TRITC) fluorescence. In this setting, amyloid clumps give a bright brick-red colour which clearly contrasts against the dark background (Plate 21). The substance may occasionally be seen as minute hyaline bodies within the cytoplasm of the tumour cells, but is more frequently found intermingled with the tumour cells as small extracellular clumps. In some places these clumps are seen to have coalesced into large masses, replacing the tumour tissue. It is frequently also seen deposited within the fibrous

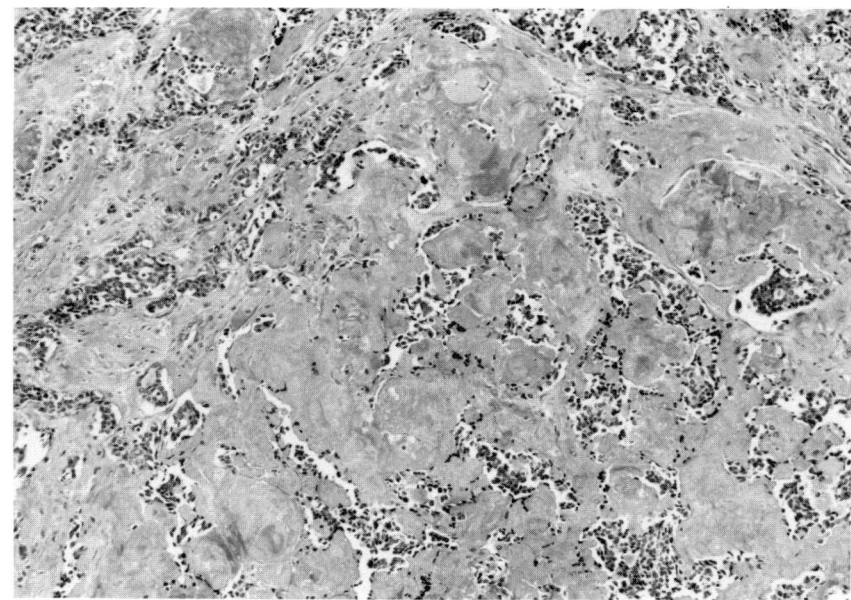

Figure 9.49 Medullary carcinoma. Trabecular growth pattern, with large amyloid deposits (H&E × 85).

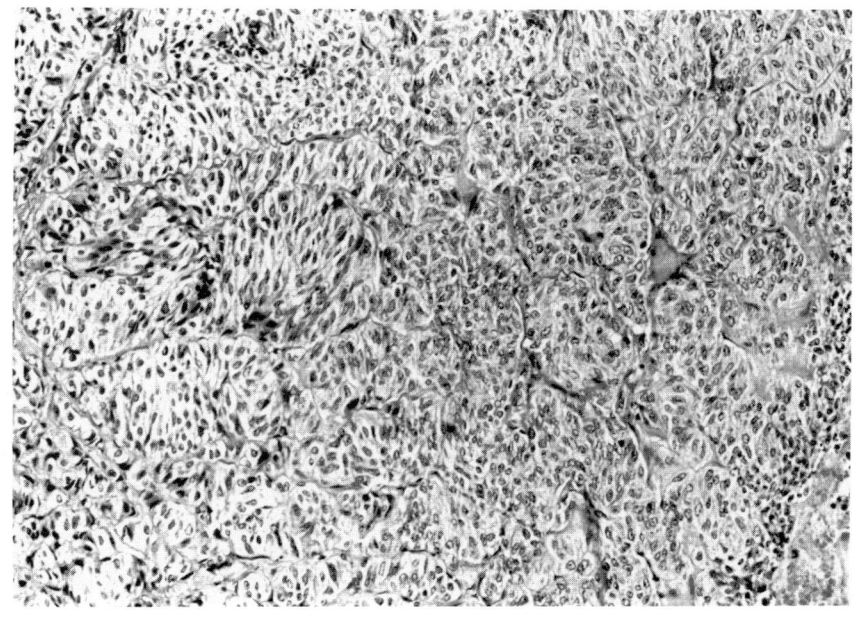

Figure 9.50 Medullary carcinoma. Alveolar growth pattern (H&E, × 165).

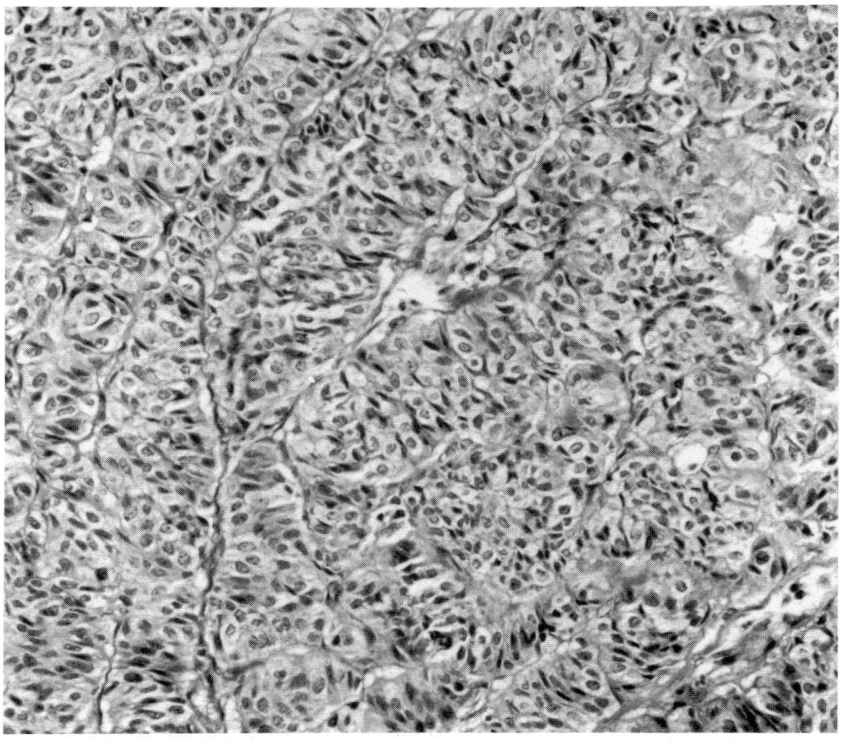

Figure 9.51 Medullary carcinoma composed of alveoli and columns of closely packed light and dark tumour cells, forming mosaic patterns. Some cells are elongated and lying in palisades (H&E, × 250).

stroma. The substance may occur in the form of spherical bodies, reminiscent of corpora amylacea and appearing as Maltese crosses when stained with alkaline Congo red and viewed in polarized light (Ljungberg, 1972b). Large masses of amyloid may have a laminated structure. Sometimes they have evoked a foreign body reaction. Old amyloid deposits may lose their stainability with Congo red and appear as amorphous hyaline material which may contain calcifications. Bone formation is seen in some cases.

Structurally, medullary carcinoma is a great imitator. The histological growth pattern may vary considerably between different parts of the tumour, and also between tumours. The most common type may be designated the trabecular growth pattern (Fig. 9.49). Globoid or trabecular amyloid deposits in varying amounts occur between the clusters of cells and in the stroma. Another, rather common type is the alveolar growth pattern formed by rounded, alveolar clusters of

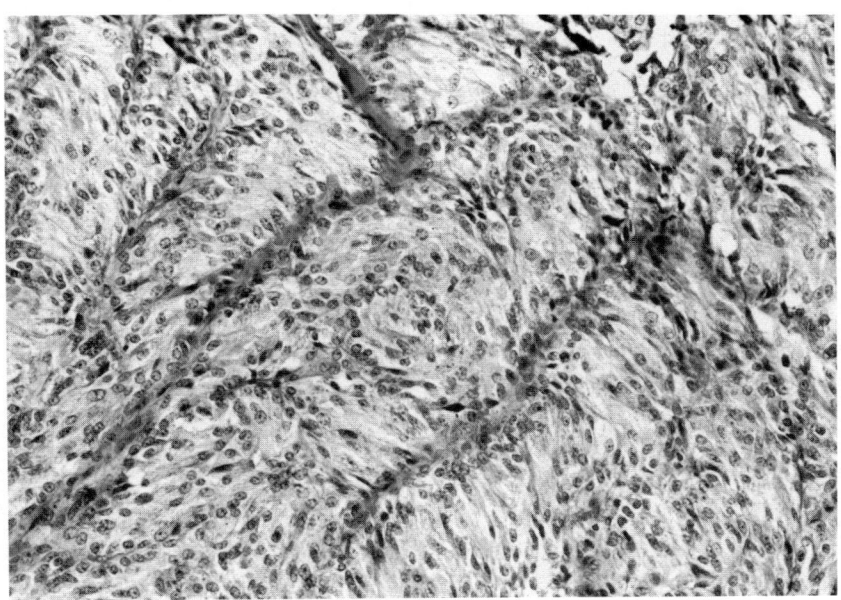

Figure 9.52 Medullary carcinoma forming columns and fascicles of elongated tumour cells, in some places with palisading nuclei, simulating a neurilemmoma (H&E, × 170).

cells, surrounded by thin fibrous stroma septa and sometimes composed of light and dark cells in mosaic patterns (Figs 9.50 and 9.51). There may be a characteristic palisading of the cells in the marginal parts of the alveoli. The tumour alveoli may form a 'Zellballen'-pattern, of the type seen in paragangliomas and phaeochromocytomas, or more irregular solid nests (insulae), as seen in a carcinoid tumour. Less frequently, a spindle-cell growth pattern may occur, usually with scanty or no amyloid deposits (Figs 9.52 and 9.53). Occasionally, medullary carcinoma may resemble an angiomatous tumour, being composed of large thin-walled channels filled with blood and lined by flattened endothelial-like tumour cells which are calcitonin-positive and argyrophilic (Figs 9.54 and 9.55).

Rarely, otherwise typical medullary carcinoma may reveal follicular and/or papillary patterns. Usually, these structures are spurious, having been formed as a result of degeneration and disappearance of tumour cells in the central parts of the cell clusters, with persistence of cells in the periphery. This phenomenon may result in irregular follicle-like structures, whose walls are formed by one or several layers of cells, or in papillary-like formations with a central stromal stalk,

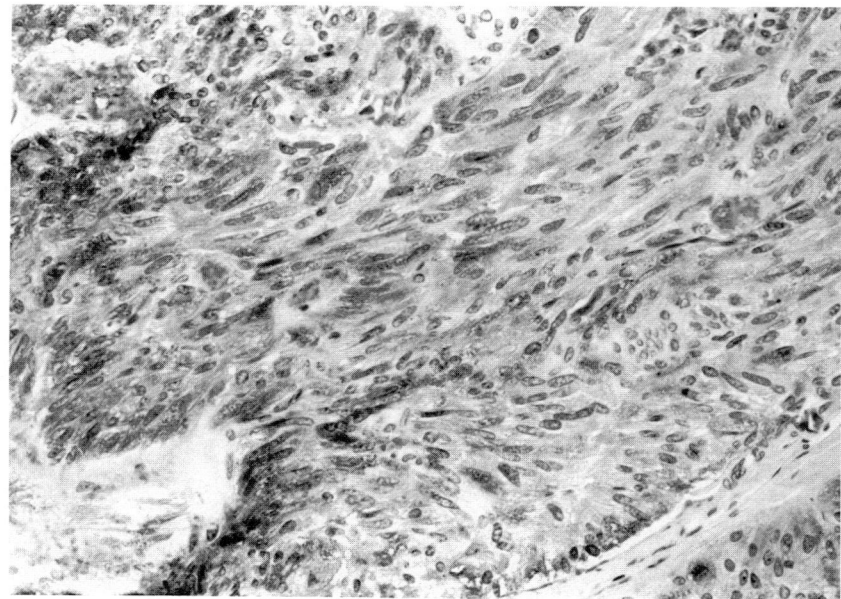

Figure 9.53 Medullary carcinoma with pronounced spindle-cell pattern (H&E, × 220).

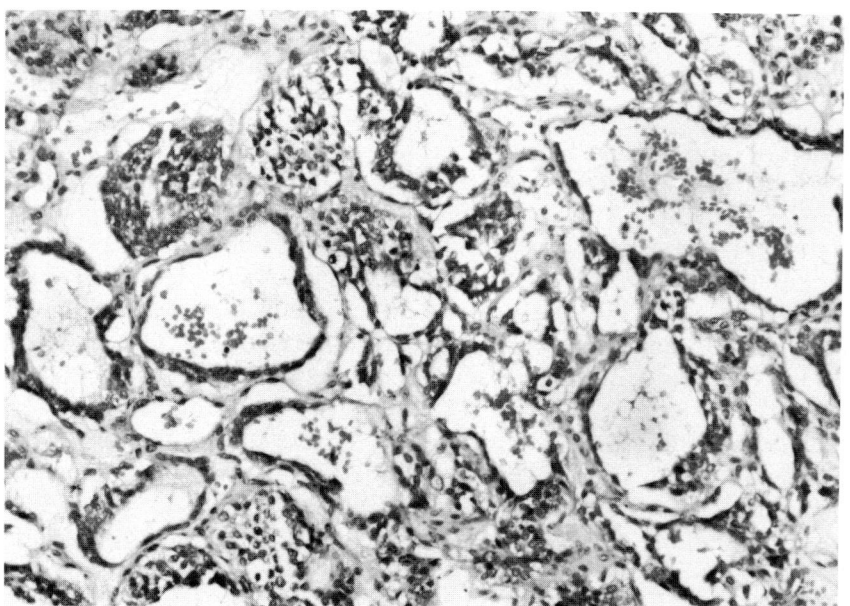

Figure 9.54 Medullary carcinoma with angiosarcoma-like features, composed of irregular spaces with red blood cells and lined by flattened atypical cells resembling endothelial cells, but in some places lying in several layers (H&E, × 160).

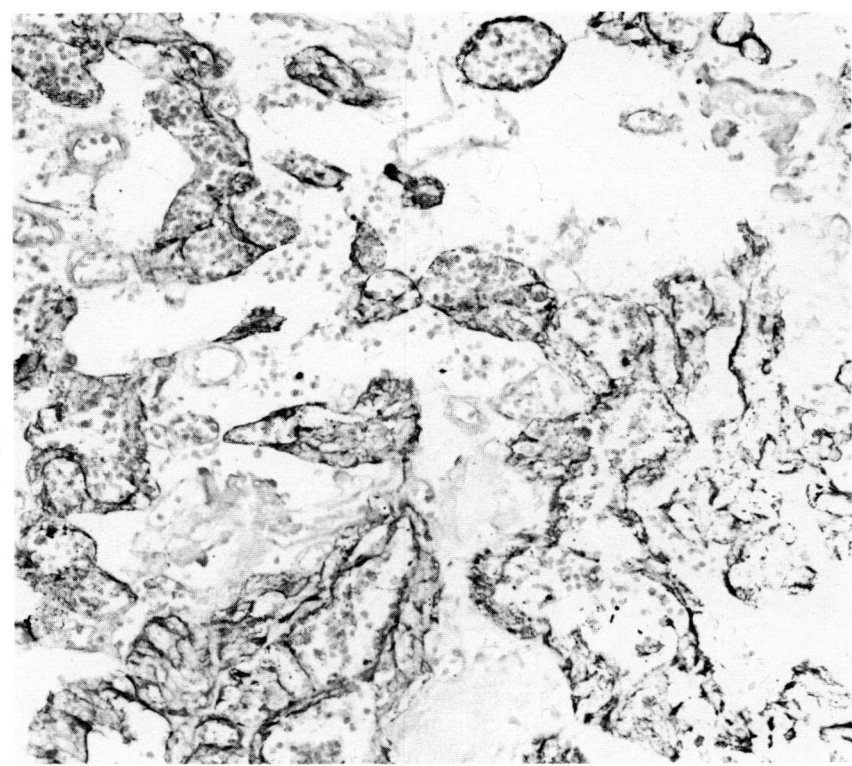

Figure 9.55 Same case as illustrated in Fig. 9.54. Cells lining the spaces are argyrophilic. By immunostaining they were also found to be calcitonin-positive (Grimelius' silver technique, × 250).

covered by one or several layers of tumour cells (Fig. 9.56) (Ljungberg, 1972b; Sambade *et al.*, 1989). Nevertheless, medullary carcinoma may occasionally form true follicles or glandular structures (Harach and Williams, 1983), as well as true papillary structures (Kakudo *et al.*, 1979). In such instances it is important to analyse the tumour ultrastructurally, as well as immunohistochemically with antibodies against thyroglobulin and neuroendocrine markers, because inter-mediate tumour forms do exist which show evidence of both follicular and parafollicular cell differentiation (p. 228). Recently, a mixed medullary–papillary carcinoma has in fact also been identified, in which the papillary component showed the cytological and nuclear features typical of classic papillary carcinoma (Albores-Saavedra *et al.*, 1990). However, when thyroglobulin-immunoreactive tumour cells

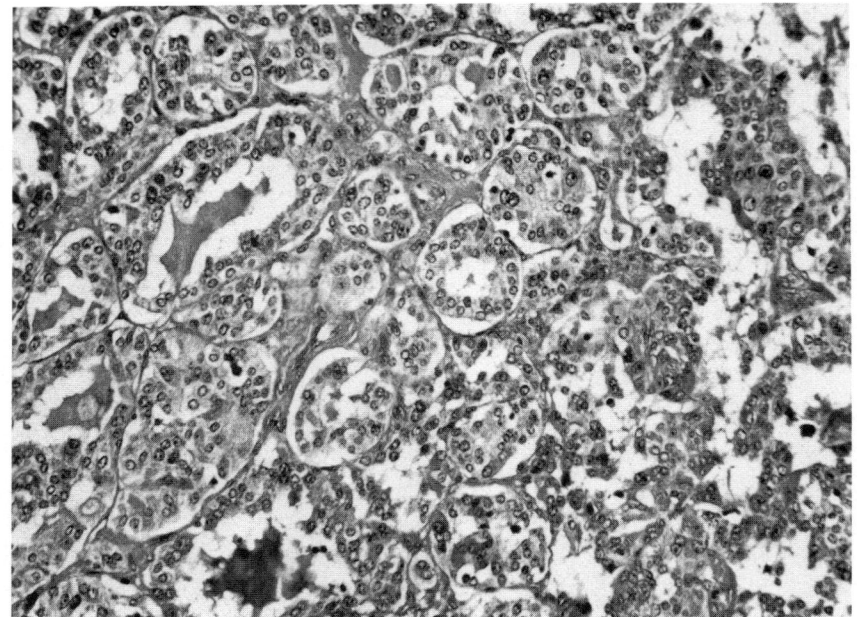

Figure 9.56 Medullary carcinoma showing pseudofollicular and pseudo-papillary growth patterns, which developed as a result of degeneration and loss of cells in the central parts of the tumour alveoli (H&E, × 180).

constitute a minority cell population (less than 25% of the total cell mass) the tumour should probably be classified within the medullary carcinoma group (Ljungberg and Nilsson, 1991; see also p. 241).

Well differentiated tumour cells are usually rather uniform, polyhedric, often with distinct plasma membranes, and have varying amounts of cytoplasm which is often finely granular and eosinophilic, sometimes more or less amphophilic (Fig. 9.57). The cells may also have markedly vacuolated or frayed cytoplasm. Nuclei usually have a fine chromatin network and small nucleoli, and they often lie eccentrically in the cells. Binucleate cells are common. Mitoses are rarely seen in the classic form of medullary carcinoma (Ljungberg, 1972b).

Although the tumour cells usually display only a slight degree of cellular and nuclear pleomorphism, focal areas may be seen where this is marked, showing highly atypical cells and even giant cells, with single or multiple bizarre, hyperchromatic nuclei (Fig. 9.58). Such features, when focal, do not seem to reflect changes in the degree of

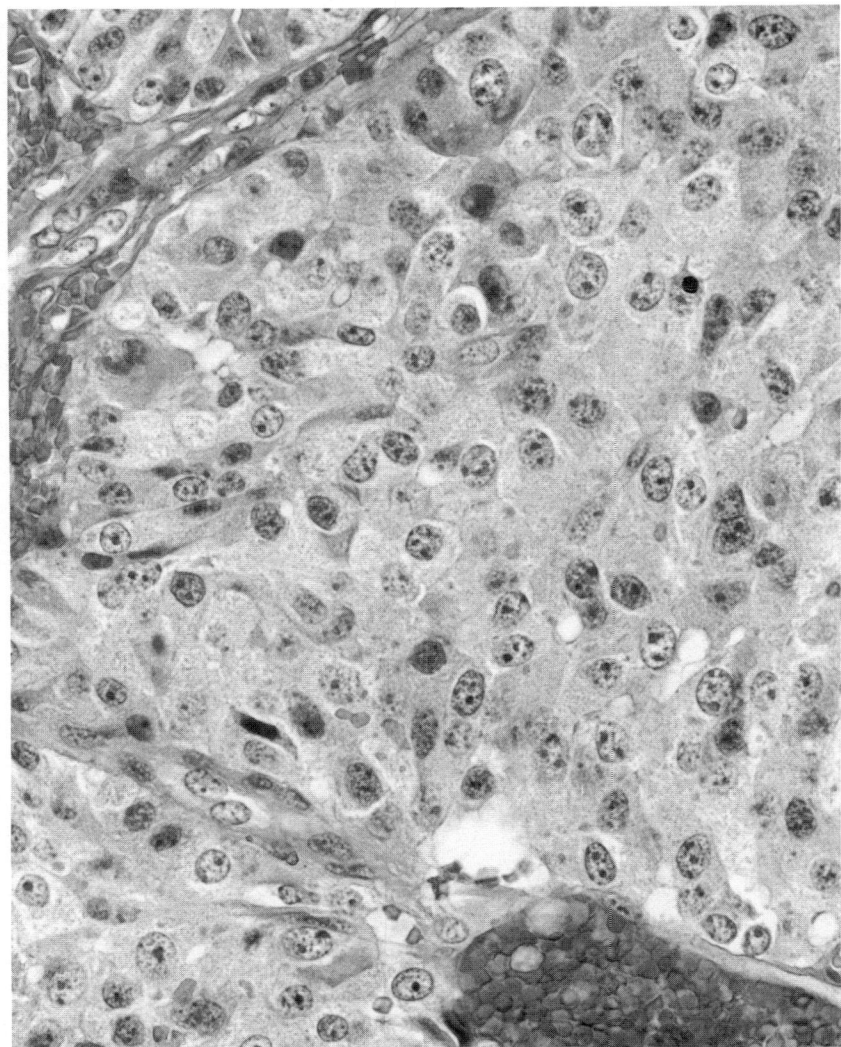

Figure 9.57 Highly differentiated medullary carcinoma composed of rather monomorphic, large polygonal cells with light, granular cytoplasm and round or oval nuclei with fine chromatin network and distinct nucleoli (H&E, × 500).

differentiation but are more probably due to degeneration in the tumour, and thus, should not be interpreted as a sign of increased aggressiveness of the lesion (Ljungberg, 1972b). Typical for such areas, as is the case for other parts of the tumour, is the scantiness of mitotic figures. However, it should be noted, that a rare, anaplastic giant

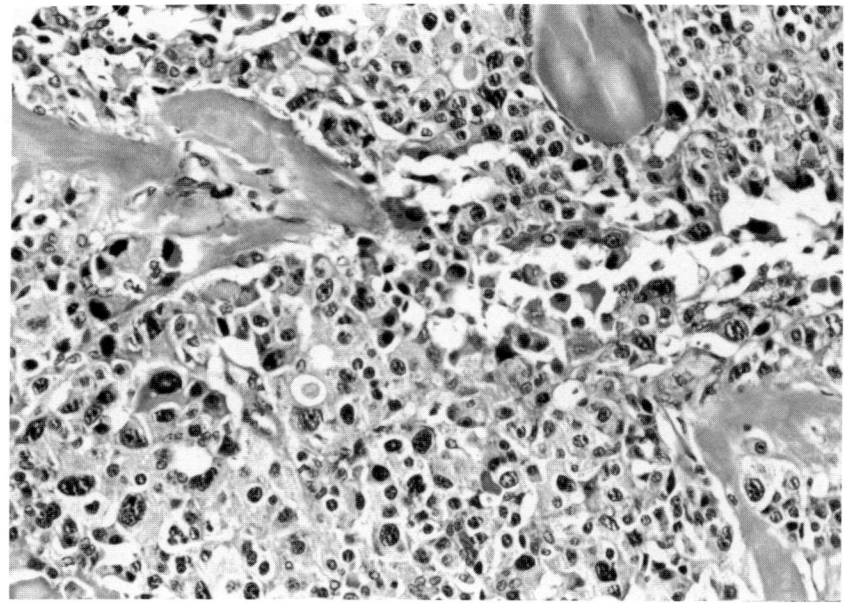

Figure 9.58 Well-differentiated medullary carcinoma with a focal area showing marked nuclear pleomorphism (H&E, × 215).

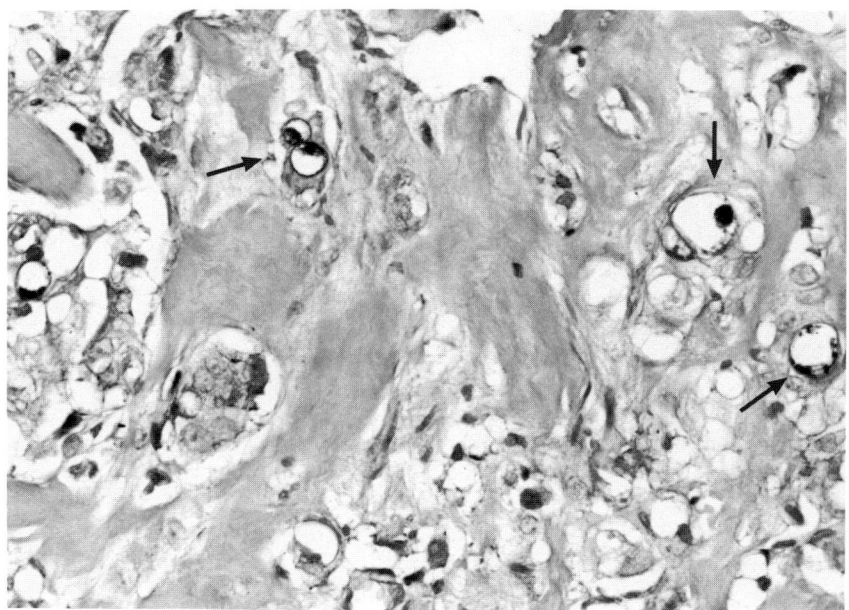

Figure 9.59 Medullary carcinoma with signet-ring cells, having PAS-positive intracytoplasmic material (arrows) (PAS, × 345).

cell variant of medullary carcinoma also exists, in which the pleo-morphic cells constitute the major part of the tumour. This variant shows abundant mitotic figures and areas of necrosis indicative of rapid growth. It runs a highly malignant course, approaching that of the classic undifferentiated giant cell carcinoma.

Medullary carcinoma may show clear cell change, even with the presence of signet-ring cells, similar to that sometimes seen in follicular cell tumours (pp. 125 and 178). Such cells may contain intracytoplasmic droplets which are stained with mucin stains, such as PAS–Alcian blue (Fig. 9.59) and mucicarmine (Fernandes *et al.*, 1982; Zaatari *et al.*, 1983; Landon and Ordoñez, 1985; Mlynek *et al.*, 1985; Uribe *et al.*, 1985). Similar material has also been observed in the interstitial tissue. In addition, occasional examples of medullary carcinoma have been reported showing oxyphil cell change. Ultrastructurally the oxyphil cells have showed increased numbers of large, often abnor-mal, dilated mitochondria, in addition to neurosecretory granules (Dominguez-Malagon *et al.*, 1989). The presence of clear or oxyphil cells does not seem to influence the clinical behaviour of the tumour.

Foci of squamous cell carcinoma have also been observed in other-wise conventional medullary carcinoma (Schröder *et al.*, 1988b; Dominguez-Malagon *et al.*, 1989). Whether such a change has the same ominous prognostic consequence as is the case when it occurs in differentiated follicular cell carcinomas cannot be judged from the small number of reported cases.

Medullary carcinoma, with otherwise typical histological and cytological features, may contain focal areas with elongated or spider-shaped cells with long cytoplasmic extensions, mimicking dendritic processes. Their cytoplasmic extensions can be seen ramifying between neighbouring tumour cells of ordinary type, and they contain brownish granules which are argentaffin (Ljungberg, 1970, 1972a). The nature of these cells is uncertain. They may correspond to the cells later described by Marcus *et al.* (1982) and Eng and Chen (1989), with similar features and found to form melanin-like substances. This melanin-like material was found to be stored in intracellular, membrane-limited melanosomes, co-existing with the calcitonin-containing dense-core granules. Medullary carcinoma, being composed of cells related to the diffuse neuroendocrine cell system, may develop monoamine-handling properties (the APUD-mechanism), which was found by Pearse (1969, 1977) to be a characteristic of this system. Although normal parafollicular cells of man, in contrast to those of certain other mammals, are incapable of producing the final monoamine products, their neoplastic counterpart, the medullary carcinoma, may achieve the full monoamine-handling capacity. Monoamines, such as 5-hydroxytryptamine and primary catecholamines, have been detected in a small proportion of tumour cells with the sensitive formaldehyde-induced fluorescence method (Falck *et al.*, 1968). In addition, various enzymes involved in monoamine metabolism have been reported to occur in medullary carcinoma, i.e. dopa decarboxylase, histaminase, monoamine-oxidase and catechol-*o*-methyltransferase (Baylin *et al.*, 1979; Feldman and Wells, 1981; Lippman *et al.*, 1982). It is not surprising that medullary carcinoma cells may also attain the amine-handling qualities that are specific to the melanocyte, another member of the neuroendocrine cell system.

9.7.3 Giant and small cell variants of medullary carcinoma

Unusual examples of medullary carcinoma composed predominantly of giant cells have been reported (Kakudo *et al.*, 1978; Nieuwenhuijzen Kruseman *et al.*, 1982). They are characterized by argyrophilia and positive staining reactions for calcitonin and other markers of parafollicular cells, present in the majority of the tumour cell population. Reported cases have contained little or no amyloid. It has been suggested that most classic undifferentiated (anaplastic) giant cell carcinomas are in fact anaplastic variants of medullary carcinoma. However, although a few undifferentiated thyroid carcinomas may contain a minority cell population with the characteristics of neuro-endocrine cells, such as immunoreactivity for calcitonin and/or somatostatin, the majority are completely non-reactive (Ljungberg *et al.*, 1984). Furthermore, some undifferentiated carcinomas may contain a component of well-differentiated follicular or papillary carcinoma showing unequivocal evidence of follicular cell differentiation, including immunoreactivity for thyroglobulin. Therefore, undifferentiated thyroid carcinoma should be looked upon rather as a heterogeneous group of tumours consisting of the anaplastic counterparts of all the differentiated forms of thyroid carcinoma (Ljungberg, 1984).

Poorly differentiated forms of thyroid carcinoma have also been recorded (Calmettes *et al.*, 1982; Mendelsohn *et al.*, 1980), composed of small, sometimes spindle-shaped, basophilic cells showing argyrophilia and immunoreactivity for chromogranin A and calcitonin (Fig. 9.60). They have behaved aggressively with a great tendency for local and metastatic spread. Structurally, these neoplasms may resemble small cell bronchial carcinomas, growing in solid nests or trabeculae in a sometimes hyalinized stroma. The author has also seen a variant composed of a giant cell as well as a small cell component, both revealing neuroendocrine features. The patient in this case was a 37-year-old woman, having had a goitre for about 6 years when the goitre suddenly began to grow. She presented with a large, hard

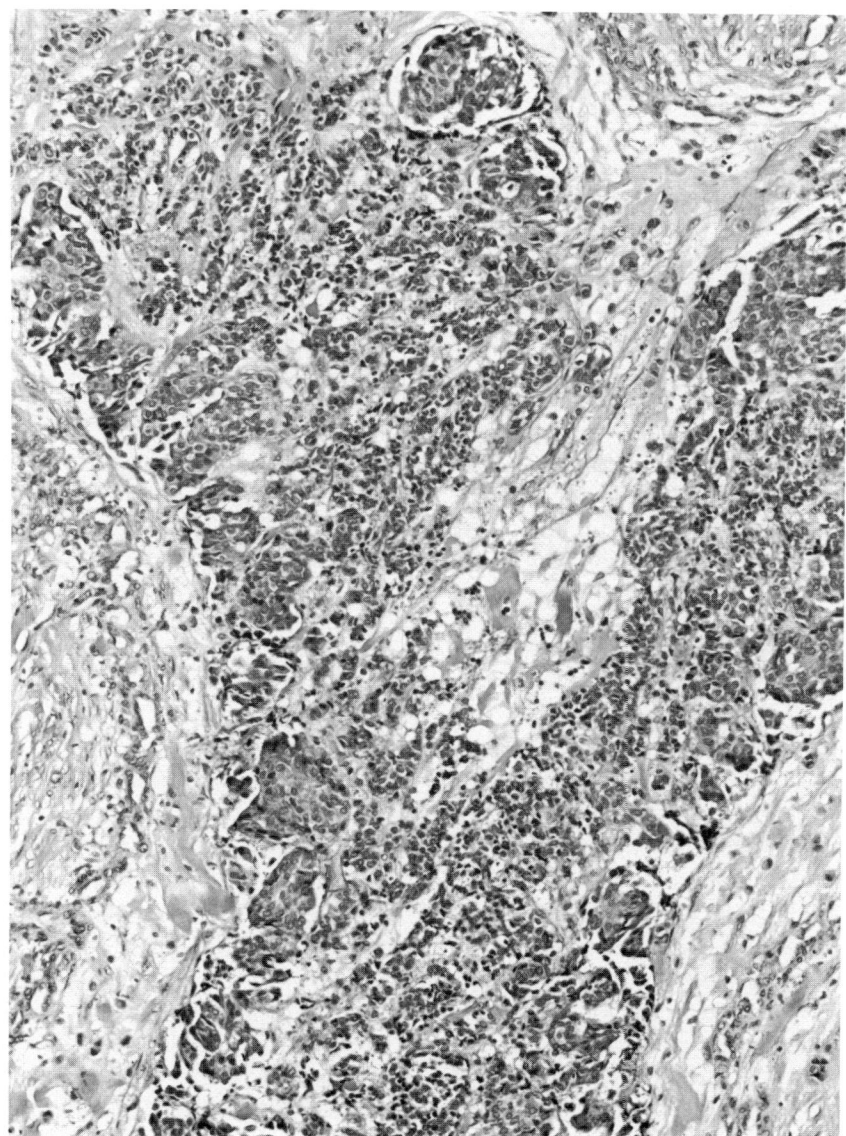

Figure 9.60 Poorly differentiated medullary carcinoma. A 77-year-old woman presented with a large thyroid mass, hoarseness and difficulties in swallowing for 2 months. The large thyroid tumour had invaded extrathyroidal structures, including the tracheal wall and extended into the mediastinum. The histological structure of the tumour is composed of haphazardly arranged, small basophilic, highly atypical cells, growing diffusely, in nests or trabeculae. Tumour cells were found to be calcitonin and chromogranin A positive, but non-reactive with thyroglobulin antibodies. No amyloid was demonstrated (H&E, × 125).

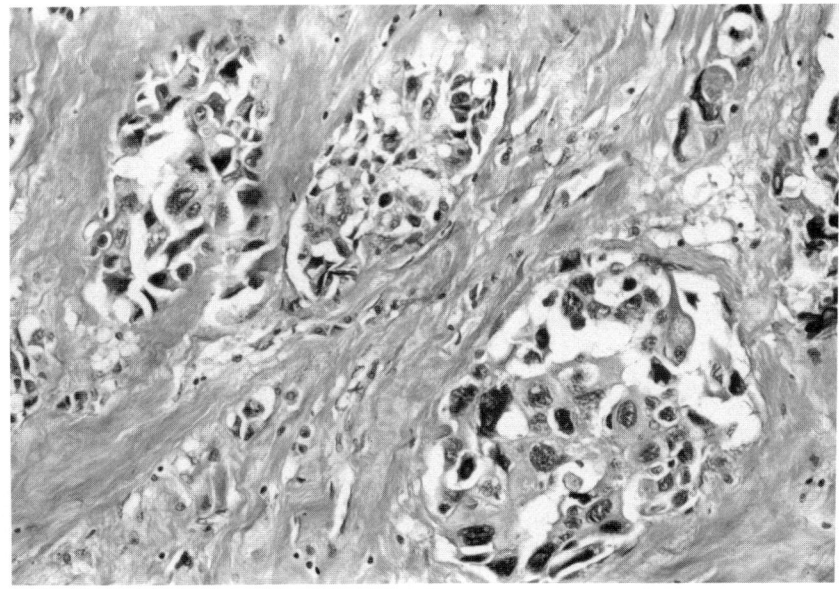

Figure 9.61 Same case as in Plate 22. Histological features of the giant cell component of the tumour. Clusters of grotesque giant tumour cells infiltrate in the abundant, sclerotic stroma (H&E, × 160).

goitre and several large lymph node metastases in the neck. Fine-needle aspiration from the thyroid lesion revealed highly polymorphic epithelial cells including many atypical giant cells (Plate 22). Cytologically, the tumour was diagnosed as an undifferentiated giant cell carcinoma of the thyroid. Surgery was not considered possible and the patient was given combined irradiation and chemotherapy. After a transient remission, the disease progressed rapidly with signs of skeletal and liver metastases, and she died 12 months after admission to the hospital. Autopsy revealed a large hard and partly necrotic tumour in the region of the right thyroid lobe and metastases were found in cervical and mediastinal lymph nodes, as well as in the lungs, liver, adrenal glands and bones. Histologically, the primary tumour and its metastases revealed a major component composed of solid nests of highly pleomorphic giant cells with bizarre nuclei (Fig. 9.61). A minor component consisted of nests and anastomosing trabeculae of rather uniform, small basophilic tumour cells with chromatin-dense nuclei (Fig. 9.62). In some areas the cells were spindle-shaped, growing in fascicle-like patterns reminiscent of a sarcomatous tumour. The stroma was

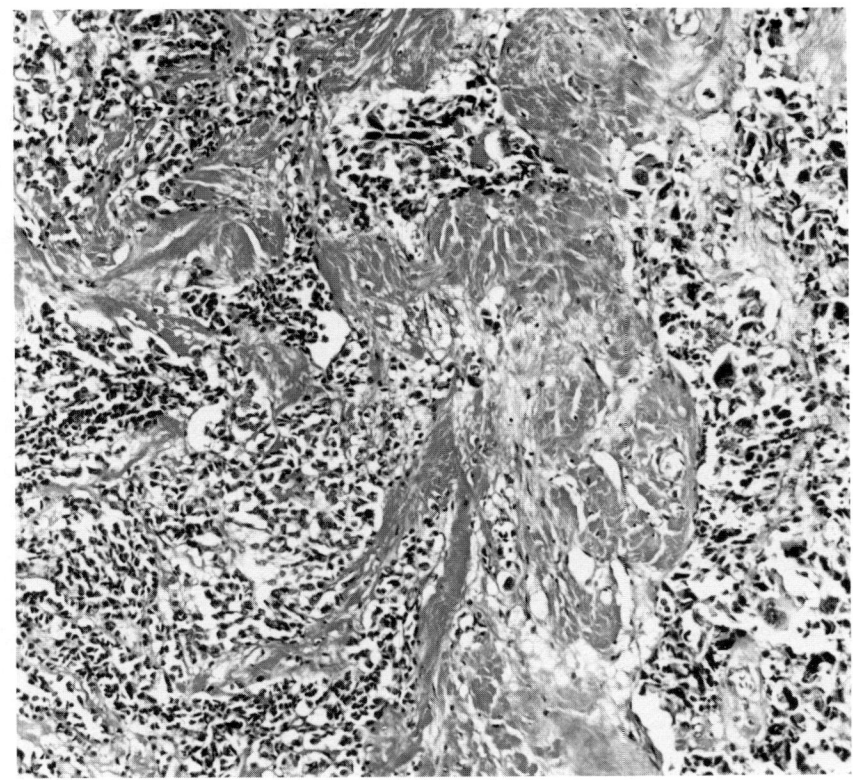

Figure 9.62 Same case as shown in Plate 22 and Fig. 9.61. Small cell component of the tumour showing loose clusters of small tumour cells with scanty cytoplasm and dark nuclei. On the right are some giant tumour cells (H&E, × 115).

abundant and often hyalinized, but amyloid was not found. Both components showed many mitoses, and extensive necrosis and haemorrhage. The majority of the tumour cells of both components contained argyrophilic granules and showed immunostaining for chromogranin A and calcitonin (Fig. 9.63), but were completely negative for thyroglobulin. Ultrastructurally, the tumour cells revealed typical intracytoplasmic neurosecretory granules. These findings indicate that the tumour, by definition, should be regarded as a medullary carcinoma, with poorly differentiated and undifferentiated components accounting for the highly malignant clinical course.

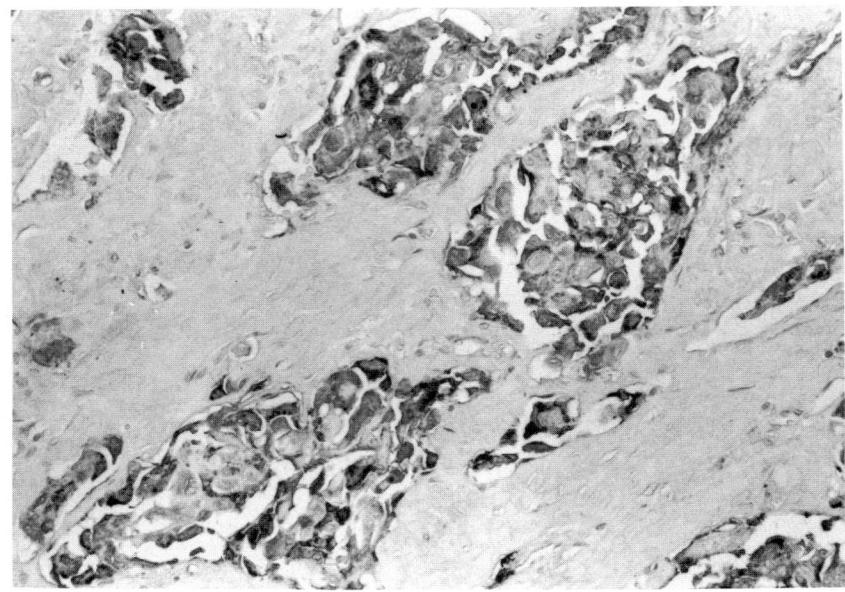

Figure 9.63 Same tumour as shown in Fig. 9.62. Area with calcitonin immuno-reactive giant tumour cells (PAP technique, with antibodies against human calcitonin, × 180).

9.7.4 Hormonal and non-hormonal products as markers of medullary carcinoma

Not only is medullary carcinoma a great structural imitator, it is also a multifunctional tumour capable of producing a variety of hormonal and non-hormonal substances, some of which may be used, with varying degrees of specificity, as markers for its diagnosis. Immunohistochemically, the tumour cells are reactive for general neuroendocrine markers, such as the chromogranins (Schmid *et al.*, 1987), synaptophysin (Gould *et al.*, 1987), neuron-specific enolase (Lloyd *et al.*, 1983), and for calcitonin; which is the most important specific marker of parafollicular cells and a prerequisite for the diagnosis of medullary carcinoma (DeLellis *et al.*, 1978). Besides calcitonin and other fragments of the calcitonin precursor (Roos *et al.*, 1983), the tumour may also elaborate the calcitonin gene related peptide (CGRP) (Amara *et al.*, 1982; Steenbergh *et al.*, 1984). Medullary carcinoma also shows various degrees of histochemical and/or biochemical immunoreactivity for somatostatin (Sundler *et al.*, 1977; Sano *et al.*, 1980), substance P (Skrabanek *et al.*, 1979; Kakudo and Vacca, 1983) and neurotensin (Ljungberg *et al.*, 1984), some of which are also constituents of normal parafollicular cells in many species. However, these latter products are not invariably present in the tumour cells and, when demonstrable, may only occur in minority cell populations (Ljungberg *et al.*, 1984). Other peptides reported to occur in medullary carcinoma include gastrin-releasing peptide/bombesin (Kameya *et al.*, 1983; Bostwick and Bensch, 1985; Ghatei *et al.*, 1985), ACTH and other fragments of its precursor molecule, pro-opiomelanocortin (Bussolati *et al.*, 1973; Ohashi *et al.*, 1987; Roth *et al.*, 1987), gastrin, VIP, insulin, glucagon (Uribe *et al.*, 1985), pancreatic polypeptide (O'Hare *et al.*, 1986), alpha-HCG (Krisch *et al.*, 1985), helodermin (Sundler *et al.*, 1988), and nerve growth factor (Cramer *et al.*, 1979).

In diagnostic practice, positive immunostaining for calcitonin, combined with negative staining for thyroglobulin, is the most reliable histopathological method for establishing the diagnosis of classic medullary carcinoma (Fig. 9.64). Carcinoembryonic antigen (CEA) appears to be less useful as a marker, because staining patterns may differ with various antibodies used. This may be due to variability between tumours in the expression of some of its epitopes (Dasovic-Knezevic *et al.*, 1989). Chromogranin A immunoreactivity is invariably present in medullary carcinoma. Exceptionally however, thyroid carcinomas may occur, which lack amyloid and show negative immunoreactions for calcitonin and thyroglobulin, but which otherwise show histological and ultrastructural features typical of medullary carcinoma. Such tumours have been argyrophilic with Grimelius' silver stain and shown positive immunoreactions for general neuroendocrine markers (e.g. chromogranin A), and sometimes also for serotonin, neuron-specific enolase and CEA (Uribe *et al.*, 1985). Whether such tumours should be included within the medullary carcinoma group, or should be regarded as a distinct group of neuroendocrine thyroid carcinomas, remains to be elucidated.

Several of these various products, including calcitonin, CEA, somatostatin, and chromogranin A (DeLellis *et al.*, 1978; O'Connor and Deftos, 1986; Pacini *et al.*, 1989), may, by means of their serum levels, be valuable as markers for the clinical diagnosis and/or for monitoring recurrence of the disease. CEA is said to correlate better with the extent of disease than do serum calcitonin levels (Kodama *et al.*, 1979; Saad *et al.*, 1984a). It should be noted, however, that none of these compounds is absolutely specific for medullary carcinoma, since their levels in serum may be increased in patients with other neuroendocrine neoplasms. Elevated serum calcitonin levels, for instance, have been demonstrated in patients with small cell carcinoma of the lung, carcinoma of the breast, hepatocellular carcinoma, malignant lymphoma and leukaemia (Body *et al.*, 1983).

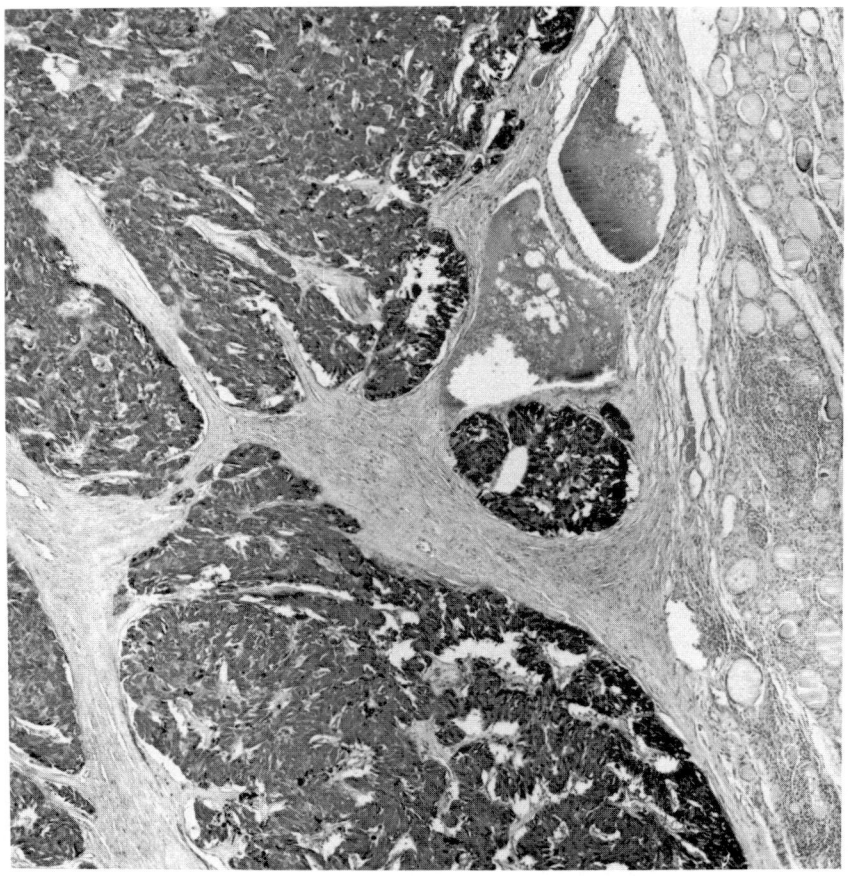

Figure 9.64 Classic medullary carcinoma showing extensive and strong immuno-reactivity for calcitonin (PAP technique, with antibodies against human calcitonin, × 47).

Although immunoreactivity for a great variety of hormones can be demonstrated in the tumour cells, and even in the serum, patients with medullary carcinoma seldom present with obvious endocrine symptoms. Very rarely, inappropriate hormonal syndromes such as Cushing's syndrome may be seen (Williams *et al.*, 1968). Prominent diarrhoea may occur in as many as 30% of patients with medullary carcinoma, and appears to be related to the total mass of the tumour (Williams, 1966; Ljungberg, 1972b). It has been variously ascribed to prostaglandins, 5-hydroxytryptamine, bradykinin or vasoactive intestinal peptide, but no common hormonal mediator for the diarrhoea has so far been identified.

Although the amyloid is stained with common amyloid stains, it appears to be unrelated to the immunoamyloid. Antisera produced against amyloid from medullary carcinoma do not react with immuno-amyloid (Sletten *et al.*, 1976). On the other hand, there is evidence that amyloid proteins of medullary carcinoma might be related to calcitonin production. Fibrils synthesized *in vitro* from calcitonin fragments have a structure that resembles the amyloid fibrils (Keydar *et al.*, 1976). Amyloid in medullary carcinoma may show reactivity with antibodies against calcitonin, but it is uncertain whether this is due to calcitonin from secretory granules that have been entrapped between the fibrils, or represents a reaction with the fibrils themselves (Uribe *et al.*, 1985).

9.7.5 Ultrastructural features of medullary carcinoma

The multihormonal potential of medullary carcinoma is also suggested by its ultrastructural features. The tumour cells contain typical neuroen-docrine secretory granules (Fig. 9.65), representing the storage site of calcitonin and other hormonal products elaborated by the tumour (DeLellis *et al.*, 1977). The great variation in shape, size, and struc-ture of the granules suggests the presence of multiple neuroendocrine types of tumour cells (Horvath *et al.*, 1972; Kameya *et al.*, 1977; Capella *et al.*, 1978). Functionally, however, some tumour cells may be multihormonal, since several secretory products have been shown to coexist in the same cell (Uribe *et al.*, 1985). Granules range from 60 to 550 nm in diameter (DeLellis and Wolfe, 1981). Larger granules have moderately electron dense, granular material, closely applied to the limiting membranes, whereas smaller granules tend to be more elec-tron dense, the central core sometimes being separated from the limiting membranes by an electron lucent space. Amyloid has a typical microfibrillary appearance, structurally indistinguishable from immunoamyloid and other forms of amyloid. It is mainly located in the intercellular stroma but is sometimes also observed within the cytoplasm of tumour cells. Occasionally, the microfilaments are seen to be intermingled with granular structures resembling disintegrating secretory granules (Huang and McLeish, 1968). This may account for the reported positive immunoreaction of amyloid with antibodies against calcitonin (Butler and Khan, 1986).

9.7.6 Cytological diagnosis of medullary carcinoma

Fine-needle aspiration biopsy of classic medullary carcinoma reveals polygonal, plasmacytoid, spindle-shaped cells often with eccentric nuclei (Plate 23). Some cells are binucleate. The cells occur singly or

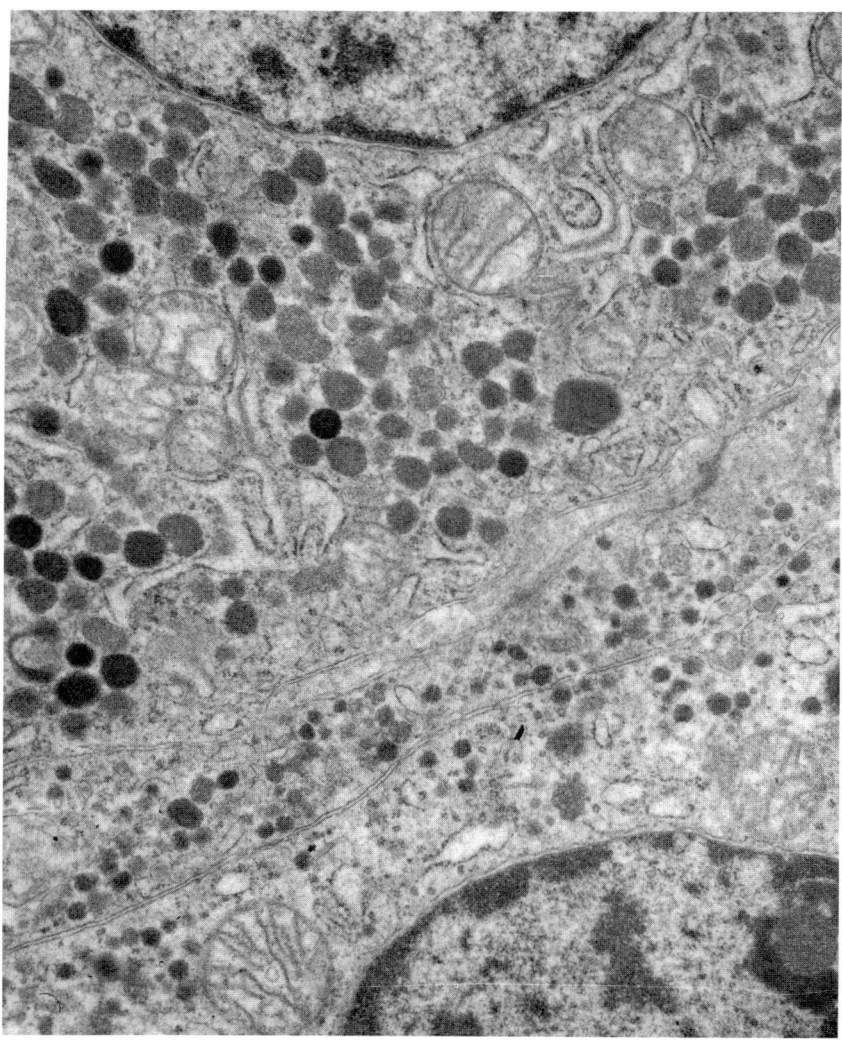

Figure 9.65 Ultrastructure of medullary carcinoma. Portions of different cells, showing neurosecretory granules of different morphology (× 20 000).

in loosely cohesive groups. Intracytoplasmic red or bluish granules, similar to those observed in tumours from other members of the diffuse neuroendocrine cell system (e.g. carcinoids), may be seen in smears stained with May–Grünwald Giemsa. Clumps of amorphous amyloid material may occur, which show a yellow and apple-green bire-fringence in polarized light when stained with alkaline Congo red (Ljungberg, 1972c). They also stand out clearly when the same slides are viewed in a fluorescence microscope provided with a filter setting for rhodamine fluorescence. In some instances, intranuclear cytoplasmic inclusions are found (Lew et al., 1984) which may cause confusion with papillary carcinoma. Tumour cells show positive immunoreaction with antibodies against calcitonin and chromogranin A (Walts et al., 1985; Ljungberg, unpublished observations). As pointed out on p.131 confusion with the trabecular hyalinizing adenoma is another diagnostic trap. This tumour may yield smears showing spindle cells with a tendency for poor cell cohesion, and some nuclei may exhibit intranuclear cytoplasmic inclusions. In addition, the presence of amorphous hyaline material may suggest amyloid. How-ever, amyloid staining is negative and intracytoplasmic purple or bluish granulations are not found. Immunocytochemically, tumour cells of the trabecular hyalinizing adenoma are non-reactive with antibodies against calcitonin but may show positive thyroglobulin immunoreactivity.

9.7.7 Familial medullary carcinoma

Familial medullary carcinoma is a component of multiple endo-crine neoplasia (MEN) syndrome, type 2A and type 2B, both of which are also associated with the development of multiple and bilateral adrenal phaeochromocytomas. Some patients show clinical evidence of primary hyperparathyroidism and nodular parathyroid hyperplasia (Chapter 17). Type 2B, in addition, is associated with multiple mucosal neuromas in the conjunctiva, lips, and oral mucosa, and ganglioneuromatosis of the gastrointestinal tract. Furthermore, these patients often show a marfanoid habitus and various skeletal and soft-tissue abnormalities, including thick lips, high-arched palate, pectus excavatus and muscular weakness. Some patients with type 2B have more generalized involvement of the peripheral nervous system, including proliferation of autonomic nerves and neuroma-like lesions of the posterior spinal nerve roots, and, occa-sionally, greatly thickened peripheral somatic nerve trunks (Bartlett et al., 1971). A substantial number of families has also been reported in which medullary carcinoma has occurred in the absence of the other components of the syndrome (Block et al., 1980; Noll et al., 1984; Farndon et al., 1986; Sobol et al., 1989). Familial medullary carcinoma is inherited as an autosomal dominant lesion with vir-tually complete penetrance. The thyroid tumours in type 2B become clinically apparent at a young age (mean age 25 years), and may even occur in children. This differs markedly from the average

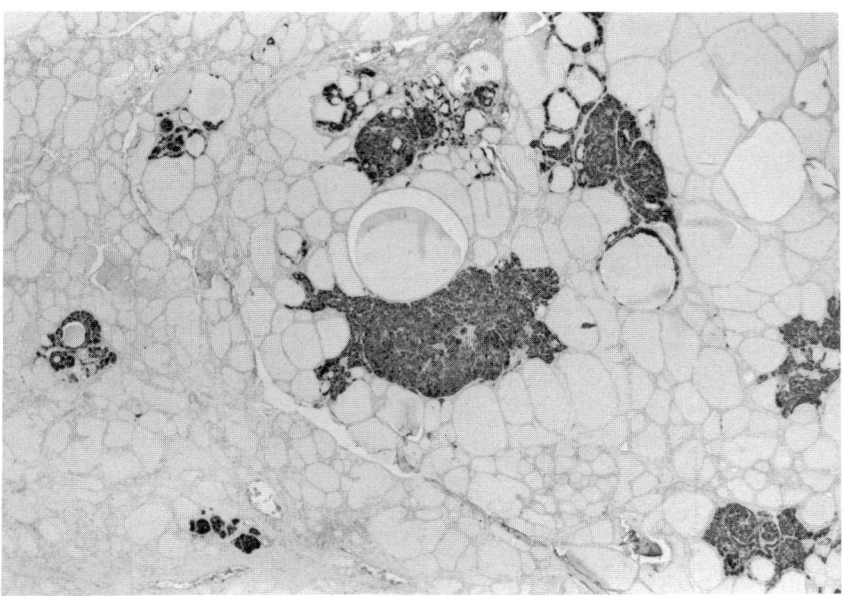

Figure 9.66 A case of familial medullary carcinoma. Immunostaining for calcitonin shows multifocal hyperplasia of parafollicular cells, of diffuse and nodular type. The larger nodule in the centre has replaced several normal follicles and may possibly represent early neoplastic change (PAP technique, with antibodies against human calcitonin, × 22).

age at presentation for medullary carcinoma in general (40–50 years), and also for the familial type 2A variant (35–40 years).

As pointed out above, medullary carcinoma in its familial setting is a multifocal lesion, which is accompanied by a diffuse or multinodular hyperplasia of the entire parafollicular cell system of the gland (Ljungberg, 1972b; Wolfe *et al.*, 1973). It is believed that Knudson's (1971) two-mutational-theory is applicable in the initiation of the various tumours in MEN 2. Thus it is suggested that inherited medullary carcinoma is the result of at least two mutational events, the first mutation being an inherited abnormality which gives rise to a hyperplasia of the parafollicular cells, and the second event being an acquired somatic mutation which causes a neoplastic change in the parafollicular cell. The MEN 2A gene has recently been mapped to the centromeric region of chromosome 10 (Mathew *et al.*, 1987; Simpson *et al.*, 1987). Observations in MEN 2 indicate that the hyperplastic change is caused by a first, heritable chromosome 10 mutation, whereas hyperplastic parafollicular cells may be converted into cancer cells by a second series of mutations at any of many possible loci (Jackson and Norum, 1989; Okazaki *et al.*, 1989).

Parafollicular cell hyperplasia, as seen in the familial syndromes, is a progressive process (Ljungberg, 1972b; DeLellis and Wolfe, 1981). In its early stage, the parafollicular cells show evidence of a diffuse proliferation within the follicles. It is usually not possible to recognize this stage on routine haematoxylin–eosin sections. Immunolocalization techniques, using antibodies against calcitonin or chromogranin A, however, will clearly visualize the proliferating cells (Fig. 9.66). As these cells proliferate, they form crescentic, annular or micronodular clusters of cells, which progressively replace the follicular cells. Finally, the entire follicle may consist of parafollicular cells obliterating the follicular lumen, and thus transforming the follicle into a compact nodule which is still encircled by an intact basement membrane. At this stage, the hyperplasia has been referred to as nodular (Fig. 9.66) (DeLellis and Wolfe, 1981).

It is not easy to make the distinction between nodular hyperplasia and an early stage of neoplasia. Morphologically, there appears to be a continuous series of changes towards an unequivocal neoplastic stage and it seems to be impossible, on the basis of morphology or immunohistochemistry, to decide exactly when the transformation to neoplastic cells has occurred. According to the author's experience, signs of early neoplastic change include destruction of the basement membrane around the nodules, as evidenced by irregular outline of the nodules, and signs of invasion of parafollicular cells into surrounding interstitial stroma (Fig. 9.67). Sometimes an increase in the amount of stroma, with a characteristic sclerosis is seen (Fig. 9.68). Later, a definite destruction of the basic architecture of the parenchyma occurs; at this stage the neoplastic nature of the lesion is beyond any doubt (Fig. 9.68). With immunohistochemical staining for calcitonin and CEA, hyperplastic foci typically stain more intensely than do clearly neoplastic lesions (Plate 24). In diagnostic practice, this finding may be helpful in distinguishing between the two types of change, although it is not known whether the change in staining intensity occurs simultaneously with the event of neoplastic transformation.

The diagnosis of parafollicular cell hyperplasia in cases belonging to families with the MEN syndromes is complicated by the fact that the parafollicular cell density appears to vary considerably in the normal gland (Chapter 1, p. 13). Autopsy studies have shown that large clusters of parafollicular cells resembling the hyperplastic nodules seen in MEN patients, may occur in cases lacking abnormalities of calcium metabolism, as well as family histories of medullary carcinoma. They tend to be most common in elderly persons, and predominate

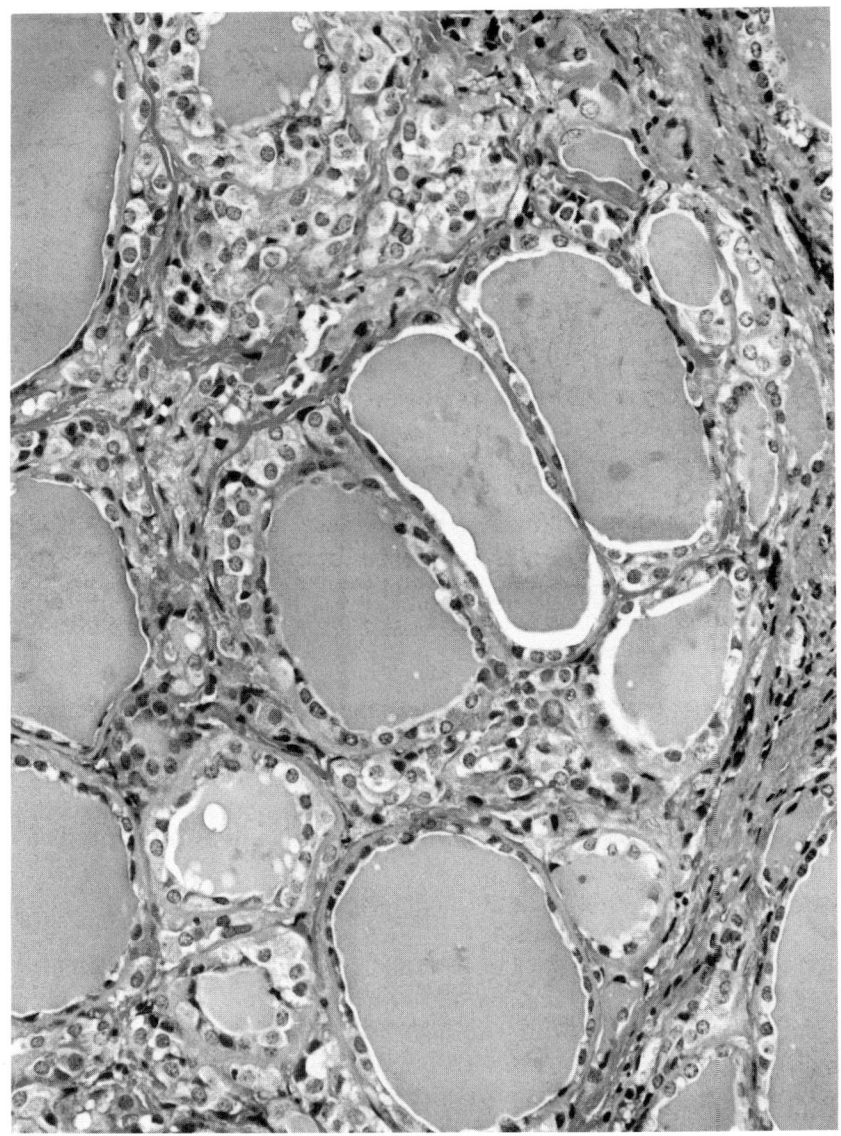

Figure 9.67 Case of familial medullary carcinoma. A microscopic focus of proliferating light, cytoplasm-rich parafollicular cells. Interstitial growth with some fibrotic stromal reaction, probably indicating early neoplastic change. In addition, some follicles show diffuse parafollicular cell hyperplasia, i.e. an increased number of light cytoplasm-rich cells in their walls (H&E, × 330).

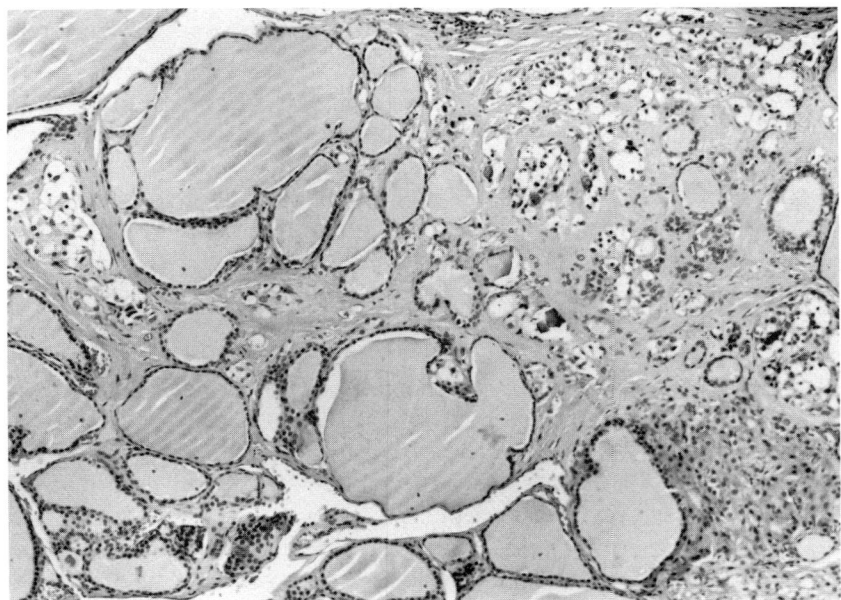

Figure 9.68 Case of familial medullary carcinoma. A minute focus of marked interstitial stromal sclerosis, destruction of follicles and infiltration of neoplastic parafollicular cells. This focus is undoubtedly an early medullary carcinoma (H&E, × 90).

in males. This suggests that parafollicular cell changes in these situations are not premalignant but rather a normal phenomenon, possibly related to the human ageing process (O'Toole *et al.*, 1985). It should be pointed out that parafollicular cell hyperplasia may also accompany various thyroid disorders, unrelated to familial medullary carcinoma. It has occasionally been observed in cases of sporadic medullary carcinoma (Ulbright *et al.*, 1981), intermediate type of thyroid carcinoma (Ljungberg and Nilsson, 1991), follicular cell tumours (Albores-Saavedra *et al.*, 1988), and Hashimoto's thyroiditis with hypothyroidism (Libbey *et al.*, 1989). In addition, it has also been demonstrated in long-standing hypercalcaemic states (Ljungberg and Dymling, 1972; LiVolsi and Feind, 1979).

9.7.8 Differential diagnosis of medullary carcinoma

The histopathological appearance of classic medullary carcinoma is well recognized and seldom causes differential diagnostic problems. The most distinguishing feature is the presence of amyloid, which is practically never found in any of the other forms of primary thyroid neoplasia. Tumours devoid of amyloid, on the other hand, especially those with atypical histological patterns, may be diagnostically problematic. Immunostaining can be critical in order to identify these structural variants of medullary carcinoma. This applies also to the differential diagnosis versus trabecular hyalinizing adenoma (p. 131), as well as poorly differentiated, small cell thyroid carcinomas (p. 187) and carcinomas with anaplastic giant cell features (p. 212). Tumours with combined solid and follicular structures should also be analysed immunohistochemically for markers of follicular and parafollicular cells, to establish their relationship to the intermediate type of carcinoma, discussed below.

Rarely, the patient presents clinically with a cervical lymph node metastasis, which will be found to have originated from an occult medullary thyroid carcinoma, sometimes of microscopic size (Fig. 9.48). In this situation, it should be remembered however, that although a metastatic lesion may show structural characteristics compatible with medullary carcinoma – and even amyloid deposits and calcitonin-immunoreactive tumour cells – one cannot, on the basis of these features only, be absolutely certain of its thyroidal site of origin. It is well known that calcitonin levels may be increased both in the serum and in tumour extracts from patients with a variety of non-thyroidal neoplasms. These include certain neuroendocrine tumours, particularly in the lung, thymus and pancreas (Deftos and Burton, 1980). Amyloid deposits as well as calcitonin immunoreactive tumour cells may also be present (Gordon et al., 1973; DeLellis and Wolfe, 1974). This author has seen one patient presenting with a large metastasis in the chest wall, initially interpreted as a secondary from medullary thyroid carcinoma, but later found to have developed from a peripheral neuroendocrine lung carcinoma (also referred to as atypical carcinoid of the lung). Although the metastasis contained calcitonin immunoreactive cells, these cells constituted only a small part of the tumour cell population, unevenly distributed among the non-reactive cells. This may be a discriminating feature, since immunoreaction for calcitonin is generally more extensive and evenly distributed in medullary thyroid carcinoma. It should also be mentioned that, vary rarely, ovarian strumal carcinoids may contain calcitonin-producing endocrine cells indistinguishable from medullary thyroid carcinoma cells, and such tumours have also shown an amyloid stroma (Dayal et al., 1979).

9.7.9 Clinical behaviour and prognostic aspects in medullary carcinoma

Medullary carcinoma invades locally and tends to metastasize early to cervical and mediastinal lymph nodes. Cervical lymph node metastases are found in more than 50% of cases at first operation. Its behaviour is unpredictable and variable and the course is sometimes protracted. Occasionally, the tumour may show evidence of spontaneous regression (Ljungberg, 1972b). In advanced cases metastases are most often found in lymph nodes, liver, lungs, pleurae, and bones (Ljungberg, 1972b).

High mitotic rates appear to be associated with a worse prognosis and the rare, poorly differentiated small or giant cell variants with this feature tend to be highly aggressive. It has been claimed that the spindle-call pattern is associated with a poor prognosis (Williams et al., 1966) but this has been refuted by others (Saad et al., 1984b; Schröder et al., 1988b). The histological architecture of the highly differentiated forms, the amount of amyloid present, or the pattern of peptide immunoreactivity, however, does not appear to be related to prognosis (Ljungberg, 1972b; Schröder et al., 1988b). Instead, survival has been found to be correlated with age, sex and stage of disease. The best prognosis is seen in young women with a tumour confined within the gland (Saad et al., 1984b). Familial medullary carcinoma in MEN 2A generally has a better prognosis than the sporadic form, possibly because it is mostly diagnosed at an earlier stage. Screening of members of families with MEN 2A with provocative tests for calcitonin secretion is important in order to identify the thyroid lesions in their preclinical phase of development (Ponder, 1984). This is especially important in MEN 2B patients in whom thyroid tumours may become clinically evident in childhood and behave aggressively (Norton et al., 1979). It is likely that the great advances that have been made in recent years regarding the identification of the gene(s) responsible for the familial medullary carcinoma, will soon result in still more effective methods for the screening of affected families.

The value of nuclear DNA ploidy pattern as a prognostic variable of medullary carcinoma is disputed. (Tangen et al., 1983; Schröder et al., 1988b; Galera-Davidson et al., 1990).

9.8 Thyroid carcinoma of intermediate type (follicular–parafollicular cell carcinoma)

Follicular and medullary thyroid carcinoma have for years been regarded as biologically distinct entities. In recent years, however, evidence has been accumulating for the existence of a tumour group with traits of both follicular and medullary carcinoma, indicating that the distinction between these two categories is not as clear as was previously believed (Hales et al., 1982; Ljungberg et al., 1983, 1984;

Pfaltz *et al.*, 1983; Parker *et al.*, 1985; Holm *et al.*, 1986; Adkins and Hartley, 1987; Burt *et al.*, 1987; Ljungberg and Nilsson, 1991). This tumour type has so far been included within the medullary or follicular main groups of current thyroid tumour classifications. In the revised WHO classification of 1988, this type has been separated as a subgroup under medullary carcinoma. Probably, however, most cases are hiding within the subgroup of follicular carcinoma which comprises less well differentiated or less solid growth patterns. At least some of the tumours classified as poorly differentiated thyroid carcinoma (so-called insular type) are probably also of the intermediate type.

Structurally, this intermediate type of thyroid carcinoma has a position which is intermediate between the follicular and medullary tumour groups. It may be defined as a complex neoplasm with evidence of a multidirectional differentiation into follicular and parafollicular, as well as intermediate, cell forms and sometimes also immature structures of the type seen in remnants of the ultimobranchial bodies of the human thyroid gland (Ljungberg and Nilsson, 1991). The following description is based on the findings in the author's own series of 18 cases.

9.8.1 Macroscopic features

The tumours are mostly well defined and at least partially encapsulated, measuring form 1.6 to 6 cm in diameter. Some are ill-defined massive growths, up to 9 cm in largest diameter, occupying the major part of the gland and having multinodular or diffuse growth patterns, often with extrathyroidal infiltration.

9.8.2 Microscopic, and immunohistochemical ultrastructural features

The tumours characteristically show solid as well as follicular or cribriform structures in varying proportions, the solid growth pattern usually being the most conspicuous component (Fig. 9.69). The solid areas consist of trabeculae or nests of epithelial cells, embedded in various amounts of connective stroma which is sometimes hyalinized. The solid areas may develop well-differentiated cells with light granulated cytoplasm arranged in patterns showing close structural similarity to classic medullary carcinoma. Occasionally, amyloid deposits may be found in such areas. There is usually a gradual transition towards formation of small follicles lined by follicular cells and containing small clumps of colloid, but larger follicles with flattened epithelial cells may also occur

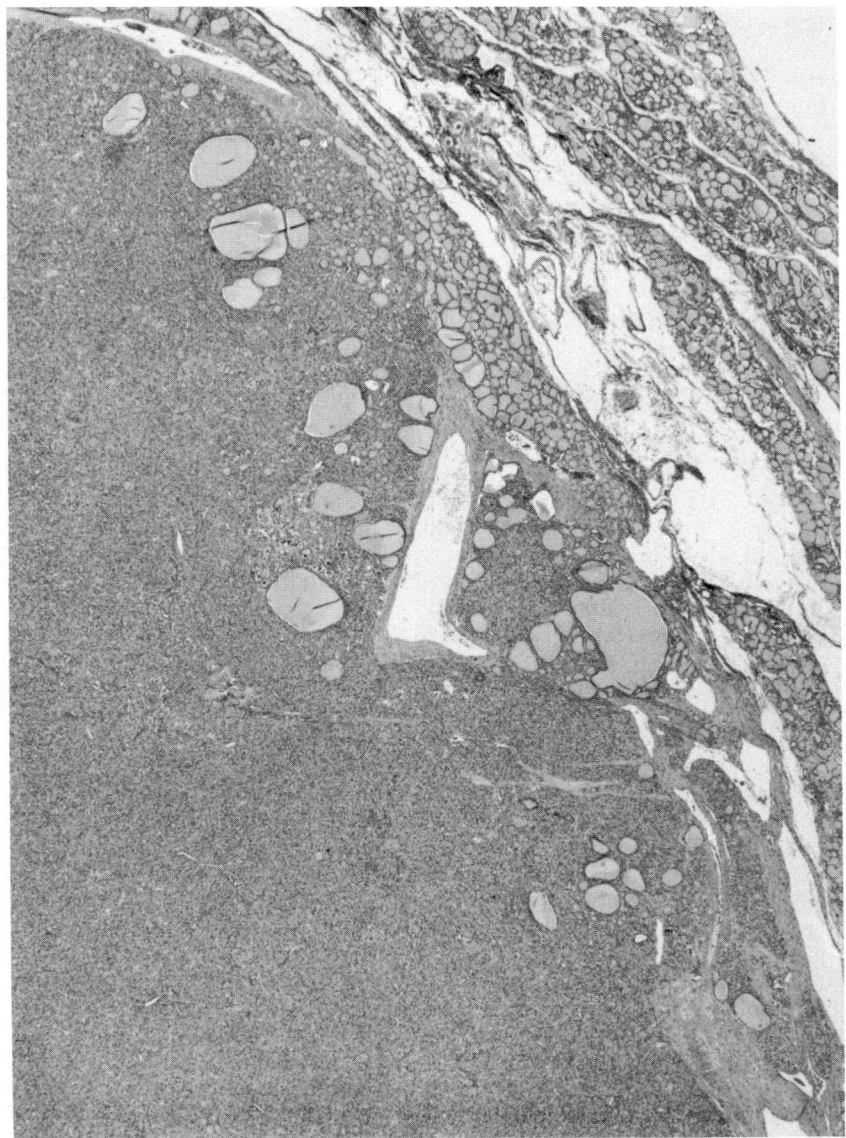

Figure 9.69 Thyroid carcinoma of intermediate type. Chiefly solid in structure. The tumour has no capsule and infiltrates between the surrounding normal follicles, some of which have been entrapped within its peripheral parts (H&E, × 12.5).

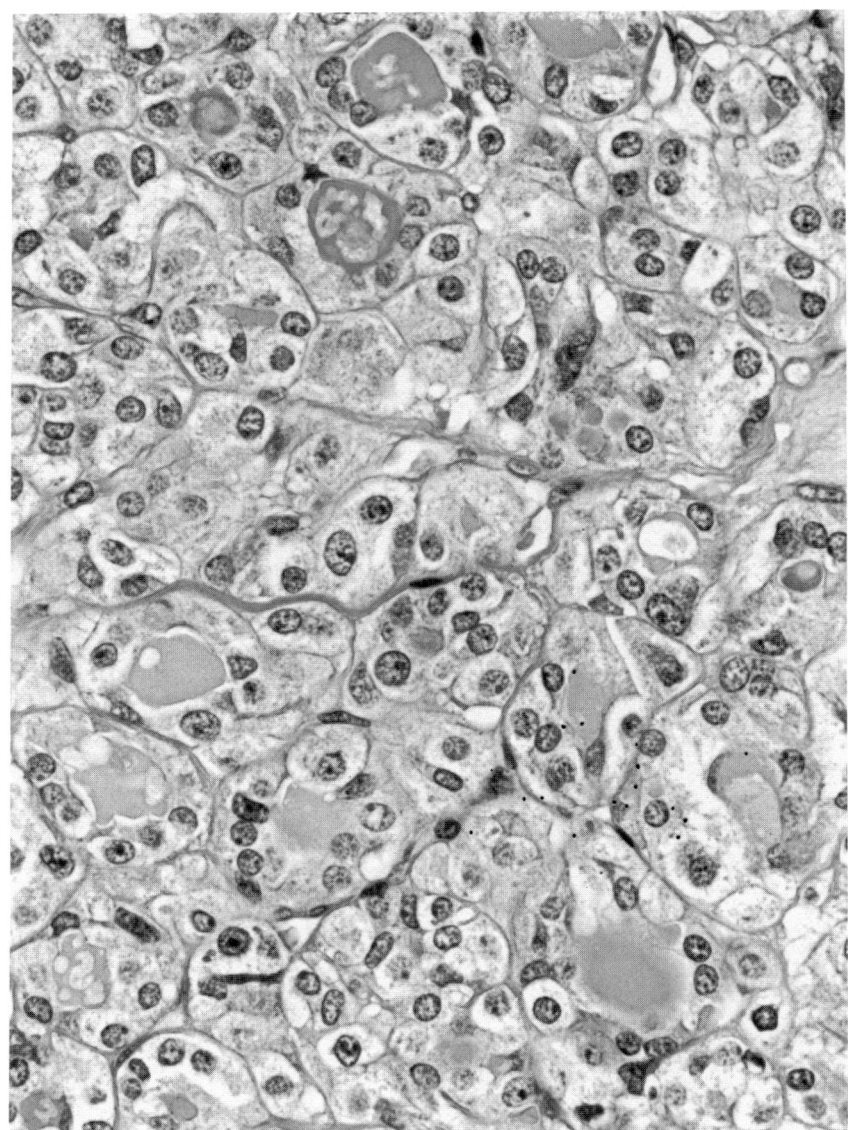

Figure 9.70 Thyroid carcinoma of intermediate type. Well-differentiated variant. Tumour cells are arranged in small solid nests or form follicular structures with colloid. Most tumour cells have abundant, light and granulated cytoplasm and resemble neoplastic parafollicular cells (H&E, × 500).

(Fig. 9.70). The well-defined tumours usually show little cytological atypia and few mitoses. Some tumours, however, have less differentiated areas, with more atypical cells, which may show a nest-like pattern mimicking that of the so-called insular type of poorly differentiated carcinoma (Fig. 9.71), described by Carcangiu *et al.* (1984). This pattern is most often seen in the massively invasive growths showing an aggressive growth pattern, a high mitotic rate and areas of necrosis.

Less frequently, tumours may contain immature looking areas, with small basophilic, sometimes spindle-shaped, cells with scanty cytoplasm and chromatin-dense nuclei, arranged in solid sheets or nests, embedded in varying amounts of stroma. Such parts may

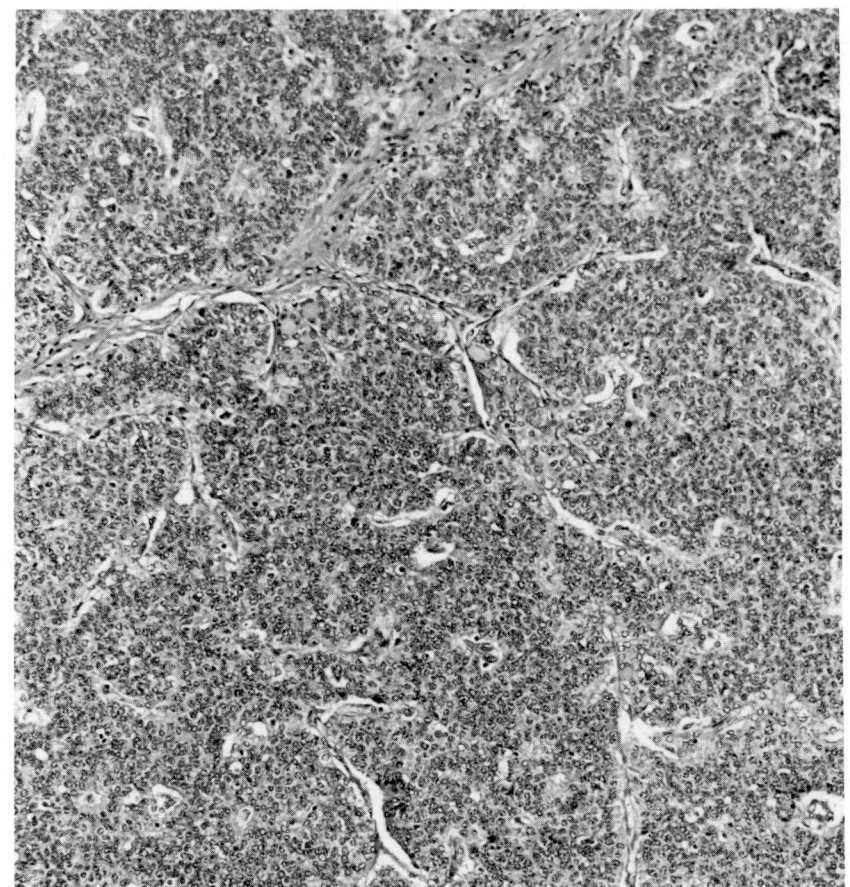

Figure 9.71 Thyroid carcinoma of intermediate type. Poorly differentiated variant, with small basophilic, atypical cells in a nest-like pattern (H&E, × 125).

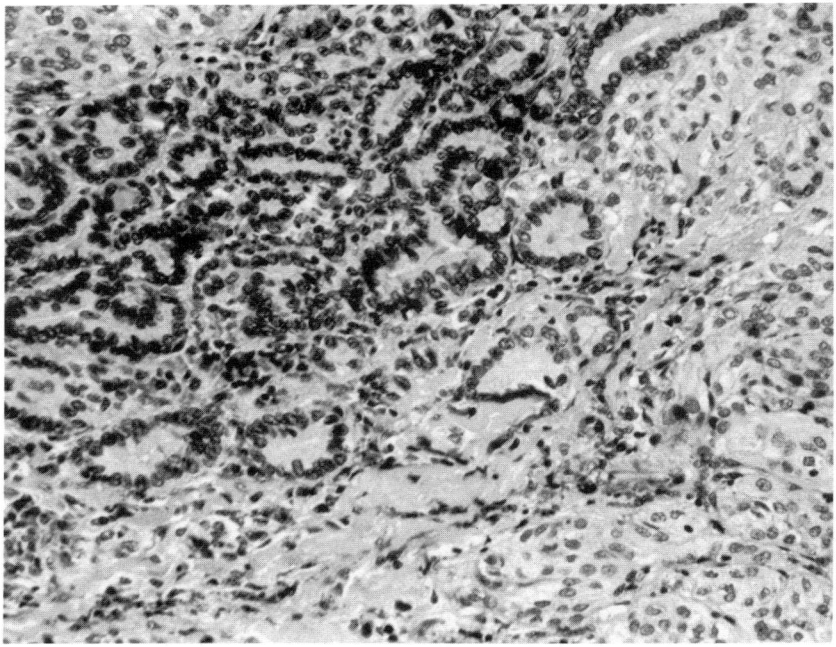

Figure 9.72 Thyroid carcinoma of intermediate type. Less differentiated variant developing irregular tubules lined by cylindrical cells with chromatin-dense nuclei (H&E, × 160).

develop tubular structures usually lined by tall cylindrical cells (Figs 9.72 and 9.73). They often contain weakly eosinophilic material which stains with Alcian blue and also small clumps positive for periodic acid–Schiff (PAS). There are also various-sized cysts lined by cylindrical or flattened cells, the latter sometimes arranged in several layers and resembling epidermoid cells. The cysts appear empty or contain detritus with degenerated cells, sometimes also Alcian blue- and PAS-positive substances. Occasionally, marked squamous cell metaplasia has been observed.

The tumour, by definition, shows immunohistochemical evidence of the presence of both thyroglobulin (Fig. 9.74) which is typical of follicular cells, and neuropeptides characteristic of parafollicular cells, such as calcitonin, somatostatin and neurotensin, as well as general neuroendocrine markers (e.g. chromagranin A) (Figs 9.75 and 9.76). Using double immunolocalization techniques, it has been shown that some tumour cells are thyroglobulin-positive, others exhibit immunoreactivity for one or several neuropeptides, whereas a third category

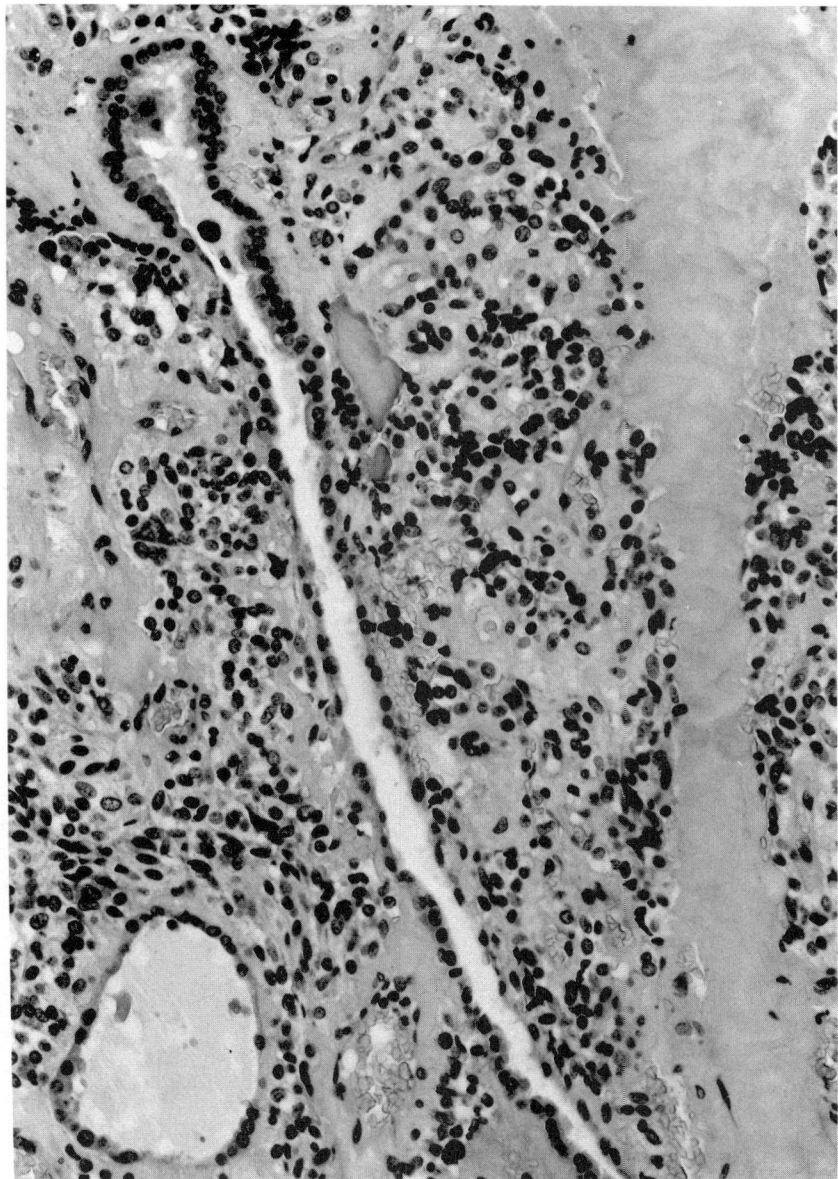

Figure 9.73 Thyroid carcinoma of intermediate type. Poorly differentiated variant composed of loose clusters of immature-looking tumour cells and a slender tubular structure lined by partly columnar, partly flattened epithelium with dark nuclei (H&E, × 250).

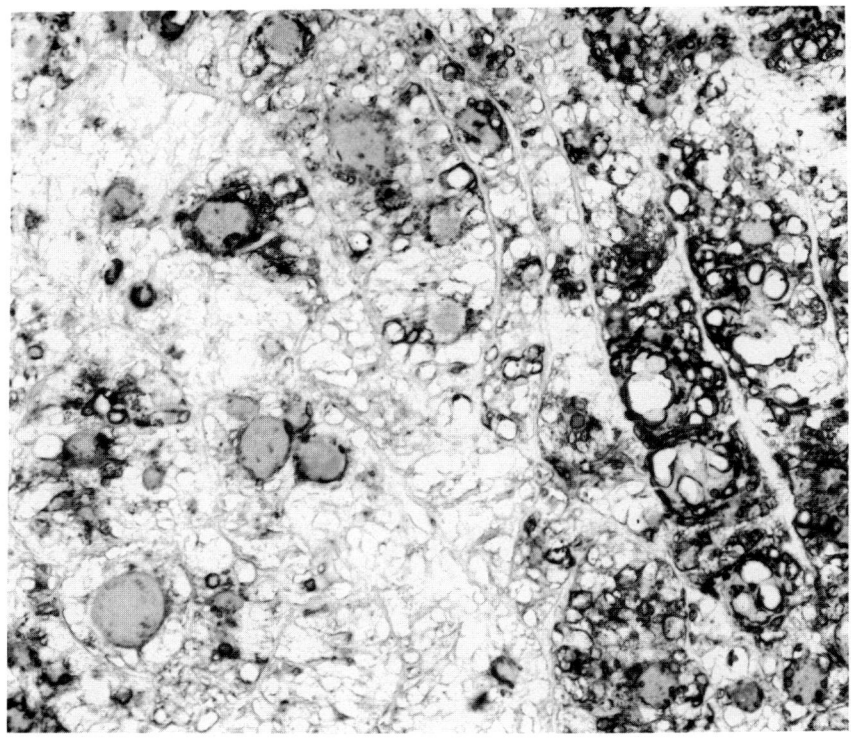

Figure 9.74 Thyroid carcinoma of intermediate type. A little less than half the tumour cell population shows immunoreactivity for thyroglobulin, most intensely in areas of follicular differentiation (ABC technique with antibodies against human thyroglobulin, × 165).

shows a staining pattern indicating co-existence of both thyroglobulin and peptide hormones. Finally, a fourth component of cells appears indifferent, being non-reactive with the antibodies used (Ljungberg *et al.*, 1983; Ljungberg and Nilsson, 1991). Immunoreaction for thyroglobulin is usually most conspicuous in follicular and cribriform areas of the tumour. Tumour cells showing immunoreactivity for chromogranin A and peptide hormones are usually most prominent in the solid parts. Not infrequently, however, they are seen to be intermingled with thyroglobulin-positive cells and may also occur together with such cells in the walls of neoplastic follicles.

Electron microscopic studies are compatible with the immunohistochemical results, suggesting the presence of several types of

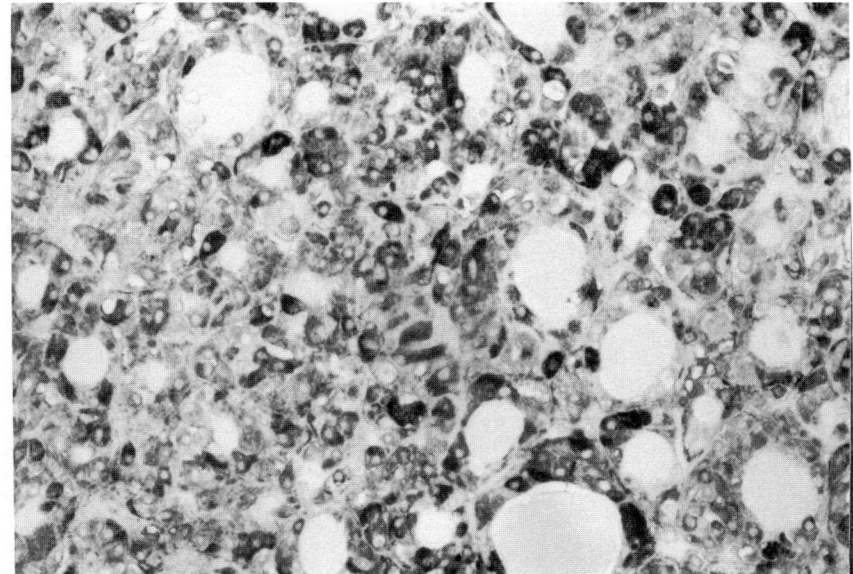

Figure 9.75 Same case as shown in Fig. 9.74. The majority of the tumour cells show positive staining for chromogranin A (ABC technique with antibodies against human chromogranin A, × 165).

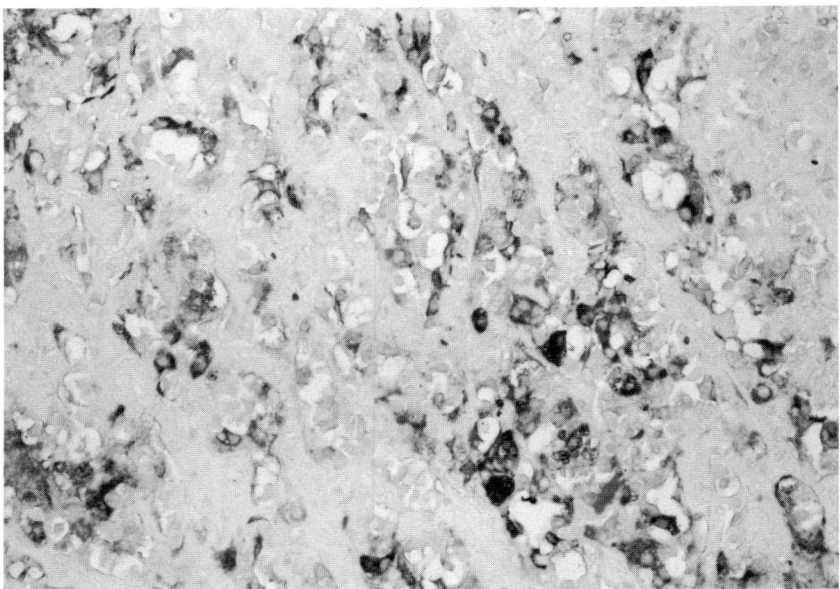

Figure 9.76 Same case as in Figs 9.74 and 9.75. Immunoreactive neurotensin was demonstrated in a minor portion of the tumour cell population. (Besides this peptide, calcitonin and somatostatin were also demonstrated immunohistochemically in the tumour.) (ABC technique, with antibodies against bovine neurotensin, × 165).

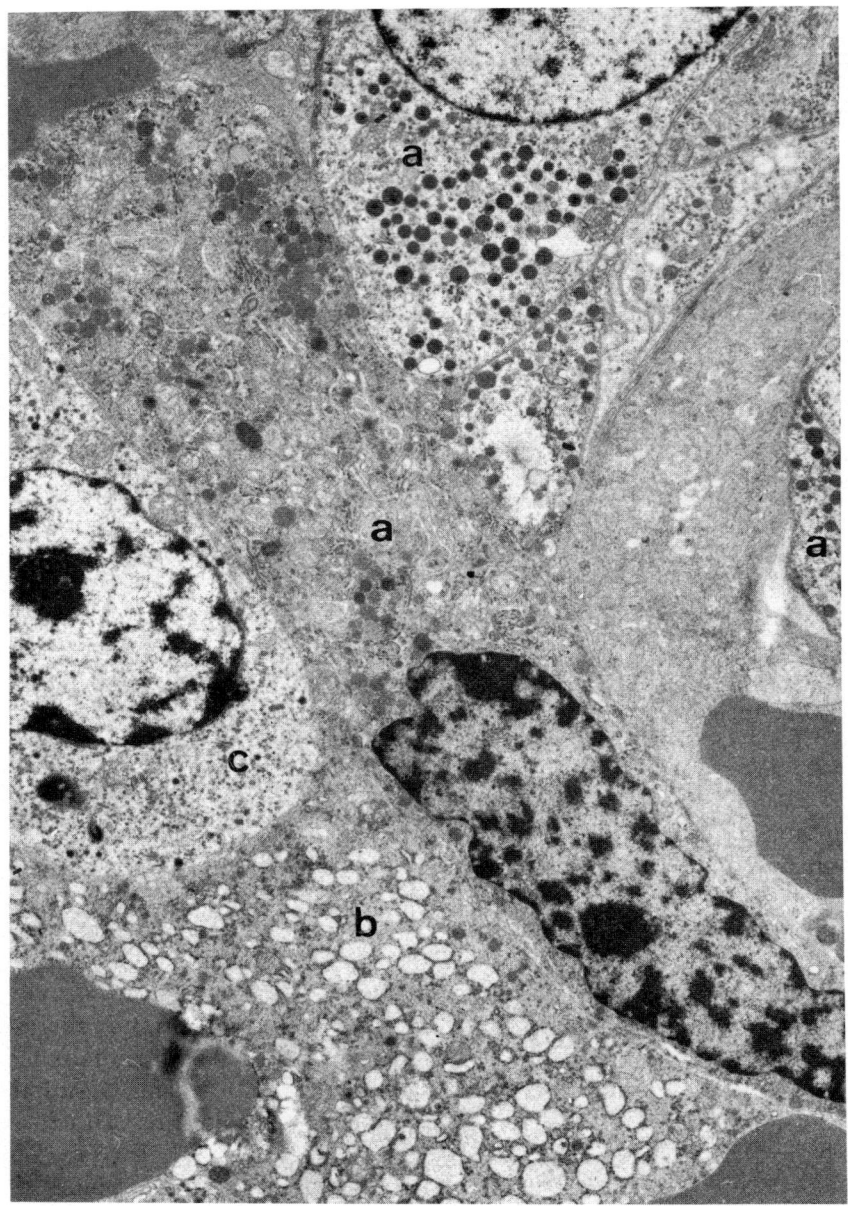

Figure 9.77 Thyroid carcinoma of intermediate type, well-differentiated form. Electron micrograph showing tumour cells with secretory granules typical of parafollicular cells (a), a portion of a cell with numerous dilated cisternae of rough endoplasmic reticulum as found in follicular cells (b), and a part of an indifferent cell lacking specific organelles (c) (× 7 500). Reprinted from Ljungberg, O. and Nilsson, P.O. (1991), Intermediate thyroid carcinoma in humans and ultimobranchial tumors in bulls: a comparative morphological and immunohistological study. *Endocrine Pathology*, **2**, 24–39, by permission of Blackwell Scientific Publications, Inc.

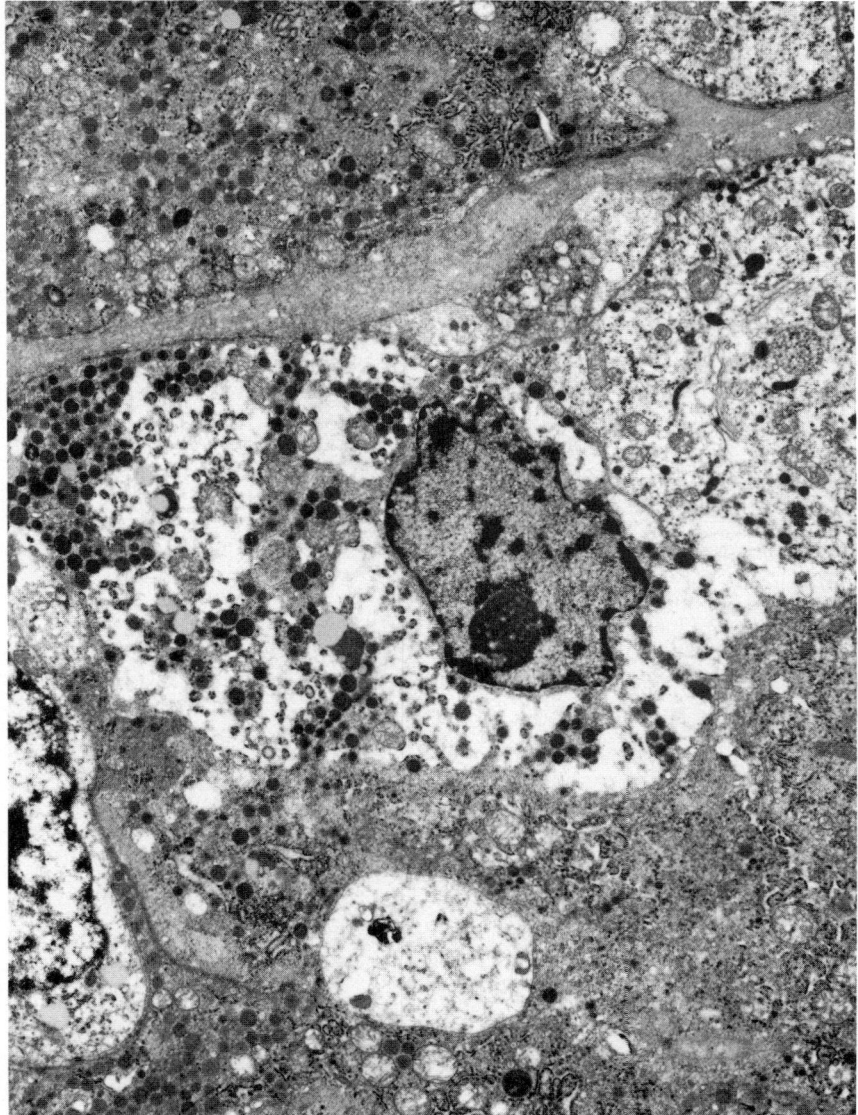

Figure 9.78 Same case as illustrated in Fig. 9.77. In the centre of the picture an intermediate type of tumour cell is seen having features of both a parafollicular cell and a follicular cell (× 8 000).

cells in the tumour (Figs 9.77 and 9.78). One type of cell reveals neurosecretory granules similar to those seen in normal and neoplastic parafollicular cells. Another type shows numerous dilated cisternae of endoplasmic reticulum as found in follicular cells. A third type is intermediate, with features of both of the aforementioned types, and finally, a fourth type looks indifferent being devoid of specific organelles (Ljungberg and Nilsson, 1991).

9.8.3 Pitfalls in interpretation of immunohistochemical staining

When using commercial antisera against the neuropeptides of interest in this context, special attention should be paid to the fact that certain antisera may have been raised against the peptide bound to thyroglobulin, in order to enhance the antigenicity. Such antisera may therefore be contaminated with antibodies against thyroglobulin, which may cause serious misinterpretation of the staining, unless they are absorbed with thyroglobulin. These antisera should be avoided in the immunohistochemical analysis of thyroid neoplasms in general. Instead, antisera raised against immunogens coupled with large molecules unrelated to thyroid tissue (e.g. haemocyanin) should be used.

Other pitfalls when interpreting immunohistochemical sections from cases suspected of being intermediate type carcinomas, may be related to entrapment of residual normal follicles or even single normal follicular cells showing thyroglobulin immunoreactivity, within a tumour which is otherwise composed of cells with neuroendocrine features (p. 30). Unless these thyroglobulin-positive elements show unequivocally malignant features, such a tumour should not be regarded as being of the intermediate type, but referred to the medullary carcinoma group. Another diagnostic problem may be the possibility of secondary uptake of one hormone into tumour cells of a neoplasm producing the opposite hormone, e.g. transfer of thyroglobulin into medullary carcinoma cells from the normal follicles which the tumour has invaded. Electron microscopic examination may be helpful solving this problem, by identifying the characteristic spectrum of tumour cell types present in the intermediate carcinoma. The demonstration of the co-existence of both hormones in metastatic tumour tissue strongly supports the diagnosis of intermediate carcinoma.

9.8.4 Nature of intermediate type of thyroid carcinoma

There is indirect evidence that the intermediate type of thyroid carcinoma may be histogenetically related to the ultimobranchial body, considered to give rise to the lateral thyroid anlage during early embryonic life (see also p. 19). Studies in certain mammals indicate that during embryonic life, undifferentiated stem cells in the ultimobranchial bodies are capable of differentiating not only into follicular cells but also into parafollicular cells, which diffusely colonize the developing thyroid parenchyma (Calvert and Isler, 1970; Calvert, 1972; Kameda *et al.*, 1980). The so-called solid cell nests, sometimes incidentally found at microscopy of the adult human gland, and believed to represent the vestiges of the ultimobranchial bodies, have been shown to contain a complex arrangement of various cell types also found in tumours of the intermediate type. They include calcitonin-immunoreactive parafollicular cells, thyroglobulin-positive follicular cells, scattered within the solid nests of immunohisto-chemically indifferent, immature-looking cells. Cystic structures lined by epidermoid cells, sometimes with an admixture of follicular and parafollicular cells may accompany the solid nests. It is of interest that cysts and mixed follicles have also been described in these remnants, containing mucinous material and having an admixture of mucinous cells in their walls (Harach, 1985a). Mixed follicles of similar type have also been observed in the rat (Calvert and Isler, 1970). From a histogenetic point of view, it may be relevant to mention that in one

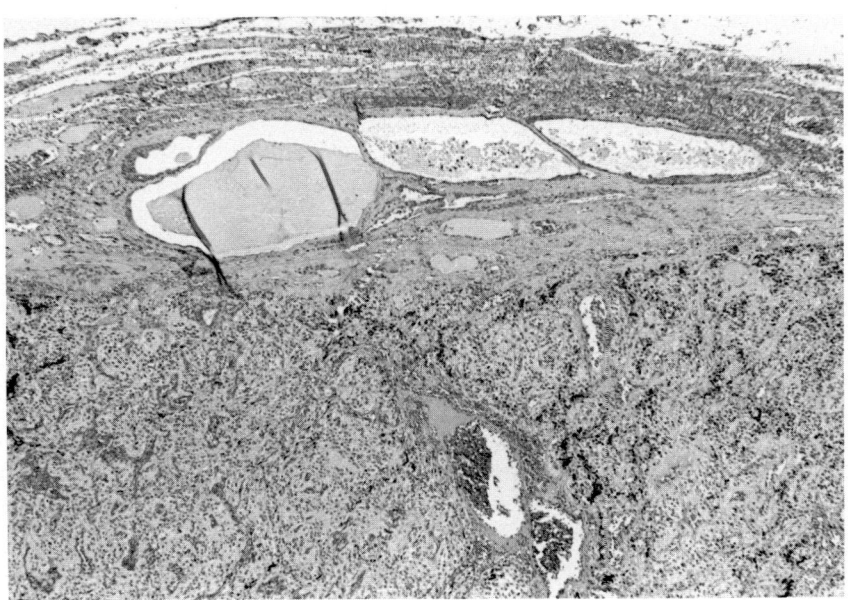

Figure 9.79 Thyroid carcinoma of intermediate type. In the fibrous zone between the tumour and the adjacent thyroid parenchyma there is is a benign-looking structure composed of cysts lined by multilayered, flat, squamoid epithelial cells and containing detritus with degenerated cells. This structure closely resembles ultimobranchial remnants in human thyroid (H&E, × 36).

of the cases of intermediate type thyroid studied by the author, several benign-looking cysts lined by squamoid cells, and small tubules of ultimobranchial type were found in the fibrous zone between the tumour and the surrounding thyroid parenchyma (Fig. 9.79).

In addition, human intermediate carcinoma shows close structural and immunohistochemical similarities to a common thyroid neoplasm in the bull, which is generally considered to be of ultimobranchial origin (Ljungberg and Nilsson, 1985).

Interestingly, bull ultimobranchial tumours are usually associated with a marked hyperplasia of the parafollicular cell system present in the non-tumorous parts of the gland, and sometimes even pure parafollicular cell tumours (i.e. medullary carcinomas) may be seen co-existing with the ultimobranchial tumours. A similar parafollicular cell hyperplasia, has also been noticed in human cases, but less frequently, and frank medullary carcinomas have not been observed (Ljungberg and Nilsson, 1991). However, the biological significance of this finding in human cases must be judged with caution, in the light of the fact that parafollicular cell hyperplasia may also occur in normal glands, especially in older individuals (p. 13).

It is thus believed that, although both medullary carcinoma and intermediate type of carcinoma appear to be histogenetically related to the ultimobranchial body, the former shows evidence of a pure parafollicular cell differentiation, whereas the latter develops both follicular and parafollicular cells, as well as intermediate cell forms, and sometimes even immature structures of the type seen in the ultimobranchial vestiges of the adult gland. It should be mentioned here that a biphasic differentiation has recently been observed in ovarian strumal carcinoids in which intermediate tumour cell forms, similar to those reported in the intermediate thyroid carcinoma, have also been identified (Kimura et al., 1986; Snyder and Tavassoli, 1986).

9.8.5 Aspects of classification and clinical features

The recognition of the intermediate type of thyroid carcinoma as a separate entity also raises new problems for the classification of thyroid tumours. The term intermediate type has been suggested because it is the author's belief that it is to be positioned in a broad intermediate zone of a continuous spectrum of differentiated thyroid carcinomas for which the pure follicular cell tumours (whether papillary carcinoma should also be included here is uncertain) and parafollicular cell carcinoma (medullary carcinoma) constitute the two extremes. It is also conceivable that poorly differentiated forms of thyroid carcinoma (p. 187) and the undifferentiated (anaplastic) group of carcinomas are heterogeneous groups comprising the less differentiated and undifferentiated counterparts not only of follicular and papillary carcinomas, but also of the intermediate type, and of medullary carcinoma.

From a prognostic and therapeutic point of view it is debatable whether it is justified to separate the intermediate type of thyroid carcinoma as a distinct entity. The author's series indicates that the tumour behaves aggressively with a tendency to lymphatic as well as haematogenous dissemination. Of 18 patients six developed metastases and four died from malignant disease, 1 month to 15 years after the primary operation. There have been occasional reports of thyroid carcinomas that had been classified as medullary carcinomas on histopathological grounds, but found to be able to concentrate radioiodine and also to regress on treatment with iodine–131 (Nusynowitz et al., 1982; Spencer et al., 1982). It is possible that such tumours are in fact examples of the intermediate category. However, more clinical knowledge is needed to evaluate the behaviour and prognosis of intermediate carcinoma and to reach a consensus on its proper classification. Until then, for practical purposes, it would seem reasonable to reserve such an intermediate group only for those neoplasms in which each of the two differentiated cell types (i.e. follicular and parafollicular cells) comprise a considerable part (perhaps more than 25% each) of the total tumour cell population. Those cases in which one of the two functioning cell types predominates, the other forming a minority cell population, may then be included within the classic follicular or medullary group, respectively. The very rare tumour which is mainly composed of immature, immunohistochemically indifferent cells and develops tubules and cysts of ultimobranchial type may be regarded as an immature variant within the intermediate group. So defined, the intermediate type of thyroid carcinoma is probably a very rare neoplasm, less common than medullary carcinoma. However, since the series studied by the author is partly composed of selected cases the exact frequency of the neoplasm among thyroid malignancies in general cannot be stated.

9.9 Undifferentiated (anaplastic) carcinoma

Undifferentiated thyroid carcinoma, also referred to as anaplastic carcinoma, may be defined as a neoplasm composed in part or exclusively of undifferentiated cells. It belongs to the most aggressive malignant tumours occurring in humans. It accounts for approximately 5–10% of all thyroid carcinomas and mainly presents in the elderly (Meissner and Warren, 1969). There is some evidence that the incidence of differentiated carcinoma is decreasing (Rossi et al., 1978). This may be due to both a genuine decrease in occurrence as well as improved diagnosis with exclusion of tumours found not to be related to this category (see below). Most cases occur during the seventh and eighth decades of life, with a slight female preponderance. Patients with this tumour have been seen in the second and third decades but these are exceptions.

Undifferentiated carcinoma typically presents as a rapidly growing, firm or hard mass in the region of the thyroid gland. Typical local symptoms include hoarseness, dysphagia and dyspnoea. The further course is very rapid, and nodal as well as distant metastases are common. Involvement of vital structures in the neck is usually the primary cause of death. The mortality rate is almost 100%, mean survival being less than a year. Commonly, local disease is so advanced at initial presentation that surgery cannot be carried out, and treatment is restricted to radiation and chemotherapy. The diagnosis is often made only by fine-needle or surgical biopsy; some cases are diagnosed at autopsy. Occasional patients have a somewhat longer survival. In these instances the primary tumour has usually been small and resectable (Aldinger *et al.*, 1978).

9.9.1 Pathological features

Grossly, the tumour is usually ill-defined, and replaces large parts of the gland. It tends to infiltrate into adjacent soft tissues of the neck and neighbouring structures, such as the trachea and oesophagus or the overlying skin.

The three major histological growth patterns seen in undifferentiated thyroid carcinoma are giant-cell, spindle-cell and squamoid patterns (Hedinger *et al.*, 1988). The clinical course of tumours having any of these patterns appears to be the same, and moreover, they are often seen in combination. Therefore, division of undifferentiated carcinoma into subtypes seems unjustified (Rosai *et al.*, 1985). Recent studies including immunohistochemical analysis have shown that small cell tumours, previously considered as a subgroup of undifferentiated carcinoma are, for the most part malignant lymphomas (Aldinger *et al.*, 1978; Rossi *et al.*, 1978; Burt *et al.*, 1985; Ralfkiaer *et al.*, 1985; Myskow *et al.*, 1986), the rest probably representing poorly differentiated (insular) carcinomas (Carcangiu *et al.*, 1984), poorly differentiated medullary carcinomas (Mendelsohn *et al.*, 1980; Nieuwenhuijzen Kruseman *et al.*, 1982; Mambo and Iriwn, 1984; Myskow *et al.*, 1986; Rosai and Carcangiu, 1987) and possibly immature variants of the intermediate type of carcinoma (p. 187). These tumour forms should not be included within the group of undifferentiated carcinoma.

Microscopically, most cases are composed of solid masses of irregularly arranged, highly pleomorphic cells with marked nuclear

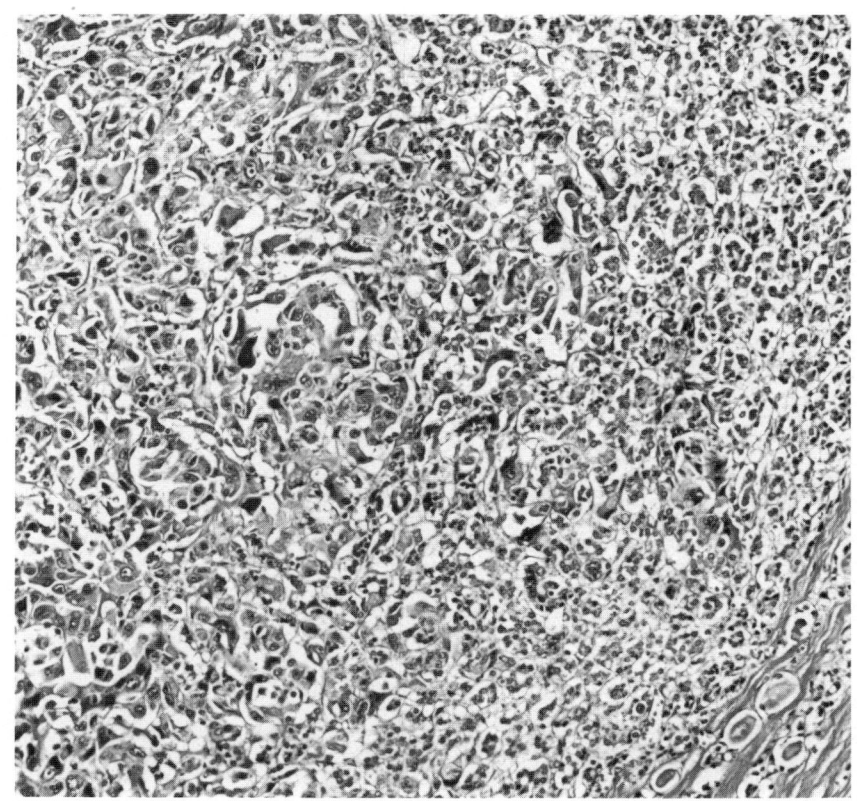

Figure 9.80 Undifferentiated carcinoma. Diffuse growth of irregularly and loosely arranged, large, highly pleomorphic tumour cells. To the right, there is a portion showing some degree of differentiation with development of small follicles lined by small uniform cells (H&E, × 125).

atypia and a high mitotic rate (Fig. 9.80). Sometimes foci of a well-differentiated component, usually of papillary or follicular type, may be observed (Figs 9.80 and 9.81). Tumour cells may also show oxyphil change. Necrosis and an inflammatory reaction with extensive infiltration of neutrophil granulocytes are common features, as is a marked invasive tendency. Foci of necrosis may have a geographic configuration, surrounded by palisading tumour cells. Neoplastic, multinucleate giant cells with bizarre nuclei and many mitoses are often seen. In addition, some lesions show a component of multinucleated giant cells with uniform nuclei, which morphologically, histochemically and ultrastructurally are analogous to normal osteoclasts (Fig. 9.83) (Gaffey *et al.*, 1991). The spindle-cell pattern often shows fascicular or storiform arrangements of elongated cells,

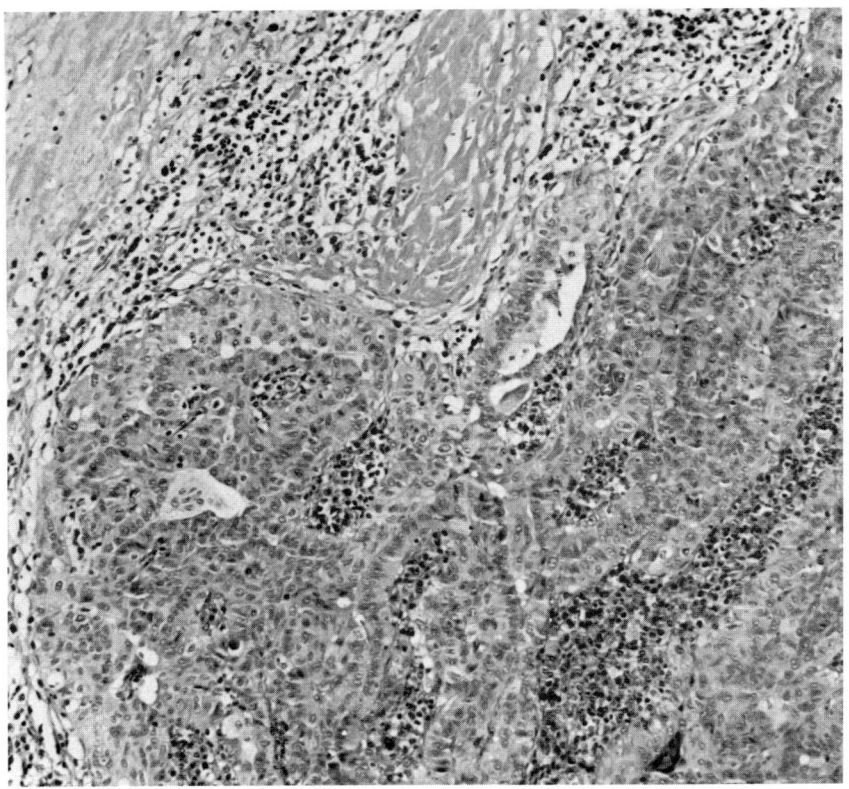

Figure 9.81 Undifferentiated carcinoma with a focus of differentiated papillary carcinoma (H&E, × 125).

marked vascularization, and, rarely, even areas with cartilaginous and/or osseous metaplasia. The frequent admixture of sarcoma-like features may cause diagnostic problems, the tumour being mistaken for fibrosarcoma, malignant fibrous histiocytoma (Fig. 9.82), giant cell tumour of bone, haemangiosarcoma (Fig. 9.83) or even osteosarcoma or chondrosarcoma. It should be pointed out, however, that thyroid tumours with sarcomatous features are nearly always undifferentiated carcinomas. Diagnostically useful features in the carcinomas (Carcangiu *et al.*, 1985a) include the palisading of tumour cells at the margins of necrotic areas and an 'angiotropic quality', i.e. a tendency of tumour cells to invade the walls of veins and arteries within the tumour.

Some undifferentiated carcinomas show evidence of squamous metaplasia (Fig. 9.84), with a component indistinguishable from

Plate 14 The small specimen is a slice from a thyroid lobe showing a papillary microcarcinoma, 7×8 mm in size, removed from a 19-year-old woman. The larger specimen represents large ipsilateral cervical lymph node metastases, by which the tumour had revealed itself clinically.

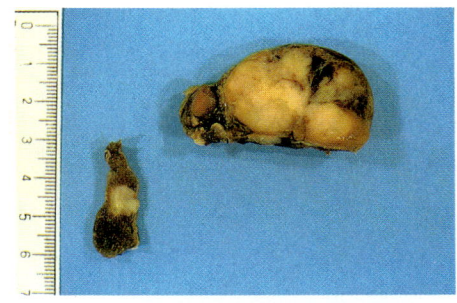

Plate 15 Typical papillary thyroid carcinoma with ill-defined borders, some white streaks in the centre and small cysts in its marginal parts.

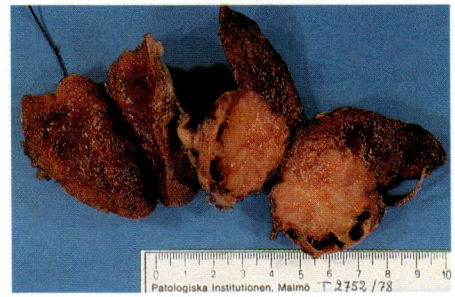

Plate 16 Papillary carcinoma with follicular structures. Same technique as in Fig. 9.17. The neoplastic follicles show a positive immunoreaction for low-molecular-weight cytokeratins.

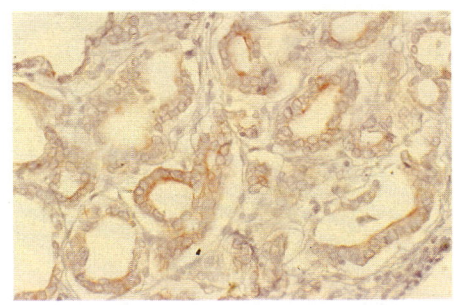

Plate 17 Minimally invasive follicular carcinoma. The fibrous capsule is partly thickened with calcifications and its inner contour is irregular with suspicious areas of tumour infiltration. Central scarring is also seen.

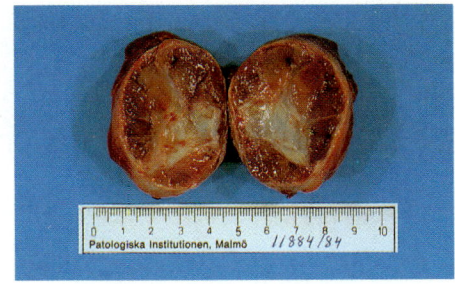

Plate 18 Widely invasive follicular carcinoma showing extensive multinodular growth, which may simulate a benign multinodular goitre.

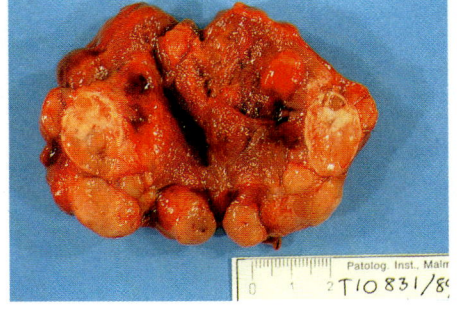

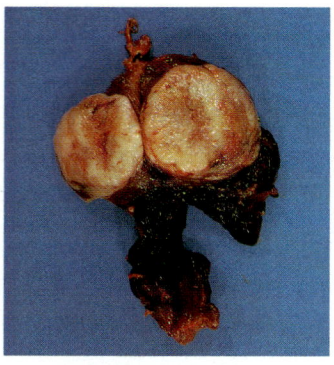

Plate 19 Sporadic medullary carcinoma. Well-defined, partly encapsulated tumour with slightly lobulated outer contour and bulging, greyish-white, finely granular cut surface.

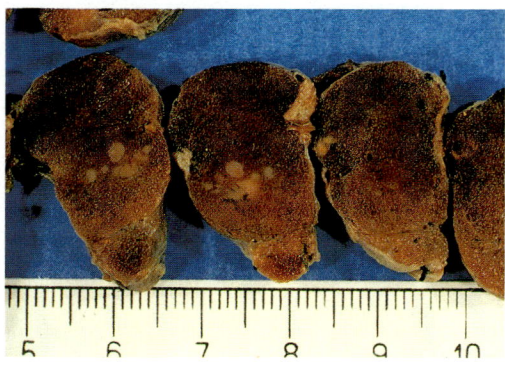

Plate 20 Familial medullary thyroid carcinoma at an early stage of development. A cluster of discrete tiny, greyish nodules is seen in the central part of the lobe. Similar changes were present in the opposite lobe.

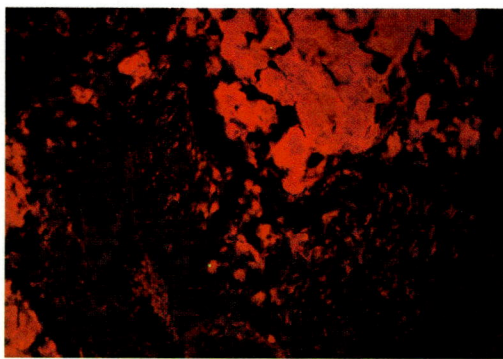

Plate 21 Amyloid clumps in medullary carcinoma, stained with alkaline Congo red and examined in a fluorescence microscope with filters for rhodamine fluorescence. The amyloid clumps show a bright brick-red fluorescence against the dark background.

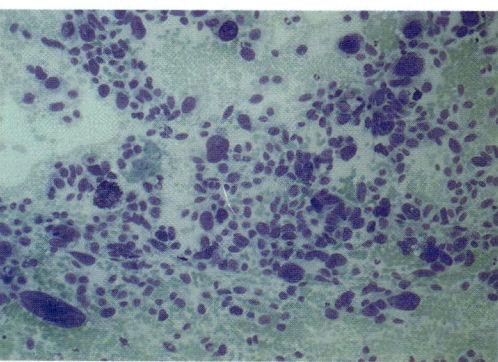

Plate 22 Fine needle aspirate from a medullary carcinoma with anaplastic component. Extremely pleomorphic cells with huge, grotesque nuclei, admixed with small undifferentiated cells with chromatin dense rounded or oval nuclei. Red cytoplasmic granules, typical of classic medullary carcinoma, could not be found (May–Grünwald Giemsa stain).

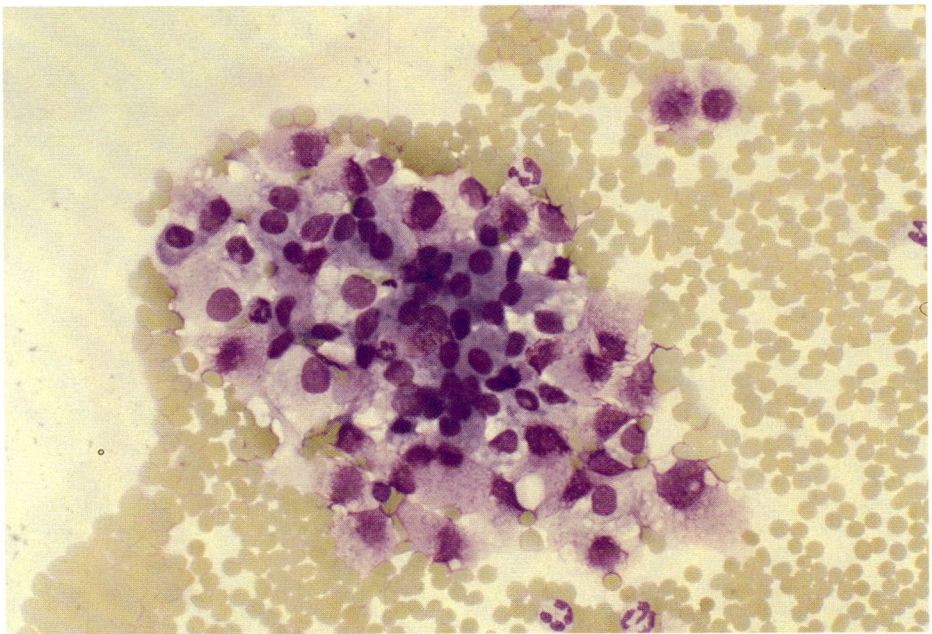

Plate 23 Fine needle aspirate of medullary carcinoma. Loose cluster of cytoplasm-rich, polygonal tumour cells, often with eccentric nuclei and purple granulation of the cytoplasm (May–Grünwald Giemsa stain).

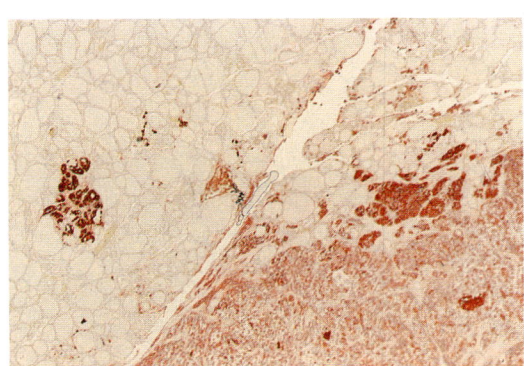

Plate 24 Case of familial medullary carcinoma. Immunostaining for calcitonin. Foci of parafollicular cell hyperplasia stain more intensely than does the medullary carcinoma seen in the lower right of the picture (ABC technique with antibodies against human calcitonin).

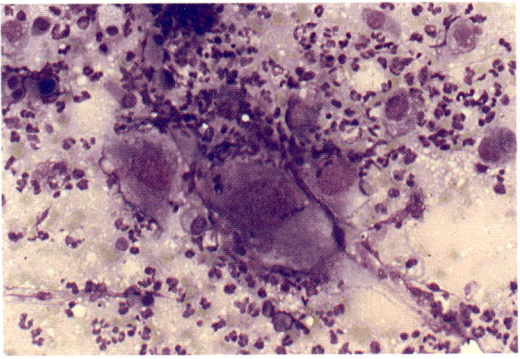

Plate 25 Fine needle aspirate from undifferentiated thyroid carcinoma, giant cell variant. The picture shows a group of large, highly pleomorphic tumour cells, necrotic material and leukocytes (May–Grünwald Giemsa stain).

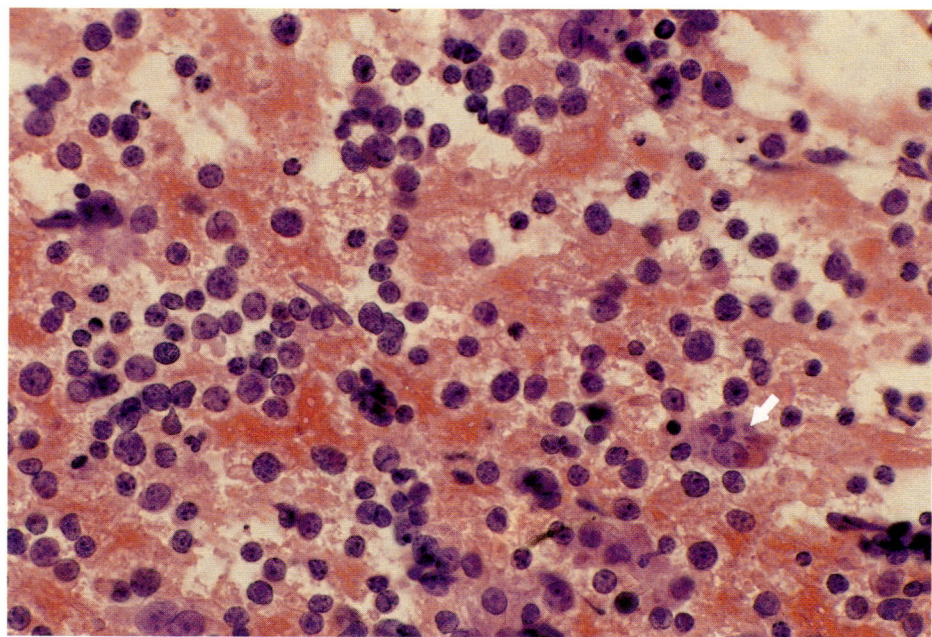

Plate 26 Fine needle aspirate of malignant lymphoma, low grade. The picture shows numerous discrete, atypical small lymphoid cells with prominent nuclei. In contrast to the dense nuclei of normal lymphocytes, these nuclei have an open chromatin pattern and often multiple nucleoli and indented membranes. To the right (arrow) a histiocyte with phagocytized nuclear debris is seen.

Plate 27 Parathyroid adenoma with remnant of normal parenchyma. Fat staining of frozen section with oil-red O. The remnant (at the top and on the right) shows a pronounced accumulation of fat droplets, evenly distributed in the chief cells. At the bottom and on the left a portion of the adenoma with a reduced amount of fat in the epithelial cells.

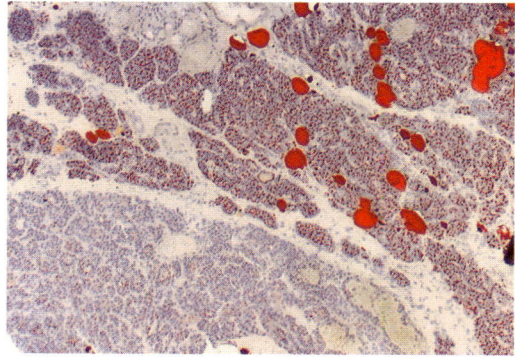

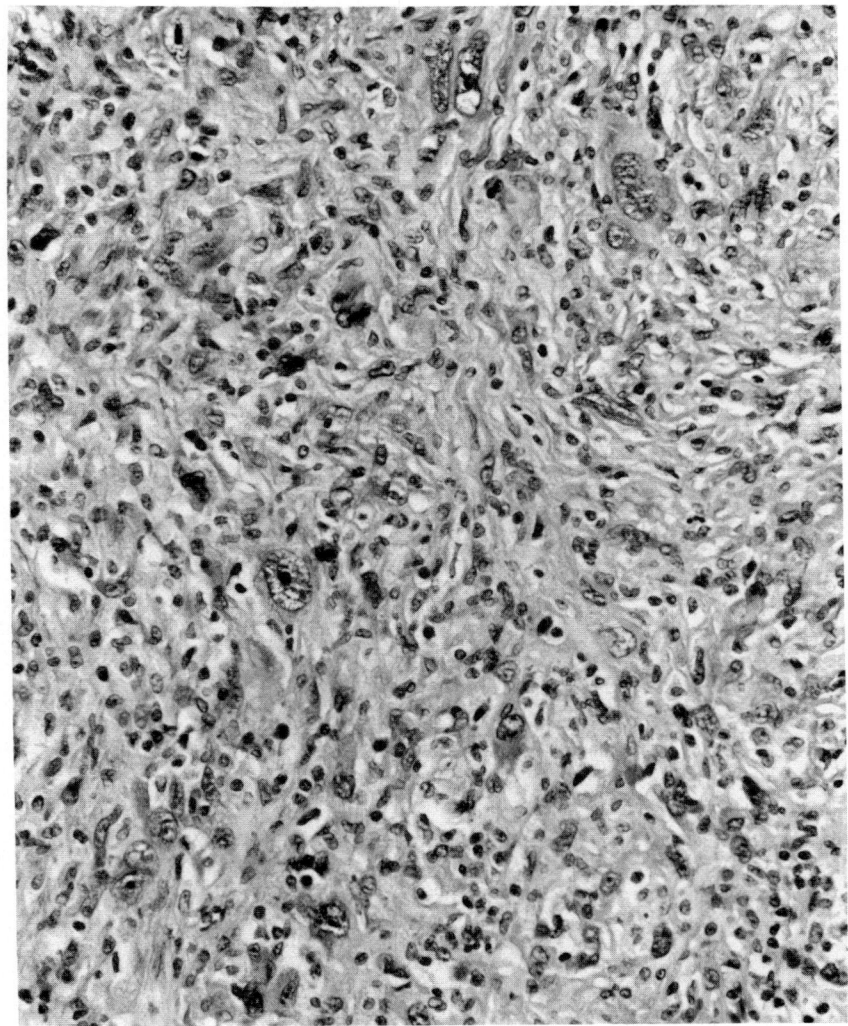

Figure 9.82 Another part of the tumour illustrated in Fig. 9.81, showing sarcoma-like features, simulating a malignant fibrous histiocytoma (H&E, × 250).

squamous cell carcinoma. Tumours with squamoid foci, but otherwise showing the features of undifferentiated carcinoma should be included in the major group of undifferentiated carcinoma on the basis of a similar clinical behaviour. On the other hand, according to the WHO classification (Hedinger *et al.*, 1988), the rare examples of tumours having a pure squamous cell pattern throughout should be

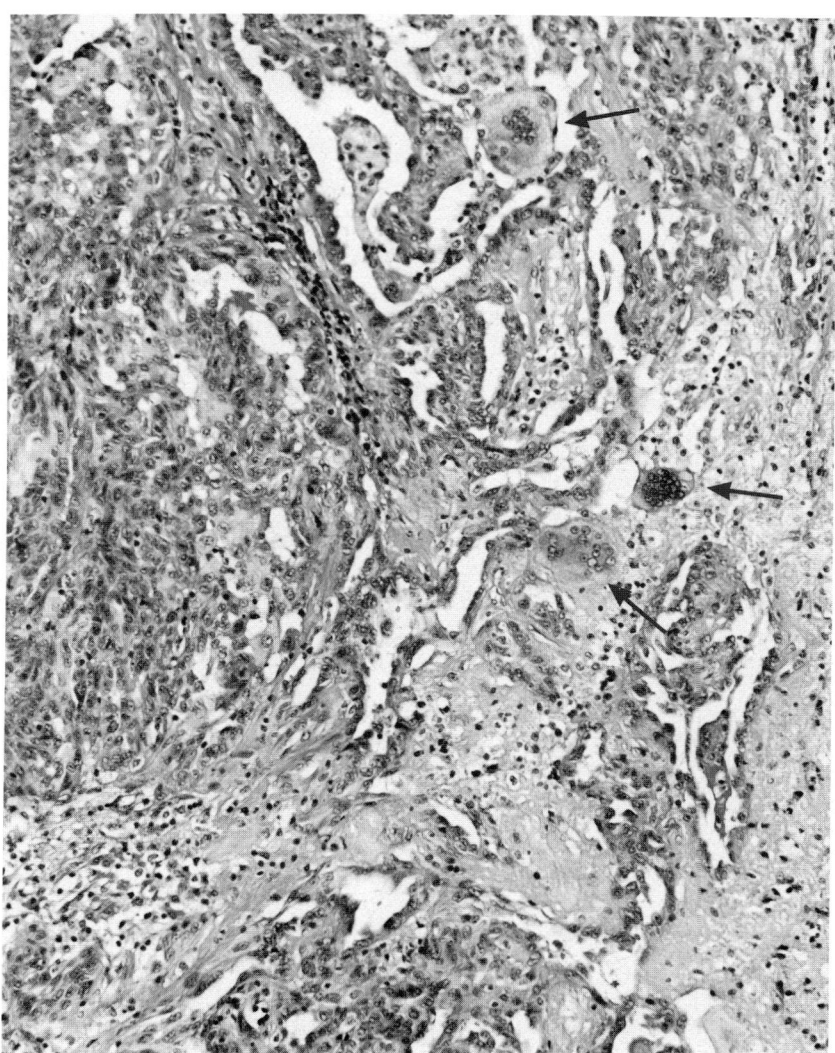

Figure 9.83 Undifferentiated carcinoma; same case as shown in Figs 9.81 and 9.82. A portion of the tumour resembling an angiosarcoma. A few benign-looking osteoclast-like multinucleate giant cells are also present in the stroma (arrows) (H&E, × 125).

separated as a subgroup under the general category of 'other carcinomas'. Pure squamous cell carcinomas of the thyroid tend to behave aggressively, although their growth rate may be somewhat lower than that of undifferentiated tumours (Huang and Assor, 1971).

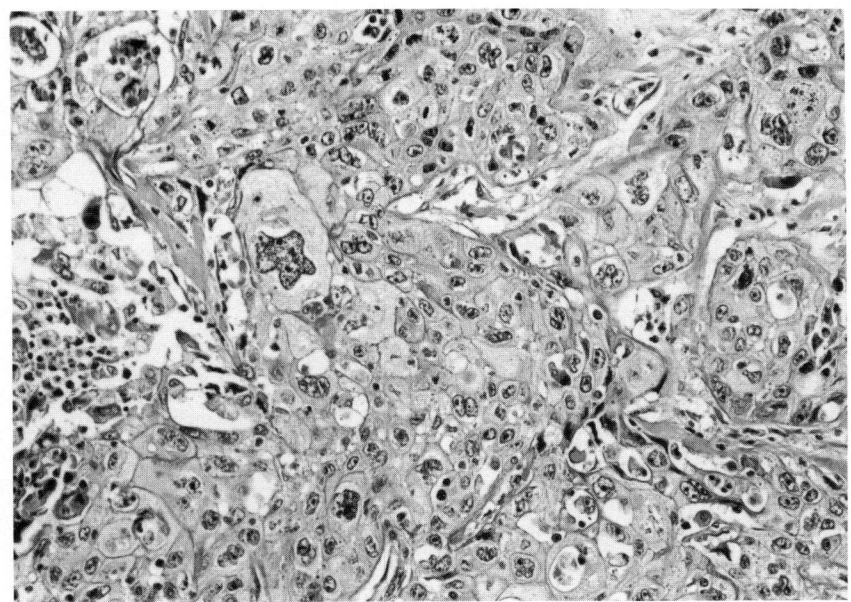

Figure 9.84 Undifferentiated carcinoma with squamous metaplasia (H&E, × 165).

9.9.2 Fine-needle aspiration cytology

When there is a rapidly expanding mass in the neck in an old person, clinically suspected of being an undifferentiated thyroid carcinoma, fine-needle aspiration cytology is of particular value. This procedure may give a rapid confirmation of the diagnosis, spare the patient from surgery and allow palliative radiotherapy or chemotherapy to be started without delay. Smears from undifferentiated carcinomas contain characteristic pleomorphic cells with bizarre nuclei and abnormal nucleoli (Plate 25). Mitotic figures and multinucleated giant cells are common. Also present are spindle cell elements showing fusiform cytoplasm and elongated, hyperchromatic nuclei which may resemble the cells of a sarcoma. Malignant squamous cells may also occur. Frequently there is an admixture of polymorphonuclear leukocytes and necrotic material. This component may be conspicuous and a diligent search for malignant cells in such smears is essential.

9.9.3 Immunohistochemical and ultrastructural features

The most useful immunohistochemical marker of undifferentiated carcinoma is cytokeratin, an epithelial marker which is present in

50–100% of cases (Carcangiu *et al.*, 1985a). However, tumours with spindle cell (sarcoma-like) patterns may lose their affinity for antibodies against cytokeratins. Beltrami *et al.* (1989) have recently recommended the application of antibodies to desmoplakins, which are antigens that are strictly related to the epithelial phenotype and represent desmosomal-related proteins that also seem to persist in cells that have lost the ability to synthesize cytokeratin; their use however is limited to cryostat sections. Vimentin is usually present in the spindle cell component, but vimentin may be present in both epithelial and stromal cells. Therefore, this marker is of little help in distinguishing carcinoma from mesenchymal neoplasms (Schröder *et al.*, 1986).

Thyroglobulin immunoreactivity has been demonstrated in focal areas showing follicular or papillary differentiation, but is usually not detected in truly undifferentiated parts of the tumours (Ljungberg *et al.*, 1984; Harach and Franssila, 1988). Likewise, calcitonin and other neuro-peptides characteristic of medullary carcinoma are absent in the majority of undifferentiated thyroid carcinomas. Nevertheless, immunoreactivity for such peptides has been described in a small number of undifferen-tiated carcinomas, usually occurring in a minority cell population (Mendelsohn *et al.*, 1980; Ljungberg *et al.*, 1984; Logmans and Jobsis, 1984). Another case, also found to have ultrastructural features of a neuroendocrine neoplasm, was described on p. 212. Undifferentiated thyroid carcinoma may thus be considered as a heterogeneous group of neoplasms including occasional examples that are histogenetically related to medullary carcinoma and the intermediate type of carcinoma.

Ultrastructurally, undifferentiated carcinomas may show epithelial markers, such as specialized cell junctions and microvilli, follicular remnants and transitional forms between undifferentiated and differentiated follicular tumour cells. However, these characteristics are not invariably present; in about half of the cases the tumours appear totally undifferentiated, and are devoid of such markers (Newland *et al.*, 1981; Carcangiu *et al.*, 1985a).

9.9.4 *Nature of undifferentiated carcinoma*

Most cases of undifferentiated thyroid carcinoma appear to arise from pre-existing, well-differentiated tumours. Many patients have had a long history of nodular goitre or well-differentiated thyroid carcinoma, prior to the development of the undifferentiated tumour. In addition, careful sampling of undifferentiated carcinomas often reveals foci of a well-differentiated component (Figs 9.80 and 9.81).This is often a papillary carcinoma, but the other main forms, including medullary carcinoma and the poorly differentiated form, may also be encoun-tered. This suggests that undifferentiated carcinoma arises as a result of anaplastic transformation of a pre-existing well-differentiated neoplasm (Harada *et al.*, 1977; Carcangiu *et al.*, 1985a). Anaplastic transformation may also occur in a distant metastasis of a well-differentiated thyroid carcinoma and it has even been reported in a median cyst of thyroglossal duct remnant (Nussbaum *et al.*, 1981).

9.10   Other forms of carcinoma

9.10.1 *Squamous cell carcinoma of the thyroid gland*

Squamous cells may occur normally in the ultimobranchial remnants of the thyroid gland or in association with normal follicular cells (Harcourt-Webster, 1966; LiVolsi and Merino, 1978). They may also be seen in remnants of the thyroglossal duct and often occur as a component of the epithelium lining cysts developing from such remnants (p. 41).

Benign squamous cells have been observed as an expression of squamous cell metaplasia in various thyroid lesions. It may thus occur in Hashimoto's thyroiditis, especially in its fibrous variant (Fig. 4.9), and has been reported in nodular goitres and benign follicular neoplasms (Huang and Assor, 1971). Squamous cell metaplasia may be an accompanying feature in some follicular and papillary carcinomas, most frequently in the latter (Klinck and Merik, 1951; Meissner and Warren, 1969; Tscholl-Ducommun and Hedinger, 1982; Carcangiu *et al.*, 1985c), undifferentiated thyroid carcinoma (Carcangiu *et al.*, 1985a) and, rarely, in medullary (Schröder, 1988; Dominguez-Malagon *et al.*, 1989), and intermediate type carcinoma (p. 232).

The benign squamous metaplasia found as an accompanying feature in thyroid neoplasms is typically non-keratinizing, lacks intercellular bridges (Shimaoka and Tsukado, 1980) and is composed of benign-looking squamous cells. Such a change does not alter the prognosis of the neoplasm (Huang and Assor, 1971; Carcangiu *et al.*, 1985c; Schröder, 1988).

However, squamous foci may occasionally show varying degrees of atypia. There seems to be a spectrum of changes towards the development of foci of unequivocal squamous cell carcinoma with obviously malignant features, such as cytological atypia and/or frank keratinization or intercellular bridges. It is important to recognize the malignant features of the squamoid component in these cases, because they strongly influence the prognosis. The rare examples of combined follicular cell and squamous cell carcinoma appear to run a highly malignant course. This also applies to the extremely rare pure squamous cell carcinoma of the thyroid. Such tumours have a dismal prognosis and resemble undifferentiated carcinoma clinically.

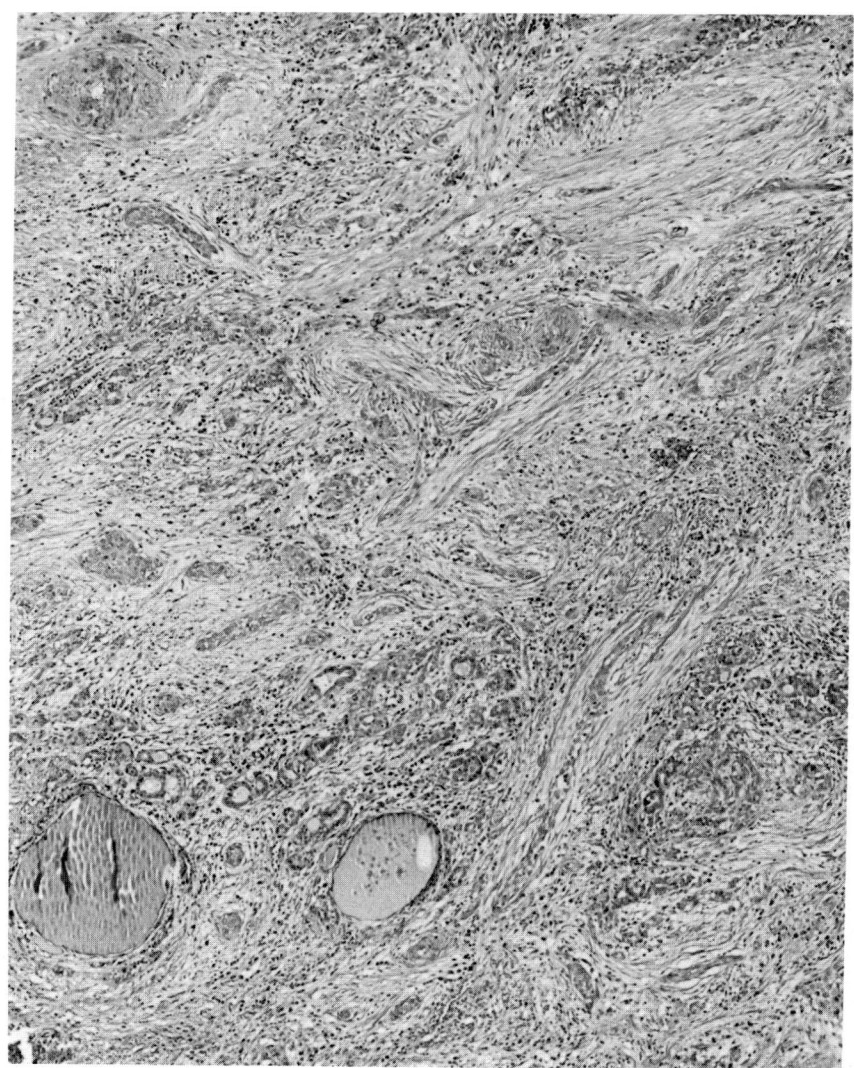

Figure 9.85 Squamous cell carcinoma in a 74-year-old woman presenting with dyspnoea and dysphagia. The thyroid parenchyma is almost completely destroyed and replaced by dense fibrous tissues with chronic inflammation, suggesting a pre-existing chronic thyroiditis. In the fibrous tissue, there are infiltrating islands and branched trabeculae of highly differentiated squamous carcinoma. A few remaining foci of very atrophic follicles are also seen (H&E, × 50).

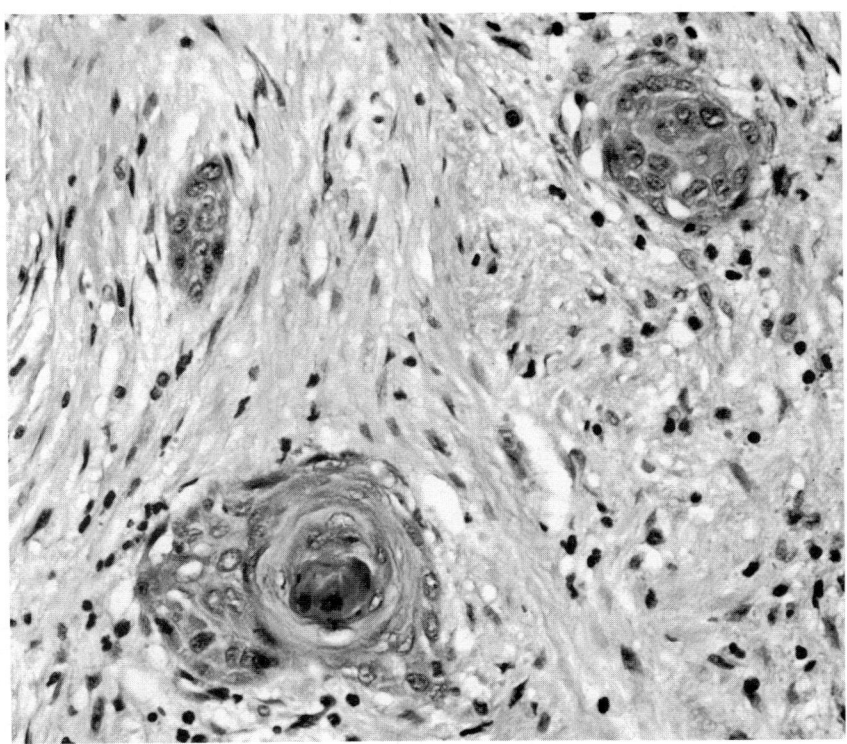

Figure 9.86 Same tumour as shown in Fig. 9.85. Tumour islands are well differentiated with evidence of keratinization and epithelial pearl formation. Despite its high degree of differentiation, the tumour behaved aggressively and the patient died 5 months postoperatively from local recurrence (H&E, × 330).

Squamous cell carcinoma, whether combined or pure, occurs in older patients, often with a long previous history of goitre. They grow rapidly, reach large size, and often present with tracheal or oesophageal compression.

Histologically, a spectrum from well-differentiated, keratin-forming lesions with little cytological atypia (Figs 9.85 and 9.86) to poorly differentiated, basaloid types has been described.

Regardless of the degree of differentiation however, prognosis remains very poor, with survival of less than a year (Huang and Assor, 1971; Bahuleyan and Ramachandran, 1972; Shimaoka and Tsukado, 1980). The tumours are often unresectable and treatment is usually restricted to palliation. The tumours tend to be radioresistant (Shimaoka and Tsukado, 1980; O'Brien, 1988).

It should be stressed here that primary thyroid squamous cell carcinoma should be distinguished from secondary involvement of the gland due to local extension from a carcinoma originating in the larynx, trachea or oesophagus, or from a metastatic carcinoma (e.g. from the lung) (Harrison, 1973; LiVolsi and Merino, 1978).This is not always possible, especially when the tumour is a pure squamous cell carcinoma, when clinical information on a primary outside the thyroid is lacking, and only excisional or fine-needle biopsies are available for microscopic examination.

9.10.2 Mucoepidermoid carcinoma

Mucoepidermoid carcinoma of the thyroid is even rarer than primary squamous cell carcinoma. It has been defined as a tumour histologically characterized by the presence of both squamous features and mucin production. In this respect it resembles mucoepidermoid carcinoma of the salivary glands (Rhatigan et al., 1977; Katoh et al., 1990). In some cases, areas of papillary carcinoma have also been described, co-existing with the mucoepidermoid component (Franssila et al., 1984). Small foci of thyroglobulin-positive cells have been observed in some of these neoplasms, as well as scattered cells showing immunoreactivity for neuroendocrine markers, such as calcitonin, calcitonin gene-related peptide, somatostatin, neurotensin and bombesin (Sambade et al., 1990). Abundance of cytoplasmic tonofilaments is a characteristic ultrastructural feature. The tumour thus seems to have a complex cellular composition. It may be related to papillary carcinoma but, in addition, shows evidence of multidirectional differentiation, having mucinous, squamous, follicular and neuroendocrine features. It has been suggested that the ultimobranchial remnants, claimed to have capabilities for such multidirectional differentiation, may be the origin of mucoepidermoid carcinoma (Fialho et al., 1981; Harach, 1985b).

Clinically, mucoepidermoid carcinoma of the thyroid has been described as a neoplasm of low-grade malignancy, showing a behaviour similar to that of papillary carcinoma. Like papillary carcinoma, it tends to metastasize to the regional lymph nodes (Franssila et al., 1984). Occasionally, however, mucoepidermoid carcinoma may behave aggressively (Schröder, 1988).

References

Adkins, G.F. and Hartley, L. (1987) Differentiated thyroid carcinoma, intermediate type. *Aust. NZ J. Surg.*, **57**, 275–7.

Albores-Saavedra, J., Altamirano-Dimas, M., Alcorta-Anguizola, B. and Smith, M. (1971) Fine structure of human papillary thyroid cancers. *Cancer*, **28**, 763–74.

Albores-Saavedra, J., de la Mora, T., de la Torre-Rendon, F. and Gould, E. (1990) Mixed medullary-papillary carcinoma of the thyroid: a previously unrecognized variant of thyroid carcinoma. *Hum. Pathol.*, **21**, 1151–5.

Albores-Saavedra, J., Monforte, H., Nadji, M. and Morales, A.R. (1988) C-cell hyperplasia in thyroid tissue adjacent to follicular cell tumors. *Hum. Pathol.*, **19**, 795–9.

Albores-Saavedra, J., Nadji, M., Civantos, F. and Morales, A.R. (1983) Thyroglobulin in carcinoma of the thyroid: an immunohistochemical study. *Hum. Pathol.*, **14**, 62–6.

Aldabagh, S.M., Trujillo, Y.P. and Taxy, J.B. (1986) Occult medullary thyroid carcinoma. Unusual histological variant presenting with metastatic disease. *Am. J. Clin. Pathol.*, **85**, 247–50.

Aldinger, K.A., Samaan, N.A., Ibanez, M. and Hill, C.S., Jr (1978) Anaplastic carcinoma of the thyroid. A review of 84 cases of spindle and giant cell carcinoma of the thyroid. *Cancer.*, **41**, 2267–75.

Amara, S.G., Jonas, V., Rosenfeld, M.G. *et al.* (1982) Alternative RNA processing in calcitonin gene expression generates mRNAs encoding different polypeptide products. *Nature (Lond.)*, **298**, 240–4.

Angus, B., Purvis, J., Stock, D. *et al.* (1987) NCL-5D3: a new monoclonal antibody recognizing low molecular weight cytokeratins effective for immunohistochemistry using fixed paraffin-embedded tissue. *J. Pathol.*, **153**, 377–84.

Auer, G.U., Bäckdahl, M., Forsslund, G.M. and Askensten, U.G. (1985) Ploidy levels in nonneoplastic and neoplastic thyroid cells. *Anal. Quant. Cytol. Histol.*, **7**, 97–105.

Bahuleyan, C.K. and Ramachandran, P. (1972) Primary squamous cell carcinoma of the thyroid. *Indian J. Cancer*, **9**, 89–91.

Bartlett, R.C., Myall, R.W.T., Bean, L.R. and Mandelstam, P. (1971) A neuropolyendocrine syndrome: mucosal neuromas, pheochromocytoma and medullary carcinoma. *Oral Surg.*, **31**, 206–20.

Batsakis, J.G., Nishiyama, R.H. and Rich, C.R. (1960) Microlithiasis (calcospherites) and carcinoma of the thyroid. *Arch. Pathol.*, **69**, 493–8.

Bauman, A. and Strawbridge, H.T.G. (1983) Spontaneous disappearance of an atypical Hürthle cell adenoma. *Am. J. Clin. Pathol.*, **80**, 399–402.

Baylin, S.B., Mendelsohn, G., Weisburger, W.R. *et al.* (1979) Levels of histaminase and L-dopa decarboxylase activity in the transition from C-cell hyperplasia to familial medullary thyroid carcinoma. *Cancer*, **44**, 1315–21.

Beaugie, J.M., Brown, C.L., Doniach, I. and Richardson, J.E. (1976) Primary malignant tumours of the thyroid. The relationship between histological classification and clinical behaviour. *Br. J. Surg.*, **63**, 173–81.

Beaumont, A., Othman, S.B. and Fragu, P. (1981) The fine structure of papillary carcinoma of the thyroid. *Histopathology*, **5**, 377–88.

Beltrami, C.A., Criante, P. and Di Loreto, C. (1989) Immunocytochemistry of anaplastic carcinoma of thyroid gland. *Appl. Pathol.*, **7**, 122–33.

Bergholm, U., Adami, H-O., Bergström, R. *et al.* (1989) Clinical characteristics in sporadic and familial medullary thyroid carcinoma. A nationwide study of 249 patients in Sweden from 1959 through 1981. *Cancer*, **63**, 1196–204.

Block, M.A., Jackson, C.E., Greenawald, K.A. *et al.* (1980) Clinical characteristics distinguishing hereditary from sporadic medullary thyroid carcinoma: treatment implications. *Arch. Surg.*, **115**, 142–8.

Böcker, W., Dralle, H., Koch, G. *et al.* (1978) Immunohistochemical and electron microscope analysis of adenomas of the thyroid gland. II. Adenomas with specific cytological differentiation. *Virchows Arch.* [*A*], **380**, 205–20.

Body, J.J., Kellerman, G., Muquardt, C. and Borkowski, A. (1983) Serum immunoreactive calcitonin: useful as a general tumor marker? *Horm. Metab. Res.*, **15**, 624.

Bondeson, L. and Ljungberg, O. (1981) Occult thyroid carcinoma at autopsy in Malmö, Sweden. *Cancer*, **47**, 319–23.

Bondeson, L. and Ljungberg, O. (1984) Occult papillary thyroid carcinoma in the young and the aged. *Cancer*, **53**, 1790–2.

Bondeson, L., Bondeson, A-G., Ljungberg, O. and Tibblin, S. (1981) Oxyphil tumors of the thyroid. Follow-up of 42 surgical cases. *Ann. Surg.*, **194**, 677–80.

Bondeson, L., Azavedo, E., Bondeson A-G *et al.* (1986) Nuclear DNA content and behaviour of oxyphil thyroid tumors. *Cancer*, **58**, 672–5.

Bondeson, L., Bondeson, A-G., Lindholm, K. *et al.* (1983) Morphometric studies of nuclei in smears of fine needle aspirates from oxyphilic tumors of the thyroid. *Acta Cytol.*, **27**, 437–40.

Borup Christensen, S. and Ljungberg, O. (1983) Correspondence. Familial occurrence of papillary thyroid carcinoma. *Br. J. Surg.*, **70**, 508–9.

Borup Christensen, S. and Ljungberg, O. (1984) Mortality from thyroid carcinoma in Malmö, Sweden 1960–77. A clinical and pathologic study of 38 fatal cases. *Cancer*, **54**, 1629–34.

Borup Christensen, S., Ljungberg, O. and Tibblin, S. (1984) A clinical epidemiologic study of thyroid carcinoma in Malmö, Sweden. *Curr. Probl. Cancer*, **8**, 1–49.

Bostwick, D.G. and Bensch, K.G. (1985) Gastrin releasing peptide in human neuroendocrine tumours. *J. Pathol.*, **147**, 237–44.

Burt, A.D., Kerr, D.J., Brown, I.L. and Boyle, P. (1985) Lymphoid and epithelial markers in small cell anaplastic thyroid tumours. *J. Clin. Pathol.*, **38**, 893–6.

Burt, A.D., MacGuire, J., Lindop, G.B. and Browne M.K. (1987) Mixed follicular–parafollicular carcinoma of the thyroid. *Scott. Med. J.*, **32**, 50–1.

Bussolati, A., Van Norden, S. and Bordi, C. (1973) Calcitonin and ACTH-producing cells in medullary carcinoma. *Virchows Arch.* [*A*], **360**, 123–7.

Butler, M. and Khan, S. (1986) Immunoreactive calcitonin in amyloid fibrils of medullary carcinoma of the thyroid. *Arch. Pathol. Lab. Med.*, **110**, 647–9.

Byar, D., Green, S., Dor, P. *et al.* (1979) A prognostic index for thyroid carcinoma: a study of the E.O.R.T.C. Thyroid Cancer Cooperative Group. *Eur. J. Cancer*, **15**, 1033–41.

Cady, B. (1981) Surgery of thyroid cancer. *World J. Surg.*, **5**, 3–14.

Cady, B., Rossi, R., Silverman, M. and Wool, M. (1985) Further evidence of the validity of risk group definition in differentiated thyroid carcinoma. *Surgery*, **98**, 1171–8.

Calmettes, C., Caillou, B., Moukhtar, M.S. *et al.* (1982) Calcitonin and carcinoembryonic antigen in poorly differentiated follicular carcinoma. *Cancer*, **49**, 2342–8.

Calvert, R. (1972) Electron microscopic observations on the contribution of the ultimobranchial bodies to thyroid histogenesis in the rat. *Am. J. Anat.*, **133**, 269–90.

Calvert, R. and Isler, H. (1970) Fine structure of a third epithelial component of the thyroid gland of the rat. *Anat. Rec.*, **168**, 23–42.

Camiel, M.R., Mule, J.E., Alexander, L.L. and Benninghoff, D.L. (1968) Association of thyroid carcinoma with Gardner's syndrome in siblings. *N. Engl. J. Med.*, **278**, 1056–8.

Capella, C., Bordi, C., Monga, G. *et al.* (1978) Multiple endocrine cell types in thyroid medullary carcinoma. *Virchows Arch. [A]*, **377**, 111–28.

Carcangiu, M.L. and Bianchi, S. (1989) Diffuse sclerosing variant of papillary thyroid carcinoma. Clinicopathologic study of 15 cases. *Am. J. Surg. Pathol.*, **13**, 1041–9.

Carcangiu, M.L., Sibley, R.K. and Rosai, J. (1985d) Clear cell change in primary thyroid tumors. *Am. J. Surg. Pathol.*, **9**, 705–22.

Carcangiu, M.L. Steeper, T., Zampi, G. and Rosai, J. (1985a) Anaplastic thyroid carcinoma. A study of 70 cases. *Am. J. Clin. Pathol.*, **83**, 135–58.

Carcangiu, M.L., Zampi, G., Pupi, A. *et al.* (1985b) Papillary carcinoma of the thyroid. A clinicopathologic study of 241 cases treated at the University of Florence, Italy. *Cancer*, **55**, 805–28.

Carcangiu, M.L., Zampi, G. and Rosai, J. (1984) Poorly differentiated ('insular') thyroid carcinoma. A reinterpretation of Langhans' 'wuchernde Struma'. *Am. J. Surg. Pathol.*, **8**, 655–68.

Carcangiu, M.L., Zampi, G. and Rosai, J. (1985c) Papillary thyroid carcinoma: a study of its many morphologic expressions and clinical correlates. *Pathol. Annu*, **20**, (1), 1–44.

Carney, J.A., Ryan, J. and Goellner, J.R. (1987) Hyalinizing trabecular adenoma of the thyroid gland. *Am. J. Surg. Pathol.*, **11**, 583–91.

Chan, J.K.C. and Saw, D. (1986) The grooved nucleus. A useful diagnostic criterion of papillary carcinoma of the thyroid. *Am. J. Surg. Pathol.*, **10**, 672–9.

Chan, J.K.C. and Tse, C.C.H. (1988) Mucin production in metastatic papillary carcinoma of the thyroid. *Hum. Pathol.*, **19**, 195–200.

Chan, J.K.C., Tsui, M.S. and Tse, C.H. (1987) Diffuse sclerosing variant of papillary carcinoma of the thyroid: a histological and immunohistochemical study of three cases. *Histopathology*, **11**, 191–202.

Chen, K.T.K. and Rosai, J. (1977) Follicular variant of thyroid papillary carcinoma: a clinicopathologic study of six cases. *Am. J. Surg. Pathol.*, **1**, 123–30.

Christ, M.L. and Haja, J. (1979) Intranuclear cytoplasmic inclusions (invaginations) in thyroid aspirations: frequency and specificity. *Acta Cytol.*, **23**, 327–31.

Civantos, F., Albores-Saavedra, J., Nadji, M. and Morales, A.R. (1984) Clear cell variant of thyroid carcinoma. *Am. J. Surg. Pathol.*, **8**, 187–92.

Cohn, K., Bäckdahl, M., Forsslund, G. *et al.* (1984) Prognostic value of nuclear DNA content in papillary thyroid carcinoma. *World J. Surg.*, **8**, 474–80.

Cooke, W.M. Jr and Carrera, G.M. (1964) Mixed squamous cell carcinoma and papillary adenocarcinoma (adenocacanthoma) of the thyroid gland. *Am. J. Surg.*, **10**, 432–3.

Cramer, S.F., Bradshaw, R.A., Baglan, N.C. and Meyer, J.A. (1979) Nerve growth factor in medullary carcinoma of the thyroid. *Hum. Pathol.*, **10**, 731–6.

Crile, G., Jr and Hazard, J.B . (1953) Relationship of the age of the patient to the natural history and prognosis of carcinoma of the thyroid. *Ann. Surg.*, **138**, 33–8.

Cuello, C., Correa, P. and Eisenberg, H. (1969) Geographic pathology of thyroid carcinoma. *Cancer*, **23**, 230–9.

Cushman, P. (1962) Familial endocrine tumors. Report of two unrelated kindreds affected with pheochromocytomas, one also with multiple thyroid carcinomas. *Am. J. Med.*, **32**, 352–60.

Dasovic-Knezevic, M., Børmer, O., Holm, R. *et al.* (1989) Carcinoembryonic antigen in medullary thyroid carcinoma: an immunohistochemical study applying six novel monoclonal antibodies. *Mod. Pathol.*, **2**, 610–17.

Dayal, Y., Tashjian, A.H., Jr and Wolfe, H.J. (1979) Immunocytochemical localization of calcitonin-producing cells in a strumal carcinoid with amyloid stroma. *Cancer*, **43**, 1331–8.

Deftos, L.J. and Burton, D.W. (1980) Immunohistological studies on non-thyroidal calcitonin-producing tumors. *J. Clin. Endocrinol. Metab.*, **50**, 1041–5.

DeGroot, L.J. and Paloyan, E. (1973) Thyroid carcinoma and radiation: a Chicago endemic. *JAMA*, **225**, 487–91.

DeGroot, L.J., Kaplan, E.L., McCormick, M. and Straus, F.H. (1990) Natural history, treatment and course of papillary thyroid carcinoma. *J. Clin. Endocrinol. Metab.*, **71**, 414–24.

DeLellis, R.A. and Wolfe, H.J. (1974) Correspondence. *Hum. Pathol.*, **5**, 500–1.

DeLellis, R.A. and Wolfe, H.J. (1981) The pathobiology of the human calcitonin (C)-cell: a review. *Pathol. Annu.* **16**, (2), 25–52.

DeLellis, R.A., Nunnenmacher, G. and Wolfe, H.J. (1977) C-cell hyperplasia. An ultrastructural analysis. *Lab. Invest.*, **36**, 237–48.

DeLellis, R.A., Rule, A.H., Spiler, I. *et al.* (1978) Calcitonin and carcino-embryonic antigen as tumor markers in medullary thyroid carcinoma. *Am. J. Clin. Pathol.*, **70**, 587–94.

Dickersin, G.R., Vickery, A.L., Jr and Smith, S.B. (1980) Papillary carcinoma of the thyroid, oxyphil cell type, 'clear cell' variant. A light- and electron-microscopic study. *Am. J. Surg. Pathol.*, **4**, 501–9.

Dominguez-Malagon, H., Delgado-Chavez, R., Torres-Najera, M. *et al.*, (1989) Oxyphil and squamous variants of medullary thyroid carcinoma. *Cancer*, **63**, 1183–8.

Eng, H-L. and Chen, W-J. (1989) Melanin-producing medullary carcinoma of the thyroid gland. *Arch. Pathol. Lab. Med.*, **113**, 377–80.

Evans, H.L. (1984) Follicular neoplasms of the thyroid. A study of 44 cases followed for a minimum of 10 years, with emphasis on differential diagnosis. *Cancer*, **54**, 535–40.

Evans, H.L. (1986) Columnar-cell carcinoma of the thyroid. A report of two cases of an aggressive variant of thyroid carcinoma. *Am. J. Clin. Pathol.*, **85**, 77–80.

Evans, H.L. (1987) Encapsulated papillary neoplasms of the thyroid. A study of 14 cases followed for a minimum of 10 years. *Am. J. Surg. Pathol.*, **11**, 592–7.

Falck, B., Ljungberg, O. and Rosengren, E. (1968) On the occurrence of mono-amines and related substances in familial medullary thyroid carcinoma with pheochromocytoma. *Acta Pathol. Microbiol. Scand.*, **74**, 1–10.

Farndon, J.R., Leight, G.S., Dilley, W.G. *et al.* (1986) Familial medullary thyroid carcinoma without associated endocrinopathies. A distinct clinical entity. *Br. J. Surg.*, **73**, 278–81.

Feldman, J.M. and Wells, S.A. (1981) Case report: tissue levels of monoamines and monoamine metabolizing enzymes in medullary carcinoma and other thyroid diseases. *Am. J. Med. Sci.*, **283**, 34–40.

Feldman, P.S., Horvath, E. and Kovacs, K. (1972) Ultrastructure of three Hürthle cell tumors of the thyroid. *Cancer*, **30**, 1279–85.

Fernandes, B.J., Bedard, Y.C. and Rosen, I. (1982) Mucus-producing medullary cell carcinoma of the thyroid. *Am. J. Clin. Pathol.*, **78**, 536–40.

Fialho, F., Basílio-de-Oliveira, C.A., Azevedo, N. *et al.* (1981) Carcinoma mucoepidermoide da tireoide. *Folha Méd. (BR)*, **82**, 397–402.

Fisher, E.R. and Kim, W.S. (1977) Primary clear cell thyroid carcinoma with squamous features. *Cancer*, **39**, 2497–502.

Fjälling, M., Tisell, L.E., Carlsson, S. *et al.* (1986) Benign and malignant thyroid nodules after neck irradiation. *Cancer*, **58**, 1219–24.

Flint, A. and Lloyd, R.V. (1990) Hürthle-cell neoplasms of the thyroid gland. *Pathol. Annu.* **25**, (1), 37–52.

Flint, A., Davenport, R.D., Lloyd, R.V. *et al.* (1988) Cytophotometric measurements of Hürthle cell tumors of the thyroid gland. Correlation with pathologic features and clinical behaviour. *Cancer*, **61**, 110–13.

Franklin, W.A., Mariotto, S., Kaplan, D. and DeGroot, L.J. (1982) Immuno-fluorescence localization of thyroglobulin in metastatic thyroid cancer. *Cancer*, **50**, 939–45.

Franssila, K.O. (1971) Value of histologic classification of thyroid cancer. *Acta. Pathol. Microbiol. Scand.* [A], (Suppl 225) 1–76.

Franssila, K.O. (1973) Is the differentiation between papillary and follicular thyroid carcinoma valid? *Cancer*, **32**, 853–64.

Franssila, K.O. (1975), Prognosis in thyroid carcinoma. *Cancer*, **36**, 1138–46.

Franssila, K.O., Harach, H.R. and Wasenius, V-M (1984) Mucoepidermoid carcinoma of the thyroid. *Histopathology*, **8**, 847–60.

Franssila, K.O., Ackerman, L.V., Brown, C.L. and Hedinger, C.E. (1985) Session II: Follicular carcinoma. *Semin. Diagn. Pathol.*, **2**, 101–22.

Franssila, K.O., Saxén, E., Teppo, L. *et al.* (1981) Incidence of different morphological types of thyroid cancer in the Nordic countries. *Acta Pathol. Microbiol. Scand.* [A], **89**, 49–55.

Frauenhoffer, C., Patchefsky, A. and Cobanoglu, A. (1979) Thyroid carcinoma: a clinical and pathologic study of 125 cases. *Cancer*, **43**, 2414–21.

Friedell, G.H., Carey, R.J. and Rosen, H. (1962) Familial thyroid cancer. *Cancer*, **15**, 241–5.

Fukanaga, F.H. and Yatani, R. (1975) Geographic pathology of occult thyroid carcinomas. *Cancer*, **36**, 1095–9.

Gaffey, M.J., Lack, E.E., Christ, M.L. and Weiss, L.M. (1991) Anaplastic thyroid carcinoma with osteoclast-like giant cells. A clinicopathologic, immunohistochemical, and ultrastructural study. *Am. J. Surg. Pathol.*, **15**, 160–8.

Galera-Davidson, H., González-Cámpora, R., Mora-Martín, J.A. *et al.* (1990) Cytophotometric DNA measurements in medullary thyroid carcinoma. *Cancer*, **65**, 2255–60.

Ghatei, M.A., Springall, D.R., Nicholl, C.G. *et al.* (1985) Gastrin-releasing peptide-like immunoreactivity in medullary thyroid carcinoma. *Am. J. Clin. Pathol.*, **84**, 581–6.

Gherardi, G. (1987) Signet ring cell 'mucinous' thyroid adenoma: a follicle cell tumor with abnormal accumulation of thyroglobulin and a peculiar histochemical profile. *Histopathology*, **11**, 317–26.

Goellner, J.R. and Carney, J.A. (1989) Cytologic features of fine-needle aspirates of hyalinizing trabecular adenoma of the thyroid. *Am. J. Clin. Pathol.*, **91**, 115–9.

González-Cámpora, R., Montero, C., Martin-Lacave, I. and Galera, H. (1986a) Demonstration of vascular endothelium in thyroid carcinomas using *Ulex europaeus* I agglutinin. *Histopathology*, **10**, 261–6.

González-Cámpora, R., Herrero-Zapatero, A., Lerma, E. *et al.* (1986b) Hürthle cell and mitochondrion-rich cell tumors. A clinicopathologic study. *Cancer*, **57**, 1154–63.

Goodman, M.T., Yoshizawa, C.N. and Kolonel, L.N. (1988) Descriptive epidemiology of thyroid cancer in Hawaii. *Cancer*, **61**, 1272–81.

Gordon, H.W., Miller, R. and Mittman, C. (1973) Medullary carcinoma of the lung with amyloid stroma: a counterpart of medullary carcinoma of the thyroid. *Hum. Pathol.*, **4**, 431–6.

Gosain, A.K. and Clark, O.H. (1984) Hürthle cell neoplasms: malignant potential. *Arch. Surg.*, **119**, 515–9.

Gould, V.E., Wiedenmann, B., Lee, I. *et al.* (1987) Synatophysin expression in neuroendocrine neoplasms as determined by immunochytochemistry. *Am. J. Pathol.*, **126**, 243–57.

Gundry, S.R., Burney, R.E., Thompson, N.W. and Lloyd, R. (1983) Total thyroidectomy for Hürthle cell neoplasm of the thyroid. *Arch. Surg.*, **118**, 529–32.

Hakulinen, T., Andersen, A.A., Malker, B. *et al.* (1986) Trends in cancer incidence in the Nordic countries. A collaborative study of the five Nordic Cancer Registries. *Acta Pathol. Microbiol. Immunol. Scand.* [*A*], **94**, (Suppl. 228), 1–51.

Hales, M., Rosenau, W., Okerlund, M.D. and Galante, M. (1982) Carcinoma of the thyroid with a mixed medullary and follicular pattern. *Cancer*, **50**, 1352–9.

Hanson, G.A., Komorowski, R.A., Cerletty, J.M. and Wilson, S.D. (1983) Thyroid gland morphology in young adults: normal subjects versus those with prior low-dose neck irradiation in childhood. *Surgery*, **94**, 984–8.

Hapke, M.R. and Dehner, L.P. (1979) The optically clear nucleus. A reliable sign of papillary carcinoma of the thyroid? *Am. J. Clin. Pathol.*, **3**, 31–8.

Har-El, G., Hadar, T., Segal, K. *et al.* (1986) Hürthle cell carcinoma of the thyroid gland. A tumor of moderate malignancy. *Cancer*, **57**, 1613–7.

Harach, H.R. (1985a) Thyroid follicles with acid mucins in man: a second kind of follicle? *Cell Tissue Res.*, **242**, 211–5.

Harach, H.R. (1985b) A study on the relationship between solid cell nests and mucoepidermoid carcinoma of the thyroid. *Histopathology*, **9**, 195–207.

Harach, H.R. and Franssila, K.O. (1988) Thyroglobulin immunostaining in follicular thyroid carcinoma: a relationship to the degree of differentiation and cell type. *Histopathology*, **13**, 43–54.

Harach, H.R. and Williams, E.D. (1983) Glandular (tubular and follicular) variants of medullary carcinoma of the thyroid. *Histopathology*, **7**, 83–97.

Harach, H.R., Franssila, K.O. and Wasenius, V.M. (1985) Occult papillary carcinoma of the thyroid. A 'normal' finding in Finland. A systematic autopsy study. *Cancer*, **56**, 531–8.

Harada, T., Ito, K., Shimaoka, K. *et al.* (1977) Fatal thyroid carcinoma.

Anaplastic transformation of adenocarcinoma. *Cancer*, **39**, 2588–96.

Harber, M.H. and Lipkovic, P. (1970) Thyroid cancer in Hawaii. *Cancer*, **25**, 1224–7.

Harcourt-Webster, J.N. (1966) Squamous epithelium in the human thyroid gland. *J. Clin. Pathol.*, **19**, 384–8.

Harrison, D.F.N. (1973) Thyroid gland in the management of laryngo-pharyngeal cancer. *Arch. Otolaryngol.*, **97**, 301–2.

Hawk, W.A. and Hazard, J.B. (1976) The many appearances of papillary carcinoma of the thyroid. *Cleve. Clin. Q.*, **43**, 207–16.

Hazard, J.B. (1960) Small papillary carcinoma of the thyroid. *Lab. Invest.*, **9**, 86–97.

Hazard, J.B. (1964) Neoplasia. In *The Thyroid* (eds. J.B. Hazard and D.C. Smith), Williams and Wilkins, Baltimore, pp. 239–55.

Hazard, J.B. (1968) Nomenclature of thyroid tumours. In *Thyroid Neoplasia* (eds. S. Young and D.R. Inman), Academic Press, London and New York, pp. 3–37.

Hazard, J.B. and Kenyon, R. (1954) Encapsulated angioinvasive carcinoma (angioinvasive adenoma) of thyroid gland. *Am. J. Clin. Pathol.*, **24**, 755–66.

Hazard, J.B., Crile, G. Jr and Dempsey, W.S. (1949) Nonencapsulated sclerosing tumors of the thyroid. *J. Clin. Endocrinol.*, **9**, 1216–31.

Hazard, J.B., Hawk, W.A. and Crile, G. (1959) Medullary (solid) carcinoma of the thyroid – a clinicopathologic entity. *J. Clin. Endocrinol. Metab.*, **19**, 152–61.

Hedinger, C. (1981) Geographic pathology of thyroid disease. *Pathol. Res. Pract.*, **171**, 285–92.

Hedinger, C., Williams, E.D. and Sobin, L.H. (1988) Histological typing of thyroid tumours. 2nd edn, in: *International Histological Classification of Tumours*, no. 11, World Health Organization, Springer-Verlag, Berlin, Heidelberg, New York, London, Paris, Tokyo and Hong Kong.

Hedinger, C., Williams, E.D. and Sobin, L.H. (1989) The WHO histological classification of thyroid tumours: a commentary on the second edition. *Cancer*, **63**, 908–11.

Heitz P., Moser, H. and Staub, J.J. (1976) Thyroid cancer. A study of 573 thyroid tumors and 161 autopsy cases observed over a thirty-year period. *Cancer*, **37**, 2329–37.

Hempelmann, L.H. (1977) Thyroid neoplasms following irradiation in infancy. In *Radiation-Associated Thyroid Carcinoma* (eds L.J. DeGroot, L.A. Frohman, E.L. Kaplan and S. Refetoff), Grune and Stratton, New York, San Francisco and London, pp. 221–31.

Hempelmann, L.H., Pifer, J.W., Burke, G.J. *et al.* (1967) Neoplasms in persons treated with X-rays in infancy for thymic enlargement. A report of the third follow-up study. *J. Natl Cancer Inst.*, **38**, 317–41.

Hirabayashi, R.N. and Lindsay, S. (1961) Carcinoma of the thyroid gland: a statistical study of 390 patients. *J. Clin. Endocrinol. Metab.*, **21**, 1596–1610.

Høie, J., Stenwig, A.E., Kullman, G. and Lindegaard, M. (1988) Distant metastases in papillary thyroid cancer. A review of 91 patients. *Cancer*, **61**, 1–6.

Holm, L.-E., Lundell, G. and Walinder, G. (1980a) Incidence of malignant thyroid tumours in humans after exposure to diagnostic doses of iodine-131. I. Retrospective cohort study. *J. Natl. Cancer Inst.*, **64**, 1055–9.

Holm, L.-E., Dahlqvist, I., Israelsson, A. and Lundell, G. (1980b) Malignant thyroid tumors after iodine-131 therapy: a retrospective cohort study. *N. Engl. J. Med.*, **303**, 188–91.

Holm, R., Sobrino-Simões, M. and Johannessen, J.V. (1986) Concurrent production of calcitonin and thyroglobulin by the same neoplastic cells. *Ultrastruct. Pathol.*, **10**, 241–8.

Horvath, E., Kovacs, K. and Ross, R.C. (1972) Medullary cancer of the thyroid gland and its possible relations to carcinoids. An ultrastructural study. *Virchows. Arch. [A]*, **356**, 281–92.

Huang, S-N. and McLeish, W.A. (1968) Pheochromocytoma and medullary carcinoma of the thyroid. *Cancer*, **21**, 302–11.

Huang, T.Y. and Assor, D. (1971) Primary squamous cell carcinoma of the thyroid gland. A report of four cases. *Am. J. Clin. Pathol.*, **55**, 93–8.

Hubert, J.P., Kiernan, P.D., Beahrs, O.H. *et al.* (1980) Occult papillary carcinoma of the thyroid. *Arch. Surg.*, **115**, 394–8.

Hutter, R.V.P., Tollefsen, H.R., DeCosse, J.J. *et al.* (1965) Spindle and giant cell metaplasia in papillary carcinoma of the thyroid. *Am. J. Surg.*, **110**, 660–8.

Idia, F., Yonekura, M. and Miyakawa, M. (1969) Study of intraglandular dissemination of thyroid cancer. *Cancer*, **24**, 764–71.

Ishimaru, Y., Fukuda, S., Kurano, R., *et al.* (1988) Follicular thyroid carcinoma with clear cell change showing unusual ultrastructural features. *Am. J. Clin. Pathol.*, **12**, 240–6.

Ito, J., Noguchi, S., Murakami, N. and Noguchi, A. (1980) Factors affecting the prognosis of patients with carcinoma of the thyroid. *Surg. Gynecol. Obstet.*, **150**, 539–44.

Jackson, C.E. and Norum, R.A. (1989) Genetic mechanisms of neoplasia in MEN 2. *Henry Ford Hosp. Med. J.*, **37**, 116–9.

Joensuu, H., Klemi, P. and Eerola, E. (1986) DNA aneuploidy in follicular adenomas of the thyroid gland. *Am. J. Pathol.*, **124**, 373–6.

Johannessen, J.V. and Sobrinho-Simões, M. (1980) The origin and significance of thyroid psammoma bodies. *Lab. Invest.*, **43**, 287–96.

Johannessen, J.V., Gould, V.E. and Jao, W. (1978) The fine structure of human thyroid cancer. *Hum. Pathol.*, **9**, 385–400.

Johannessen, J.V., Sobrinho-Simões, M., Lindmo, T. and Tangen, K.O. (1982) The diagnostic value of flow cytometric DNA measurements in selected disorders of the human thyroid. *Am. J. Clin. Pathol.*, **77**, 20–5.

Johannessen, J.V., Sobrino-Simões, M., Tangen, K.O. and Lindmo, T. (1981) A flow-cytometric deoxyribonucleic acid analysis of papillary thyroid carcinoma. *Lab. Invest.*, **45**, 336–41.

Johnson, T.L., Lloyd, R.V., Burney, R.E. and Thompson, N.W. (1987) Hürthle cell thyroid tumors. An immunohistochemical study. *Cancer*, **59**, 107–12.

Johnson, T.L., Lloyd, R.V., Thompson, N.W. *et al.* (1988) Prognostic implications of the tall cell variant of papillary thyroid carcinoma. *Am. J. Surg. Pathol.*, **12**, 22–7.

Jónasson, J.G., Hrafnkelsson, J. and Björnsson, J. (1989) Tumours in Iceland. 11. Malignant tumours of the thyroid gland. A histological classification and epidemiological considerations. *APMIS*, **97**, 625–30.

Kahn, N.F. and Perzin, K.H. (1983) Follicular carcinoma of the thyroid: an evaluation of the histological criteria used for diagnosis. *Pathol. Annu.*, **18**, (1), 221–53.

Kakado, K. and Vacca, L.L. (1983) Immunohistochemical study of substance P-like immunoreactivity in human thyroid and medullary carcinoma of the thyroid. *J. Submicrosc. Cytol.*, **15**, 563–8.

Kakudo, K., Miyauchi, A., Ogihara, T. *et al.* (1978) Medullary carcinoma of the thyroid. Giant cell type. *Arch. Pathol. Lab. Med.*, **102**, 445–7.

Kakudo, K., Miyauchi, A., Takai S. *et al.* (1979) C cell carcinoma of the thyroid. Papillary type. *Acta Pathol. Jpn*, **29**, 653–9.

Kameda, Y., Shigemoto, H. and Ikeda, A. (1980) Development and cyto-differentiation of C cell complexes in dog fetal thyroids. An immuno-histochemical study using anti-calcitonin, anti-C thyroglobulin and anti-19 S thyroglobulin antisera. *Cell Tissue Res.*, **206**, 403–15.

Kameya, T., Bessho, T., Tsumuraya, M. *et al.* (1983) Production of gastrin releasing peptide by medullary carcinoma of the thyroid. An immuno-histochemical study. *Virchows Arch [A]*, **401**, 99–108.

Kameya, T., Shimosato, Y., Adachi, I. *et al.* (1977) Immunohistochemical and ultrastructural analysis of medullary carcinoma of the thyroid in relation to hormone production. *Am. J. Pathol.*, **89**, 555–74.

Katoh, R., Sugai, T., Ono, S. *et al.* (1990) Mucoepidermoid carcinoma of the thyroid gland. *Cancer*, **65**, 2020–7.

Keydar, I., Ravid, M. and Sohar, E. (1976) *In vitro* synthesis of amyloid fibrils from insulin, calcitonin and parathormone. *Isr. J. Med. Sci.*, **12**, 1137–40.

Keyhani-Rofagha, S., Kooner, D.S., Keyhani, M. and O'Toole, R.V. (1990) Necrosis of a Hürthle cell tumor of the thyroid following fine needle aspiration. Case report and literature review. *Acta Cytol.*, **34**, 805–8.

Kimura, N., Sasano, N. and Namiki, T. (1986) Evidence of hybrid cell of thyroid follicular cell and carcinoid cell in strumal carcinoid. *Int. J. Gynecol. Pathol.*, **5**, 269–77.

Klinck, G.H. and Merik, K.F. (1951) Squamous cells in the human thyroid. *Milit. Surg.*, **109**, 406–14.

Klinick, G.H. and Winship, T. (1959) Psammoma bodies and thyroid cancer. *Cancer*, **12**, 656–62.

Knudson, A.G., Jr (1971) Mutation and cancer: statistical study of retinoblastoma. *Proc. Natl. Acad. Sci. USA*, **68**, 820–3.

Kodama, T., Fujino, M., Endo, Y. *et al.* (1979) Family study of serum carcino-embryonic antigen in inherited medullary carcinoma of the thyroid. *Cancer*, **44**, 661–4.

Krayenbühl, J.C. and Hedinger, C. (1985) Grosszellig-eosinophile Tumoren der Schilddrüse. *Schweiz. Med. Wochenschr.*, **115**, 512–22.

Krisch, K., Krisch, I., Horvat, G. *et al.* (1985) The value of immunohisto-chemistry in medullary thyroid carcinoma: a systematic study of 30 cases. *Histopathology*, **9**, 1077–89.

Landon, G. and Ordoñez, N.G. (1985) Clear cell variant of medullary carcinoma of the thyroid. *Hum. Pathol.*, **16**, 844–7.

Lang, W., Choritz, H. and Hudeshagen, H. (1986) Risk factors in follicular thyroid carcinomas. A retrospective follow-up study covering a 14-year period with emphasis on morphological findings. *Am. J. Surg. Pathol.*, **10**, 246–55.

Lang, W., Georgii, A., Stauch, G. and Kienzle, E. (1980) The differentiation of atypical adenomas and encapsulated follicular carcinomas in the thyroid. *Virchows Arch. [A]*, **385**, 125–41.

Langhans, T. (1907) Über die epithelialen Formen der malignen Struma. *Virchows Arch. [A]*, **189**, 69–188.

Layfield, L.L. and Lones, M.A. (1991) Necrosis in thyroid nodules after fine needle aspiration biopsy. Report of two cases. *Acta Cytol.*, **35**, 427–30.

Lee, F.I. and MacKinnon, M.D. (1981) Papillary thyroid carcinoma associated with polyposis coli. A case of Gardner's syndrome. *Am. J. Gastroenterol.,* **76**, 138–40.

Lew, S., Orell, S. and Henderson, D.W. (1984) Intranuclear vacuoles in non-papillary carcinoma of the thyroid. A report of three cases. *Acta Cytol.,* **28**, 581–6.

Libbey, N.P., Nowakowski, K.J. and Tucci, J.R. (1989) C-cell hyperplasia of the thyroid in a patient with goitrous hypothyroidism and Hashimoto's thyroiditis. *Am. J. Surg. Pathol.,* **13**, 71–7.

Lindsay, S. (1960) *Carcinoma of the Thyroid Gland. A Clinical and Pathologic Study of 293 Patients at the University of California Hospital.* C.C. Thomas, Springfield, p. 43.

Linell, F., Ljungberg, O. and Andersson, I. (1980) Breast carcinoma. Aspects of early stages, progression and related problems. *Acta Pathol. Microbiol. Scand., [A],* (Suppl. 272), 1–233.

Lippman, S.M., Mendelsohn, G., Trump, D.L. *et al.* (1982) The prognostic and biological significance of cellular heterogeneity in medullary thyroid carcinoma: a study of calcitonin, L-dopa decarboxylase, and histaminase. *J. Clin. Endocrinol. Metab.,* **54**, 233–40.

LiVolsi, V.A. (1987) Editorial. Mixed thyroid carcinoma: a real entity? *Lab. Invest.,* **57**, 237–9.

LiVolsi, V.A. and Feind, C.R. (1979) Incidental medullary carcinoma in sporadic hyperparathyroidism. An expansion of the concept of C-cell hyperplasia. *Am. J. Clin. Pathol.,* **71**, 595–9.

LiVolsi, V.A. and Merino, M.J. (1978) Squamous cells in the human thyroid gland. *Am. J. Surg. Pathol.,* **2**, 133–40.

LiVolsi, V.A. and Merino, M.J. (1981) Histopathologic differential diagnosis of the thyroid. *Pathol. Annu..* **16**, (2), 357–406.

Ljungberg, O. (1966) Medullary carcinoma of the thyroid and phaeochromo-cytoma: a hereditary syndrome. *Acta Pathol. Microbiol. Scand.,* **67**, 589.

Ljungberg, O. (1970) Two cell types in familial medullary thyroid carcinoma. A histochemical study. *Virchows Arch. [A],* **349**, 312–22.

Ljungberg, O. (1972a), Argentaffin cells in human thyroid and parathyroid and their relationship to C cells and medullary carcinoma. *Acta Pathol. Microbiol. Scand. [A],* **80**, 589–99.

Ljungberg, O. (1972b) On medullary carcinoma of the thyroid. *Acta Pathol. Microbiol. Scand. [A],* (Suppl. 231), 1–57.

Ljungberg (1972c) Cytologic diagnosis of medullary carcinoma of the thyroid gland. With special regard to the demonstration of amyloid in smears of find needle aspirates. *Acta Cytol.,* **16**, 253–5.

Ljungberg, O. (1984) Histogenesis of thyroid carcinoma. *Med. Hypotheses,* **14**, 253–7.

Ljungberg, O. and Dymling, J-F. (1972), Pathogenesis of C-cell neoplasia in thyroid gland. C-cell proliferation in a case of chronic hypercalcaemia. *Acta Pathol. Microbiol. Scand. [A],* **80**, 577–88.

Ljungberg, O. and Nilsson, P.O. (1985) Hyperplastic and neoplastic changes in ultimobranchial remnants and in parafollicular (C) cells in bulls: a histologic and immunohistochemical study. *Vet. Pathol.,* **22**, 95–103.

Ljungberg, O. and Nilsson, P.O. (1991) Intermediate thyroid carcinoma in humans and ultimobranchial tumors in bulls: a comparative morphological and immunohistochemical study. *Endocr. Pathol.,* **2**, 24–39.

Ljungberg, O., Bondeson, L. and Bondeson, A-G. (1984) Differentiated thyroid carcinoma, intermediate type: a new tumor entity with features of follicular and parafollicular cell carcinoma. *Hum. Pathol.,* **15**, 218–28.

Ljungberg, O., Cederquist, E. and von Studnitz, E. (1967) Medullary thyroid carcinoma and phaeochromocytoma – a familial chromaffinomatosis. *Br. Med. J.*, **i**, 279–81.

Ljungberg, O., Ericsson, U-B., Bondeson, L. and Thorell, J. (1983) A compound follicular–parafollicular cell carcinoma of the thyroid: a new tumor entity? *Cancer*, **52**, 1053–61.

Lloyd, R.V., Sisson, J.C. and Marangos, P.J. (1983) Calcitonin, carcinoembryonic antigen and neuron-specific enolase in medullary thyroid carcinoma. An immunohistochemical study. *Cancer*, **51**, 2234–9.

Logman, S.C. and Jobsis, A.C. (1984) Thyroid-associated antigens in routinely embedded carcinomas. Possibilities and limitations studied in 116 cases. *Cancer*, **54**, 274–9.

Lote, K., Andersen, K., Nordal, E. and Brennhovd, I.O. (1980) Familial occurrence of papillary thyroid carcinoma. *Cancer*, **46**, 1291–7.

Mambo, N.C. and Irwin, S.M. (1984) Anaplastic small cell neoplasms of the thyroid; an immunoperoxidase study. *Hum. Pathol.*, **15**, 55–60.

Marcus, J.N., Dise, C.A. and LiVolsi, V.A. (1982) Melanin production in a medullary thyroid carcinoma. *Cancer*, **49**, 2518–26.

Maruchi, N., Furihata, R. and Makiuchi, M. (1971) Population surveys on the prevalence of thyroid cancer in a non-endemic region, Nagano, Japan. *Int. J. Cancer*, **7**, 575–83.

Massin, J-P., Savoie, J-C., Garnier, H. *et al.* (1984) Pulmonary metastases in differentiated thyroid carcinoma. Study of 58 cases with implications for the primary tumor treatment. *Cancer*, **53**, 982–92.

Mathew, C.G.P., Chin, K.S., Easton, D.F. *et al.* (1987) A linked genetic marker for multiple endocrine neoplasia type 2A on chromosome 10. *Nature*, **328**, 527–8.

Mazzaferri, E.L., Young, R.L., Oertel, J.E. *et al.* (1977) Papillary thyroid carcinoma: the impact of therapy in 576 patients. *Medicine (Baltimore)*, **56**, 171–96.

McConahey, W.M., Hay, I.D., Woolner, L.B. *et al.* (1986) Papillary thyroid cancer treated at the Mayo Clinic, 1946 through 1970. Initial manifestations, pathologic findings, therapy, and outcome. *Mayo Clin. Proc.*, **61**, 978–96.

McIntosh, C.A., Ewen, S.W.B. and Matheson, N.A. (1986) Nuclear DNA and prognosis in papillary thyroid carcinoma. *J. Pathol.*, **148**, 80A.

Meissner, W.A. and Warren, S. (1969) Tumors of the thyroid gland. In *Atlas of Tumor Pathology*, Fascicle 4, Second Series, Armed Forces Institute of Pathology, Washington, DC.

Mendelsohn, G. (1984) Signet-cell-stimulating microfollicular adenoma of the thyroid. *Am. J. Surg. Pathol.*, **8**, 705–8.

Mendelsohn, G., Baylin, S.B., Bigner, S.H. *et al.* (1980) Anaplastic variants of medullary thyroid carcinoma. A light-microscopy and immunohistochemical study. *Am. J. Surg. Pathol.*, **4**, 333–41.

Miller, R.H., Estrada, R., Sneed, W.F. and Mace, M.L. (1983) Hürthle cell tumors of the thyroid. *Laryngoscope*, **93**, 884–8.

Mlynek, M-L., Richter, H.J. and Leder, L-D. (1985) Mucin in carcinomas of the thyroid. *Cancer*, **56**, 2647–50.

Motoyama, T. and Watanabe, H. (1983) Simultaneous squamous cell carcinoma and papillary adenocarcinoma of the thyroid gland. *Hum. Pathol.*, **14**, 1009–10.

Myskow, M.W., Krajewski, A.S., Dewar, A.E. *et al.* (1986) The role of immunoperoxidase techniques on paraffin embedded tissue in determining the histogenesis of undifferentiated thyroid neoplasms. *Clin. Endocrinol.*, **24**, 335–41.

Nesland, J.M., Sobrino-Simões, M.A., Holm, R. *et al.* (1985), Hürthle cell lesions of the thyroid. A combined study using transmission electron microscopy and immunocytochemistry. *Ultrastruct. Pathol.*, **2**, 121–9.

Nielsen, B. and Zetterlund, B. (1985) Malignant thyroid tumors at autopsy in a Swedish goitrous population. *Cancer*, **55**, 1041–3.

Nieuwenhuijzen Kruseman, A.C., Bosman, F.T., van Bergen Henegouw, J.D. *et al.* (1982) Medullary differentiation of anaplastic thyroid carcinoma. *Am. J. Clin. Pathol.*, **77**, 541–7.

Noll, W.W., Maurer, L.H., Herzberg, V.L. *et al.* (1984) Familial medullary carcinoma of the thyroid: clinical studies in northern New England. *Henry Ford Hosp. Med. J.*, **32**, 244–5.

Norton, J.A., Froome, L.J., Farrell, R.E. and Wells, S.A., Jr (1979) Multiple endocrine neoplasia type IIb: the most aggressive form of medullary thyroid cancer. *Surg. Clin. North Am.*, **59**, 109–18.

Nourok, D.S. (1964) Familial pheochromocytoma and thyroid carcinoma. *Ann. Intern. Med.*, **60**, 1028–40.

Nussbaum, M., Buchwald, R.P., Ribner, A. *et al.* (1981) Anaplastic carcinoma arising from median ectopic thyroid (thyroglossal duct remnant). *Cancer*, **48**, 2724–8.

Nusynowitz; M.L., Pollard, E., Benedetto, A.R. *et al.* (1982) Treatment of medullary carcinoma of the thyroid with I-131. *J. Nucl. Med.*, **23**, 143–6.

O'Brien, C.J. (1988) Uncommon tumours of the thyroid gland. In *Textbook of Uncommon Cancer* (eds C.J. Williams, J.G. Krikorian, M.R. Green and D. Raghavan), Wiley, New York, pp. 707–23.

O'Connor, D.T. and Deftos, L.J. (1986) Secretion of chromogranin A by peptide-producing endocrine neoplasms. *N. Engl. J. Med.*, **314**, 1145–51.

O'Hare, M.M.T., Shaw, C., Johnston, C.F. *et al.* (1986) Pancreatic polypeptide immunoreactivity in medullary carcinoma of the thyroid: identification and characterisation by radioimmunoassay, immunocytochemistry and high performance liquid chromatography. *Regul. Pept.* 14, 169–80.

O'Toole, K., Fenoglio-Preisler, C. and Pushparaj, N. (1985) Endocrine changes associated with the human aging process: III. Effect of age on the number of calcitonin immunoreactive cells in the thyroid gland. *Hum. Pathol.*, **16**, 991–1000.

Ohashi, M., Yanase, T., Fujio, N. *et al.* (1987) alpha-neoendorphin-like immunoreactivity in medullary carcinoma of the thyroid. *Cancer*, **59**, 277–80.

Okazaki, M., Miya, A., Tanaka, N. *et al.* (1989) Allele loss on chromosome 10 and point mutation of ras oncogenes are infrequent in tumors of MEN 2A. *Henry Ford Hosp. Med. J.*, **37**, 112–5.

Ott, R.A., Calandra, D.B., McCall, A. *et al.* (1985) The incidence of thyroid carcinoma in patients with Hashimoto's thyroiditis and solitary cold nodules. *Surgery*, **98**, 1202–6.

Pacini, F., Elisei, R., Anelli, S. *et al.* (1989) Somatostatin in medullary thyroid carcinoma. In vitro and in vivo studies. *Cancer*, **63**, 1189–95.

Parker, LN., Kollin, J., Wu, S-Y. *et al.* (1985) Carcinoma of the thyroid with a mixed medullary, papillary, follicular, and undifferentiated pattern. *Arch. Intern. Med.*, **145**, 1507–9.

Patchefsky, A.S. and Hoch, W.S. (1972) Psammoma bodies in diffuse toxic goiter. *Am. J. Clin. Pathol.*, **57**, 551–6.

Pearse, A.G.E. (1969) The cytochemistry and ultrastructure of polypeptide hormone-producing cells of the APUD series and the embryologic, physiologic and pathologic implications of the concept. *J. Histochem. Cytochem.*, **17**, 303–13. .

Pearse, A.G.E. (1977) The diffuse neuroendocrine system and the APUD concept: related 'endocrine' peptides in brain, intestine, pituitary, placenta and anuran cutaneous glands. *Med. Biol.*, **55**, 115–25.

Pfaltz, M., Hedinger, Chr. E. and Mühlethaler, J.P. (1983) Mixed medullary and follicular carcinoma of the thyroid. *Virchows Arch. [A]*, **400**, 53–9.

Phade, V.R., Lawrence, W.R. and Max, M.H. (1981) Familial papillary carcinoma of the thyroid. *Arch. Surg.*, **116**, 836–7.

Ponder, B.A.J. (1984) Screening for familial medullary thyroid carcinoma: a review. *J. R. Soc. Med.*, **77**, 585–94.

Puchtler, H., Sweat, F. and Levine, M. (1962) On the binding of Congo red by amyloid. *J. Histochem. Cytochem.*, **10**, 355–64.

Ralfkiaer, N., Gatter, K.C., Alcock, C. *et al.* (1985) The value of immuno-cytochemical methods in the differential diagnosis of anaplastic thyroid tumours. *Br. J. Cancer*, **52**, 167–70.

Rhatigan, R.M., Roque, J.L. and Bucher, R.L. (1977) Mucoepidermoid carcinoma of the thyroid gland. *Cancer*, **39**, 210–4.

Rigaud, C. and Bogomoletz, W.V. (1987) 'Mucin secreting' and 'mucinous' primary thyroid carcinomas: pitfalls in mucin histochemistry applied to thyroid tumours. *J. Clin. Pathol.*, **40**, 890–5.

Rigaud, C., Peltier, F. and Bogomoletz, W.V. (1985) Mucin producing microfollicular adenoma of the thyroid. *J. Clin. Pathol.*, **38**, 277–80.

Roos, B.A., Huber, M.B., Birnbaum, R.S. *et al.* (1983) Medullary thyroid carcinomas secrete a noncalcitonin peptide corresponding to the carboxyl-terminal region of preprocalcitonin. *J. Clin. Endocrinol. Metab.*, **56**, 802–7.

Rosai, J. and Carcangiu, M.L. (1984) Pathology of thyroid tumors: some recent and old questions. *Hum. Pathol.*, **15**, 1008–12.

Rosai, J. and Carcangiu, M.L. (1987) Pitfalls in the diagnosis of thyroid neoplasms. *Pathol. Res. Pract.*, **182**, 169–79.

Rosai, J., Saxén, E.A. and Woolner, L. (1985) Session III: Undifferentiated and poorly differentiated carcinoma. *Semin. Diagn. Pathol.*, **2**, 123–36.

Rosai, J., Zampi, G. and Carcangiu, M.L. (1983) Papillary carcinoma of the thyroid. A discussion of its several morphologic expressions, with particular emphasis on the follicular variant. *Am. J. Surg. Pathol.*, **7**, 809–17.

Ross, R.C. (1947) Mixed squamous cell carcinoma and papillary adeno-carcinoma (adenocanthoma) of the thyroid gland. *Arch. Pathol.*, **44**, 192–7.

Rossi, R., Cady, B., Meissner, W.A. *et al.* (1978) Prognosis of undifferentiated carcinoma and lymphoma of the thyroid. *Am. J. Surg. Pathol.*, **135**, 589–95.

Roth, K.A., Bensch, K.G. and Hoffman, A.R. (1987) Characterization of opioid peptides in human thyroid medullary carcinoma. *Cancer*, **59**, 1594–8.

Roudebush, C.P. and DeGroot, L.J. (1977) The natural history of radiation-associated thyroid cancer. In *Radiation-Associated Thyroid Carcinoma* (eds L.J. DeGroot, L.A. Frohman, E.L.Kaplan and S. Refetoff), Grune and Stratton, New York, San Francisco and London, pp. 97–118.

Russell, W.O., Ibanez, M.L., Clark, R.L. and White, E.C. (1963) Thyroid carcinoma. Classification, intraglandular dissemination, and clinico-pathological study based upon whole organ sections of 80 glands. *Cancer*, **16**, 1425–60.

Ryff-de Léche, A., Staub, J.-J., Kohler-Faden, R. *et al.* (1986) Thyroglobulin production by malignant thyroid tumors. An immunocytochemical and radioimmunoassay study. *Cancer*, **57**, 1145–53.

Saad, M.F., Fritsche, H.A. and Samaan, N.A. (1984a) Diagnostic and prognostic values of carcinoembryonic antigen in medullary carcinoma of the thyroid. *J. Endocrinol. Metab.*, **58**, 889–94.

Saad, M.F., Ordoñez, N.G., Rashid, R.K. *et al.* (1984b) Medullary carcinoma of the thyroid. A study of the clinical features and prognostic factors in 161 patients. *Medicine (Baltimore)*, **63**, 319–42.

Sakamoto, A., Kasai, N. and Sugano, H. (1983) Poorly differentiated carcinoma of the thyroid. A clinicopathologic entity for a high-risk group of papillary and follicular carcinomas. *Cancer*, **52**, 1849–55.

Saleiro, J.V., Faria, V. and Oliveira, M.C. (1981) Clear cell tumor of the thyroid gland. *J. Submicrosc. Cytol.*, **13**, 75–7.

Sambade, C., Baldaque-Faria, A., Cardoso-Oliveira, M. and Sobrinho-Simões, M. (1989) Follicular and papillary variants of medullary carcinoma of the thyroid. *Pathol. Res. Pract.*, **184**, 98–103.

Sambade, C., Franssila, K., Basílo-de-Oliviera, C.A. and Sobrinho-Simões, M. (1990) Mucoepidermoid carcinoma of the thyroid revisited. *Surg. Pathol.*, **3**, 271–80.

Sampson, R.J. (1977) Prevalence and significance of occult thyroid cancer. In *Radiation-Associated Thyroid Carcinoma* (eds L.J. DeGroot, L.A. Frohman, E.L. Kaplan and S. Refetoff), Grune and Stratton, New York, San Francisco and London, pp. 137–53.

Sampson, R.J., Key, C.R., Buncher, C.R. and Iijima, S. (1971) Smallest forms of papillary carcinoma of the thyroid. A study of 141 microcarcinomas less than 0.1 cm in greatest dimension. *Arch. Pathol.*, **91**, 334–9.

Sancho-Garnier, H. (1977) L'epidèmiologie des cancers de la thyroide. *Ann. Radiol.*, **20**, 715–21.

Sano, T., Kagawa, N., Hizawa, K. *et al.* (1980) Demonstration of somatostatin production in medullary carcinoma of the thyroid. *Jpn. J. Clin. Oncol.*, **10**, 221–8.

Saxén, E., Franssila, K., Bjarnason, O. *et al.* (1978) Observer variation in histologic classification of thyroid cancer. *Acta Pathol. Microbiol. Scand. [A]*, **86**, 483–6.

Schelfhout, L.J.D.M., Van Muijen, G.N.P. and Fleuren, G.J. (1989) Expression of keratin 19 distinguishes papillary thyroid carcinoma from follicular carcinoma and follicular thyroid adenoma. *Am. J. Clin. Pathol.*, **92**, 654–8.

Schimke, R.N. and Hartmann, W.H. (1965) Familial amyloid-producing medullary thyroid carcinoma and pheochromocytoma. A distinct genetic entity. *Ann. Intern. Med.*, **63**, 1027–39.

Schlumberger, M., Tubiana, M., De Vathaire, F. *et al.* (1986) Long-term results of treatment of 283 patients with lung and bone metastases from differentiated thyroid carcinoma. *J. Clin. Endocrinol. Metab.*, **63**, 960–7.

Schmid, K.W., Fischer-Colbrie, R., Hagn, C. *et al.* (1987) Chromogranin A and B and secretogranin II in medullary carcinomas of the thyroid. *Am. J. Surg. Pathol.*, **11**, 551–6.

Schottenfeld, D. and Gershman, S.T. (1977) Epidemiology of thyroid cancer – part I. *Clin. Bull.*, **7**, 47–54.

Schröder, S. (1988) Pathologie und Klinik maligner Schilddrüsentumoren. Klassifikation, Immunohistologie, Prognosekriterien. *Veröff. Pathol.*, **30**, 1–163.

Schröder, S. and Böcker, W. (1986) Clear-cell carcinomas of thyroid gland: a clinicopathologic study of 13 cases. *Histopathology*, **10**, 75–89.

Schröder, S., Böcker, W., Baisch, H. *et al.* (1988b) Prognostic factors in medullary thyroid carcinomas. Survival in relation to age, sex, stage, histology, immunohistochemistry, and DNA content. *Cancer*, **61**, 806–16.

Schröder, S., Böcker, W., Dralle, H. *et al.* (1984a) The encapsulated papillary carcinoma of the thyroid. A morphologic subtype of the papillary thyroid carcinoma. *Cancer*, **54**, 90–3.

Schröder, S., Dockhorn-Dworniczak, B., Kastendieck, H. *et al.* (1986) Intermediate-filament expression in thyroid gland carcinomas. *Virchows Arch. [A]*, **409**, 751–66.

Schröder, S., Hüsselmann, H. and Böcker, W. (1984b) Lipid-rich cell adenoma of the thyroid gland. Report of a peculiar thyroid tumour. *Virchows Arch. [A]*, **404**, 105–8.

Schröder, S., Pfannschmidt, N., Dralle, H. *et al.* (1984c) The encapsulated follicular carcinoma of the thyroid. A clinicopathologic study of 35 cases. *Virchows Arch. [A]*, **402**, 259–73.

Schröder, S., Schwarz, W., Rehpenning, W. *et al.* (1988a) Dendritic/ Langerhans cells and prognosis in patients with papillary thyroid carcinomas. Immunocytochemical study of 106 thyroid neoplasms correlated to follow-up data. *Am. J. Clin. Pathol.*, **89**, 295–300.

Shimaoka, K. and Tsukado, Y. (1980) Squamous cell carcinomas and adenosquamous carcinomas originating from the thyroid gland. *Cancer*, **46**, 1833–42.

Simpson, N.E., Kidd, K.K., Goodfellow, P.J. *et al.* (1987) Assignment of multiple endocrine neoplasia type 2A to chromosome 10 by linkage. *Nature*, **328**, 528–30.

Sipple, J.H. (1961) The association of pheochromocytoma with carcinoma of the thyroid. *Am. J. Med.*, **31**, 163–6.

Skrabanek, P., Cannon, D., Dempsey, J. *et al.* (1979) Substance P in medullary carcinoma of the thyroid. *Experientia*, **35**, 1259–60.

Sletten, K., Westermark, P. and Natung, J.B. (1976) Characterization of amyloid fibril protein from medullary carcinoma of the thyroid. *J. Exp. Med.*, **143**, 993–8.

Snyder, R.R. and Tavassoli, F.A. (1986) Ovarian strumal carcinoid: immunohistochemical, ultrastructural, and clinicopathologic observations. *Int. J. Gynecol. Pathol.*, **5**, 187–201.

Sobol, H., Narod, S.A. Schuffenecker, I. *et al.* (1989) Hereditary medullary thyroid carcinoma: genetic analysis of three related syndromes. *Henry Ford Hosp. Med. J.*, **37**, 109–11.

Sobrinho-Simões, M., Nesland, J.M. and Johannessen, J.V. (1988) Columnar-cell carcinoma. Another variant of poorly differentiated carcinoma of the thyroid. *Am. J. Clin. Pathol.*, **89**, 264–7.

Sobrinho-Simões, M.A., Nesland, J.M., Holm, R. and Johannessen, J.V. (1985) Transmission electron microscopy and immunocytochemistry in the diagnosis of thyroid tumours. *Ultrastruct. Pathol.*, **9**, 255–75.

Sobrinho-Simões, M., Soares, J., Carneiro, F. and Limbert, E. (1990) Diffuse follicular variant of papillary carcinoma of the thyroid: report of eight cases of a distinct aggressive type of thyroid tumor. *Surg. Pathol.*, **3**, 189–203.

Spencer, R.P., Garg, V., Raisz, L.,G. *et al.* (1982) Radioiodine uptake and turnover in a pseudo-medullary thyroid carcinoma. *J. Nucl. Med.*, **23**, 1006–10.

Steenbergh, P.H. Höppener, J.W.M., Zandberg, J. *et al.* (1984) Calcitonin gene related peptide coding sequence is conserved in the human genome and is expressed in medullary thyroid carcinoma. *J. Clin. Endocrinol. Metab.*, **59**, 358–60.

Stephenson, T.J., Griffiths, D.W.R. and Mills, P.M. (1986) Comparison of *Ulex europaeus* I lectin binding and factor VIII-related as antigen markers of vascular endothelium in follicular carcinoma of the thyroid. *Histopathology*, **10**, 251–60.

Strate, S.M., Lee, E.L. and Childers, J.H. (1984) Occult papillary carcinoma of the thyroid with distant metastases. *Cancer*, **54**, 1093–100.

Straus, F.H. and Spitalnik, P.F. (1977) Histologic parenchymal changes in the human thyroid gland after low dose childhood irradiation. In *Radiation-Associated Thyroid Carcinoma* (eds L.J. DeGroot, L.A. Frohman, E.L. Kaplan and S. Refetoff), Grune and Stratton, New York, San Francisco and London, pp. 183–7.

Sundler, F., Alumets, J., Håkanson, R. *et al.* (1977) Somatostatin-immunoreactive cells in medullary carcinoma of the thyroid. *Am. J. Pathol.*, **88**, 381–5.

Sundler, F., Christophe, J., Robberecht, P. *et al.* (1988) Is helodermin produced by medullary thyroid carcinoma cells and normal C cells? Immunocytochemical evidence. *Regul. Pept.*, **20**, 83–9.

Tangen, K.O., Lindmo, T., Sobrinho-Simões, M. and Johannessen, J.V. (1983) A flow cytometric DNA analysis of medullary thyroid carcinoma. *Am. J. Clin. Pathol.*, **79**, 172–7.

Tennvall, J., Biörklund, A., Möller, T. *et al.* (1986) Is the EORTC prognostic index of thyroid cancer valid in differentiated thyroid carcinoma? *Cancer*, **57**, 1405–14.

Thompson, J.S., Harned, R.K., Anderson, J.C. and Hodgson, P.E. (1983) Papillary carcinoma of the thyroid and familial polyposis coli. *Dis. Colon Rectum*, **26**, 583–5.

Thompson, N.W., Dunn, E.L., Batsakis, J.G. and Nishiyama, R.H. (1974) Hürthle cell lesions of the thyroid. *Surg. Gynecol. Obstet.*, **139**, 555–60.

Tollefsen, H.R., Shah, J.P. and Huvos, A.G. (1975) Hürthle cell carcinoma of the thyroid. *Am. J. Surg.*, **130**, 390–4.

Tscholl-Ducommun, J. and Hedinger, Chr. E. (1982) Papillary thyroid carcinoma. Morphology and prognosis. *Virchows Arch. [A]*, **396**, 19–39.

Ulbright, T., Kraus, F.T. and O'Neal, L.W. (1981) C-cell hyperplasia developing in residual thyroid following resection for sporadic medullary carcinoma. *Cancer*, **48**, 2076–9.

Uribe, M., Fenoglio-Preiser, C.M., Grimes, M. and Feind, C. (1985) Medullary carcinoma of the thyroid gland. Clinical, pathological, and immunohistochemical features with review of the literature. *Am. J. Surg. Pathol.*, **9**, 577–94.

Valenta, L.J. and Bechet, M.M. (1977) Ultrastructure and biochemistry of thyroid carcinoma. *Cancer*, **40**, 284–300.

Variakojis, D., Getz, M.L., Paloyan, E. and Straus, F.H. (1975) Papillary clear cell carcinoma of the thyroid gland. *Hum. Pathol.*, **6**, 384–90.

Viale, G., Dell'Orto, P., Coggi, G. and Gambacorta, M. (1989) Coexpression of cytokeratins and vimentin in normal and diseased thyroid glands. *Am. J. Surg. Pathol.*, **13**, 1034–40.

Vickery, A.L., Jr. (1981) Needle biopsy pathology. *Clin. Endocrinol. Metab.*, **10**, 275–93.

Vickery, A.L., Jr (1983) Thyroid papillary carcinoma. Pathological and philosophical controversies. *Am. J. Surg. Pathol.*, **7**, 797–807.

Vickery, A.L., Jr, Carcangiu, M.L., Johannessen, J.V. and Sobrinho-Simões, M. (1985) Papillary carcinoma. *Semin.Diagn. Pathol.*, **2**, 90–100.

Walts, A.E., Said, J.W. and Shintaku, I.P. (1985) Chromogranin as a marker of neuroendocrine cells in cytological material – an immunocytochemical study. *Am. J. Clin. Pathol.*, **84**, 273–7.

Waterhouse, J., Muir, C., Correa, P. and Powell, J. (1976) Cancer incidence in five continents. Vol. III. *International Agency for Research on Cancer Scientific Publications* no. 15. International Agency for Research on Cancer, Lyon.

Watson, R.G., Brennan, M.D., Goellner, J.R. *et al.* (1984) Invasive Hürthle cell carcinoma of the thyroid. Natural history and management. *Mayo Clin. Proc.*, **59**, 851–5.

White, I.L., Vimadalal, S.D., Catz, B. *et al.* (1981) Occult medullary carcinoma of thyroid. An unusual clinical and pathologic presentation. *Cancer*, **47**, 1364–8.

Williams, E.D. (1965) A review of 17 cases of carcinoma of the thyroid and phaeochromocytoma. *J. Clin. Pathol.*, **18**, 288–92.

Williams, E.D. (1966) Diarrhoea and thyroid carcinoma. *Proc. R. Soc. Med.*, **59**, 18–19.

Williams, E.D., Brown, C.L. and Doniach, I. (1966) Pathological and clinical findings in a series of 67 cases of medullary carcinoma of the thyroid. *J. Clin. Pathol.*, **19**, 103–13.

Williams, E.D., Morales, A.M. and Horn, R.C. (1968) Thyroid carcinoma and Cushing's syndrome. *J. Clin. Pathol.*, **21**, 129–35.

Williams, E.D., Doniach, I., Bjarnason, O. and Michie, W. (1977) Thyroid cancer in an iodine rich area. A histopathological study. *Cancer*, **39**, 215–22.

Winship, T. and Rosvoll, R.V. (1970) Thyroid carcinoma in childhood: final report on a 20 year study. *Clin. Proc. Child. Hosp.*, **26**, 327–48.

Wolfe, H.J., Melvin, K.E.W., Cervi-Skinner, S.J. *et al.* (1973) C-cell hyperplasia preceding medullary thyroid carcinoma: cytological, immunological and biological studies. *N. Engl. J. Med*, **289**, 437–41.

Woolner, L.B. (1971) Thyroid carcinoma: pathologic classification with data on prognosis. *Semin. Nucl. Med.*, **1**, 481–502.

Woolner, L.B., Beahrs, O.H., Black, B.M. *et al.* (1961) Classification and prognosis of thyroid carcinoma. A study of 885 cases observed in a thirty year period. *Am. J. Surg.*, **102**, 354–87.

Zaatari, G.S., Saigo, P.E. and Huvos, A.G. (1983) Mucin production in medullary carcinoma of the thyroid. *Arch. Pathol. Lab. Med.*, **107**, 70–4.

10 Non-epithelial thyroid tumours

10.1 Malignant lymphoma

Lymphoma involving the thyroid gland may be a primary lesion, or it may involve the gland secondarily, as a manifestation of systemic disease. Primary lymphomas of the thyroid gland are uncommon, accounting for less than 5% of thyroid malignancies. They are predominantly non-Hodgkin's lymphomas. The majority of thyroid lymphomas are of the B-cell type. It has been suggested that they represent neoplasms of mucosa-associated lymphoid tissue (MALT) (Anscombe and Wright, 1985). They comprise about 0.5% of all non-Hodgkin's lymphomas and about 6.5% of primary extranodal lymphomas (Chak *et al.*, 1981). Hodgkin's disease and plasmacytoma primarily involving the thyroid gland also occur but are extremely rare neoplasms at this site.

Primary thyroid lymphomas occur predominantly in elderly women, the sex ratio being about 2–6 females to 1 male (Burke *et al.*, 1977; Compagno and Oertel, 1980; Devine *et al.*, 1981; Rasbach *et al.*, 1985). The most common presenting signs and symptoms are a rapidly enlarging mass in the thyroid region, dysphagia, swelling of the anterior neck, hoarseness, and choking due to tracheal compression. These features bear a close resemblance to those of undifferentiated carcinoma, but, unlike the latter, malignant lymphoma is less often associated with a history of pre-existing goitre (Devine *et al.*, 1981). Most patients are euthyroid. On the scintiscan, the lesion may present as one or several cold nodules (Devine *et al.*, 1981).

10.1.1 Pathological features

Grossly, the thyroid gland is involved by bulging grey–tan fleshy, homogeneous masses, which are rubbery or firm in character. Often most of the thyroid has been replaced by tumour tissue and extrathyroidal extension is common (Compagno and Oertel, 1980). Residual thyroid tissue may be present as a compressed rim along the periphery of the expanding process. Adjacent lymph nodes may also be involved, although rather infrequently. This feature was demonstrated in 20% of cases in the large series studied by Compagno and Oertel (1980).

Microscopically, most thyroid lymphomas are diffuse, histiocytic lymphomas, using the Rappaport classification (Burke et al., 1977; Compagno and Oertel, 1980). In the series studied with the Kiel classification, the majority of cases belonged to the centrocytic/centroblastic category (Heimann et al., 1978; Anscombe and Wright, 1985) whereas in studies using Lukes' classification the most common diagnosis was large uncleaved cell lymphoma or immunoblastic lymphoma (Maurer et al., 1979). Poorly differentiated lymphomas also exist, as do variants with plasmacytoid features (Compagno and Oertel, 1980). Regardless of which classification is used, the lymphomas may be roughly divided into tumours of high-grade and low-grade malignancy, showing a good correlation with prognosis (Heimann et al., 1978).

Malignant lymphomas invade the gland diffusely, without sharp demarcation. Characteristically, the lymphoma cells are not cohesive and, in contrast to most carcinomas, lack organization into solid nests, trabecular, ribbon or follicular patterns. However, they tend to invade the lumina of pre-existing follicles which may be stuffed with tumour cells (Fig. 10.1). This phenomenon is diagnostically important; it is not seen in lymphocytic thyroiditis. Isaacson and Wright (1984) designated this phenomenon of invasion of the epithelial structures as a 'lymphoepithelial lesion', which interestingly, is also a characteristic of tumours of mucosa-associated lymphoid tissue in the gastrointestinal tract, lung and salivary glands (Anscombe and Wright, 1985). Due to secondary degeneration of the surrounding follicular cells, the tumour cells may come to lie in rounded masses held together by the surrounding reticulin network, which may imitate an alveolar growth pattern of a carcinoma. Another characteristic feature is the diffuse infiltration of the entire thickness of the walls of blood vessels by neoplastic lymphoid cells, which come to lie beneath the intima, but do not obliterate the lumen. In contrast, carcinomas involving blood vessels tend to break through their walls, forming plugs which partly or completely occlude their lumina.

In at least 75% of cases, malignant lymphomas develop on a background of chronic lymphocytic thyroiditis, especially Hashimoto's thyroiditis (Lindsay and Dailey, 1955; Woolner et al., 1966; Compagno and Oertel, 1980; Holm et al., 1985). The majority of patients with primary thyroid lymphoma show positive serum tests for thyroid autoantibodies (Rasbach et al., 1985). It has been suggested that chronic antigenic stimulation occurring in autoimmune thyroiditis may result in the development of a malignant clone and the subsequent

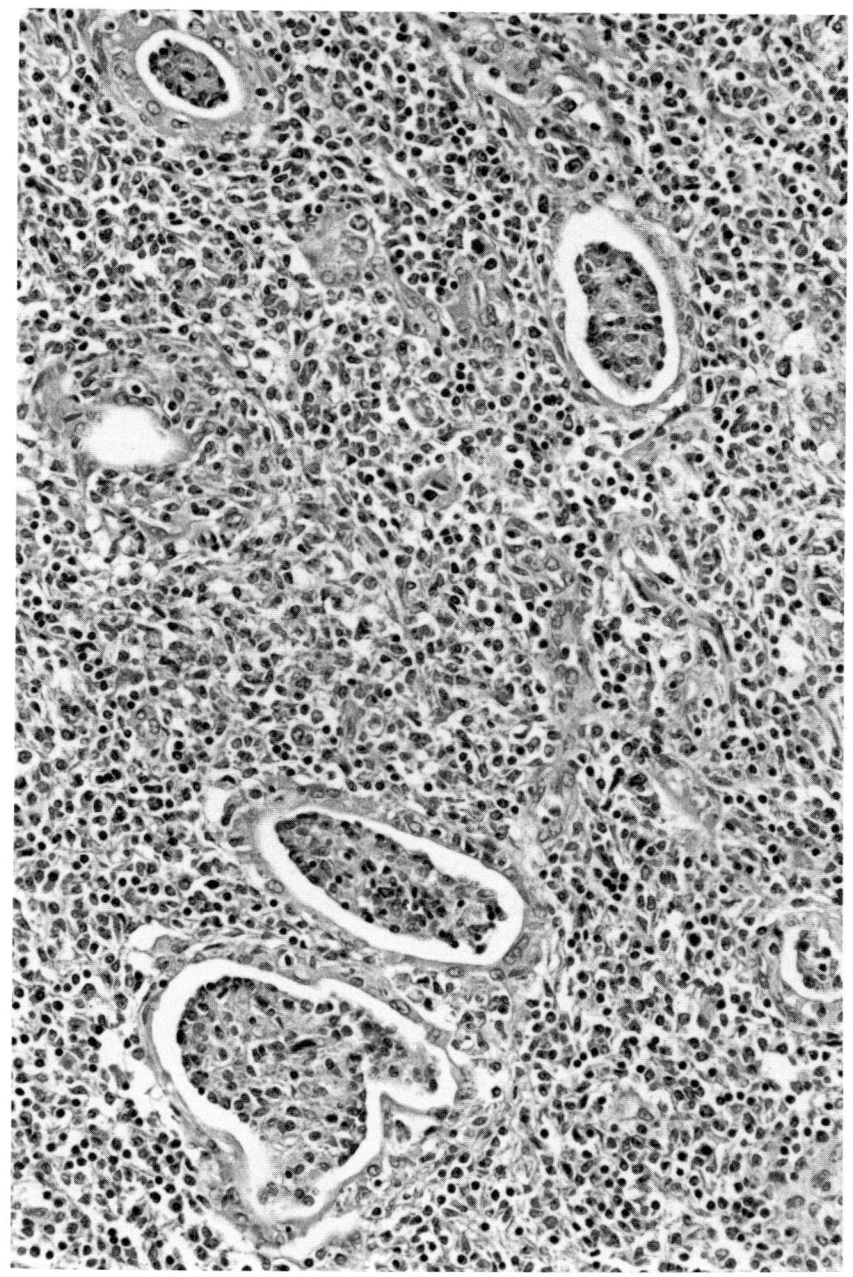

Figure 10.1 Malignant lymphoma: low grade, mixed centrocytic/centroblastic type. The parenchyma is almost completely destroyed and replaced by a diffuse infiltrate of atypical lymphoid cells. Typical lymphoepithelial lesions (H&E, × 250).

establishment of a malignant lymphoma (Burke *et al.*, 1977). The histological distinction between lymphocytic thyroiditis and malignant lymphoma is based on the typical malignant features of the lymphoma as outlined above. These include the destruction of the thyroid architecture by dense infiltrates lacking germinal centres and composed of atypical lymphoid cells, the frequent invasion of the thyroid capsule and extrathyroidal structures, the typical infiltration of vessel walls, and the characteristic 'follicular plugging' (lymphoepithelial lesion).

It is important to distinguish primary thyroid lymphomas from involvement of the thyroid by a systemic lymphoma, since the former has a better prognosis. The initial spread of primary thyroid lymphoma tends to involve the adjacent lymph nodes. Recurrence after resection is usually local. When dissemination occurs, the gastrointestinal tract is often involved (Smithers, 1970; Compagno and Oertel, 1980; Williams, 1981). The association with the gastrointestinal tract may be significant and may reflect the tendency of mucosa-associated lymphomas to spread to other sites where mucosa-associated lymphoid tissue is present (Anscombe and Wright, 1985).

As noted in the section on poorly differentiated and undifferentiated carcinoma (p. 242), thyroid lymphoma must also be distinguished from the rare cases of small cell carcinomas. Small cell tumours, previously believed to represent a subgroup of undifferentiated carcinoma, have been shown for the most part to be malignant lymphomas. However, there remains a small residuum of tumours which lack lymphoid markers but show clear evidence of epithelial differentiation. This group of small cell carcinomas probably includes poorly differentiated (insular) carcinomas, poorly differentiated medullary carcinomas, and possibly, immature variants of the intermediate of carcinoma. Immunohistochemical and ultrastructural analysis may be helpful separating these forms.

Immunohistochemically, thyroid malignant lymphomas consistently show a positive reaction for leukocyte common antigen (Fig. 10.2) but are unreactive with antigens against common epithelial markers, such as epithelial membrane antigen and cytokeratins (Burt *et al.*, 1985; Pizzolo *et al.*, 1980; Warnke *et al.*, 1984). These antibodies can all be applied to sections obtained from formalin-fixed, paraffin embedded tissue. Almost all lymphomas of the thyroid express markers characteristic of lymphomas of B-cell derivation (Aozasa *et al.*, 1987). The most useful diagnostic feature is the expression of immunoglobulin kappa or lambda light chains. However, the application of monoclonal antibodies to these antigens using paraffin section has given variable results (Mambo and Irwin, 1984; Myskow *et al.*, 1986) and it may be preferable to use frozen tissue sections or cytological techniques on cell suspensions or flow cytometry.

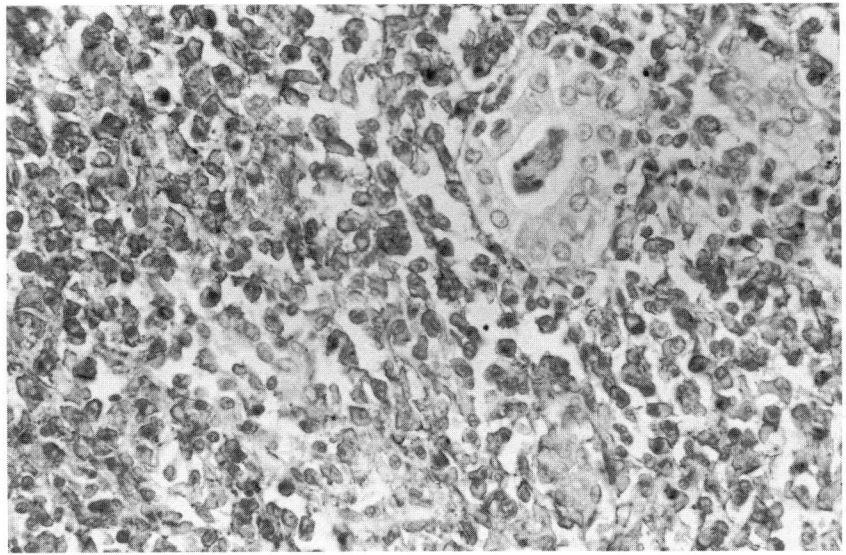

Figure 10.2 Malignant lymphoma; low grade. Immunostaining with antibodies against leukocyte common antigen shows a positive reaction in the lymphoid cell population (ABC technique, × 330).

Fine-needle aspiration diagnosis of thyroid malignant lymphoma (Plate 26) has been shown to be useful in the preoperative evaluation. Aspirates from cases with non-Hodgkin's lymphomas reveal malignant looking lymphoid cells, often lying in aggregates, with prominant nuclei, and sometimes showing indented nuclear membranes. Mitoses are frequent. Karyorrhexis is commonly seen, with fragmented nuclei lying free or as phagocytized debris within the histiocytes. Frequently, however, there may be an admixture of normal lymphoid cells, plasma cells and clusters of hypertrophic follicular cells and oxyphil cells suggestive of Hashimoto's thyroiditis. Against this background the recognition of the malignant lymphoid cells may be difficult, and requires close scrutiny of the smears by an observant and experienced cytopathologist (Willems and Löwhagen, 1981). This is especially true for cases of low-grade lymphomas derived from follicle-centre cells and composed of a spectrum of small to large lymphoid cells with less atypical features (Hamburger *et al.*, 1983). Immunocytochemical examination with antibodies against kappa- and lambda-light chains may help to solve this dilemma, by offering a possibility of assessing whether the B cells are neoplastic (monoclonal) or inflammatory (polyclonal) (Tani and Skoog, 1988).

Plasmacytoma may have a primary origin in the thyroid gland, or may be a component of multiple myeloma. It is a very rare neoplasm and sporadic cases have appeared in the literature. Compagno and Oertel (1980), however, described 21 plasmacytomas among 245 cases of lymphoproliferative thyroid tumours, and more recently, Aozasa *et al.* (1986) described a series comprising six cases. Histologically, the thyroid parenchyma is obliterated by broad sheets of mature plasma cells often

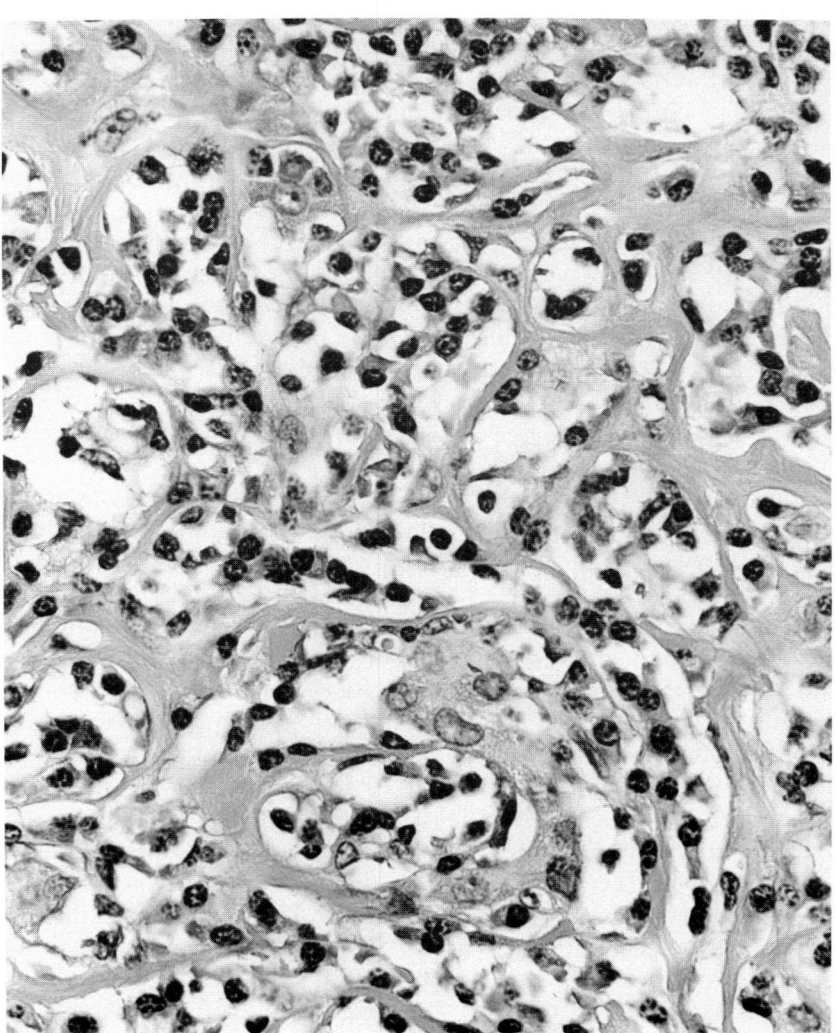

Figure 10.3 Plasmacytoma of the thyroid gland. The thyroid parenchyma is wiped out by a diffuse infiltration of atypical plasma cells. There is some stromal fibrosis (H&E, × 500).

with an admixture of lymphoid follicles, and scattered lymphocytes and histiocytes. Varying degrees of fibrosis may be seen (Fig. 10.3). Non-neoplastic parts of the gland usually show features of chronic lymphocytic thyroiditis. Typically, immunohistochemistry shows the monoclonal nature of the neoplastic plasma cell population (Macpherson et al., 1981; Aozasa et al., 1986). Plasmacytoma should be distinguished from the so-called plasma cell granuloma of the thyroid, which is a non-neoplastic, plasma cell infiltration composed of polyclonal, mature plasma cells, admixed with other inflammatory cells and histiocytes and sometimes associated with a polyclonal gammopathy (Holck, 1981; Yapp et al., 1985). Consequently, on immunohistochemical analysis, plasma cell granuloma shows a mixture of both kappa- and lambda-chain positive plasma cells.

Hodgkin's disease primarily involving the thyroid gland has been reported sporadically (DeBaets et al., 1981; Feigin et al., 1982), although Compagno and Oertel (1980) described seven cases in their series of 245 thyroid lymphomas. The patients usually present with a goitre showing cold nodules on the scintiscan. Associated cervical lymphadenopathy may be seen. Histologically, the lesions have been of the nodular sclerosing or the lymphocytic depletion type.

10.2   Sarcoma and related tumours

According to current opinion, proved sarcomas of the thyroid gland are extremely rare. Most thyroid tumours showing sarcoma-like features are undifferentiated carcinomas (Carcangiu et al., 1985). The growth pattern of such carcinomas can simulate very closely that of fibrosarcoma, malignant fibrous histiocytoma, malignant haemangiopericytoma and angiosarcoma. However, the epithelial nature of these lesions can sometimes be revealed by the microscopic demonstration of focal areas of obvious epithelial structures. When such areas cannot be found, immunohistochemical demonstration of epithelial markers, such as low-molecular-weight cytokeratins or epithelial membrane antigen will mostly establish the carcinomatous nature of the lesion. Desmoplakin has recently been suggested as a valuable specific marker for poorly or undifferentiated carcinomas, but its demonstration is restricted to cryostat sections (Beltrami et al., 1989) (p. 248). Additional support may be achieved by electron microscopic examination, showing epithelial differentiation manifested by specialized cell junctions and the presence of tonofilaments and microvilli (Newland et al., 1981; Carcangiu et al., 1985). Nevertheless, there is a small fraction of sarcoma-like tumours which totally lack epithelial markers but show definite evidence of specific sarcomatous differentiation, as exemplified by the case of osteogenic sarcoma described by Ohbu et al., (1989).

Angiosarcoma or malignant haemangioendothelioma is a rare, highly malignant tumour mostly developing in patients with a long-standing nodular goitre. The majority of cases have been reported in European alpine regions with endemic goitre, whereas this type of tumour is almost unknown in other countries. The tumour is made up of vascular channels and may show solid areas of highly polymorphous tumour cells that may have the appearance of a carcinoma. Spindle cell areas may also occur. Metastases are most frequently seen in the lungs and pleurae. They are usually haemorrhagic and may give rise to extensive bleeding (Egloff, 1983). There has been controversy as to whether these tumours are variants of undifferentiated carcinomas or lesions of truly endothelial origin (Krisch et al., 1980). Recent ultrastructural and immunohistochemical findings support the existence of a thyroid neoplasm of endothelial origin (Egloff, 1983; Ruchti et al., 1984; Tötsch et al., 1990). It should be remembered, however, that undifferentiated carcinomas, and rarely even medullary carcinomas (see Figs. 9.83 and 9.54 respectively), may focally show angiosarcoma-like features, which, on the contrary, show immunohistochemical evidence of their epithelial nature.

References

Anscombe, A.M. and Wright, D.H. (1985) Primary malignant lymphoma of the thyroid – a tumour of mucosa-associated lymphoid tissue: review of seventy-six cases. *Histopathology*, **9**, 81–97.

Aozasa, K., Onoue, A., Yoshimura, H. et al. (1986) Plasmacytoma of the thyroid gland. *Cancer*, **58**, 105–10.

Aozasa, K., Ueda, T., Katagiri, S. et al. (1987) Immunologic and immunohistologic analysis of 27 cases with thyroid lymphomas. *Cancer*, **60**, 969–73.

Beltrami, C.A., Criante, P. and Di Loreto, C. (1989) Immunocytochemistry of anaplastic carcinoma of thyroid gland. *Appl. Pathol.*, **7**, 122–33.

Burke, J.S., Butler, J.J. and Fuller, L.M. (1977) Malignant lymphomas of the thyroid. A clinical pathologic study of 35 patients including ultrastructural observations. *Cancer*, **39**, 1587–602.

Burt, A.D., Kerr, D.J., Brown, I.L. and Boyle, P. (1985) Lymphoid and epithelial markers in small cell anaplastic thyroid tumours. *J. Clin. Pathol.*, **38**, 893–6.

Carcangiu, M.L., Steeper, T., Zampi, G. and Rosai, J. (1985) Anaplastic thyroid carcinoma. A study of 70 cases. *Am. J. Clin. Pathol.*, **83**, 135–58.

Chak, L.Y., Hoppe, R.T., Burke, J.S. and Kaplan, H.S. (1981) Non-Hodgkin's lymphoma presenting as thyroid enlargement. *Cancer*, **48**, 2712–6.

Compagno, J. and Oertel, J.E. (1980) Malignant lymphoma and other lymphoproliferative disorders of the thyroid gland. A clinicopathologic study of 245 cases. *Am. J. Clin. Pathol.*, **74**, 1–11.

DeBaets, M., Vanholer, R., Eeckhaut, W. et al. (1981) Primary Hodgkin's disease of the thyroid. Report of a case demonstrated by immunoperoxidase positive Reed–Sternberg cells. *Acta. Haematol.*, **65**, 54–9.

Devine, R.M., Edis, A.J. and Banks, P.M. (1981) Primary lymphoma of the thyroid: a review of the Mayo Clinic experience through 1978. *World J. Surg.*, **5**, 33–8.

Egloff, B. (1983) The hemangioendothelioma of the thyroid. *Virchows Arch.* [*A*], **400**, 119–42.

Feigin, G.A., Buss, D.H., Paschal, B. *et al.* (1982) Hodgkin's disease manifested as a thyroid nodule. *Hum. Pathol.*, **13**, 774–6.

Hamburger, J.I., Miller, M. and Kini, S.R. (1983) Lymphoma of the thyroid. *Ann. Intern. Med.*, **99**, 685–93.

Heimann, R., Vannineuse, A., De Sloover, D., and Dor, P. (1978) Malignant lymphomas and undifferentiated small cell carcinoma of the thyroid: a clinicopathological review in the light of the Kiel classification for malignant lymphomas. *Histopathology*, **2**, 201–13.

Holck, S. (1981) Plasma cell granuloma of the thyroid. *Cancer*, **48**, 830–2.

Holm, L-E., Blomgren, H. and Löwhagen, T. (1985) Cancer risks in patients with chronic lymphocytic thyroiditis. *N. Engl. J. Med.*, **312**, 601–4.

Isaacson, P. and Wright, D.H. (1984) Extranodal malignant lymphoma arising from mucosa associated lymphoid tissue. *Cancer*, **53**, 2515–24.

Krisch, K., Holzner, J.H., Kokoschka, R. *et al.* (1980). Haemangioendothelioma of the thyroid gland – true endothelioma or anaplastic carcinoma? *Pathol. Res. Pract.*, **170**, 230–42.

Lindsay, S. and Dailey, M.E. (1955) Malignant lymphoma of thyroid gland and its relation to Hashimoto disease: a clinical and pathological study of 8 patients. *J. Clin. Endocrinol. Metab.*, **15**, 1332–53.

Macpherson, T.A., Dekker, A. and Kapadia, S.B. (1981) Thyroid-gland plasma cell neoplasm (plasmacytoma). *Arch. Pathol. Lab. Med.*, **105**, 570–2.

Mambo, N.C. and Irwin, S.M. (1984) Anaplastic small cell neoplasms of the thyroid: an immunoperoxidase study. *Hum. Pathol.*, **15**, 55–60.

Maurer, R., Taylor, C., Terry, R. and Lukes, R.J. (1979) Non-Hodgkin lymphomas of the thyroid. *Virchows Arch.* [*A*], **383**, 293–317.

Myskow, M.W., Krajewski, A.S., Dewar, E. *et al.* (1986) The role of immunoperoxidase techniques on paraffin embedded tissue in determining the histogenesis of undifferentiated thyroid neoplasms. *Clin. Endocrinol.*, **24**, 335–41.

Newland, J.R., Mackay, B., Hill, C.S., Jr and Hickey, R.C. (1981) Anaplastic thyroid carcinoma. An ultrastructural study of 10 cases. *Ultrastruct. Pathol.*, **2**, 121–9.

Ohbu, M., Kameya, T., Wada, C. *et al.* (1989) Primary osteogenic sarcoma of the thyroid gland: a case report. *Surg. Pathol.*, **2**, 67–72.

Pizzolo, G., Sloane, J., Beverley, P. *et al.*, (1980) Differential diagnosis of malignant lymphoma and nonlymphoid tumours using monoclonal anti-leukocyte antibody. *Cancer*, **46**, 2640–7.

Rasbach, D.A., Modschein, M.S., Harris, N.L. *et al.* (1985) Malignant lymphoma of the thyroid gland: a clinical and pathologic study of twenty cases. *Surgery*, **98**, 1166–70.

Ruchti, C., Gerber, H.A. and Schaffner, T. (1984) Factor VIII-related antigen in malignant hemangioendothelioma of the thyroid: additional evidence for the endothelial origin of this tumor. *Am. J. Clin. Pathol.*, **82**, 474–80.

Smithers, D.W. (1970) Malignant lymphoma of the thyroid. In *Tumours of the Thryoid Gland* (ed. D.W. Smithers), E. and S. Livingstone, Edinburgh and London, pp. 141–54.

Tani, E. and Skoog, L. (1988) Fine-needle aspiration cytology and immuno-cytochemistry in the diagnosis of lymphoid lesions of the thyroid gland. *Acta Cytol.*, **33**, 48–52.

Tötsch, M., Dobler, G., Feichtinger, H. *et al.* (1990) Malignant hemangio-endothelioma of the thyroid. Its immunohistochemical discrimination from undifferentiated thyroid carcinoma. *Am. J. Surg. Pathol.*, **14**, 69–74.

Warnke, R.A., Gatter, K.C. and Falini, B. (1984) The diagnosis of human lymphoma using monoclonal antileukocyte antibodies. *N. Engl. J. Med.*, **309**, 1275–81.

Willems, J-S. and Löwhagen, T. (1981) The role of fine-needle aspiration cytology in the management of thyroid disease. *Clin. Endocrinol. Metab.*, **10**, 267–73.

Williams, E.D. (1981) Malignant lymphoma of the thyroid. *Clin. Endocrinol. Metab.*, **10**, 379–89.

Woolner, R.L.B., McConahey, W.M., Beahrs, O.H. and Black, S.M. (1966) Primary malignant lymphoma of the thyroid. Review of forty-six cases. *Am. J. Surg.*, **111**, 502–23.

Yapp, R., Linder, J., Schenken, J.R. and Karrer, F.W. (1985) Plasma cell granuloma of the thyroid. *Hum. Pathol.*, **16**, 848–50.

11 Miscellaneous thyroid tumours

11.1 Paraganglioma

Paragangliomas are tumours derived from the extra-adrenal paraganglia (Glenner and Grimley, 1974). Occasionally, such tumours have also been reported to occur within the thyroid gland (Haegert et al., 1974; Buss et al., 1980; Mitsudo et al., 1987), or located in close proximity to the gland (Kay et al., 1975; Bronner et al., 1988). The tumours have been described as being well defined, usually encapsulated, measuring about 1.5–4 cm in diameter and having tan or greyish, sometimes haemorrhagic cut surfaces.

Microscopically, thyroid paragangliomas show features common to paragangliomas in general (Glenner and Grimley, 1974). They are composed of elongated, oval or polygonal cells with oval nuclei and eosonophilic, often finely granular cytoplasm. These cells show characteristic 'Zellballen' arrangements, embedded in a highly vascularized stroma. Some areas may show trabecular or solid patterns, sometimes with palisading tumour cells. Areas with increased amounts of hyalinized stroma may be seen.

Ultrastructurally, tumour cells show secretory granules of neuro-endocrine type. In accordance with this feature histochemical analysis reveals the presence of general markers of neuroendocrine cells. Thus, silver staining, according to Grimelius' technique generally reveals argyrophil cytoplasmic granules and tumour cells show immuno-reactivity for chromogranin A. They also give positive reaction with antibodies against neurofilaments, which has been found to be typical of paraganglioma cells in general (Bronner et al., 1988).

Thyroid paragangliomas may show close histological resemblance to medullary carcinoma. The fact that both tumours share ultrastructural and immunohistochemical markers of a neuroendocrine neoplasm adds to the confusion with medullary carcinoma. A discriminating point however, is the lack of calcitonin immunoreactivity in paragangliomas. Histologically, paragangliomas may also simulate a trabecular, hyalinizing thyroid adenoma. But in contrast to paraganglioma, the adenoma reveals its follicular differentiation by a positive immunostaining for thyroblobulin. The sustentacular cells at the periphery of the tumour cell nests have been shown to be protein S-100 positive, which may be of diagnostic importance. But it should be remembered that S-100 immunoreactivity may also be demonstrated in thyroid follicular cells (Vanstape *et al.*, 1986).

A rare case of malignant thyroid paraganglioma, invading the trachea, has been reported by Mitsudo *et al.* (1987).

11.2 Teratoma

Cervical teratomas are rare, accounting for only about 5% of all childhood teratomas (Berry *et al.*, 1969; Grosfeld *et al.*, 1976). Their origin is uncertain, but it has been proposed that they originate in the embryonic thyroid anlage. Most are found within or around the thyroid gland (Bale, 1950). They typically occur in infants or children, with equal sex distribution (Newtedt and Shirkey, 1964; Stone *et al.*, 1967). Thyroid teratomas are often congenital and, due to their large size, may complicate delivery and require caesarian section. In other instances, they may cause respiratory distress in the infant, due to airway obstruction (Dehner, 1983; Lack, 1985).

The tumours are usually well encapsulated, histologically benign and can be resected without difficulty. The teratoma may contain both solid and cystic parts and show derivatives of all three germ layers. Less than 10% of thyroid teratomas occur in adults, and then chiefly among women (Wolvos *et al.*, 1985). Teratomas in adults are usually malignant with an aggressive behaviour. Death usually occurs within a year or two. Pulmonary metastases are common. Recurrences are usually resistant to both radiation and chemotherapy (Kimler and Muth, 1978).

11.3 Thymoma

Occasional cases of an intrathyroidal tumour have been described which morphologically resembled epithelial thymoma (Miyauchi *et al.*, 1985; Asa *et al.*, 1988). The tumours were typically located in the middle or lower parts of the lobes and were slow growing. They were composed of sheets and cords of spindle cells and epidermoid cells with occasional keratinization in concentric whorls, and combined with a lymphoplasmacytic infiltration. The epithelial structures were arranged in characteristic lobular formations divided by thick fibrous bands. Mitoses were rare. Ectopic thymic tissue may occur in the thyroid, from which the thyroidal thymoma is supposed to orig-inate. The prognosis of the thymoma is much more favourable than that of squamous cell carcinoma, from which it must be separated.

In contrast to the thymoma, squamous cell carcinoma grows rapidly, shows marked cytological atypia, many mitoses and a high degree of invasiveness, and usually, inconspicuous lymphocytic infiltration.

11.4 Parathyroid adenoma

As will be discussed in Chapter 14 (p. 300), ectopic parathyroid tissue may be present within the thyroid. Consequently intrathyroidal parathyroid adenomas or hyperplastic parathyroid glands may also be encountered (p. 321). Since parathyroid adenomas may have a follicular growth pattern, their distinction from a thyroid adenoma may be problematic. Immunohistochemical analysis with antibodies against thyroglobulin, parathormone and chromogranin A, the latter two being present in parathyroid cells, is helpful for making this differentiation.

References

Asa, S.L., Dardick,I., Van Nostrand, A.W.P. et al. (1988) Primary thyroid thymoma: a distinct clinicopathologic entity. Hum. Pathol., 19, 1463–7.

Bale, G.F. (1950) Teratoma of the neck in the region of the thyroid gland. A review of the literature and report of four cases. Am. J. Pathol., 26, 565–80.

Berry, C.L., Keeling, J. and Hilton, C. (1969) Teratomata of infancy and childhood. J. Pathol., 98, 241–52.

Bronner, M.P., LiVolsi, V.A. and Jennings, T.A. (1988) Plat: paraganglioma-like adenomas of the thyroid. Surg. Pathol., 1, 383–9.

Buss, D.H., Marshall, R.B., Baird, F.G. and Myers, R.T. (1980) Paraganglioma of the thyroid gland. Am. J. Surg. Pathol., 4, 589–93.

Dehner, L.P. (1983) Gonadal and extragonadal germ cell neoplasia of childhood. Hum. Pathol., 14, 493–511.

Glenner, G.G. and Grimley, P.M. (1974) Tumours of the extra-adrenal paraganglion system (including chemoreceptors). In Atlas of Tumor Pathology, Fascicle 9, Second Series, Armed Forces Institute of Pathology, Washington, DC.

Grosfeld, J.L., Ballantine, T.V.N., Lowe, D. and Baehner, R.L. (1976) Benign and malignant teratomas in children: analysis of 85 cases. Surgery, 80, 297–305.

Haegert, D.G., Wang, N.S., Farrer, P.A. et al. (1974) Non-chromaffin paragangliomatosis manifesting as a cold thyroid nodule. Am. J. Clin. Pathol., 61, 561–70.

Kay, S., Montague, J.W. and Dodd, R.W. (1975) Nonchromaffin paraganglioma (chemodectoma) of thyroid region. Cancer, 36, 582–5.

Kimler, S.C. and Muth, W.F. (1978), Primary malignant teratoma of the thyroid: case report and literature review of cervical teratomas in adults. Cancer, 42, 311–7.

Lack, E.E. (1985) Extragonadal germ cell tumors of the head and neck region: review of 16 cases. Hum. Pathol., 16, 56–64.

Mitsudo, S.M., Grajower, M.M., Balbi, H. and Silver, C. (1987) Malignant paraganglioma of the thyroid gland. *Arch. Pathol. Lab. Med.*, **111**, 378–80.

Miyauchi, A., Kuma, K., Matsuzuka, F. *et al.* (1985) Intrathyroidal epithelial thymoma: an entity distinct from squamous cell carcinoma of the thyroid. *World J. Surg.*, **9**, 128–35.

Newstedt, J.R. and Shirkey, H.C. (1964) Teratoma of the thyroid region, report of a case with seven-year follow-up. *Am. J. Dis. Child.*, **107**, 88–95.

Stone, H.H., Henderson, W.D. and Guido, F.A. (1967) Teratomas of the neck. *Am. J. Dis. Child.*, **113**, 222–4.

Vanstapel, M.J., Gatter, K., de-Wolf-Peeters, C. *et al.* (1986) New sites of human S-100 immunoreactivity detected with monoclonal antibodies. *Am. J. Clin. Pathol.*, **85**, 160–8.

Wolvos, T.A., Chong, F.K., Razui, S.A. and Tully, G.L. III (1985) An unusual thyroid tumour: a comparison to a literature review of thyroid teratomas. *Surgery*, **97**, 613–7.

12 Metastasis to the thyroid gland

12.1 Introduction

Autopsy studies have disclosed thyroid gland involvement in 4–24% of patients dying of disseminated carcinoma (Mortensen *et al.*, 1956; Silverberg and Vidone, 1966; Meissner and Warren, 1969). The wide range in incidence may be explained by the thoroughness with which the thyroid gland has been examined at autopsy. The usual sites of origin of metastases found at autopsy include the breast, kidney, lung and skin (malignant melanoma) (Harcourt-Webster, 1965; von Goumoëns, 1968). Other types of carcinoma involve the thyroid only very rarely. Thyroid metastases discovered at autopsy however, are usually small, sometimes microscopic. Therefore, they have most often been clinically occult, the patients having had clinical signs of metastatic disease elsewhere. Although metastatic carcinoma, judging from autopsy findings, is 10 times or so more common than primary thyroid carcinoma, it rarely causes clinical or diagnostic problems, and few such lesions are confused with primary thyroid cancer (Shimaoka *et al.*, 1962). Consequently, in clinical material comprising cases operated for malignant thyroid neoplasms, metastatic lesions are reported to be uncommon. Ivy (1984) reported 30 cases with thyroid metastases among the malignant thyroid neoplasms treated at the Mayo Clinic between 1946 and 1982. Elliott and Frantz (1960) found only 14 instances among more than 500 operated cases, in which a metastatic tumour simulated primary carcinoma of the thyroid.

12.2 Metastases simulating primary thyroid carcinoma

The most common metastatic tumour to masquerade as a primary thyroid tumour is renal cell carcinoma (Figs. 12.1–12.3) (Elliott and Frantz, 1960; Ivy, 1984). Such a metastasis may be the first clinical sign of a renal carcinoma, or may occur years or even decades after the extirpation of the primary neoplasm (Carcangiu *et al.*, 1985). The histopathological distinction between a metastasis of renal clear cell carcinoma and a primary thyroid adenoma or carcinoma with clear

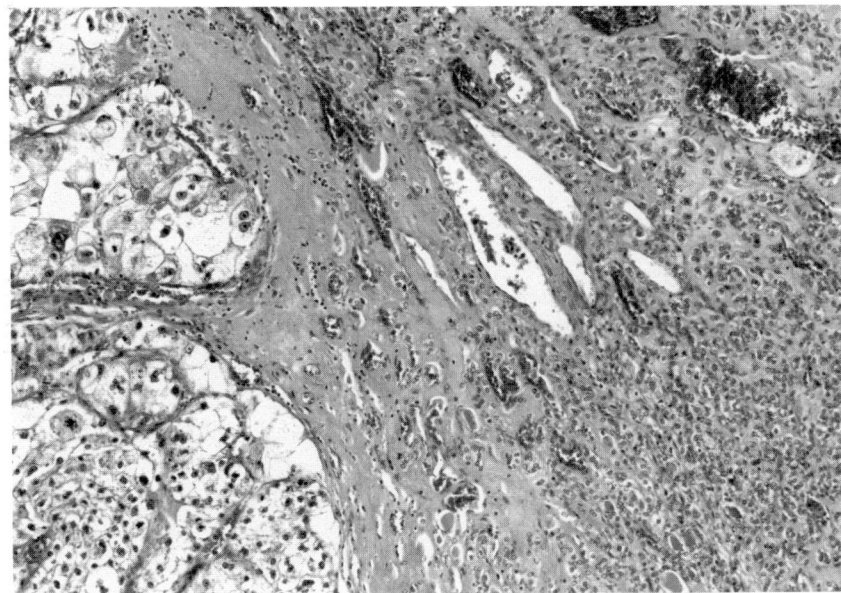

Figure 12.1 Metastasis of renal clear cell carcinoma in a hyperplastic nodule of a multi-nodular goitre. On the left, are nodules of highly atypical, vacuolated metastatic carcinoma cells. On the right, is a portion of the hyperplastic thyroid nodule (H&E, × 80).

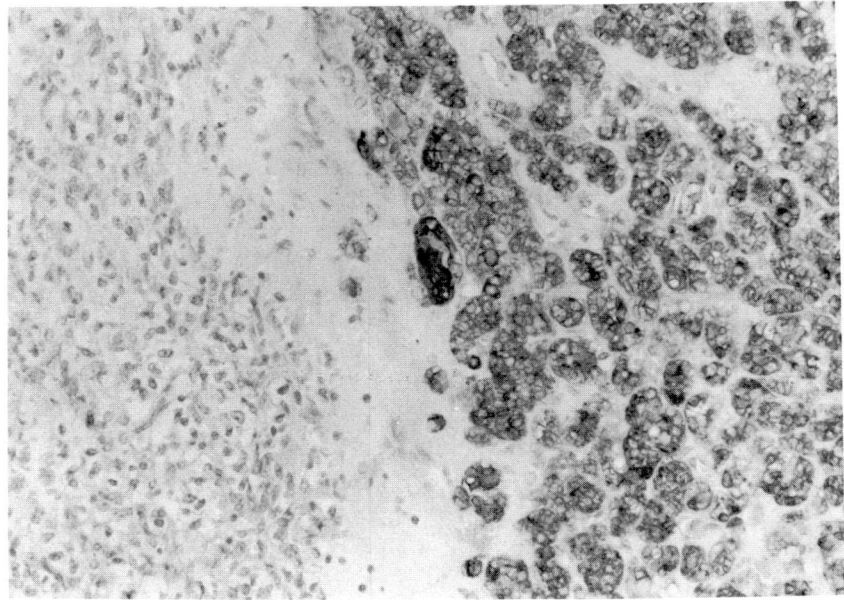

Figure 12.2 Same case as illustrated in Fig. 12.1. Immunostaining for thyroglobulin shows strong reaction in the preserved portion of the thyroid nodule in the right part of the picture, whereas metastatic carcinoma cells (on the left) are completely non-reactive (ABC technique with antibodies against human thyroglobulin, × 165).

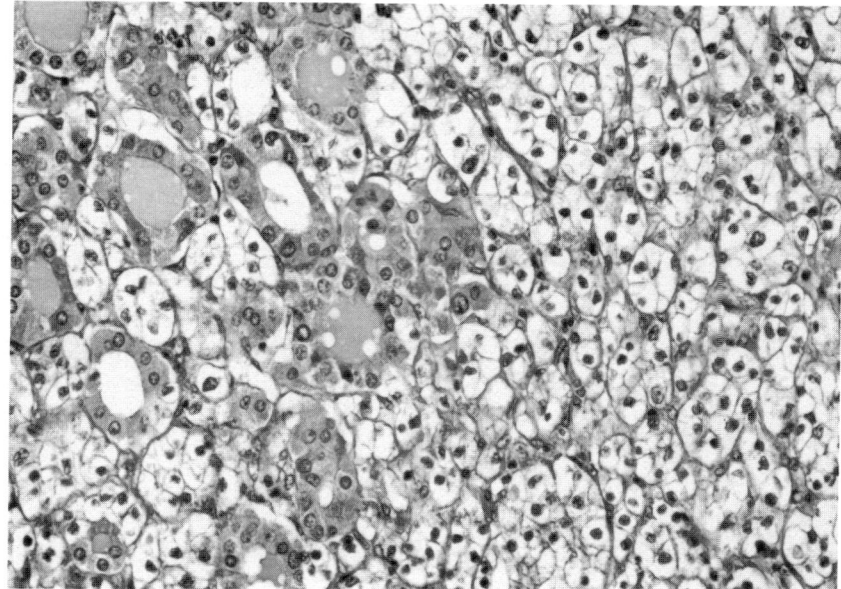

Figure 12.3 Metastasis of renal clear cell carcinoma in an oxyphil thyroid nodule. Scattered preserved follicles with oxyphil epithelium are seen on the left (H&E, × 230).

cell features may be difficult. This distinction is, nonetheless, practically important since the thyroid lesion may be the only metastatic manifestation of the disease. Hedinger *et al.* (1967) have pointed out that clear cell tumours in the thyroid are more often metastases of renal cell cacinoma than primary thyroid neoplasms. The most useful features favouring the diagnosis of metastatic renal cell carcinoma include multifocality of the lesion, formation of tubular structures lacking colloid but sometimes containing blood, vacuolated appearance of the cytoplasm of the tumour cells, which is rich in glycogen and fat, and a pronounced vascularization of the stroma, showing many sinusoidal blood vessels. Primary thyroid clear cell tumours, on the other hand, are usually solitary growths which, at least focally, form follicles with colloid, and the tumour cells tend to have a clear, but finely granulated cytoplasm, usually devoid of glycogen and fat (Stoll and Lietz, 1973). Immunohistochemical analysis with antibodies against thyroglobulin is also very helpful in distinguishing the two conditions (Fig. 12.2; see also p. 199).

Other metastases that may simulate a primary thyroid neoplasm include those from breast and lung carcinomas, especially those having

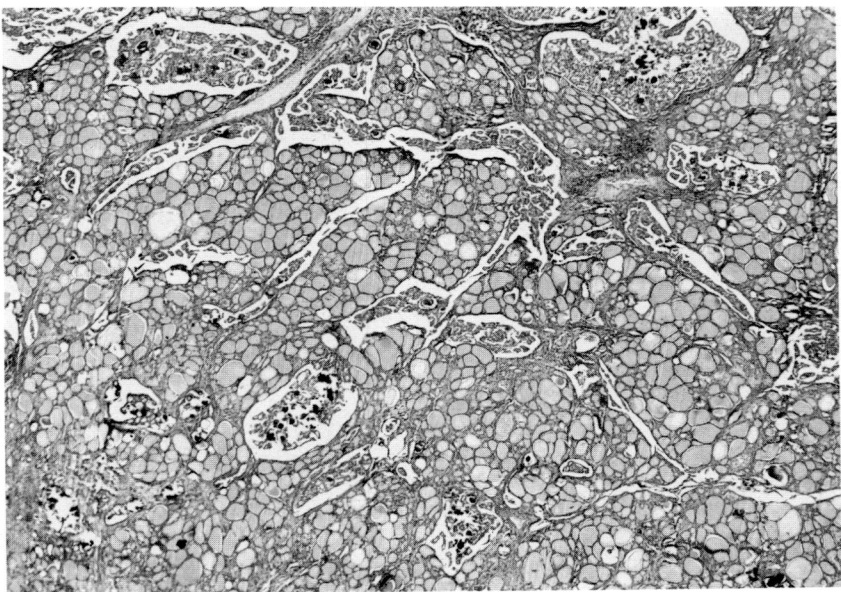

Figure 12.4 Metastasis of papillary lung carcinoma, growing extensively in the lymph vessels of the gland. The black dots represent psammoma bodies (H&E, × 18).

a papillary growth pattern which may be confused with a papillary thyroid carcinoma. In contrast to the renal cell carcinoma, which disseminates mainly by the bloodstream, metastases from these neoplasms tend to reach the thyroid gland by the lymphatics, and thus usually grow diffusely within the lymph vessels of the gland (Figs 12.4 and 12.5). An exceptional case of metastasis to the thyroid gland from a pulmonary carcinoid has been reported by Nesland *et al.* (1985), who stressed the difficulty in distinguishing such a lesion from a primary medullary carcinoma. In this context it may be recalled that neuroendocrine tumours of the lung may be calcitonin-positive, although the immunoreaction appears to be less conspicuous as compared to classic medullary carcinoma (p. 226). Also, the presence of mucin in a neoplasm in the thyroid does not necessarily indicate a metastasis; mucinous cells may occur in follicular cell tumours (p. 178), and even in medullary carcinoma (p. 211).

Laryngeal, pharyngeal and upper oesophageal carcinomas may extend directly into the thyroid gland (Gowing, 1970). These tumours are usually squamous cell carcinomas. However, primary squamous cell carcinoma of the thyroid is very rare (p. 249) and the possibility

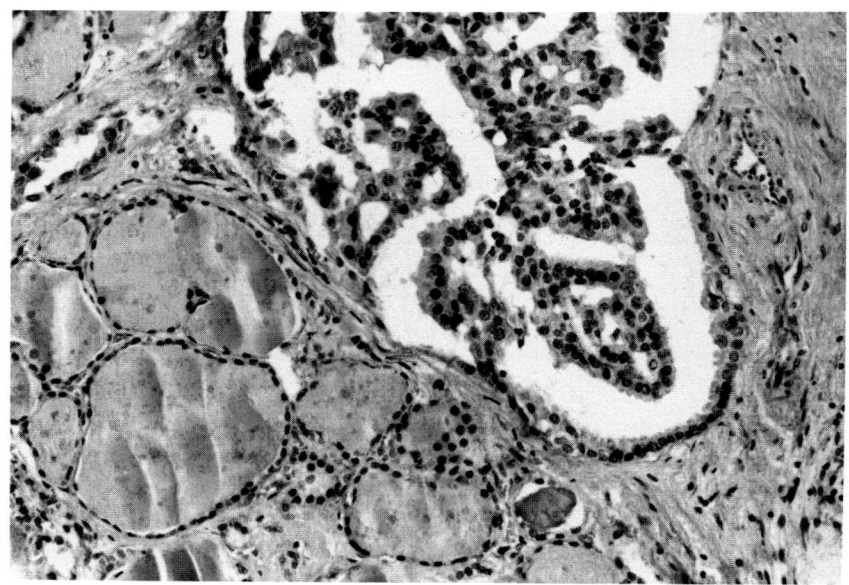

Figure 12.5 From same case as in Fig. 12.4. Close-up view of a metastatic focus reveals papillary carcinomatous structures. Cytological features however, are not compatible with primary papillary thyroid carcinoma (H&E, × 180).

of a metastatic lesion should be considered first when a carcinoma of this histological type is encountered in the gland.

References

Carcangiu, M.L., Sibley, R.K. and Rosai, J. (1985) Clear cell changes in primary thyroid tumours. A study of 38 cases. *Am. J. Surg. Pathol.*, **9**, 705–22.

Elliott, R.H., Jr and Frantz, V.K. (1960) Metastatic carcinoma masquerading as primary thyroid cancer. A report of authors' 14 cases. *Ann. Surg.*, **151**, 551–61.

Gowing, N.F.C. (1970) The pathology and natural history of thyroid tumours. In *Tumours of the Thyroid Gland* (ed. D. Smithers), E. & S. Livingstone, Edinburgh and London, pp. 103–29.

Harcourt-Webster, J.N. (1965) Secondary neoplasm of the thyroid presenting as a goitre. *J. Clin. Pathol.*, **18**, 282–7.

Hedinger, Chr., Corbat, F. and Egloff, B. (1967) Schilddrüsenmetastasen hypernephroider Nierencarcinome. *Schweiz. Med. Wochenschr.*, **97**, 1420–6.

Ivy, H.K. (1984) Cancer metastatic to the thyroid. A diagnostic problem. *Mayo Clin. Proc.*, **59**, 856–9.

Meissner, W. A. and Warren, S. (1969) Tumors of the thyroid gland. In *Atlas of Tumor Pathology*, Fascicle 4, 2nd Series, Armed Forces Institute of Pathology, Washington, DC.

Mortensen, J.D., Woolner, L.B. and Bennett, W.A. (1956) Secondary malignant tumors of the thyroid gland. *Cancer*, **9**, 306–9.

Nesland, J.M., Sobrinho-Simões, M.A., Holm, R. and Johannessen, J.V. (1985) Organoid tumor in the thyroid gland. *Ultrastruct. Res.*, **9**, 67–70.

Shimaoka, K., Sokal, J.E. and Pickren, J.W. (1962) Metastatic neoplasms in the thyroid gland. Pathological and clinical findings. *Cancer*, **15**, 557–65.

Silverberg, S.G. and Vidone, R.A. (1966) Metastatic tumours in the thyroid. *Pacif. Med. Surg.*, **74**, 175–80.

Stoll, W. and Lietz, H. (1973) Zur Kenntnis und Problematik des hellzelligen Adenomes in der Schilddrüse. *Virchows Arch. [A]*, **361**, 163–73.

von Goumoëns, E. (1968) Sekundäre Geschwülste der Schilddrüse. *Schweiz. Med. Wochenschr.*, **98**, 19–25.

13 Thyroid carcinoma in children and adolescents

Thyroid carcinoma is rare in childhood (Buckwalter *et al.*, 1975). Currently, it has been estimated that about 10% of all thyroid carcinomas occur in patients under 21 years of age (Buckwalter *et al.*, 1981). Likewise, solitary cold nodules of the thyroid are infrequent in this age group. Rallison *et al.*, (1975) reported an incidence of 1.8% in a large series of children aged 11–15 years. Although most thyroid nodules occurring in childhood are benign adenomas or examples of nodular goitre, the proportion of malignant tumours is considerably higher than that in adults. About 40% of children with solitary nodules are reported to have thyroid carcinoma (Adams, 1968; Hoffmann *et al.*, 1972; Kirkland *et al.*, 1973), as compared to 10–20% in adults.

13.1 Radiation and thyroid carcinoma

During the second quarter of this century, American authors reported a rapid increase in the number of children with thyroid carcinoma (Winship and Rosvoll, 1969). A sharp rise began in 1945, reaching a peak in 1957; subsequently the number gradually declined. This increase in incidence was presumed to be due to the widespread use of external ionizing irradiation to the region of the thyroid, usually during infancy and childhood. This therapy was most commonly given for thymic enlargement, tonsillar and adenoidal hypertrophy, cervical adenitis, brain tumours, retinoblastomas, and various skin conditions including naevi, haemangiomas, keloids, tinea capitis and even eczema. Winship and Rosvoll (1961) found that about three-quarters of the children with thyroid carcinoma in their large sample of 476 cases had a history of previous irradiation. They observed a mean latent period of 8 years between exposure to irradiation and detection of thyroid carcinoma. According to Hayles *et al.* (1963), the latent period may be as long as 40 years. This increase in thyroid carcinoma was not observed in Europe, possibly because few patients received thymic irradiation in this part of the world (Winship and Rosvoll, 1970).

With the discontinuance of irradiation for most conditions there has been a marked decline in the incidence of clinically evident thyroid carcinoma in children (Buckwalter *et al.*, 1981). Although irradiation to

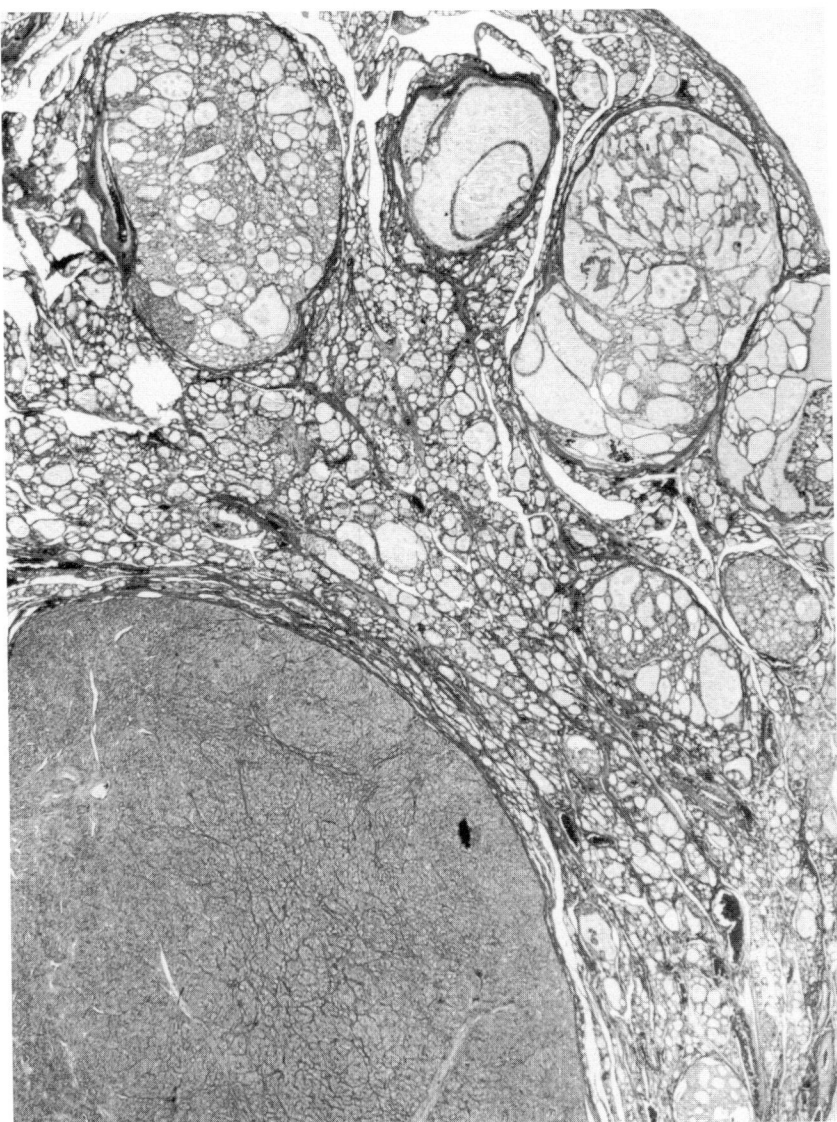

Figure 13.1 Multiple hyperplastic and involutional nodules in a gland removed from a 20-year-old man. When the patient was one-year old, he had received external irradiation to the head and neck for a fibrosarcoma in the mandibular region (Van Gieson-Hansen stain, × 12.5).

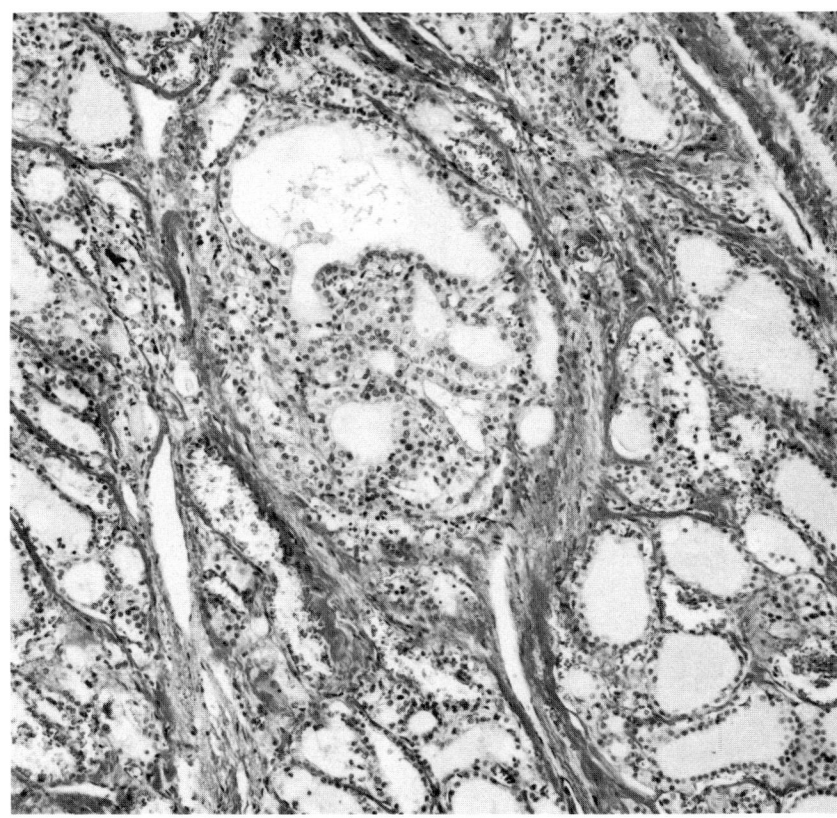

Figure 13.2 Same case as illustrated in Fig. 13.1. In some areas the thyroid paren-chyma shows interstitial fibrosis and multiple microscopic foci with marked epithelial hyperplasia (H&E, × 125).

the neck region during infancy and childhood undoubtedly leads to an increased incidence of thyroid neoplasia, such exposure more often results in various benign abnormalities in the gland, such as follicular adenomas, nodular hyperplasia, chronic lymphocytic thyroiditis, and destruction of the parenchyma and fibrosis (Figs 13.1 and 13.2). In the 1970s it was claimed that there was a persistent high prevalence of thyroid carcinoma in adults as a late consequence of irradiation given to the neck area during infancy or childhood. A considerable propor-tion of the carcinomas found in these cases, however, were very small papillary carcinomas (Fig. 13.3) (Favus *et al.*, 1976; Kumorowski and Hanson, 1977). Today, many of these tumours would undoubtedly have been regarded as occult lesions (papillary microcarcinomas),

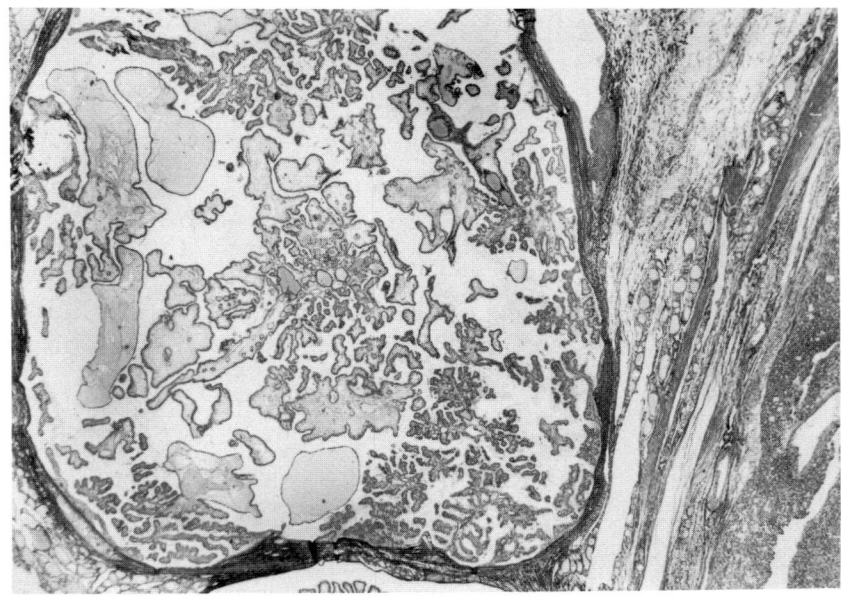

Figure 13.3 Same case as in Figs 13.1 and 13.2. A well-defined, encapsulated papillary microcarcinoma was also found in the gland (H&E, × 18).

representing incidental findings of negligible clinical importance, and occurring in a background of various radiation-induced benign lesions.

It may also be of interest to note that the administration of radioiodine, which is mostly given to adult individuals, and regardless of whether given in diagnostic or ablative doses, does not seem to be associated with the same risk of later development of clinical thyroid carcinoma (Holm *et al.*, 1980).

13.2 Clinical and pathological features

Thyroid carcinoma may occur at any age during childhood but appears to increase with each decade. Rarely, congenital tumours have been described that later were proved to be thyroid carcinomas (Winship and Rosvoll, 1961). It should be recalled here however, that tumours of the thyroid gland which present in the newborn period are for the most part benign teratomas (p. 281). Thyroid carcinoma is about twice as common in girls as in boys (Winship and Rosvoll, 1961; Buckwalter *et al.*, 1981; Tallroth *et al.*, 1986), the female preponderance thus being less marked than in the adult age group.

The presenting symptom is either a mass in the thyroid gland or in a cervical lymph node. The great majority of tumours are of the highly differentiated papillary type and are associated with a high frequency of regional lymph node metastases. Follicular carcinomas may also occur in children, but this type is rare. Undifferentiated thyroid carcinoma is virtually non-existent in children under the age of 15 years. Papillary tumours are characterized by a very slow clinical course even with metastases in the lymph nodes or the lungs and their overall prognosis is excellent (Buckwalter et al., 1975).

Medullary carcinoma is sometimes seen in children, and then usually as part of the MEN 2 B syndrome (Bartlett et al., 1971; Khairi et al., 1975). In this situation, medullary carcinoma tends to behave aggressively (Norton et al., 1979), although cases with a protracted course have also been observed (Ljungberg, 1966).

References

Adams, H.D. (1968) Carcinoma in nodular goiter of childhood. *Postgrad. Med.*, **43**, 136–9.

Bartlett, R.C., Myall, R.W.T., Bean, L.R. and Mandelstam, P. (1971) A neuropolyendocrine syndrome: mucosal neuromas, pheochromocytoma and medullary carcinoma. *Oral Surg.*, **31**, 206–20.

Buckwalter, J.A., Gurll, N.J. and Thomas, C.G. (1981) Cancer of the thyroid in the youth. *World J. Surg.*, **5**, 15–25.

Buckwalter, J.A., Thomas, S.R. and Freeman, J.B. (1975) Is childhood thyroid cancer a lethal disease? *Ann. Surg.*, **181**, 632–9.

Favus, M.J., Schneider, A.B., Stachura, M.E. et al. (1976) Thyroid cancer occurring as a late consequence of head-and-neck irradiation. Evaluation of 1056 patients. *N. Engl. J. Med.*, **294**, 1019–25.

Hayles, A.B., Johnson, L.M., Beahrs, O.H. and Woolner, L.B. (1963) Carcinoma of the thyroid gland in children. *Am. J. Surg.*, **106**, 735–43.

Hoffmann, G.L., Thompson, N.W. and Heffron, C. (1972) The solitary thyroid nodule: a reassessment. *Arch. Surg. (Chicago)*, **105**, 379–85.

Holm, L-E., Dahlqvist, I., Israelsson, A. and Lundell, G. (1980) Malignant thyroid tumors after iodine-131 therapy: a retrospective cohort study. *N. Engl. J. Med.*, **303**, 188–91.

Khairi, M.R.A., Dexter, R.N. Burzynski, N.J. and Johnston, C.C., Jr (1975) Mucosal neuroma, pheochromocytoma and medullary thyroid carcinoma: multiple endocrine neoplasia type 3. *Medicine (Baltimore)*, **54**, 89–112.

Kirkland, R.T., Kirkland, J.L., Rosenberg, H.S. et al. (1973) Solitary thyroid nodules in 30 children and report of a child with a thyroid abscess. *Pediatrics*, **51**, 85–90.

Kumorowski, R.A. and Hanson, G.A. (1977) Morphologic changes in the thyroid following low-dose childhood radiation. *Arch. Pathcl. Lab. Med.*, **101**, 36–9.

Ljungberg, O. (1966) Thyreoideacancer och pheochromocytom. *Opuscula Medica*, **2**, 57–62.

Norton, J.A., Froome, L.J., Farrell, R.E. and Wells, S.A. Jr (1979) Multiple endocrine neoplasia type IIb: the most aggressive form of medullary thyroid cancer. *Surg. Clin. North Am.*, **59**, 109–18.

Rallison, M.L., Dobyns, B.M., Keating, F.R. *et al.* (1975) Thyroid nodularity in children. *JAMA*, **233**, 1069–72.

Tallroth, E., Bäckdahl, M., Einhorn, J. *et al.* (1986) Thyroid carcinoma in children and adolescents. *Cancer*, **58**, 2329–32.

Winship, T. and Rosvoll, R.V. (1961) Childhood thyroid carcinoma. *Cancer*, **14**, 734–43.

Winship, T. and Rosvoll, R.V. (1969) Cancer of the thyroid in children. In *Thyroid Cancer* UICC Monograph Series, vol. 12 (ed. Chr. E. Hedinger), Springer-Verlag, Berlin, Heidelberg and New York, pp. 75–8.

Winship, T. and Rosvoll, R.V. (1970) Thyroid carcinoma in childhood: final report on a 20 year study. *Clin. Proc. Child. Hosp.*, **26**, 327–48.

PART TWO
The Parathyroid Glands

14 The normal parathyroid glands

14.1 Anatomy

The parathyroid glands are usually four in number. Reported frequencies of supernumerary glands (usually five) varies between 2.5% and 13% (Gilmour and Martin, 1937; Alveryd, 1968; Wang, 1976; Åkerström et al., 1984). Fewer than four is extremely rare. The glands are arranged in two pairs. The upper pair (parathyroid IV) is usually located at the middle third of the posterolateral border of the thyroid gland, at the cricothyroid junction and about 1 cm above the site of intersection of the inferior thyroid artery and the recurrent laryngeal nerve. The lower pair of glands (parathyroid III) are located near the lower thyroid poles, in close proximity to the inferior thyroid artery. Supernumerary glands are usually encountered near the lower poles of the thyroid gland, often embedded in thymic tissue. The glands usually have a symmetrical distribution. Anomalous positions may occur. Glands may be found as high as the carotid bifurcation, lateral to the inferior pole of the thyroid, or within the pericardium. A gland may also be found in the capsule or within the parenchyma of the thyroid gland, in the pharyngeal wall, or in the retropharyngeal and retro-oesophageal spaces.

The shape of the glands varies. They are usually oval, or shaped like a bean or droplet; more rarely they are lobulated or flattened out like a leaf. Their weight increases from 5–9 mg at birth until the third or fourth decade of life, when the total glandular weight reaches about 120 and 130 mg in men and women respectively (Gilmour and Martin, 1937; Åkerström et al., 1984). Any gland weighing more than 60 mg should be suspected of being abnormal (Roth, 1971). Usually, they are approximately equal in size, but on occasion their weights are uneven, one gland being larger than average. Under such circumstances the remaining three glands are correspondingly smaller, the total glandular weight being within normal limits. The weight of the parenchymal cell mass is better related to endocrine function than the weight of the total gland, because the content of adipose tissue within the gland varies according to nutritional state and other factors. The parenchymal cell weight may be determined by the density gradient technique (Åkerström *et al.*, 1980).

The colour of the glands varies with the relative content of stromal fat, which in turn, depends on age and nutritional status. Before puberty, the gland contains little fat and is reddish-brown. With increasing age, up to 25–30 years, the amount of adipose tissue increases; thereafter it is mainly determined by nutritional factors. The presence of fat renders the gland a yellowish-brown colour.

14.2   Development

The embryological development of the parathyroid glands explains much of their topographic variability. It is generally accepted that the parathyroid glands are of endodermal derivation and develop from the primitive branchial pouches.

The upper pair (parathyroid IV) descends from the fourth branchial pouch, together with the ultimobranchial bodies which form the lateral thyroid anlage. These two structures are closely associated during the further development. This may explain why an upper parathyroid gland may occasionally be located within the thyroid or why accessory parathyroid tissue may appear in close proximity to ultimobranchial remnants found in or near the thyroid capsule, or within the thyroid gland (Fig. 14.1).

The lower pair (parathyroid III) develop together with thymus III from the third branchial pouch. These two structures accompany each other during the descent in the neck. When they reach the level of the developing thyroid they separate; the parathyroids find their final place near the lower thyroid poles, whereas the thymic anlage continues its migration downwards into the thorax. The separation may occur earlier which may leave the gland high up in the neck, sometimes with small amounts of thymic tissue. A small thymic extension may occasionally remain in the neck which is then closely associated with the parathyroid gland at its normal position near the lower pole of the thyroid. Rarely, the parathyroids III follow the thymus anlage during its descent into the anterior mediastinum and may even migrate into the pericardium. Supernumerary glands are most often found in association with undescended thymic tissue reaching from behind the thyroid down to the thymus. Occasionally, as many as 11 or 12 abortive glands or rudimentary nests of parathyroid tissue have been identified in this area (Grimelius *et al.*, 1981).

14.3   Histology

The glandular parenchyma is enclosed within a thin fibrous capsule.

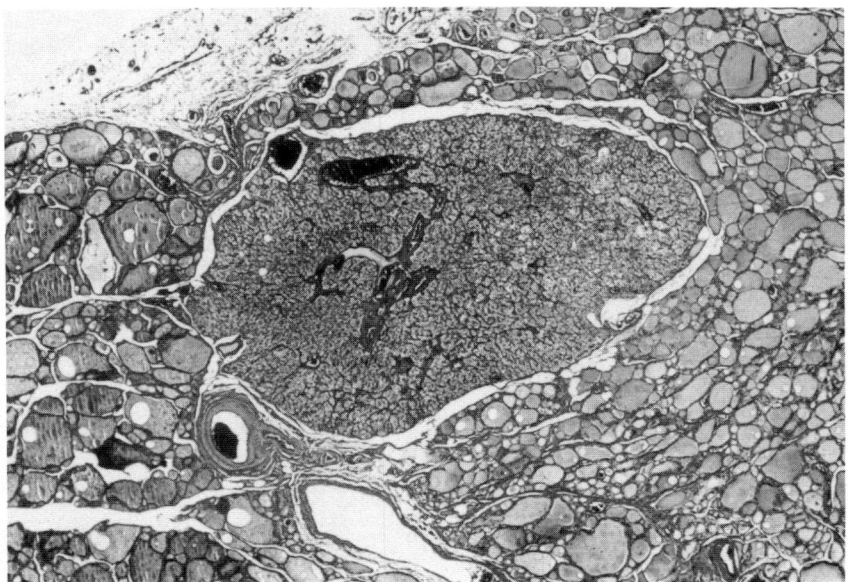

Figure 14.1 Intrathyroidal parathyroid gland, lying close to the surface of the thyroid gland (H&E, × 16).

The epithelial cells form cords and sheets which are embedded in a richly vascularized stroma. Before puberty, the stroma contains no or only a few fat cells. After puberty, the amount of adipose tissue increases with age up to 25–30 years, and thereafter it is fairly constant. However, the amount of adipose tissue is also related to nutritional status. Obese individuals usually have abundant fat cells in their parathyroid glands (Fig. 14.2) and, conversely, emaciated persons have fewer fat cells in the glands (Fig. 14.3) than is normal for their age (Gilmour, 1939; Dekker *et al.*, 1979). In glands with little adipose tissue the parenchymal cells are usually arranged in solid masses irregularly split up by connective tissue septa. In glands rich in adipose tissue on the other hand, the epithelial cells form branching or anastomosing trabeculae, often with a tendency to a lobulated arrangement.

The epithelial component comprises two principal types of cells (Fig. 14.4), the chief cell and the oxyphil cell. Chief cells predominate. They are round or polygonal, 6–10 μm in diameter with central basophilic nuclei, and pale granular cytoplasm. Based on varying stainability of the cytoplasm, chief cells have been subdivided into dark and light (vacuolated) variants. They have been thought to represent

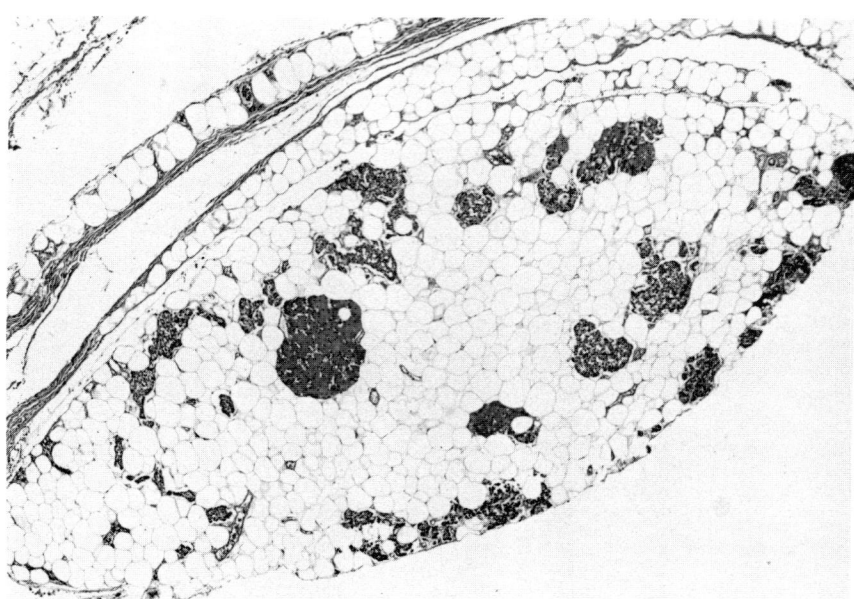

Figure 14.2 Normal parathyroid gland from an obese, normocalcaemic adult, containing abundant adipose tissue (H&E, × 32).

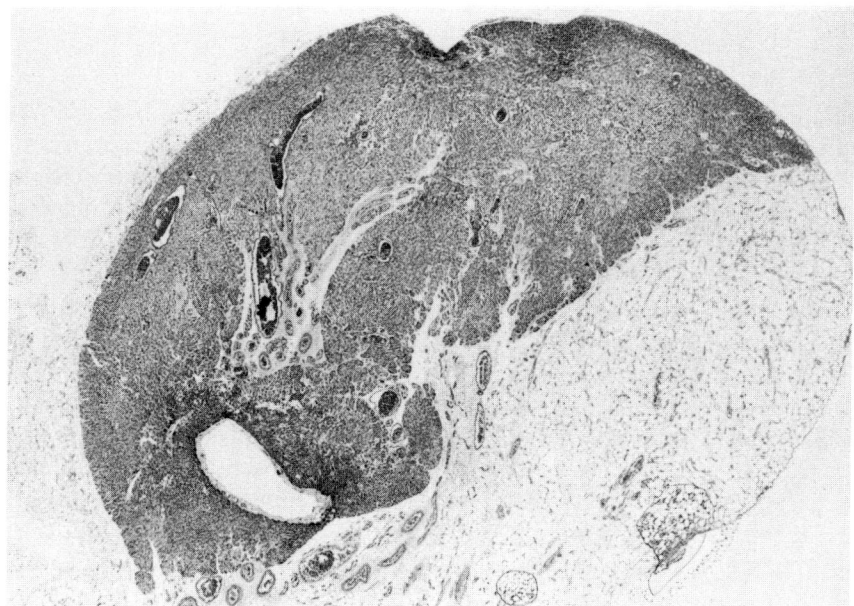

Figure 14.3 Normal parathyroid gland from an emaciated adult individual with normal serum calcium, having a rather compact structure and almost devoid of fat cells (H&E, × 17).

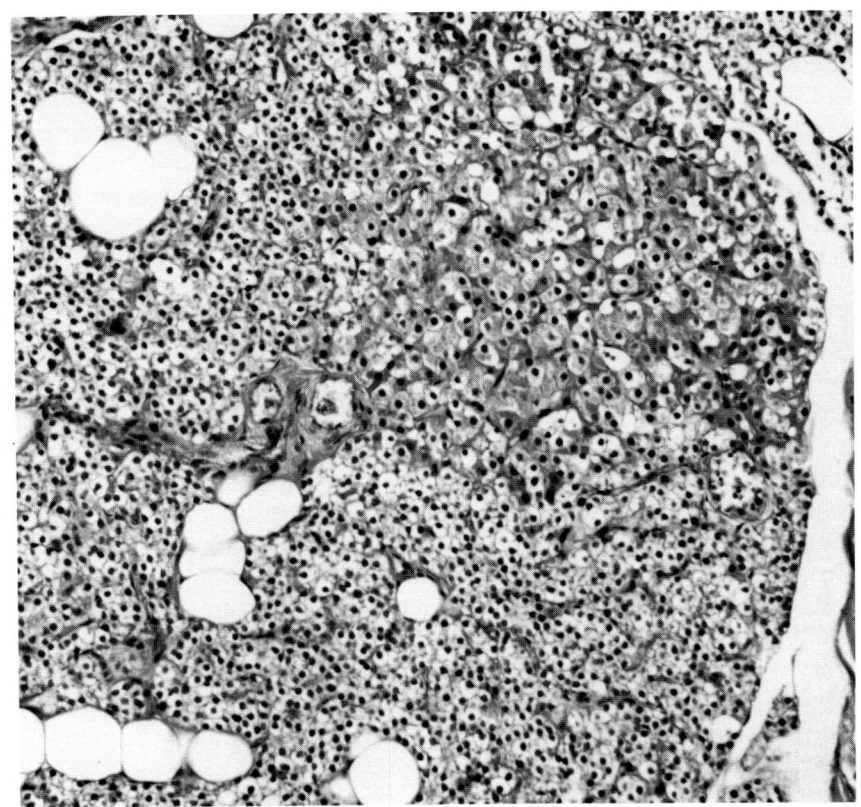

Figure 14.4 Normal parathyroid parenchyma of an adult, composed of sheets and cords of vacuolated chief cells and a nest of oxyphil cells (at the top right), admixed with scattered small groups of fat cells (H&E, × 165).

morphological variations due to differences in physiological activity, dark cells being active and light cells inactive forms. Based on ultrastructural studies it was claimed that these cell forms represent different stages in a functional cell cycle (Altenähr, 1972; Roth and Capen, 1974). Dark cells were described as having large Golgi complexes, numerous secretory granules and sparse glycogen, whereas light cells were characterized by a small Golgi complex, abundant glycogen, and few secretory granules. However, the concept of dark and light chief cells and the existence of functional cycles has been challenged (Stoeckel and Porte, 1966). Recent ultrastructural and quantitative stereological studies on experimental animals suggest that the variation in cytoplasmic staining found on histological and ultrathin sections may be due to artifacts of fixation and may also be partly explained by an asymmetrical distribution of the various cell organelles (Stoeckel and Porte, 1966; Larsson *et al.*, 1984; Svensson *et al.*, 1989). Likewise, Nilsson (1977) was unable to differentiate between active and inactive cells on the basis of ultrastructural examination of normal adult human glands. Even chief cells containing large amounts of glycogen which is lost during preparation and thus causing them to appear light and vacuolated under the microscope, have been shown to contain well-developed organelles associated with hormone synthesis and secretion (Nilsson, 1977).

Normal chief cells contain varying amounts of lysosomal lipid bodies. However, this cytoplasmic component appears to be related to the functional state of the cell. Ultrastructural studies of 'normal' glands from patients with hypercalcaemia have demonstrated that the majority of chief cells are chronically suppressed and contain large amounts of lipid. This fat tends to coalesce, forming a large lipid body ranging from 0.5 to 5 µm. Hyperfunctioning chief cells as seen in adenomas and chief cell hyperplasia, on the other hand, have only a few fat droplets (Roth and Munger, 1962; Altenähr, 1972; Roth and Capen, 1974; Nilsson, 1977).

Oxyphil cells were first mentioned by Sandström (1880) but were clearly described by Welsh in 1898. They are few in children and young individuals; their number increases with age (Gilmour, 1939). In the elderly they may form irregular nests or even nodules. Such accumulations may become large and mimic adenomas or nodular hyperplasia and they may cause differential diagnostic problems in cases of primary hyperparathyroidism caused by other parathyroid lesions (p. 333). The cells are large, up to 20 µm in diameter, and have strongly eosinophilic, granular cytoplasm (Fig 14.5). Ultrastructurally, their cytoplasm is seen to be filled with numerous large, sometimes bizarre mitochondria but few secretory granules and other components associated with hormone production and secretion. The same features also apply to the oxyphil cells (oncocytes) of other organs, such as the thyroid and salivary glands. They are rich in oxidative enzymes (Tremblay and Pearse, 1959). They appear to be derived from the chief cells. This is supported by the finding of cell forms with features intermediate between chief cells and fully developed oxyphil cells, i.e. fewer mitochondria than fully developed oxyphil cells, and a well-developed endoplasmic reticulum and Golgi complex, as well as secretory granules; such cells are often referred to as transitional oxyphil cells. The nature of the oxyphil cells is unknown. The fully developed oxyphil cell appears to have a markedly reduced capability for protein synthesis. Boquist (1975) and McGregor *et al.* (1978) reported *in vitro* transformation of chief cells into oxyphil cells during long-term culture in high calcium ion concentrations. This raises the

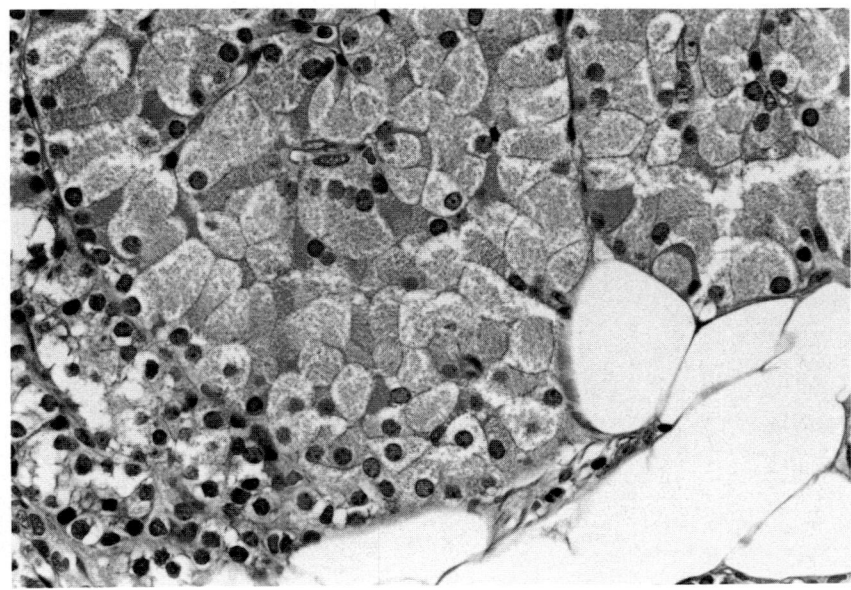

Figure 14.5 Normal parathyroid parenchyma. Close-up view showing a cluster of polygonal oxyphil cells with abundant, finely granular cytoplasm. Cords of chief cells are present at the lower left (H&E, × 350).

possibility of hypercalcaemia acting as the stimulus for oxyphil cell transformation.

Acini and follicle-like structures with colloid-like material may be seen in normal glands (Boquist, 1973). Sometimes follicles may contain material which is positive with amyloid stains (Lieberman and DeLellis, 1973; Anderson and Ewen, 1974; Antonutto *et al.*, 1975). Varied-sized cysts are sometimes also found (Castleman and Mallory, 1935; Black and Watts, 1949; Haid *et al.*, 1967; Hoehn *et al.*, 1969). Glands with follicular patterns may mimic thyroid tissue. However, unlike follicular epithelial cells, parathyroid cells of normal and diseased glands are argyrophilic (Frigerio *et al.*, 1982) and react with antibodies against chromogranin A (Wilson and Lloyd, 1984; Weiler *et al.*, 1988) and are thyroglobulin-negative.

14.4 Physiology

The parathyroid glands synthesize and secrete parathyroid hormone (PTH), which together with the active metabolite(s) of vitamin D and calcitonin, maintains the concentration of ionized calcium in

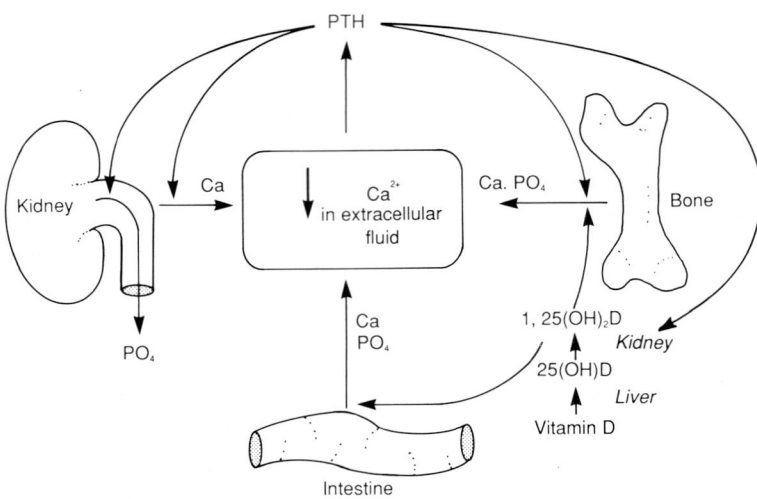

Figure 14.6 Schematic diagram of the role of parathormone (PTH) and vitamin D (1,25(OH)$_2$D) in calcium homeostasis. A fall in the concentration of extracellular Ca^{2+} stimulates the release of PTH. Increased circulating PTH levels enhance the renal tubular reabsorption of Ca^{2+}, promotes phosphaturia, and stimulates the synthesis of 1,25(OH)$_2$D. The latter, in turn, promotes the intestinal absorption of Ca^{2+}, and, in conjunction with PTH, increases the release of skeletal Ca^{2+}. This results in a restoration of extracellular Ca^{2+} levels to normal and a reduction of PTH secretion back to normal.

extracellular fluids within a very narrow range (Brown *et al.*, 1986; Mallette, 1989). A schematic diagram of the role of PTH and vitamin D in calcium homeostasis is given in Fig. 14.6. PTH brings about calcium homeostasis primarily by acting on bone and kidney. The active form of vitamin D primarily acts on the intestinal mucosa, influencing its absorption of calcium and phosphate, but may also, in synergy with PTH, stimulate calcium resorption in bone. The precursor molecule of vitamin D (7-dehydrocholesterol), present in the dermis, is converted into cholecalciferol (vitamin D$_3$) by ultraviolet light. Cholecalciferol may also be obtained from dietary sources. Vitamin D from either source is hydroxylated in the liver to 25-hydroxyvitamin D, which is then converted into 1,25-dihydroxyvitamin D in the kidney. This is the most active form of the hormone. The renal 1-alpha-hydroxylase, which is responsible for this hydroxylation, is stimulated by PTH and low serum phosphate concentration. Although calcitonin may play an important role in mineral metabolism in some species, its importance in man is still controversial (Austin and Heath, 1981). The parathyroid glands are sensitive to changes in the extracellular concentration of ionized calcium and respond with alterations in the secretion of PTH by a negative feed-back relationship. Consequently, a decrease in the concentration of extracellular Ca^{2+} produces an increased rate of PTH secretion. PTH interacts with specific receptors on the cell surface of target tissues. The cellular responses are thought to be mediated by several mechanisms including activation of adenyl cyclase, increased cytosolic Ca^{2+} concentration and/or changes in the metabolism of phosphatidylinositides (Caulfield and Rosenblatt, 1990). The most rapid effects of PTH include increased release of Ca^{2+} from bone by inhibition of bone formation and stimulation of bone resorption, and increased distal tubular reabsorption of Ca^{2+} by the kidney. PTH also promotes the renal excretion of phosphate, mobilized from bone together with calcium. These are the short-term effects of PTH which occur within a few minutes to an hour. If the decrease of extracellular fluid Ca^{2+} concentration persists for several hours or longer, there is a sustained increase in PTH which stimulates renal synthesis of $1,25(OH)_2D$. This hormone stimulates intestinal absorption of calcium and phosphate, and further promotes the release of Ca^{2+} ions from bone (Reichel et al., 1989). An even more prolonged hypocalcaemia (over days or weeks) produces further increases in PTH secretion through parathyroid cellular hyperplasia and hypertrophy. In bone, PTH causes an increase in the number of osteoclasts and stimulates their phagocytic activity resulting in increased bone resorption. It causes retraction of the osteoblatic cell layer and thereby produces free bone surfaces where the osteoclasts may attach and attack the bone (Rodan and Martin, 1981).

PTH is a single-chain polypeptide hormone consisting of 84 amino acid residues with a molecular weight of about 9600. Residues 2–25 constitute the minimum sequence required for biological activity and a fragment consisting of the first 34 amino acid residues is probably as active as the intact hormone (Rosenblatt et al., 1989). The carboxyl (C)-terminal fragment has generally been considered biologically inactive. It appears that in fact the major part of newly synthesized PTH may be degraded into various fragments within the chief cell itself (Cohn et al., 1986). This degradation seems to be a mechanism for controlling the amount of secreted hormone. Calcium may influence this degradation mechanism; the ratio of inactive fragments to intact PTH increases with hypercalcaemia, and, conversely, decreases in hypocalcaemic states (Mayer et al., 1979; Bienkowski, 1983).

In addition to PTH, the parathyroid glands also synthesize another major protein, secretory protein I (SP-I) (Kemper et al., 1974). This is an acidic glycoprotein which appears to be similar if not identical to chromogranin A, which was initially demonstrated in the adrenal medulla (Cohn *et al.*, 1982; Greeley *et al.*, 1989), but later found to be widely distributed in cells of the neuroendocrine system and its tumours (Lloyd *et al.*, 1989). SP-I/chromogranin A is present in secretory granules containing PTH and is cosecreted with this hormone. It is also cosecreted with adrenaline and noradrenaline from the adrenal medulla and probably with other endocrine products with which it coexists in secretory granules (Cohn *et al.*, 1986). This suggests that it plays a general role in hormone secretion and storage mechanisms. Recent evidence suggests that pancreastatin, which has been demonstrated in porcine pancreatic islets, is a peptide with a sequence identical to a part of the porcine SP-I/chromogranin A molecule. SP-I/chromogranin A and pancreastatin have both been demonstrated to inhibit insulin secretion in the pancreatic islets of the rat suggesting that circulating SP-I plays a physiological role in the regulation of insulin secretion (Greeley *et al.*, 1989). It may also act as a precursor of peptide hormones (Eiden, 1987).

The parathyroid cells are typically argyrophilic with the Grimelius **std** silver technique (Frigerio *et al.*, 1982) and show chromogranin A immunoreactivity (Wilson and Lloyd, 1984), features that may be related to their content of SP-I/chromogranin. PTH can also be demonstrated immunohistochemically (Ordoñez *et al.*, 1983; Pesce *et al.*, 1989).

References

Åkerström, G., Grimelius, L., Johansson, H. *et al.* (1980) Estimation of the parathyroid parenchymal cell mass by density gradients. *Am. J. Pathol.*, **99**, 685–94.

Åkertröm, G., Malmaeus, J. and Bergström, R. (1984) Surgical anatomy of human parathyroid glands. *Surgery*, **95**, 14–21.

Altenähr, E. (1972) Ultrastructural pathology of parathyroid glands. *Curr. Top. Pathol.*, **56**, 1–54.

Alveryd, A. (1968) Parathyroid glands in thyroid surgery. I. Anatomy of parathyroid glands. *Acta Chir. Scand.*, Suppl. **389**, 9–49.

Anderson, T.J. and Ewen, S.W.B. (1974) Amyloid in normal and pathological parathyroid glands. *J. Clin. Pathol.*, **27**, 656–63.

Antonutto, G., Bartoli, M. and Melato, M. (1975) Histochemical study on systematic amyloid microdeposits with special references to parathyroid intrafollicular deposits. *Virchows. Arch. [A]*, **368**, 23–34.

Austin, L.A. and Heath, H. (1981) Calcitonin: physiology and pathophysiology. *N. Engl. J. Med.*, **304**, 269–78.

Bienowski, R.S. (1983) Review article. Intracellular degradation of newly synthesized secretory proteins. *Biochem. J.*, **214**, 1–10.

Black, B.M. and Watts, C.F. (1949) Cysts of parathyroid origin: report of two cases and study of the incidence and pathogenesis of cysts in parathyroid glands. *Surgery*, **25**, 941–9.

Boquist, L. (1973) Follicles in human parathyroid glands. *Lab. Invest.*, **28**, 313–20.

Boquist, L. (1975) Occurrence of oxyphil cells in suppressed parathyroid glands. *Cell Tiss. Res.*, **163**, 465–70.

Brown, E.M., LeBoff, M.S., Oetting, M. *et al.* (1986) Secretory control in normal and abnormal parathyroid tissue. *Rec. Prog. Horm. Res.*, **43**, 337–82.

Castleman, B. and Mallory, T.B. (1935) The pathology of the parathyroid gland in hyperparathyroidism. A study of 25 cases. *Am. J. Pathol.*, **11**, 1–72.

Caulfield, M.P. and Rosenblatt, M. (1990) Parathyroid hormone–receptor interactions. *TEM*, **1**, 164–8.

Cohn, D.V., Kumarasamy, R. and Ramps, W.K. (1986) Intracellular processing and secretion of parathyroid gland proteins. *Vitam. Horm.*, **43**, 283–316.

Cohn, D.V., Zangerle, R., Fischer-Colbrie, R. *et al.* (1982) Similarity of secretory protein I from parathyroid gland to chromogranin A from adrenal medulla. *Proc. Natl Acad. Sci.*, **79**, 6056–9.

Dekker, A., Dunsford, H.A. and Geyer, S.J. (1979) The normal parathyroid gland at autopsy: the significance of stromal fat in adult patients. *J. Pathol.*, **128**, 127–32.

Eiden, E. (1987) Is chromogranin a prohormone? *Nature*, **325**, 301.

Frigerio, B., Capella, C., Wilander, E. and Grimelius, L. (1982) Argyrophil reaction in parathyroid glands. *Acta Pathol. Microbiol. Scand. [A]*, **90**, 323–6.

Gilmour, J.R. (1939) The normal histology of the parathyroid glands. *J. Pathol. Bacteriol.*, **48**, 187–222.

Gilmour, J.R. and Martin, W.J. (1937) The weight of the parathyroid glands. *J. Pathol. Bacteriol.*, **44**, 431–62.

Greeley, G.H., Thompson, J.C., Ishizuka, J. *et al.* (1989) Inhibition of glucose-stimulated insulin release in the perfused rat pancreas by parathyroid secretory protein-I (Chromogranin A). *Endocrinology*, **124**, 1235–8.

Grimelius, L., Åkerström, G., Johansson, H. and Bergström, R. (1981) Anatomy and histopathology of human parathyroid glands. *Pathol. Annu.*, **16**(2), 1–24.

Haid, S.P., Method, H.L. and Beal, J.M. (1967) Parathyroid cysts. Report of two cases and a review of the literature. *Arch. Surg.*, **94**, 421–6.

Hoehn, J.G., Beahrs, O.H. and Woolner, L.B. (1969) Unusual surgical lesions of the parathyroid gland. *Am. J. Surg.*, **118**, 770–8.

Kemper, B., Habener, J.F., Rich, A. and Potts, J.T. Jr (1974) Parathyroid secretion: discovery of a major calcium-dependent protein. *Science*, **184**, 167–9.

Larsson, H-O., Lorentzon, R. and Boquist, L. (1984) Structure of the parathyroid glands, as revealed by different methods of fixation. A quantitative light- and electron-microscopic study in untreated Mongolian gerbils. *Cell Tissue Res.*, **235**, 51–8.

Lieberman, A. and DeLellis, R.A. (1973) Intrafollicular amyloid in normal parathyroid glands. *Arch. Pathol.*, **95**, 422–3.

Lloyd, R.V., Iacangelo, A., Eiden, L.E. *et al.* (1989) Chromogranin A and B messenger ribonucleic acids in pituitary and other normal and neoplastic human endocrine tissues. *Lab. Invest.*, **60**, 548–56.

Mallette, L.E. (1989) Regulation of blood calcium in humans. *Endocrinol. Metab. Clin. North Am.*, **18**, 601–10.

Mayer, G.P., Keation, J.A., Hurst, J.G. and Habener, J.F. (1979) Effects of plasma calcium concentration on the relative proportion of hormone and carboxyl fragments in parathyroid venous blood. *Endocrinology*, **104**, 1778–84.

McGregor, D.H., Lotuaco, L.G., Rao, M.S. and Chu, L.L.H. (1978) Functioning oxyphil adenoma of parathyroid gland. An ultrastructural and biochemical study. *Am. J. Pathol.*, **92**, 691–712.

Nilsson, O. (1977) Studies on the ultrastructure of the human parathyroid glands in various pathological conditions. *Acta Pathol. Microbiol. Scand. [A]* (Suppl. 263), 1–88.

Ordoñez N.G., Ibanez, M.L., Samaan, N.A. and Hickey, R.C. (1983) Immunoperoxidase study of uncommon parathyroid tumors. Report of two cases of nonfunctioning parathyroid carcinoma and one intrathyroid parathyroid tumor producing amyloid. *Am. J. Surg. Pathol.*, **7**, 535–42.

Pesce, C., Tobia, F., Carli, F., and Antoniotti, G.V. (1989) The sites of hormone storage in normal and diseased parathyroid glands: a silver impregnation and immunohistochemical study. *Histopathology*, **15**, 157–66.

Reichel, H., Koeffler, H. Ph. and Norman, A.W. (1989) The role of the vitamin D endocrine system in health and disease. *N. Engl. J. Med.*, **320**, 980–91.

Rodan, G.A. and Martin, J.T. (1981) Role of osteoblasts in hormonal control of bone resorption – a hypothesis. *Calcif. Tissue Int.*, **33**, 349–51.

Rosenblatt, M., Kronenberg, H.M. and Potts, J.T. Jr (1989) Parathyroid hormone. Physiology, chemistry, biosynthesis, secretion, metabolism and mode of action. In *Endocrinology*, Vol. 2 (ed. L.J. DeGroot), W.B. Saunders, Philadelphia, pp. 848–91.

Roth, S.I. (1971) Recent advances in parathyroid gland pathology. *Am. J. Med.*, **50**, 612–22.

Roth, S.I. and Capen, C.C. (1974) Ultrastructural and functional correlations of the parathyroid gland. *Int. Rev. Exp. Pathol.*, **13**, 161–221.

Roth, S.I. and Munger, B.L. (1962) The cytology of adenomatous, atrophic, and hyperplastic parathyroid glands of man. *Virchows Arch. Pathol. Anat.*, **335**, 389–410.

Sandström, I. (1880), Om en ny körtel hos menniskan och åtskilliga däggdjur. *Upsala Läkareförenings Förhandlingar*, **15**, 441–71.

Stoeckel, M.E. and Porte, A. (1966) Observations ultrastructurales sur la parathyroide de souris. I. Etude chez la souris normale. *Z. Zellforsch.*, **73**, 488–502.

Svensson, O., Wernerson, A. and Reinholt, F.P. (1989) The parathyroid glands in the rat as seen by ultrathin step and serial sectioning. *Bone Miner.*, **6**, 237–48.

Tremblay, G. and Pearse, A.G.E. (1959) A cytochemical study of oxidative enzymes in the parathyroid oxyphil cell and their functional significance. *Br. J. Exp. Pathol.*, **40**, 66–70.

Wang, C.A. (1976) The anatomic basis of parathyroid surgery. *Ann. Surg.*, **183**, 271–5.

Weiler, R., Fisher-Colbrie, R., Schmid, K.W. *et al.* (1988) Immunological studies on the occurrence and properties of chromogranin A and B and secretogranin II in endocrine tumours. *Am. J. Surg. Pathol.*, **12**, 877–84.

Welsh, D.A. (1898) Concerning the parathyroid glands – a critical anatomical and experimental study. *J. Anat. Physiol.*, **32**, 292–307.

Wilson, B.S. and Lloyd, R.V. (1984) Detection of chromogranin in neuroendocrine cells with a monoclonal antibody. *Am. J. Pathol.*, **115**, 458–68.

15 Pathological examination

The kind of specimens that the surgical pathologist is concerned with and that call for his diagnostic service are mainly derived from patients operated for primary hyperparathyroidism and, rarely, from patients with tertiary hyperparathyroidism. There are few areas where the theoretical and practical experience of the surgeon and the pathologist, as well as their smooth and close cooperation are as crucial for the therapeutic result as in the field of parathyroid disease.

15.1 General aspects

In parathyroid surgery there has been controversy over whether primary hyperparathyroidism should be treated with a conservative or a radical approach. Today, the prevailing strategy follows the principle not to remove more of the total parathyroid organ than is necessary to establish a correct diagnosis and to cure the patient. Conventionally, the two main parathyroid lesions that are responsible for primary hyperparathyroidism, adenoma and hyperplasia, are treated differently. Adenoma, nearly always being a single gland disease, is generally cured by extirpation of the gland involved by the adenoma. In order to verify the diagnosis, however, it is necessary to extirpate and examine microscopically at least one additional gland and establish that this gland is microscopically normal. In fact, the diagnostically crucial specimen in a case presumed to be an adenoma is not the enlarged gland but rather one of the associated glands, all of which should be normal (with the very rare exception of double adenomas). Hyperplasia, on the other hand, is a multiglandular disease and consequently, all submitted glands should show microscopic signs of hyperplasia. This condition is generally treated by subtotal parathyroidectomy, i.e. removal of three glands and partial excision of the fourth.

Since the diagnosis will determine the extent of the surgical procedure, and hence, its therapeutic result, the pathologist plays an important role in the intraoperative frozen section examination of the tissues removed. In the search for parathyroid glands the pathologist is often required to differentiate parathyroid tissue from thyroid, thymus, lymph node and adipose tissue. He is also required to decide whether a parathyroid gland is microscopically normal or abnormal and, if abnormal, whether it is involved by adenoma or hyperplasia. The differential features between these two types of lesion are discussed in detail in Chapter 16 (p. 333).

15.2   Intraoperative examination and frozen section

It is important that the exact anatomical location of all tissue specimens submitted to the pathologist is identified by the surgeon and carefully recorded. It should be specified whether resected parathyroid tissue represents a whole gland or a biopsy, and whether it refers to right or left and upper or lower gland. This is especially important if, for some reason or another, reoperation should be necessary at a later stage, and then perhaps performed by a different surgeon.

The excised tissue specimens should be weighed and measured. In the case of glands with clearly identified outlines, any extraneous tissue (usually fat or thymus tissue) should be trimmed off, before weighing.

The specimens should be carefully examined macroscopically. All submitted fat or thymic tissue is scrutinized for any divergent structures suspected of being parathyroid tissue, and, when present, specifically marked and selected separately for histological examination by frozen section. When practically feasible, all the rest of the specimen should also be investigated with frozen sections. Otherwise it is fixed in formalin and embedded in paraffin for later study with permanent sections.

Submitted parathyroid glands should also be carefully examined macroscopically, before frozen sectioning. In the case of a grossly enlarged gland, the pathologist's task is to decide whether this is due to hyperplasia or adenoma. When an adenoma is present, careful inspection will often reveal a minute cap of different colour lying somewhere at the periphery of the tumour but sometimes clearly separated from the latter. This represents a remnant of the original normal gland. It is often of softer consistency than the adenoma, having a yellowish colour, that contrasts with the somewhat darker, tan or orange–yellow colour of the adenoma (Plate 29). Although a similar macroscopic appearance may occasionally be seen in the nodular form of parathyroid hyperplasia, its presence favours the diagnosis of adenoma. In the author's experience the glandular remnant is usually found at the vascular hilus of the lesion. Many adenomas are elongated and the vascular hilus region tends to be localized at the upper pole of the lesion, probably as a result of gravitational influences during development of the tumour. The hilus region is much more easily identified by the surgeon during the exploration, when the lesion is still *in situ*. It is recommended that the surgeon mark the hilus region with a suture. This measure facilitates selecting the tissue sample for frozen section which has the greatest chance of including the remnant. It is important that the suture is applied with care and fine suture material is used, in order to avoid mechanical destruction of the tiny remnant.

In cases of hyperplasia, all glands, by definition, show more or less pronounced hyperplasia and, needless to say, all glandular tissue submitted should be examined by frozen section. Details about microscopic features of hyperplasia are given in Chapter 16 (p. 333).

15.2.1 Fat staining of cryostat sections

The introduction of fat staining of cryostat sections has been an important contribution to the intraoperative evaluation of parathyroid disease (Roth and Gallagher, 1976), and should always be performed, together with sections conventionally stained with haematoxylin–eosin. The diagnostic value of fat staining is based on the observations originally made in ultrastructural studies of normal and diseased parathyroid glands (Roth and Munger, 1962; Black, 1969; Altenähr, 1972; Paloyan *et al.*, 1973; Roth and Capen, 1974; Nilsson, 1977). They showed that chief cells in 'normal' glands of patients with hyper-calcaemia are chronically suppressed, have small inconspicuous Golgi apparatus, dispersed endoplasmic reitculum, few secretory granules, abundant glycogen, and large amounts of lipid. In contrast, chief cells in adenomas and hyperplastic glands show only few fat droplets and abundant organelles involved in hormone synthesis and secretion, reflecting an increase of the functional activity. In 1976, Roth and Gallagher showed that this difference in lipid content of chief cells could also be demonstrated at the light microscopic level.

The use of fat staining adds a functional dimension to the struc-tural evaluation of the tissue (Bondeson *et al.*, 1985). It is especially valuable in the microscopic interpretation of glands that are not appreciably enlarged, but may be hyperplastic. Such glands may show a diffusely compact structure microscopically, with reduction of the number of fat cells. Since the amount of stromal fat varies with age and body constitution and may be normally absent in young and/or lean individuals, lack of stromal fat as such cannot be considered as reliable evidence of hyperplasia. However, the demonstration of a reduced amount of intraepithelial fat in such glands, indicating increased functional activity, is supportive evidence of hyperplasia. Likewise, fat staining may be of great diagnostic assistance in cases of suspected hyperplasia of glands which microscopically show minute compact foci of chief cells of uncertain significance, embedded in otherwise normal-looking parenchyma. Fat staining may provide valuable information as to whether these minute foci show reduced intra-epithelial fat content suggesting hyperfunction and thus represent an early stage of hyperplasia, or contain increased amounts of such fat, like the surrounding normal-looking parenchyma, and thus should be interpreted as elements of a normal gland.

Fat staining may also be helpful in the identification of the rim of a residual normal gland of a lesion suspected to be an adenoma. This remnant is more easily appreciated with fat staining than in haematoxylin–eosin stained frozen sections (Plate 27). However, the diagnosis of adenoma cannot be based on the demonstration of a normal glandular remnant alone, since nodular hyperplasia may also show such features. It requires examination of at least one additional gland and the demonstration of abundant intraepithelial fat, evenly distributed within its entire parenchyma.

Since the first report by Roth and Gallagher (1976) of the clinical application of the fat-staining technique, its practical value has been debated and divergent opinions have been expressed by other investigators (Dekker et al., 1979; King and Hirose, 1979; Ljungberg and Tibblin, 1979; Saffos and Rhatigan, 1979; Kasdon et al., 1981; Dufour and Durkowski, 1982; Farnebo and von Unge, 1984; Bondeson et al., 1985). Although there has usually been agreement about the value of the method in the diagnosis of parathyroid adenoma, some authors report discrepant staining results in cases of hyperplasia, especially in glands with minute or equivocal structural changes (cf Dekker et al., 1979; Dufour and Durkowski, 1982). One important reason for this divergence in staining results may be methodological differences. The choice of fat dye and its solvent is of great importance for obtaining reliable staining results. A popular fat stain is Sudan IV in Herxheimer's mixture (equal parts of acetone and 70% ethanol). In the author's hands, this technique produced a weak staining of the intraepithelial fat droplets and the difference in the amount of fat between hyperfunctioning and suppressed chief cells was found to be subtle in many instances, resulting in difficulties in the interpretation of the sections. This is probably due to the fact that Herxheimer's solution, containing only 15% water, readily dissolves small fat droplets from the tissue (Pearse, 1968). To reduce extraction of the smallest fat globules we have used a weaker alcoholic solution with 40% water, i.e. Lillie's supersaturated isopropanol method, with oil-red O as the fat dye (Ljungberg and Tibblin, 1979). This technique gives a deeper and clearer staining and greater contrast between hyperfunctioning and normal tissue with respect to stainable fat. In addition, oil-red O is considered to give more stable supersaturated solutions and better colour effects than Sudan IV. A description of the technique is given below.

1. Prepare a stock solution of saturated dye solution in 99% isopropanol using 250–500 mg of dye (Oil Red O, C.I. No. 26125) per 100 ml.
2. Add 4 ml distilled water to 6 ml stock solution.
3. Allow to stand for 5–10 min and then filter.
4. Stain frozen sections for 2–5 min.
5. Wash sections in water.
6. Stain for 15–20 s in Harris' haematoxylin.
7. Wash in tap water.
8. Mount in glycerin or Lillie's gum syrup.

It is important that the working solution is used fresh.

15.3 Parathyroid immunohistochemistry

PTH has been localized immunohistochemically in normal, hyperplastic, and neoplastic parathyroid glands. Such immunoreactivity has been demonstrated in all cell types, including transitional and fully developed oxyphil cells, and has been found in cases of hyperplasia, adenoma, as well as carcinoma (Ordoñez et al., 1982, 1983; Pesce et al., 1989). The immunoreaction has been demonstrated in cells which are also argyrophilic with Grimelius' silver technique, supporting the view that argyrophilic granules correspond to the PTH harbouring secretory granules (Pesce et al., 1989). As noted in Chapter 14 (p. 307) parathyroid cells also show immunoreactivity for chromogranin A, which is closely related to, or identical with a second major parathyroid secretory protein, SP-I, and known to coexist with PTH in the same granules. PTH immunostaining, particularly in conjunction with thyroglobulin immunostaining, can help to differentiate parthyroid tissue from thyroid tissue, for example in the case of intrathyroidal parathyroid adenoma or parathyroid adenoma with a follicular structure, when this distinction cannot be made in routine histological sections (see also p. 327). PTH immunohistochemistry may be crucial in the identification of metastatic parathyroid carcinoma (Ordoñez et al., 1983), and in the diagnosis of the rare instances of clinically non-functional parathyroid carcinoma, which may resemble poorly differentiated follicular carcinoma or medullary carcinoma of the thyroid.

15.2.4   Fine-needle aspiration diagnosis in parathyroid disease

The need for fine-needle aspiration diagnosis has increased in recent years, due to the introduction of various radiological imaging modalities for the preoperative localization of enlarged parathyroid glands in patients with primary hyperparathyroidism (Eisenberg *et al.*, 1989). The possibility of ultrasonically guided percutaneous inactivation of hyperactive parathyroid glands (Solbiati *et al.*, 1985; Karstrup *et al.*, 1988) also relies on parathyroid identification, usually by means of aspiration cytology.

However, the diagnosis of parathyroid disease in smears of fine-needle aspirates has generally been considered difficult because of the great problem of discriminating parathyroid from thyroid epithelial cells. Although the specificity of the cytological diagnosis has been high in some reports, the sensitivity, is generally low (Glenthøj and Karstrup, 1989). By aspiration cytology, parathyroid lesions have often been confused with cellular thyroid adenomas as well as papillary carcinomas (Löwhagen, 1974; Friedman *et al.*, 1983). Likewise, oxyphil cells of parathyroid and thyroid lesions are cytologically indistinguishable. In recent studies the use of silver staining and immunocytochemical demonstration of PTH have been advocated to confirm a parathyroid origin of the aspirates (Gutekunst *et al.*, 1986; Winkler *et al.*, 1987). Chromogranin A may also be a useful marker in this respect. However, it should be noted that medullary thyroid carcinoma is also argyrophilic and chromogranin positive.

References

Altenäher, E. (1972) Ultrastructural pathology of parathyroid glands. *Curr. Top. Pathol.*, **56**, 2–54.

Black, W.C. (1969) Correlative light and electron microscopy in primary hyperparathyroidism. *Arch. Pathol.*, **88**, 225–41.

Bondeson, A-G., Bondeson, L., Ljungberg, O. and Tibblin, S. (1985) Fat staining in parathyroid disease. Diagnostic value and impact on surgical strategy. Clinicopathologic analysis of 191 cases. *Hum. Pathol.*, **16**, 1255–63.

Dekker, A., Watson, C.G. and Barnes, E.L. (1979) The pathologic assessment of primary hyperparathyroidism and its impact on therapy. A prospective evaluation of 50 cases with oil-red-O stain. *Ann. Surg.*, **190**, 671–5.

Dufour, D.R. and Durkowski, C. (1982) Sudan IV stain. Its limitations in evaluating parathyroid functional status. *Arch. Pathol. Lab. Med.*, **106**, 224–7.

Eisenberg, H., Pallotta, J., Sacks, B. and Brickman A.S. (1989) Parathyroid localization, three-dimension modeling, and percutaneous ablation techniques. *Endocrinol. Metab. Clin. North Am.*, **18**, 659–700.

Farnebo, L-O. and von Unge, H. (1984) Peroperative evaluation of parathyroid glands using fat stain on frozen sections. Advances and limitations. *Acta Chir. Scand.*, (Suppl. 520), 17–24.

Friedman, M., Shimaoka, K., Lopez, C.A. and Shedd, D.P. (1983) Parathyroid adenoma diagnosed as a papillary carcinoma on needle aspiration smears. *Acta Cytol.*, **27**, 337–40.

Glenthøj, A. and Karstrup, S. (1989) Parathyroid identification by ultrasonically guided aspiration cytology. Is correct cytological identification possible? *APMIS*, **97**, 497–502.

Gutekunst, R., Valesky, A., Borisch, B. *et al.* (1986) Parathyroid localization *J. Clin. Endocrinol. Metab.*, **63**, 1390–3.

Karstrup, S., Lohela, P., Apaja-Sarkkinen, M. *et al.* (1988) Case report. Nonoperative inactivation of a parathyroid tumour in a patient with hypercalcaemic crisis. *Acta Med. Scand.*, **224**, 187–8.

Kasdon, E.J., Cohen, R.B., Rosen, S. and Silen, W. (1981) Surgical pathology of hyperparathyroidism. Usefulness of fat stain and problems in interpretation. *Am. J. Surg. Pathol.*, **5**, 381–4.

King, D.T. and Hirose, F.M. (1979) Chief cell intracytoplasmic fat used to evaluate parathyroid disease by frozen section. *Arch. Pathol. Lab. Med.*, **103**, 609–12.

Ljungberg, O. and Tibblin, S. (1979) Preoperative fat staining of frozen sections in primary hyperparathyroidism. *Am. J. Pathol.*, **95**, 633–42.

Löwhagen, T. (1974) Thyroid. In *Aspiration Biopsy Cytology* (ed. J. Zajicek), S. Karger, Basel, pp. 87–8.

Nilsson, O. (1977) Studies on the ultrastructure of the human parathyroid glands in various pathological conditions. *Acta Pathol. Microbiol. Scand. [A]*, (Suppl. 263), 1–88.

Ordoñez, N.G., Ibañez, M.L., Samaan, N.A. and Hickey, R.C. (1983) Immunoperoxidase study of uncommon parathyroid tumours. Report of two cases of nonfunctioning parathyroid carcinoma and one intrathyroid parathyroid tumour – producing amyloid. *Am. J. Surg. Pathol.*, **7** 535–42.

Ordoñez, N.G., Ibañez, M.L., Mackay, B. *et al.*, (1982) Functioning oxyphil cell adenomas of parathyroid gland: immunoperoxidase evidence of hormonal activity in oxyphil cells. *Am. J. Clin. Pathol.*, **78**, 681–9.

Paloyan, E., Lawrence, A.M. and Straus, F.H. (1973) *Hyperparathyroidism*. Grune and Stratton, New York, London.

Pearse, A.G.E. (1968) *Histochemistry: Theoretical and Applied*, Vol. 1, 3rd edn, J. & A. Churchill, London, pp. 411–4.

Pesce, C., Tobia, F., Carli, F. and Antoniotti, G.V. (1989) The sites of hormone storage in normal and diseased parathyroid glands: a silver impregnation and immunohistochemical study. *Histopathology*, **15**, 157–66.

Roth, S.I. and Capen, C.C. (1974) Ultrastructural and functional correlations of the parathyroid gland. *Int. Rev. Exp. Pathol.*, **13**, 161–21.

Roth, S.I. and Gallagher, M.J. (1976) The rapid identification of 'normal' parathyroid glands by the presence of intracellular fat. *Am. J. Pathol.*, **84**, 521–7.

Roth, S.I. and Munger, B.L. (1962) The cytology of the adenomatous, atrophic, and hyperplastic parathyroid glands of man: a light and electron-microscopic study. *Virchows Arch. [A]*, **335**, 389–410.

Saffos, R.O. and Rhatigan, R.M. (1979) Intracellular lipid in parathyroid glands. *Hum. Pathol.*, **10**, 483–5.

Solbiati, L., Giangrande, A., De Pra, L. *et al.* (1985) Percutaneous ethanol injection of parathyroid tumours under US guidance: treatment of secondary hyperparathyroidism. *Radiology*, **155**, 607–10.

Winkler, B., Gooding, G.A.W., Montgomery, C.K. *et al.*, (1987) Immuno-peroxidase confirmation of parathyroid origin of ultrasound-guided fine needle aspirates of the parathyroid glands. *Acta Cytol.*, **31**, 40–4.

16 Primary hyperparathyroidism

16.1 General aspects

Primary hyperparathyroidism is now documented to be the most common cause of hypercalcaemia in non-hospitalized patients and is said to be present in about 2.5 per 1000 individuals, occurring in one in every 500 women over 40 years of age and one in every 2000 men (Boonstra and Jackson, 1971; Heath *et al.*, 1980). Most patients have sporadic disease but primary hyperparathyroidism may also be familial, occurring as the sole disorder (Marx *et al.*, 1973; Allo and Thompson, 1982) or as a component in multiple endocrine neoplasia syndromes, i.e. MEN type 1 (Wermer, 1963; Ballard *et al.*, 1964; Cutler *et al.*, 1964) and MEN type 2 (Sipple, 1961; Steiner *et al.*, 1968). Hyperparathyroidism associated with the MEN syndromes is discussed in section 17.3. Rarely, the disorder affects children (Randall and Lauchlan, 1963; Mühlethaler *et al.*, 1967) and even neonates (Goldbloom *et al.*, 1972; Rhone, 1975) and may then occur in a sporadic as well as familial form (Bradford *et al.*, 1973; Rapaport *et al.*, 1986). Neonatal hyperparathyroidism seems to occur more frequently when both parents have benign familial hypocalciuric hypercalcaemia, suggesting a relationship between these two disorders (Marx *et al.*, 1982). There is an increased incidence of primary hyperparathyroidism in patients exposed in the past to low dose radiation (Rao *et al.*, 1980; Netelenbos, 1983; Tisell *et al.*, 1985).

Primary hyperparathyroidism is biochemically characterized by an excess of circulating PTH which produces hypercalcaemia, hypophosphataemia and an increased excretion of calcium and phosphate in the urine. The serum concentration of $1,25(OH)_2D$ is elevated and there is an increased excretion of nephrogenous cyclic AMP. In cases having osseous manifestations (osteitis fibrosa cystica), the serum alkaline phosphatase level is elevated. Today, however, less than 10% of patients with primary hyperparathyroidism have osteitis fibrosa cystica (Lafferty, 1981). Hypercalcaemia is the key to the diagnosis of primary hyperparathyroidism. Nevertheless, there are also well-documented cases of normocalcaemic primary hyperparathyroidism with clinical manifestations similar to the classic form of the disease (Wills, 1971; Broadus et al.,1980).

Formerly, patients with primary hyperparathyroidism presented with osteitis fibrosa cystica and symptoms of painful bone, renal stones, abdominal pain and psychic disorders (Clark, 1985). Today, the majority of patients are initially discovered because of a raised serum calcium level, noted on routine laboratory testing or at biochemical screening. With the advent of biochemical screening during the early 1970s, an increase in the number of cases of primary hyperpara-thyroidism has been noted, and there has been a change in symptoma-tology. In patients below the age of 40 the incidence has remained low, but above the age of 60 the number of cases has risen markedly, especially among women (Christensson et al., 1976). Most cases in this new group of patients have a non-renal and non-osseous presenta-tion. Not infrequently, these patients are initially considered to be asymptomatic. On further questioning, however, many reveal vague or non-specific symptoms such as fatigue, lethargy and depression which often disappear or improve following successful parathyroid-ectomy (Clark, 1989; Heath, 1989).

Conditions associated with primary hyperparathyroidism include renal stones and nephrocalcinosis (Deaconson et al., 1987), chondro-calcinosis (McGill et al., 1984), osteitis fibrosa cystica and osteoporosis (Leppla et al., 1982), hypertension (Heath et al., 1980; Kristoffersson et al., 1987), gout and pseudogout (Kellett et al., 1982; Duh et al., 1986), peptic ulcer disease (Hellström and Ivemark, 1962; Linos, 1978), pancreatitis (Mixter et al., 1962; Clark, 1980), psychiatric disorders (Joborn et al., 1988) and muscle weakness (Patten et al., 1974).

The pathological changes occurring in parathyroid glands in primary hyperparathyroidism include adenoma, hyperplasia and, rarely, carcinoma. By definition, parathyroid adenoma (and carcinoma) is a neoplastic process, which nearly always involves a single parathyroid gland. Hyperplasia, on the other hand, is a diffuse or multifocal proliferation of epithelial cells involving the parathyroid organ as a whole. Hyperplasia, therefore, involves all glands present, although the severity of the pathological process may vary considerably between the glands. Consequently, a reliable distinction between adenoma and hyperplasia requires a microscopic examination, not only of the presumably abnormal gland, but also of at least one additional gland, to establish whether the latter is normal or hyperplastic. If normal, the diagnosis of the abnormal gland will be adenoma, and the operation can then be terminated. If hyperplastic, the condition is regarded as hyperplasia, and the operation must be extended to include also extirpation of a third whole gland, as well as a part of the fourth.

Data on the relative incidence of the parathyroid lesions have varied in different reports (Hellström and Ivemark, 1962; Utley and Black, 1967; Black and Utley, 1968; Haff et al., 1970; Palmer et al., 1975; Johansson et al., 1979; Ghandur-Mnaymneh and Kimura, 1984; Smith and Coombs, 1984). In some of these series, hyperplasia was the commonest pathological lesion whereas in others adenoma was claimed to be the dominating change. This variability may be partly due to differences in the composition of the material, for instance, some included many familial cases, which tend to have hyperplastic changes in their parathyroids. Other, probably more important, reasons may be inconsistent histopathological diagnostic criteria (see above), and subjective interpretation by the investigating pathologist (Highman and Esselstyn, 1979). However, adenoma is now considered to be the most common pathological change, being held responsible for about 80% of cases (Castleman and Roth, 1978; Smith and Coombs, 1984). In a series of 165 patients operated for primary hyperparathyroidism in Malmö, Sweden during the period 1977–1983, 106 were considered to have parathyroid adenoma and 14 probably adenoma (together comprising 73% of cases), 42 cases (24%) to have chief cell hyperplasia, two water clear cell hyperplasia, and one parathyroid carcinoma (Bondeson et al., 1985).

16.2 Parathyroid adenoma

As pointed out previously, parathyroid adenoma is nearly always a single gland disorder, the other glands being macroscopically and microscopically normal. Parathyroid adenomas occur most often in the middle decades of life and are two to three times as frequent in women as in men. In most current series they account for 70–80% of all cases of primary hyperparathyroidism.

Adenomas may develop in any of the parathyroid glands, although in some series, they more frequently involve the lower pair (Norris, 1947; Mayor, 1967; Kay, 1976). The location of the adenoma naturally depends on the site of the normal gland and, consequently, an adenoma may also occur at ectopic sites (p. 300). It is estimated that about 10% of adenomas occur in anomalous positions, including the thyroid (Black and Haynes, 1949; Spiegel et al., 1975; Wheeler et al., 1987; Sawady et al., 1989), mediastinum (Nathaniels et al., 1970), thymus, and the soft tissues behind the oesophagus or pharynx, or even within their walls (Sloane and Moody, 1978). Intrathyroidal adenomas usually develop from the upper glands (parathyroid IV) and are then usually found within the upper parts of the thyroid lobes, although some may also be found in the lower parts of the lobes, being derived from parathyroid III. Some adenomas are not truly intrathyroidal, but have developed from a parathyroid gland which has only been partly invaginated into the capsule of the thyroid gland or hidden in a deep cleft on its surface, giving the impression that the adenoma is within the gland. Similarly, a parathyroid adenoma developing from a gland that has been enveloped by multiple thyroid nodules, may appear to be intrathyroidal. Interestingly, however, parathyroid adenoma has occasionally been reported to occur within thyroid nodules (Herbst et al., 1976).

Adenomas in the anterior mediastinum may have developed from an aberrant lower parathyroid gland (parathyroid III), but may also be the result of a gravitational descent of an adenoma which has developed from a lower parathyroid gland with an initially normal location. Similarly, a large adenoma that involves an upper gland may cause the gland to descend in the neck because of its weight alone, and thus appear at the level of, or even below the lower glands (Castlemann and Roth, 1978).

Rare cases of double and triple adenomas have been reported (Woolner et al., 1952; Schwindt, 1967; Verdonk and Edis, 1981). However, knowing that in primary chief cell hyperplasia each gland may be grossly and microscopically identical to an adenoma, and with adequate follow-up studies, most of the so-called multiple adenomas have been found to represent asymmetrical and asynchronous chief cell hyperplasia (Wang, 1976; Castleman and Roth, 1978; Harness et al., 1979; Thompson et al., 1982).

Studies based on the relative contents of glucose-6-phosphate dehydrogenase isoenzymes have suggested that parathyroid adenoma, like primary hyperplasia, may have a multicellular (not clonal) origin (Fialkow et al., 1977). Accordingly, it has been hypothesized that adenomatous disease may be a form of hyperplastic disease and that both conditions may have a common pathogenesis. Recent analyses which involve modern molecular genetic techniques, however, have given conflicting results, indicating that some (and possibly many) parathyroid adenomas are monoclonal neoplasms. This suggests that there is, in fact, a fundamental biological difference between parathyroid adenomas and primary hyperplasia (Arnold et al., 1988).

16.2.1 Macroscopic features

Parathyroid adenomas commonly weigh 200–1000 mg, but may range from 60 mg to 300 g (Castleman and Mallory, 1935). There is a positive correlation between the size of the tumour and serum levels of calcium and PTH (Purnell et al., 1971; Castleman and Roth, 1978), although there are exceptions to this rule (Nilsson, 1977). In addition, adenomas in cases with the osseous form of the disease tend to be larger than in patients lacking skeletal manifestations (Lloyd 1968; Purnell *et al.*, 1971). Since primary hyperparathyroidism is now usually diagnosed at an early stage, and many patients have a non-osseous and non-renal presentation, the majority of parathyroid adenomas operated upon today tend to be small.

Adenomas are usually soft and pliable and have a tan to orange–brown colour; oxyphil adenomas are typically friable and tan or chocolate-brown. Larger tumours may show focal haemorrhage, fibrosis, cystic degeneration and even calcification. They are often embedded in fat tissue, sometimes with adjacent thymic or thyroid tissue.

Small adenomas often have the ellipsoid shape of the normal gland, with minimal swelling and rounding of its edges; larger tumours are round, tear-drop shape or oval, often slightly lobulated or multinodular (Plates 28 and 29). They are usually surrounded by a thin capsule and a vascular stalk can often be identified on the surface. When the tumour is located *in situ*, the stalk is often found on the upper part of the mass, probably due to the effect of gravity, the adenoma thus hanging down on its anchoring vascular stalk. It is important to identify this stalk, because when a remnant of the normal parathyroid gland is present, it is nearly always found surrounding the blood vessels at the attachment of the stalk to the lesion, i.e. its vascular hilus. The surgeon is recommended to mark it with a suture (Plate 29). This facilitates selecting adequate material for frozen section examination and enables the pathologist to identify the remnant, which is helpful in the differential diagnosis between adenoma and hyperplasia (p. 313).

16.2.2 Microscopic features

Parathyroid adenomas are surrounded by a thin fibrous capsule (Fig. 16.1). At the periphery of the adenoma a rim of a compressed remnant of the original, uninvolved parathyroid gland is sometimes identified. As noted above, such a remnant is nearly always found in the vascular hilar region. Here it often forms a wedge-shaped structure which together with the blood vessels penetrates into the adenomatous tissue, from which it is more or less clearly demarcated (Figs 16.2 and 16.3). Therefore, adequate sampling from the hilus region is important in order to enhance the possibility of identifying this remnant. Adenomas have a compact structure, being composed of masses of epithelial cells, principally of the same type as seen in the normal gland. The cells may be arranged in varying patterns, such as nodular, trabecular, alveolar, rosette-like, follicular or even papillary-like formations, and, frequently, there is a mixture of several patterns (Figs 16.4 and 16.5).

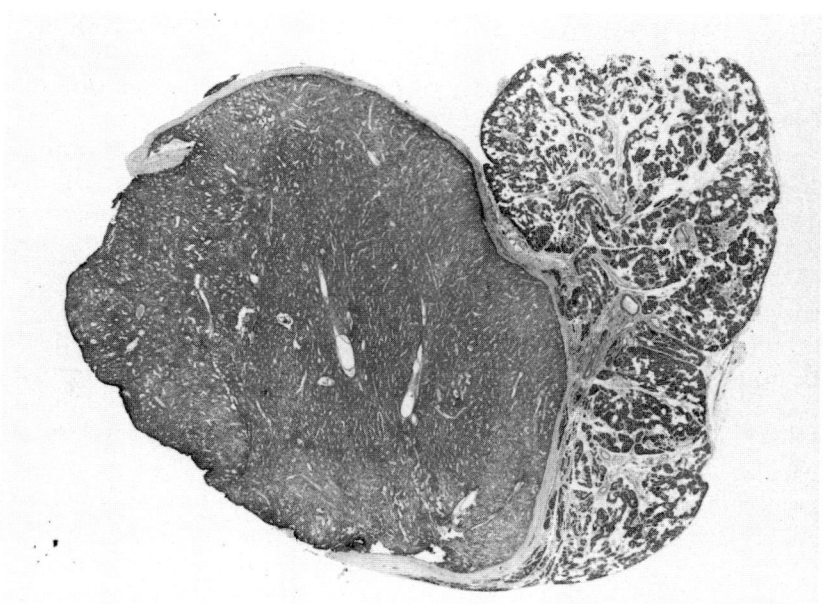

Figure 16.1 Small parathyroid adenoma clearly separated from the residual portion of the gland by a fibrous capsule (H&E, × 8.5).

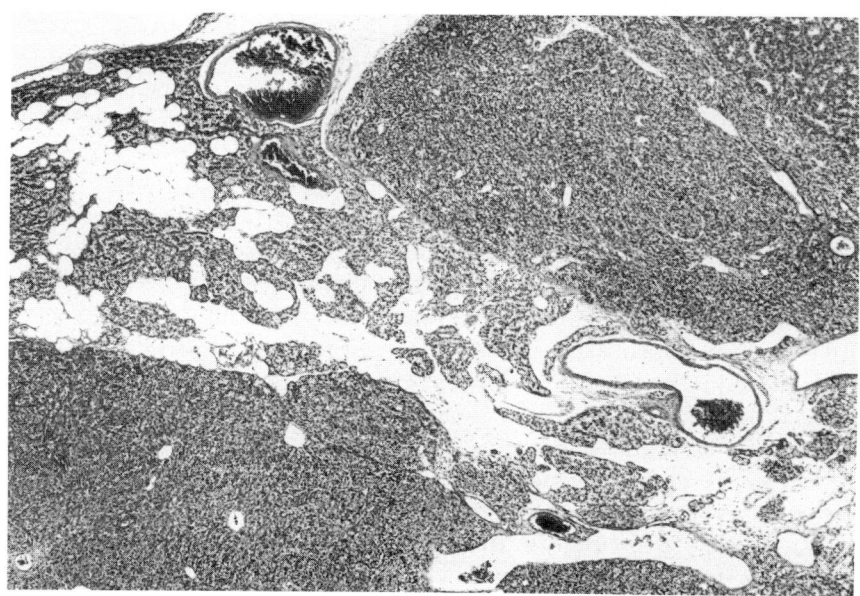

Figure 16.2 Parathyroid adenoma with a wedge-shaped remnant penetrating into the tumour. It is accompanied by the hilar blood vessels (H&E, × 35).

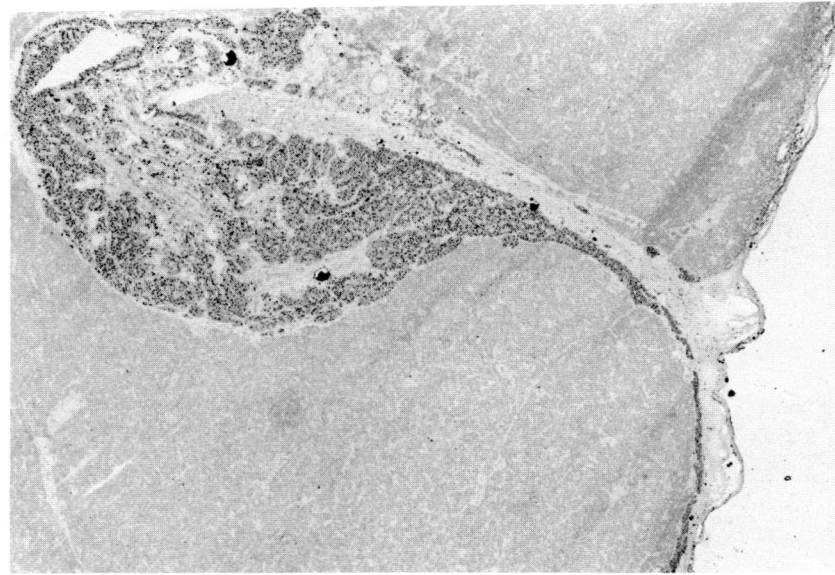

Figure 16.3 Parathyroid adenoma with vascular hilus and parathyroid remnant. Frozen section stained with oil-red O. The remnant is clearly distinguishable from the adenomatous tissue by its abundant fat content and is seen to continue as a thin rim on the surface of the adenoma (at the lower right) (× 35).

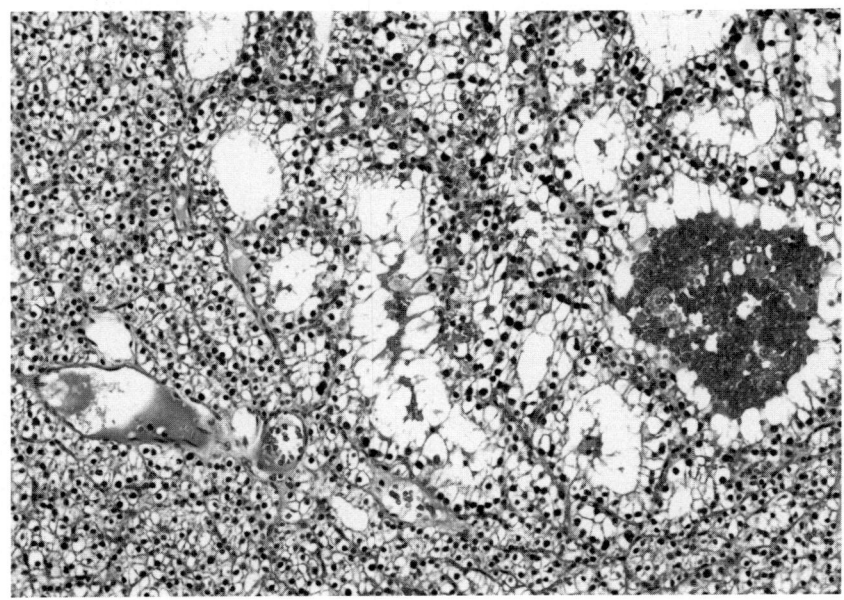

Figure 16.4 Parathyroid chief cell adenoma, mainly composed of vacuolated chief cells, which in some nodules resemble water-clear cells in acinar arrangements (H&E, × 125).

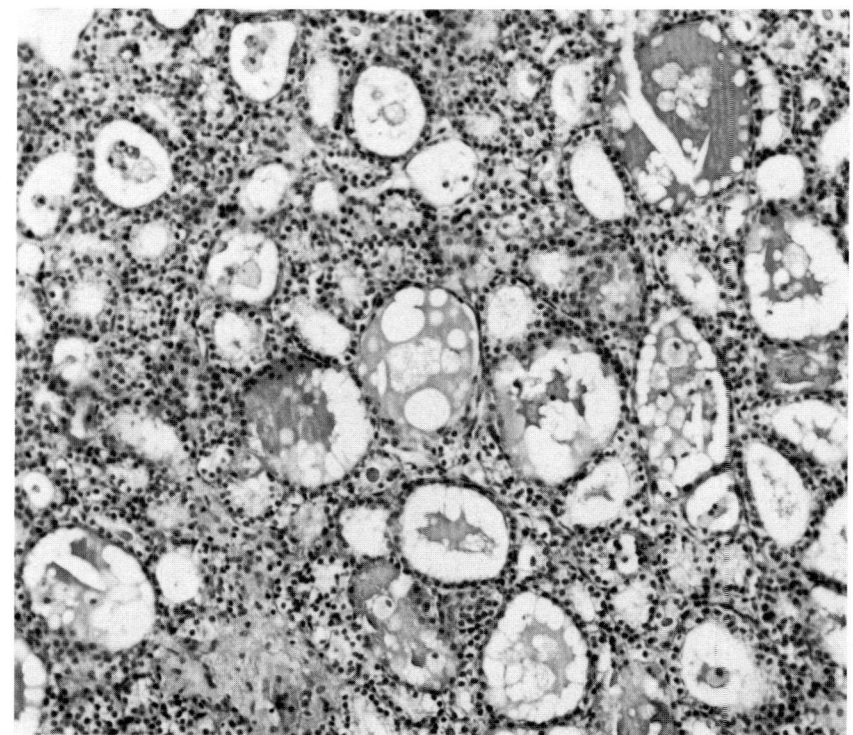

Figure 16.5 Parathyroid chief cell adenoma with follicular structure (H&E, × 125).

Adenomas may be composed of pure masses of dark or vacuolated chief cells, oxyphil cells or transitional cell forms. Water-clear cells with typical cytoplasmic vacuoles have, vary rarely, also been reported to occur in adenomas (Mitrovic *et al.*, 1967); mostly however, cells resembling water-clear cells have, on electron microscopic examination, been found to represent chief cells with large glycogen deposits, accounting for the cytoplasmic clearing (Nilsson, 1977).

More frequently, however, adenomas contain mixtures of several cell types, the chief cells usually being the dominating element. Although these cell types may be individually intermingled, they usually form discrete clusters or nodules. Most adenomas contain no or only a few fat cells. This is in contrast to the normal parathyroid remnant which usually has a lobulated architecture and intermingling clusters of fat cells, like the normal parathyroid gland. In addition, the remnant appears atrophic; the individual epithelial cells look smaller than the cells in the adenoma, and they have smaller and more uniform nuclei.

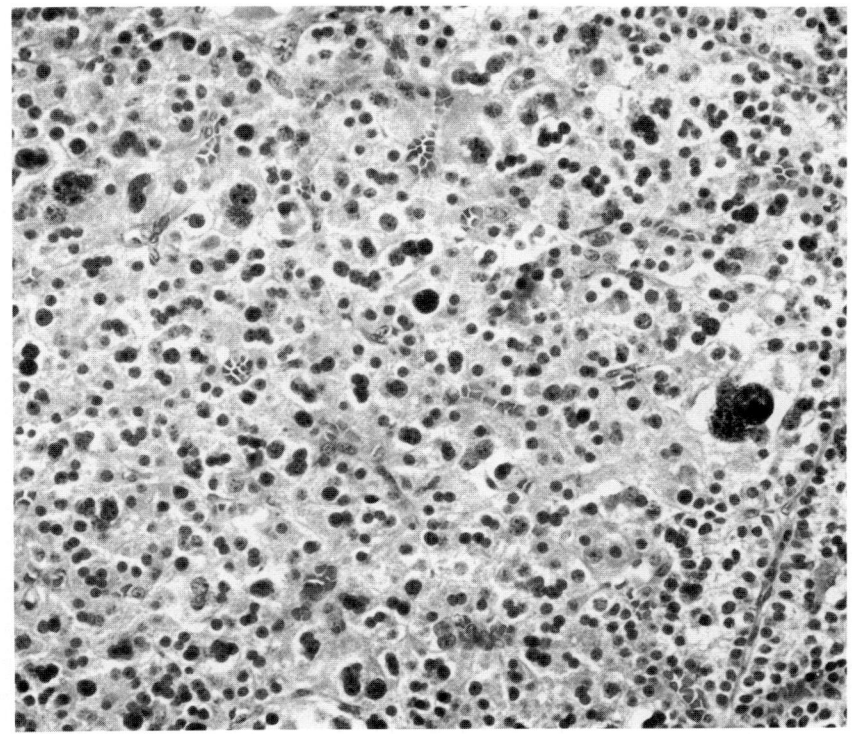

Figure 16.6 Parathyroid chief cell adenoma. Areas of chief cells with chromatin-dense and highly pleomorphic nuclei (H&E, × 230).

Parathyroid adenomas show marked variation in nuclear size and shape with focal areas of cells with bizarre, hyperchromatic giant nuclei, some multilobulated or occurring in clusters (Fig. 16.6). In spite of pronounced nuclear pleomorphism, mitotic figures are hard to find. This nuclear abnormality should not be regarded as a sign of malignancy; on the contrary, it favours a benign diagnosis.

In some adenomas a follicular pattern predominates. The formation of follicular structures appears to be the result of cell degeneration in the central parts of pre-existing solid cell nests. The lumina of such acini may be filled with eosinophilic proteinous fluid that resembles thyroid colloid. In some cases this material is stained with amyloid stains such as alkaline Congo red (Leedham and Pollock, 1970; Anderson and Ewen, 1974; Hansson *et al.*, 1974). Occasionally, the follicular pattern may be indistinguishable from that of a thyroid follicular adenoma. This may present a particularly difficult differential diagnostic problem in the rare case of an intrathyroidal parathyroid adenoma. Immunohistochemical investigation may be very helpful in solving this dilemma. Parathyroid cells show no thyroglobulin immunoreactivity, whereas thyroid follicular adenomas invariably do. Conversely, parathyroid cells are argyrophilic and show a positive reaction with antibodies against PTH and chromogranin A, which are characteristics not exhibited by the thyroid follicular cells.

Rarely, parathyroid adenomas may show areas with distinctive papillary formations (Sahin and Robinson, 1988). Such features may cause confusion with thyroid papillary carcinoma, especially in fine-needle aspirates (Löwhagen and Sprenger, 1974; Friedman et al., 1983). However, the nuclear features typical of papillary thryoid carcinoma are absent (p. 148). In addition, immunohistochemical studies as mentioned above, should be helpful in establishing a correct diagnosis.

The stroma of adenomas is usually sparse but richly vascularized. It is mainly confined to thin connective tissue septa, surrounding the epithelial structures. Some adenomas, especially when large, show areas of sclerosis and hyalinization which may contain macrophages with haemosiderin as a sign of earlier haemorrhage. Occasionally, there are areas of recent bleeding and necrosis. Massive infarction has occurred, causing hypercalcaemic crisis. As a result of degeneration, microscopic cysts may develop in the adenoma; large cysts associated with hyperparathyroidism, however, are rare (Shields and Staley, 1961). Calcification and bone formation, especially in the capsule, may also be seen, and may be visible roentgenographically (Polga and Balikian, 1971).

With the oil-red O fat stain, the majority of adenomas show a marked reduction in fat content of the chief cells. In a small proportion of tumours fat droplets may occur in some parenchymatous cells. Their distribution is usually patchy, mostly confined to very small granules, and generally, does not reach such proportions as seen in the normal accompanying glands or in the glandular remnant (Ljungberg and Tibblin, 1979). In contrast, the uninvolved glands, and the glandular remnant, show a marked accumulation of fat droplets in the chief cells, typically with an even distribution throughout the parenchyma. All chief cells usually contain one or a few large globules 2–7μm in diameter, and a varying number of minute fat granules. It should be pointed out that the difference in fat content between adenomatous and normal glandular tissue is valid only for chief cells, whereas the oxyphil cells, in our experience, contain almost the same amount of fat regardless of whether they occur in adenomatous, hyperplastic, or structurally normal glands.

16.2.3 Oxyphil adenoma

Oxyphil adenomas are composed largely or exclusively of oxyphil cells (Fig. 16.7). The cytoplasm of the neoplastic oxyphil cells contains numerous mitochondria, and the other ultrastructural elements characteristic of their normal counterparts (p. 304). Most oxyphil adenomas are not clinically active and are usually incidental findings at autopsy. Occasionally they may be hyperfunctioning and cause hyperparathyroidism. In some recently reported clinical series, they accounted for 3–7.8% of all parathyroid adenomas (Poole *et al.*, 1982; Bedetti *et al.*, 1984; Wolpert *et al.*, 1989). Hyperfunctioning oxyphil adenomas have been considered to contain an admixture of chief cells or transitional forms which are not fully developed oxyphil cells and which are thought to be responsible for the hormonal over-production (Heimann *et al.*, 1971; Altenähr, 1972; McGregor *et al.*, 1978; Chaudhry *et al.*, 1979). Although there is general agreement that transitional oxyphil cells have functional activity, there is controversy as to whether neoplastic, fully developed oxyphil cells have such a capability. Ordoñez *et al.* (1982), in an immunohistochemical study of clinically active oxyphil adenomas however, found positive immunostaining for PTH in all the neoplastic cell types, including the fully developed oxyphil cells.

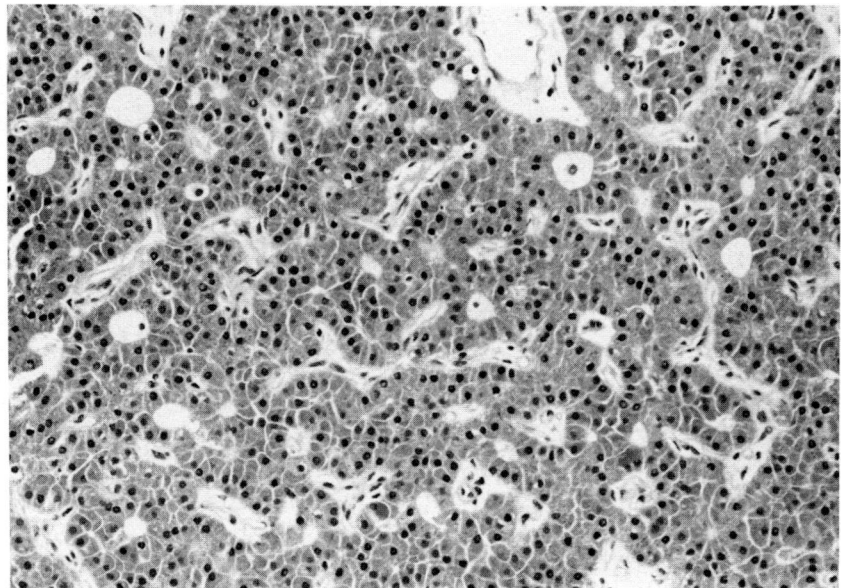

Figure 16.7 Oxyphil adenoma composed of anastomosing cords of exclusively oxyphil cells (H&E, × 175).

16.2.4 Lipoadenoma

This is a rare benign parathyroid tumour, composed of abundant mature adipose tissue, sometimes admixed with myxoid areas, and associated with glandular elements. Like ordinary parathyroid adenomas, the lipoadenoma involves only one parathyroid gland.

Lipoadenomas are soft, yellow–tan, somewhat lobulated tumours which are usually surrounded by a thin capsule. Due to their abundant adipose tissue, they may mimic lipomas. Microscopically, the tumour consists of delicate, anastomosing trabeculae of chief cells, sometimes in a palisade arrangement and embedded in mature adipose tissue (Fig. 16.8). Oxyphil cells and vacuolated chief cells may occur as isolated elements or in small nests. Some lesions may, in addition, contain an admixture of a loose myxomatous stroma with many blood capillaries and some connective tissue septa. The myxomatous stroma contains stellate or spindle-shaped stromal cells scattered in a finely fibrillar, sometimes vacuolated myxoid matrix which is Alcian blue-positive and stains metachromatically with toluidine blue. Sometimes there may be an infiltrate of lymphocytes and plasma cells, scattered throughout the stroma

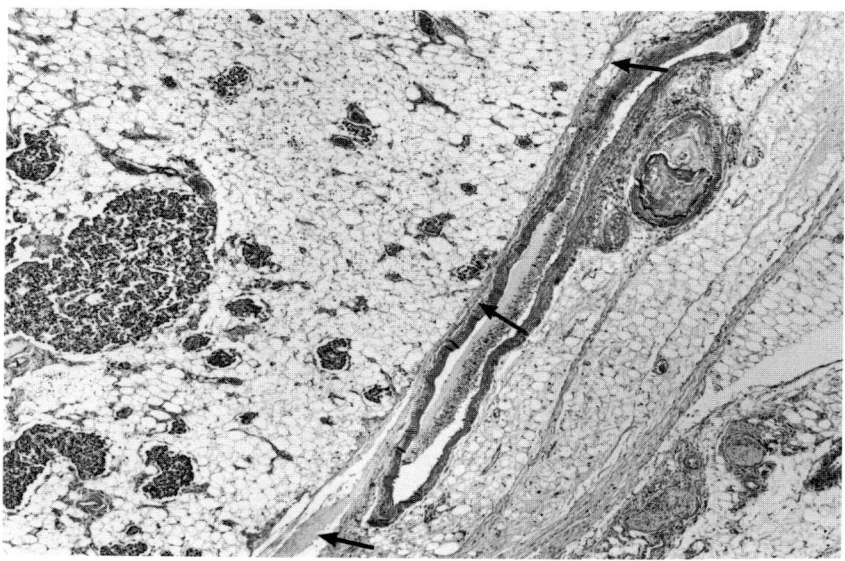

Figure 16.8 Parathyroid lipoadenoma removed from a 54-year-old woman with primary hyperparathyroidism and marked hypercalcaemia (3.0 mmol/l). The lesion consists of abundant adipose tissue associated with islands of glandular elements. The tumour has a thin fibrous capsule (arrows) separating it from the adjacent soft tissues (H&E, × 34).

(Fig. 16.9). A remnant of the normal parathyroid gland may be found adjacent to the lesion.

Staining of frozen sections with oil-red O gives the same results as in ordinary adenomas, i.e. a marked reduction of lipid droplets in

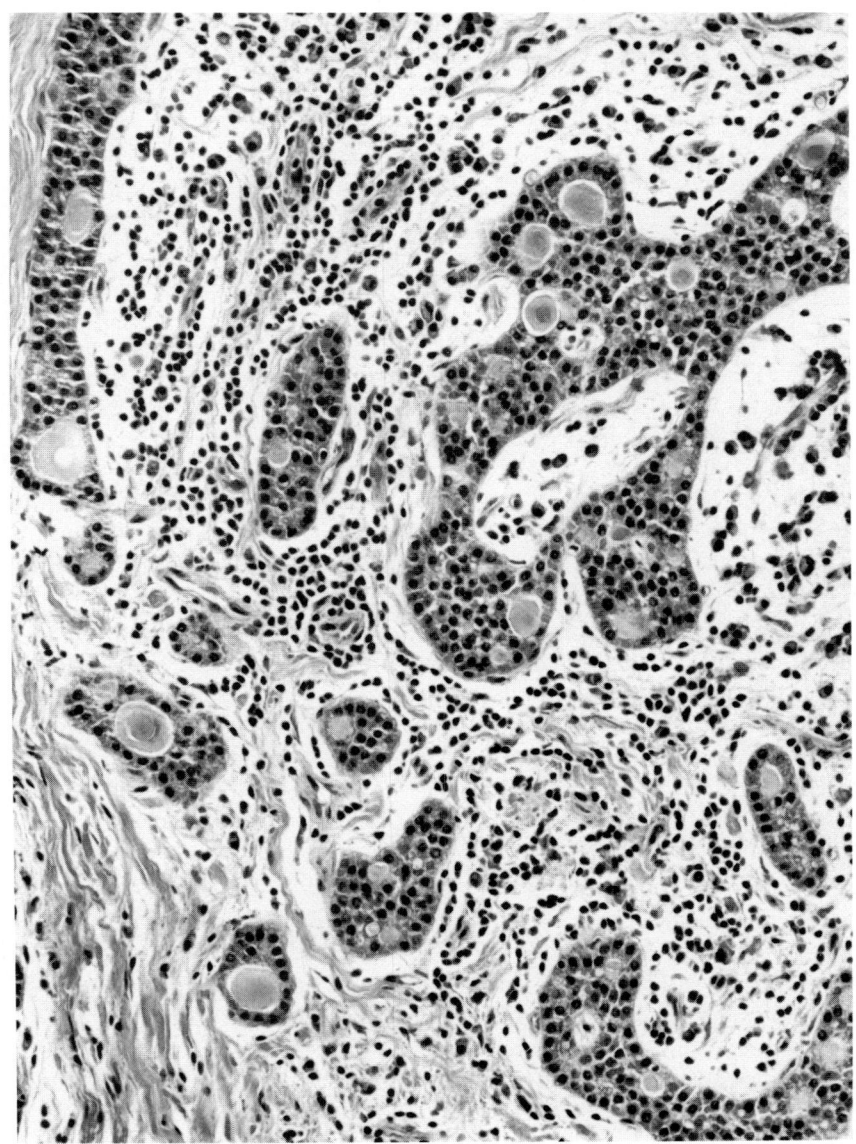

Figure 16.9 Lipoadenoma with predominant loose, myxomatous stroma infiltrated by lymphocytes and plasma cells. Thin, ramifying trabeculae of chief cells, with some acinar structures are also seen (H&E, × 250).

332 Primary hyperparathyroidism

the chief cells of the tumour, and accumulation of such droplets in the parenchymal cells of the glandular remnant (Udén *et al.*, 1987).

The first case was described in 1958 by Ober and Kaiser, and since then about 20 cases have been reported (Udén *et al.*, 1987). The majority of lipoadenomas have been associated with primary hyper-parathyroidism and found at surgical exploration for this condition. A few have been incidental findings at autopsy, and others have been encountered at operation for goitre or a cervical mass. They tend to be larger and heavier than ordinary parathyroid adenomas: the median size in hitherto published cases has been 3.0 cm in largest diameter, with a median weight of 3.9 g (range 0.25–420 g). Whereas ordinary parathyroid adenomas show a marked female preponderance, the distribution of lipoadenomas between the sexes appears to be about equal (Udén *et al.*, 1987). The tumour is usually located at the normal sites of the parathyroid glands, but larger tumours may, due to their weight, descend in the neck and reach substernal locations (Ober and Kaiser, 1958). Ectopic lipoadenomas have been described in the thymus (Wolff and Goodman, 1980), posterior mediastinum (Hargreaves and Wright, 1981), posterior neck (Geelhoed, 1982), and the thyroid gland (Aukee and Meuerman, 1971).

In some reported cases, myxomatous tissue has been the dominant or sole mesenchymal component (Ober and Kaiser, 1958; Grimelius *et al.*, 1972; Gabbert *et al.*, 1980). Frequently myxoid areas also contain lymphocytes and plasma cells the significance of which is unknown. The complex histological pattern of the lesion, which suggests a simultaneous proliferation of an epithelial component and a mesen-chymal component of fat cells and stromal cells, has led some authors (Ober and Kaiser, 1958; Legolvan *et al.*, 1977) to regard the process as a tumour-like malformation, i.e. a hamartoma.

16.3 Primary parathyroid hyperplasia

In contrast to parathyroid adenoma, which traditionally involves a single gland (or occasionally two), primary parathyroid hyperplasia is a multiple gland disorder (Roth, 1971; Sicard, 1981). There are two major types of primary parathyroid hyperplasia, i.e. chief cell hyperplasia, which is the most common form, and the rare water-clear cell hyperplasia. A third, rare form, referred to as hyperplastic chronic parathyroiditis has also been described.

16.3.1 Chief cell hyperplasia

Chief cell hyperplasia accounts for 10–25% of cases with primary hyper-parathyroidism (Wang, 1971; Castleman and Roth, 1978; Bruining *et al.*, 1981; Bondeson *et al.*, 1985). In about half the cases, all four parathyroid glands are almost equally enlarged. In the remaining cases, the hyperplastic process is uneven, and in some instances one gland is greatly enlarged, whereas the other three are only slightly so, or may appear normal-sized. On the basis of macroscopic features, such cases may easily be misinterpreted as parathyroid adenomas. There are also cases in which all four glands appear grossly normal to the surgeon, but are revealed as hyperplastic on microscopic examination (Black and Haff, 1970). The combined weight of the hyperplastic glands varies greatly and may reach as high as 10 g or more (Castleman and Roth, 1978). Rudimentary glands or supernumerary nests of parathyroid cells at ectopic sites are also involved by the process and may become large and attain clinical significance, thus accounting for cases with more than four glands.

Macroscopically, the enlarged glands are yellow–brown to red–brown. Slightly enlarged glands may be shaped as normal glands but show rounded edges and bulging surfaces. Large glands are often nodular in appearance (Plate 30), sometimes with pseudopodal extensions.

Microscopically, chief cell hyperplasia may affect the different glands to a varying extent, and sometimes only parts of the glands. In classical cases, there is a reduction in the number of fat cells, the parenchyma looks compact with closely packed masses of predominantly chief cells, but often with an admixture of other cell types, and the normal lobulated architecture is deranged throughout the gland (diffuse hyperplasia) (Fig. 16.10). In contrast to cases of adenoma, there is no rim of normal-looking parathyroid tissue at the periphery of a diffusely hyperplastic gland.

However, the hyperplastic process often has a multifocal distribution, giving rise to multiple compact nodules within the glands (nodular hyperplasia) (Figs 16.11 and 16.12). Between these nodules the glandular tissue may have a more or less normal-looking appearance, with collections of fat cells as seen in normal glands. To complicate the matter further, a gland may occasionally contain only a single hyperplastic nodule, surrounded by normal-looking parenchyma, whereas the other three glands, often being of roughly normal size, show only discrete foci of hyperplasia. It is evident that without microscopic examination of at least one of the other glands, such a nodule may be misinterpreted as an adenoma. A correct diagnosis in these cases, therefore, requires examination of at at least one (preferably more) of the other three glands and careful search for hyperplastic foci. Another difficulty in interpretation is the fact that nodules of oxyphil cells may occur as a normal phenomenon in the elderly (p. 304). Such an oxyphil nodule could be misinterpreted as

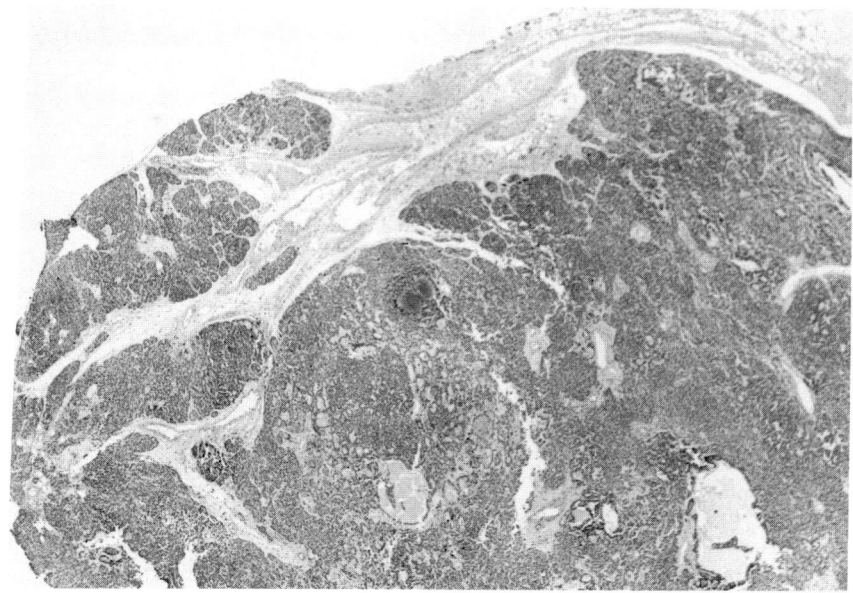

Figure 16.10 Primary chief cell hyperplasia of diffuse type. The gland is enlarged with a compact structure, reduction of fat cells and partial derangement of the lobular architecture (H&E. × 17).

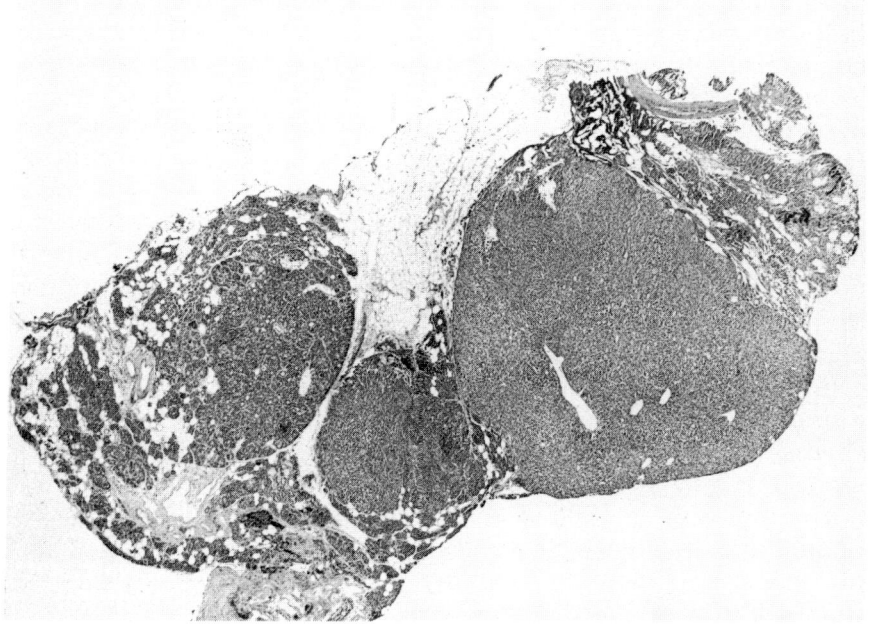

Figure 16.11 Primary chief cell hyperplasia of nodular type. The gland is occupied by small compact chief cell nodules surrounded by normal-looking parenchyma (H&E, × 17).

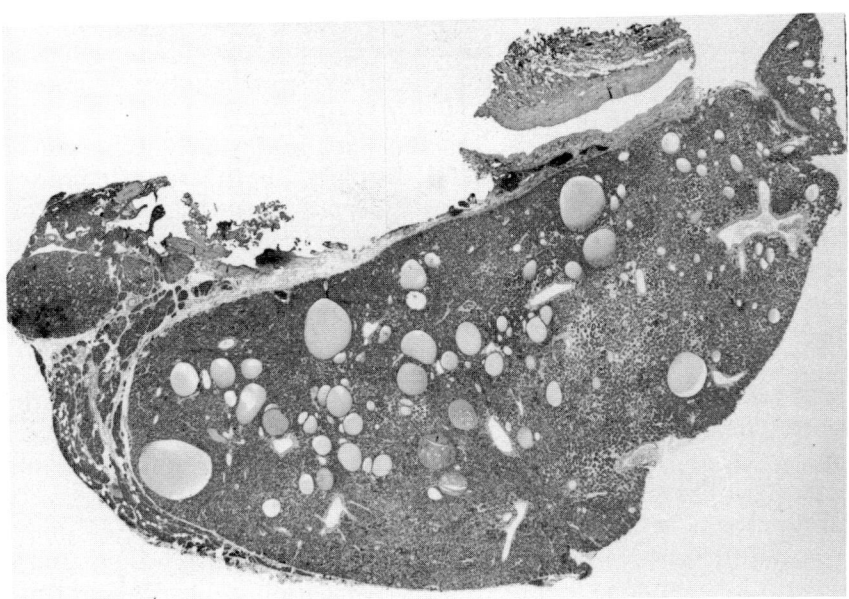

Figure 16.12 Another gland from the case as shown in Fig. 16.11, showing similar pathological changes (H&E, × 11).

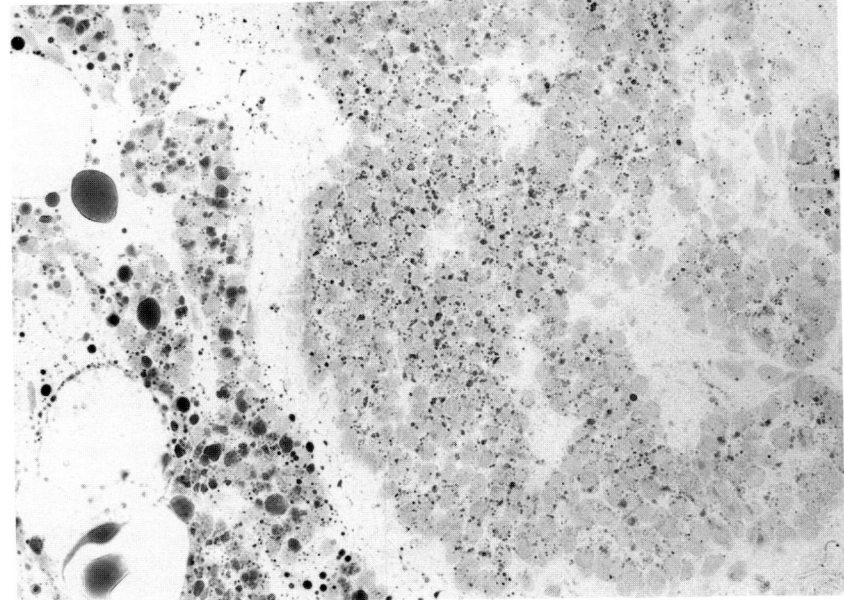

Figure 16.13 Nodular chief cell hyperplasia. On the right is a portion of a hyperplastic nodule with reduced amount of fat in the chief cells. On the left, the intervening glandular parenchyma shows accumulation of fat droplets (Oil-red O stain, × 330).

an adenoma or as a manifestation of primary nodular hyperplasia. In this situation, microscopic examination of additional parathyroid glands is needed to disclose the changes responsible for the clinical disorder.

In fat staining with oil-red O, chief cells in diffusely hyperplastic glands are usually devoid of fat droplets. In the more common multifocal or multinodular form of hyperplasia, the hyperplastic foci show marked reduction or complete absence of fat droplets in the chief cells, indicating secretory activity (Fig. 16.13). The intervening parts of the gland, which may appear morphologically more or less normal, however, usually contain fairly large numbers of fat droplets, sometimes in amounts comparable to those seen in normal, suppressed glands from cases with an adenoma. This suggests that whereas the hyperplastic nodules are autonomously hyperfunctioning, the inter-vening parts are still more or less responsive to the serum calcium level (Ljungberg and Tibblin, 1979). These findings agree with immunohistochemical results obtained with monoclonal antipara-thyroid antibodies, which are believed to interact with certain surface receptor mechanisms involved in the regulation of hormone release (Juhlin et al., 1989). With these antibodies, adenomas and hyperplastic nodules showed a reduced immunostaining indicating a secretory abnormality. However, normal and suppressed glands, as well as strands of glandular tissue lying between hyperplastic nodules and containing abundant cytoplasmic fat stained intensely, suggesting that the sensitivity of hormone release to extracellular calcium is preserv-ed in these tissues.

The cellular arrangement varies. The cells may form trabecular, alveolar or even follicular structures. In nodular hyperplasia, the individual nodule may contain only one cell type, whereas a neighbouring nodule may contain cells of a different type. The stroma is sparse and degenerative changes are uncommon although, occa-sionally, cystic changes may occur (Fallon et al., 1982). Marked cytological and nuclear pleomorphism with giant, bizarre nuclei that can be seen in adenomas, are rare features in hyperplasia and, when present, may be of some assistance in the differential diagnosis.

16.3.2 Hyperplastic chronic parathyroiditis

Boyce et al. (1982) described a case of chief cell hyperplasia associated with chronic lymphocytic parathyroiditis. No evidence of parathyroid hyperfunction was available in this case, but the lesion was interpreted as an equivalent to Hashimoto's thyroiditis and hypofunction was con-sidered to be the most likely manifestation. In 1984, we reported two cases of similar parathyroid lesions which, in addition, were associated with hyperparathyroidism (Bondeson *et al.*, 1984). Both patients were operated on because of clinical signs of primary hyperparathyroidism. In both instances, three enlarged parathyroid glands (weighing 100–1580 mg each) were extirpated. Microscopically, all glands showed varying degrees of hyperplasia, involving both chief and oxyphil cells. Frozen sections stained with oil-red O revealed the chief cells to be devoid of fat droplets, indicating hyperactivity. The stroma was abundant and contained diffuse and nodular infiltrates of lymphocytes with some admixture of plasma cells; lymphoid nodules with germinal centres were also present (Plate 31). Focally, there were also signs of parenchymal destruction. The significance of the inflammatory reaction is uncertain. It may be that this reaction might represent an auto-immune process, analogous to that seen in Graves' disease, which is associated with factors acting to stimulate the parenchyma, and result in hyperplasia and hyperfunction of the glands.

16.3.3 Water-clear cell hyperplasia

Albright *et al.* (1934) gave the first morphological description of water-clear cell hyperplasia, and defined and separated the condition from chief cell hyperplasia. Formerly, water-clear cell hyperplasia comprised some 10% of all cases of primary hyperparathyroidism (Cope *et al.*, 1958; Hellström and Ivemark, 1962; Black, 1971). In the past few decades however, the incidence has dropped markedly, approaching 1% (Castleman *et al.*, 1976). In Malmö, we found two cases among 165 patients (1.2%) operated for primary hyperparathyroidism during the period 1977–1983 (Bondeson *et al.*, 1985).

Clinically, water-clear cell hyperplasia differs in some respects from the other main forms of primary hyperparathyroidism. There is a reversed sex ratio with a predominance of males. Although water-clear cells appear to have low endocrine activity (Dawkins *et al.*, 1973; Nilsson, 1977), water-clear cell hyperplasia, because of its large cell mass, tends to cause more intense symptoms (Castleman *et al.*, 1976; Tisell *et al.* 1981). It is not associated with familial hyperparathyroidism or with the syndromes of multiple endocrine neoplasia.

Water-clear cell hyperplasia involves all parathyroid glands, the upper pair usually being larger than the lower. The total glandular weight usually exceeds 5 g, and may be 100 g or more. The glands are soft and often dark or chocolate brown, sometimes pale brown. They are often irregular in shape with pseudopodal extensions. Some may be multinodular, and often show cystic degeneration.

Microscopically, the normal glandular tissue is completely effaced out and replaced by a monotonous proliferation of large, clear cells throughout the lesion. The cells are arranged in sheets, alveoli or

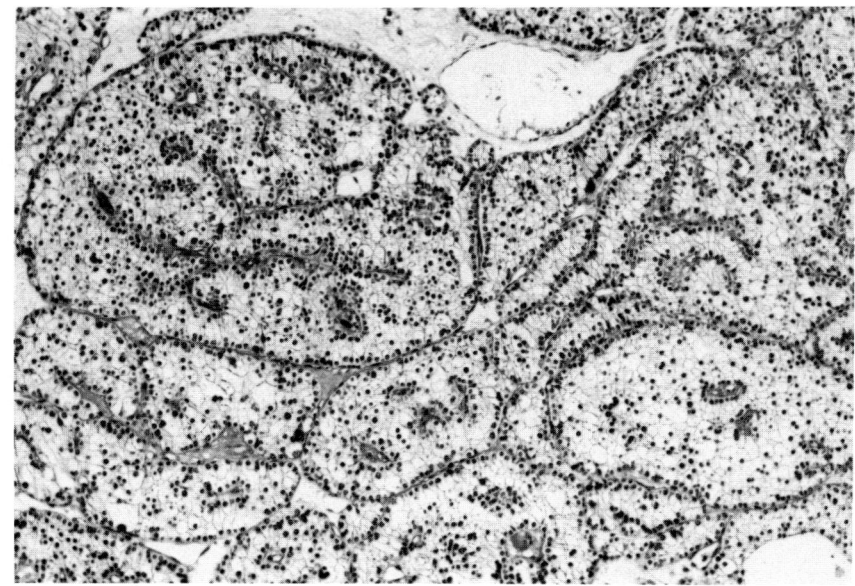

Figure 16.14 Water-clear cell hyperplasia. The normal glandular parenchyma is completely replaced by sheets and cords of uniform, large clear cells, usually with basally located, dense nuclei (H&E, × 85).

acini and the nuclei are often basal, lying close to the connective tissue septa (Fig. 16.14). Varied-sized cysts are common, and occasionally, papillary formations are also present. Water-clear cells are usually considerably larger than chief cells, measuring up to 40 μm. Strictly speaking, the term 'hyperplasia' is inadequate to describe the process, as there is also a marked hypertrophy of the cells. Their cytoplasm appears water-clear, but in thin sections and at high magnification, it is seen to contain aggregations of small rounded vauoles, surrounded by small amounts of eosinophilic cytoplasmic material. Ultrastructurally, the vacuoles comprise numerous vesicles, 0.2–2 μm in diameter, with a triple-layered membrane. They are surrounded by sparse amounts of cytoplasmic material, which are often rich in mitochondria, but contain few secretory granules (Nilsson, 1977). The vesicles are believed to be derived from the Golgi apparatus (Roth, 1970; Nilsson, 1977). On the basis of light microscopy the terms 'clear', 'transitional water-clear' or even 'water-clear' have sometimes been used to describe chief cells with large glycogen deposits. Such cells, which should preferably be called 'vacuolated chief cells', do not contain the vesicles typical of cells in water-clear hyperplasia.

Ultrastructurally, water-clear cells contain few lipid droplets and usually no aggregates of such droplets (Nilsson, 1977). In frozen sections stained with oil-red O, from a few cases of water-clear cell hyperplasia, the parenchymal cells were completely devoid of fat (Bondeson et al., 1985).

With knowledge of its characteristic microscopic features, water-clear cell hyperplasia is easily recognizable by the pathologist. It should be noted, however, that identification of the water-clear cells may be difficult in frozen sections, since freeze artifacts tend to blur the sharp outline of the cells.

Rarely, the diagnosis of water-clear cell hyperplasia involves differentiation from metastatic renal clear cell carcinoma, and clear cell tumours of the thyroid gland and female genital tract. Clinical circumstances and, if necessary, immunostaining for thyroglobulin and PTH may be helpful in establishing a correct diagnosis.

16.4   Parathyroid carcinoma

Carcinoma of the parathyroid glands is rare. It accounts for less than 3% of cases with primary hyperparathyroidism (Black, 1956; Holmes et al., 1969; Schantz and Castleman, 1973; van Heerden et al., 1979; Wang and Gaz, 1985; Shane and Bilezikian, 1989) but figures as low as 0.5–1% have also been reported (Woolner et al., 1952; Cope et al., 1953; Bondeson et al., 1985). In contrast to benign primary hyperparathyroidism, which occurs two to three times more often in women than in men, the sex ratio in parathyroid carcinoma is about equal. The average age at presentation is about 10 years younger than that of the typical patient with benign disease. Parathyroid carcinoma has also been reported to occur in familial primary hyperparathyroidism (Mallette et al., 1974; Dinnen et al., 1977).

Most cases have severe symptoms of hyperparathyroidism. Hypercalcaemia tends to be marked (> 14 mg/100 ml), serum PTH levels are often very high, and complaints such as fatigue, weakness, anorexia, weight loss, nausea, vomiting, polyuria and polydypsia are common. Kidneys and bones are more frequently affected than in benign primary hyperparathyroidism and many patients present with renal colic, bone pain or pathological fractures. The presence of severe symptoms of primary hyperparathyroidism in combination with a palpable mass in the neck are the main clinical features that distinguish this condition from its benign counterpart. Only occasional examples of non-functional parathyroid carcinoma have been reported (Sieracki and Horn, 1960; Chahinian et al., 1981; Aldinger et al., 1982).

At operation, parathyroid carcinoma usually presents as a large, hard tumour surrounded by dense fibrous tissue, adherent to adjacent structures. The tumour is irregularly outlined, and usually

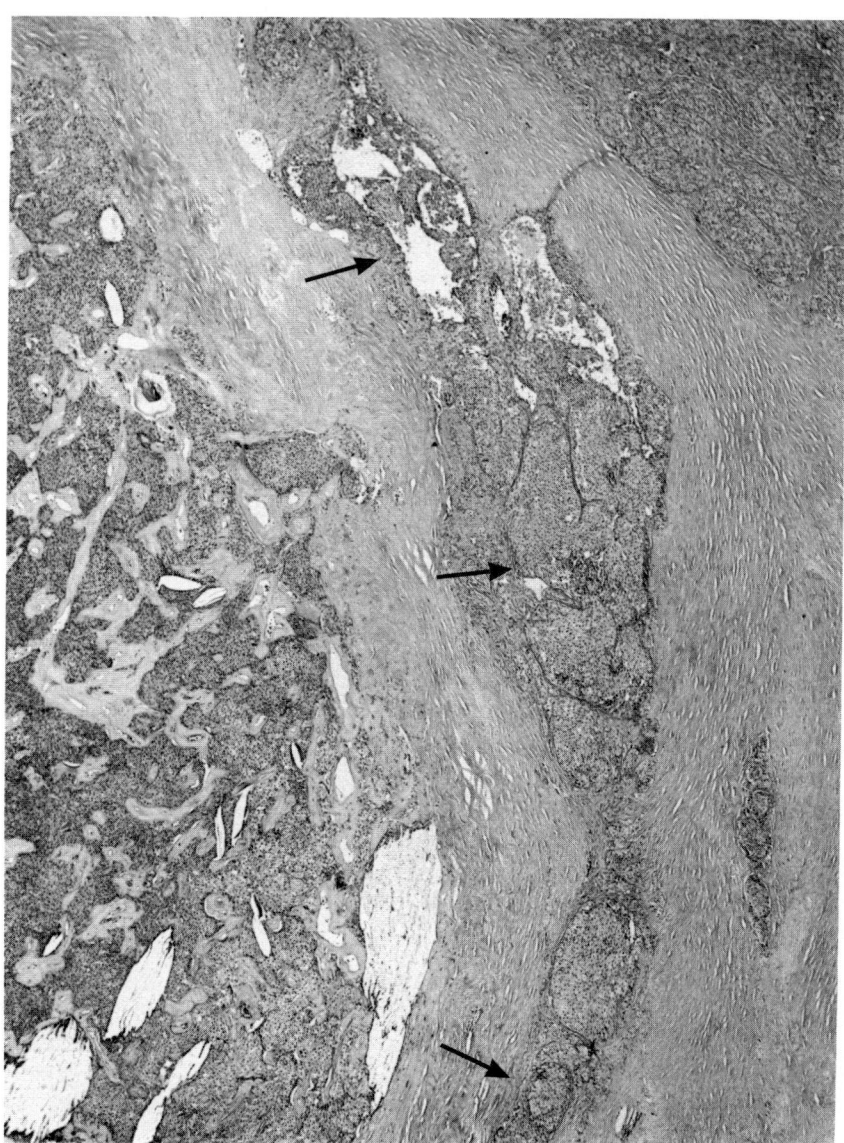

Figure 16.15 Parathyroid carcinoma. The tumour is widely infiltrating into dense, partly hyalinized, fibrous tissue. Arrows indicate tumour invading blood vessels (H&E, × 25).

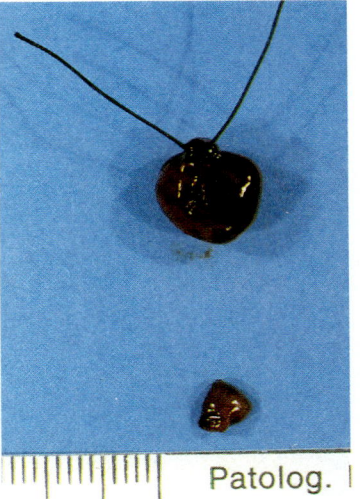

Patolog. I

Plate 28 Small parathyroid adenoma and the ipsilateral normal parathyroid gland. Note suture at the vascular stalk of the adenoma.

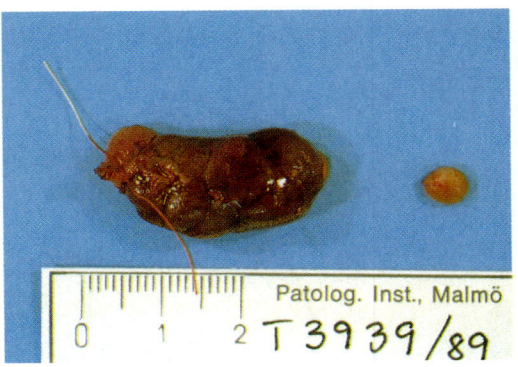

Patolog. Inst., Malmö

T 3939/89

Plate 29 Large parathyroid adenoma, accompanied by the ipsilateral normal gland. The adenoma is slightly lobulated, with a brown colour. Note the small cap of yellowish tissue at the suture, clearly distinguishable from the darker adenoma. This is a remnant of the normal gland, located at the vascular stalk at the upper pole of the lesion.

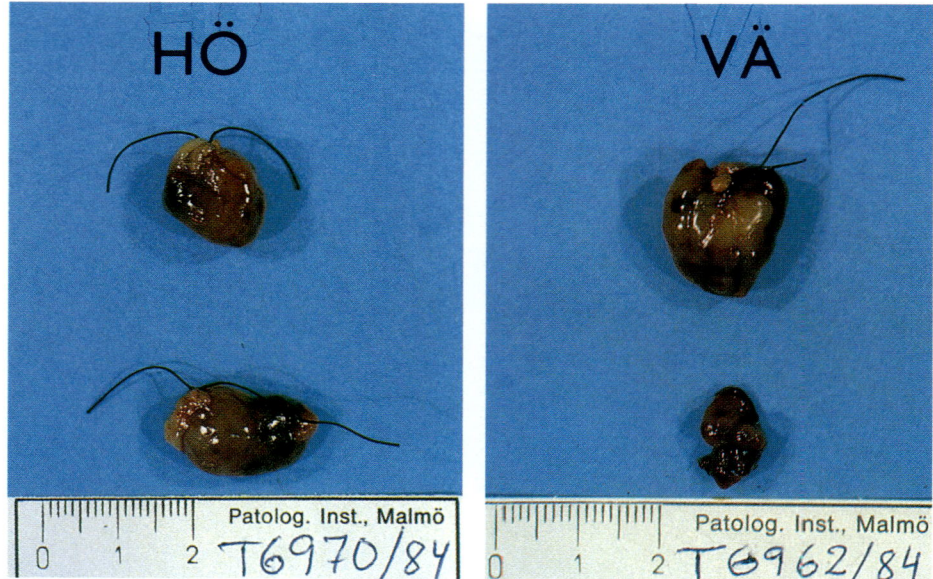

HÖ VÄ

Patolog. Inst., Malmö
T6970/84

Patolog. Inst., Malmö
T6962/84

Plate 30 Primary chief cell hyperplasia of nodular type. The picture shows three whole glands which are enlarged, with brown, knobby surfaces. The specimen to the lower right, is part of the fourth gland showing similar features.

Plate 31 Hyperplastic chronic parathyroiditis. The parenchyma shows nodular lymphoplasmacytic stromal infiltration with germinal centres.

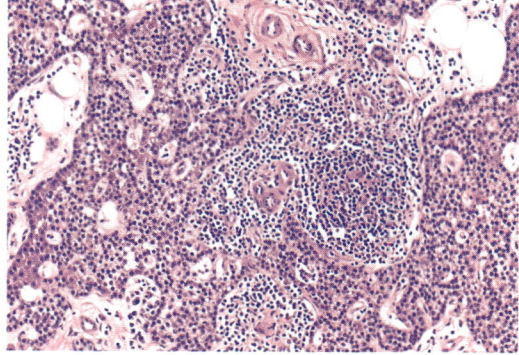

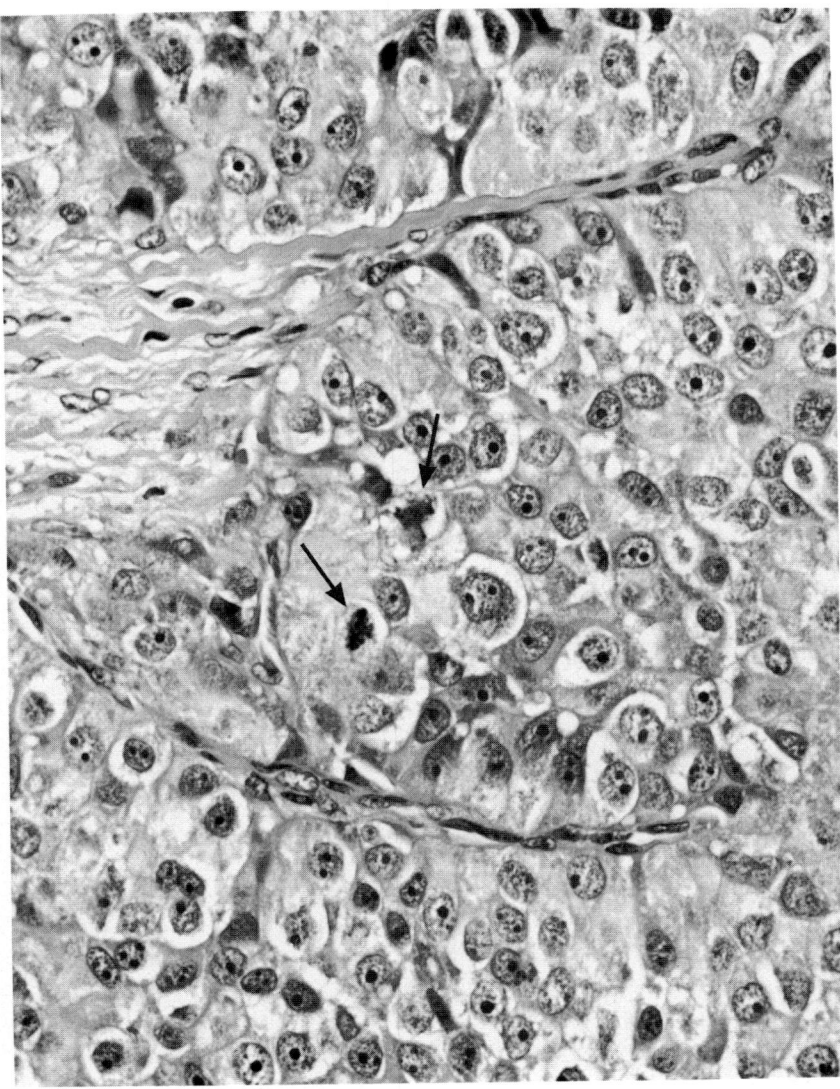

Figure 16.16 Detail of the parathyroid carcinoma shown in Fig. 16.15. Cellular area with nests and cords of closely packed atypical epithelial cells. Nuclei have a rather coarse chromatin pattern and prominent nucleoli. Arrow indicates mitotic figure (H&E, × 500).

measures 2–6 cm in diameter. A thick fibrous capsule is common and the tumour tissue usually shows a pale cut surface, sometimes with central necrosis and calcification.

Microscopically, carcinomas are characterized by invasive growth into the fibrous capsule or surrounding structures. Blood vessel invasion also occurs (Fig. 16.15). Tumour cells usually resemble dark or vacuolated chief cells, but tumours composed of oxyphil cells have also been observed (Obara *et al.*, 1985). The cells are arranged in trabecular, lobulated patterns or in sheets, often separated by thick fibrous bands. Areas with a spindle cell pattern may be seen. There is a greater tendency for a monotonous cell proliferation by comparison with parathyroid adenoma and the lesion usually lacks the highly polymorphic large cells frequently seen in adenomas. Tumour cells often have uniform nuclei with prominent nucleoli. Mean nuclear diameter tends to be greater than that in chief cell adenomas, which may be a useful discriminant (Jacobi *et al.*, 1986). Mitotic figures are usually present (Fig. 16.16) but may not be numerous, and may even be absent (Stephenson, 1950). The demonstration of occasional mitoses does not necessarily indicate that the lesion is a carcinoma. Snover and Foucar (1981), after a careful search, were able to demonstrate mitoses in about three-quarters of parathyroid adenomas and chief cell hyperplasias.

Immunohistochemical demonstration of PTH may be valuable in diagnosing non-functional tumours and in cases of metastases whose origins are in doubt (Ordoñez *et al.*, 1983). Fine-needle aspiration of tumour tissue (Guazzi *et al.*, 1982; de la Garza *et al.*, 1985) is not considered advisable, because of the risk of seeding tumour cells into the surrounding tissue (Shane and Bilezikian, 1989).

Parathyroid carcinoma usually progresses slowly. Hypercalcaemia may recur months or years after the initial operation and is often the first sign of recurrent carcinoma. The average interval between first operation and first recurrence is approximately 3 years (Gittler and Maier, 1972; Shane and Bilezikian, 1982), but much longer intervals have been reported (Ellis *et al.*, 1971; van Heerden *et al.*, 1979). The tumour has a great tendency for local invasion, usually involving adjacent structures, such as the thyroid, strap muscles, recurrent laryngeal nerve, blood vessels, oesophagus and trachea. It tends to metastasize via both the lymphatics and blood vessels, the most frequent sites of metastases being regional lymph nodes, lungs, liver, bone, pleura and pericardium.

Management of primary parathyroid carcinoma is primarily surgical. The tumour is not radiosensitive and results of chemotherapeutic regimens have been disappointing. If the carcinoma is disseminated and surgery is not possible, the ultimate prognosis is poor, the 5-year-survival of inoperable patients being less than 50% (Shane and Bilezikian, 1982). Death is usually caused by uncontrolled hyper-calcaemia. However, the course may be protracted even with known metastases. The main therapeutic approach in cases of metastatic disease is directed toward management of the hypercalcaemia by surgical removal of accessible metastases and administration of drugs capable of lowering serum calcium levels.

References

Albright, F., Bloomberg, E., Castleman, B. and Churchill, E.D. (1934) Hyper-parathyroidism due to diffuse hyperplasia of all parathyroid glands rather than adenoma of one. *Arch. Intern. Med.*, **54**, 315–29.

Aldinger, K.A., Hickey, R.C., Ibanez, M.L. *et al.* (1982) Parathyroid car-cinoma: a clinical study of seven cases of functioning and two cases of nonfunctioning parathyroid cancer. *Cancer*, **49**, 388–97.

Allo, M. and Thompson, N.W. (1982) Familial hyperparathyroidism caused by solitary adenomas. *Surgery*, **92**, 486–90.

Altenähr, E. (1972) Ultrastructural pathology of parathyroid glands. *Curr. Top. Pathol.*, **56**, 1–54.

Anderson, T.J. and Ewen, S.W.B. (1974) Amyloid in normal and pathological parathyroid glands. *J. Clin. Pathol.*, **27**, 656–63.

Arnold, A., Staunton, C.E., Kim, H.G. *et al.* (1988) Monoclonality and abnormal parathyroid hormone genes in parathyroid adenomas. *N. Engl. J. Med.*, **318**, 658–62.

Aukee, S. and Meuerman, L.O. (1971) Hormonal activity in a case of parathyroid tumour. *Acta Chir. Scand.*, **266**, 121–3.

Ballard, H.S., Frame, B. and Hartsock, R.J. (1964) Familial multiple endocrine adenoma–peptide ulcer complex. *Medicine (Baltimore)*, **43**, 481–516.

Bedetti, C.D., Dekker, A. and Watson, C.G. (1984) Functioning oxyphil cell adenoma of the parathyroid gland: a clinicopathologic study of ten patients with hyperparathyroidism. *Hum. Pathol.*, **15**, 1121–6.

Black, B.K. (1956) Carcinoma of the parathyroid. *Ann. Surg.*, **139**, 355–653.

Black, B.M. (1971) Hyperparathyroidism: some aspects of current surgical interest. *Am. Surg.*, **37**, 130–4.

Black, B.M. and Haynes, A.L. (1949) Intrathyroid hyperfunctioning parathyroid adenomas. *Proc. Mayo Clin.*, **24**, 408–13.

Black, W.C. and Haff, R.C. (1970) The surgical pathology of parathyroid chief cell hyperplasia. *Am. J. Clin. Pathol.*, **53**, 586–79.

Black, W.C. and Utley, J.R. (1968) The differential diagnosis of parathyroid adenoma and chief cell hyperplasia. *Am. J. Clin. Pathol.*, **49**, 761–75.

Bondeson, A-G., Bondseson, L. and Ljungberg, O. (1984) Chronic parathyroiditis associated with parathyroid hyperplasia and hyper-parathyroidism. *Am. J. Surg. Pathol.*, **8**, 211–5.

Bondeson, A-G., Bondeson, L., Ljungberg, O. and Tibblin, S. (1985) Fat stain-ing in parathyroid disease – diagnostic value and impact on surgical strategy: clinicopathologic analysis of 191 cases. *Hum. Pathol.*, **16**, 1255–63.

Boonstra, C.E. and Jackson, J.E. (1971) Serum calcium survey for hyperpara-thyroidism. Results in 5000 clinic patients. *Am. J. Clin. Pathol.*, **55**, 523–6.

Boyce, B.F., Doherty, V.R. and Mortimer, G. (1982) Hyperplastic parathyroiditis – a new autoimmune disease? *J. Clin. Pathol.*, **35**, 812–4.

Bradford, W.D., Wilson, J.W. and Gaede, J.T. (1973) Primary neonatal hyper-parathyroidism: an unusual cause of failure to thrive. *Am. J. Clin. Pathol.*, **59**, 265–75.

Broadus, A.E., Horst, R.L., Lang, R. *et al.* (1980) The importance of circulating 1,25-dihydroxyvitamin D in the pathogenesis of hypercalciuria and renal-stone formation in primary hyperparathyroidism. *N. Engl. J. Med.*, **302**, 421–6.

Bruining, H.A., van Houten, H., Juttmann, J.R. *et al.* (1981) Results of operative treatment of 615 patients with primary hyperparathyroidism. *World J. Surg.*, **5**, 85–90.

Castleman, B. and Mallory, T.B. (1935) The pathology of the parathyroid glands in hyperparathyroidism. A study of 25 cases. *Am. J. Pathol.*, **11**, 1–72.

Castleman, B. and Roth, S.I. (1978) Tumors of the parathyroid glands. In *Atlas of Tumor Pathology* (ed. W.H. Hartmann), Fascicle 14, 2nd series, Armed Forces Institute of Pathology, Washington, DC, pp. 1–94.

Castleman, B., Schantz, A. and Roth, S.I. (1976) Parathyroid hyperplasia in primary hyperparathyroidism. A review of 85 cases. *Cancer*, **38**, 1668–75.

Chahinian, A.P., Holland, J.F., Nieburgs, H.E. *et al.* (1981) Metastatic non-functioning parathyroid carcinoma: ultrastructural evidence of secretory granules and response to chemotherapy. *Am. J. Med. Sci.*, **282**, 80–4.

Chaudhry, A.P., Satchidanand, S., Gaeta, J.F. *et al.* (1979) A functional parathyroid gland adenoma of transitional oxyphil cells. A light and ultrastructural study. *Pathology*, **11**, 705–12.

Christensson, T., Hellström, K., Wengle, B. *et al.* (1976) Prevalence of hyper-calcemia in a health screening in Stockholm. *Acta Med. Scand.*, **200**, 131–7.

Clark, O.H. (1980) Hyperparathyroidism and pancreatitis. *JAMA*, **244**, 2413.

Clark, O.H. (1985) Hyperparathyroidism. In *Endocrine Surgery of the Thyroid and Parathyroid Glands* (ed. O.H. Clark), C.V. Mosby, St Louis, pp. 172–240.

Clark, O.H. (1989) Surgical treatment of primary hyperparathyroidism (Indications for cervical exploration and intraoperative strategy). In *Common Problems in Endocrine Surgery* (ed. J. van Heerden), Year Book Medical Publications, Chicago, pp. 164–73.

Cope, O., Keynes, W.M., Roth, S.I. and Castleman, B. (1958) Primary chief-cell hyperplasia of the parathyroid glands: a new entity in the surgery of hyperparathyroidism. *Ann. Surg.*, **148**, 375–88.

Cope, O., Nardi, G.L. and Castleman, B. (1953) Carcinoma of parathyroid glands: four cases among 148 patients with hyperparathyroidism. *Ann. Surg.*, **138**, 661–71.

Cutler, R.E., Reiss, E. and Ackerman, L.V. (1964) Familial hyper-parathyroidism: a kindred involving eleven cases, with a discussion of primary cell hyperplasia. *N. Engl. J. Med.*, **270**, 859–65.

Dawkins, R.L., Tashjian, A.H., Castleman, B. and Moore, E.W. (1973) Hyper-parathyroidism due to clear cell hyperplasia. *Am. J. Med.*, **54**, 119–26.

Deaconson, T.F., Wilson, S.D. and Lemann, J., Jr (1987) The effect of parathyroidectomy on the recurrence of nephrolithiasis. *Surgery*, **102**, 910–3.

de la Garza, S., Flores de la Garza, E. and Hernández Batres, F. (1985) Func-tional parathyroid carcinoma. Cytology, histology and ultrastructure of a case. *Diagn. Cytopathol.*, **1**, 232–5.

Dinnen, J.S., Greenwood, R.H., Jones, J.H. *et al.*(1977) Parathyroid carcinoma in familial hyperparathyroidism. *J. Clin. Pathol.*, **30**, 966–75.

Duh, Q.Y., Morris, R.C., Arnaud, C.D. and Clark, O.H. (1986) Decrease in serum uric acid level following parathyroidectomy in patients with primary hyperparathyroidism. *World J. Surg.*, **10**, 729–36.

Ellis, H.A., Floyd, M. and Herbert, F.K. (1971) Recurrent hyper-parathyroidism due to parathyroid carcinoma. *J. Clin. Pathol.*, **24**, 596–604.

Fallon, M.D., Haines, J.W. and Teitelbaum, S.L. (1982) Cystic parathyroid gland hyperplasia – hyperparathyroidism presenting as a neck mass. *Am. J. Clin. Pathol.*, **77**, 104–7.

Fialkow, P.J., Jackson, C.E., Block, M.A. and Greenawald, K.A. (1977) Multicellular origin of parathyroid 'adenomas'. *N. Engl. J. Med.*, **297**, 696–8.

Friedman, M., Shimaoka, K., Lopez, A.C. and Sheed, D.P. (1983) Parathyroid adenoma diagnosed as papillary carcinoma of thyroid on needle aspiration smears. *Acta Cytol.*, **27**, 337–40.

Gabbert, H., Rothmund, M. and Höhn, P. (1980) Myxoid lipoadenoma of the parathyroid gland. *Pathol. Res. Pract.*, **170**, 420–5.

Geelhoed, G.W. (1982) Parathyroid adenolipoma: clinical and morphologic features. *Surgery*, **92**, 806–10.

Ghandur-Mnaymneh, L. and Kimura, N. (1984) The parathyroid adenoma. A histopathologic definition with a study of 172 cases of primary hyper-parathyroidism. *Am. J. Pathol.*, **115**, 70–83.

Gittler, R.D. and Maier, H. (1972) Carcinoma of the parathyroid. *Arch. Intern. Med.*, **130**, 413–7.

Goldbloom, R.B., Gillis, D.A. and Prasad, M. (1972) Hereditary parathyroid hyperplasia: a surgical emergency of early infancy. *Pediatrics*, **49**, 514–23.

Grimelius, L., Johansson, H. and Lindquist, B. (1972) A case of unusual stromal development in a parathyroid adenoma. *Acta Chir. Scand.*, **138**, 628–9.

Guazzi, A., Gabriella, M. and Guadagni, G. (1982) Cytologic features of a functioning parathyroid carcinoma. *Acta Cytol.*, **26**, 709–13.

Haff, R.C., Black, W.C. and Ballinger, W.F. (1970) Primary hyper-parathyroidism. Changing clinical, surgical and pathological aspects. *Ann. Surg.*, **171**, 85–92.

Hansson, C.G., Säve-Söderberg, J. and Nilsson, O (1974) Corpora amylacea in parathyroids in cases of primary hyperparathyroidism. *Pathol. Eur.*, **9**, 151–8.

Hargreaves, H.K. and Wright, T.C. (1981) A large functioning parathyroid lipoadenoma found in the posterior mediastinum. *Am. J. Clin. Pathol.*, **76**, 89–93.

Harness, J.K., Ramsburg, S.R., Nishiyama, R.H. and Thompson, N.W. (1979) Multiple adenomas of the parathyroids: do they exist? *Arch. Surg.*, **114**, 468–74.

Heath, D.A. (1989) Primary hyperparathyroidism. Clinical presentation and factors influencing clinical management. *Endocrinol. Metab. Clin. North Am.*, **18**, 631–46.

Heath, H., III, Hogdson, S.F. and Kennedy, M.A. (1980) Primary hyper-parathyroidism: incidence, morbidity, and potential economic impact in a community. *N. Engl. J. Med.*, **302**, 189–93.

Heimann, P., Hansson, G. and Nilsson, O. (1971) Primary hyper-parathyroidism in a case of oxyphilic adenoma. Clinical and morphological observations. *Acta Pathol. Microbiol. Scand.* [A], **79**, 10–4.

Hellström, J. and Ivemark, B. (1962) Primary hyperparathyroidism. Clinical and structural findings in 138 cases. *Acta Chir. Scand.*, (Suppl. 294), 1–113.

Herbst, J.W., Neff, W.R. and Freeman, J.B. (1976) Hyperparathyroidism: ectopic parathyroid glands. *Am. J. Surg.*, **131**, 380–1.

Highman, L. and Esselstyn, C.B., Jr (1979) Parathyroid hyperplasia or adenoma – a reappraisal (Abstracts). *Soc. Int. Chir. Sept.*, p. 199.

Holmes, E.C., Morton, D.L. and Ketcham, A.S. (1969) Parathyroid carcinoma. A collective review. *Ann. Surg.*, **169**, 631–40.

Jacobi, J.M., Lloyd, H.M. and Smith, J.F. (1986) Nuclear diameter in parathyroid carcinomas. *J. Clin. Pathol.*, **39**, 1353–4.

Joborn, C., Hetta, J. and Johansson, H. *et al.* (1988) Psychiatric morbidity in primary hyperparathyroidism. *World J. Surg.*, **12**, 476–81.

Johansson, H., Granberg, P.O., Tibblin, S. *et al.* (1979) Scandinavian study of the parathyroid surgical activity in 1975. *Acta Chir. Scand.*, (Suppl. 493), 66.

Juhlin, C., Rastad, J., Klareskog, L. *et al.* (1989) Parathyroid histology and cytology with monoclonal antibodies recognizing a calcium sensor of parathyroid cells. *Am. J. Pathol.*, **135**, 321–8.

Kay, S. (1976) The abnormal parathyroid. *Hum. Pathol.*, **7**, 127–38.

Kellett, H.A., MacLaren, I.F. and Toft, A.D. (1982) Gout and pseudogout precipitated by parathyroidectomy. *Scott. Med. J.*, **27**, 250–1.

Kristoffersson, A., Dahlgren, K., Granstrand, B. and Järhult, J. (1987) Primary hyperparathyroidism in Northern Sweden. *Surg. Gynecol. Obstet.*, **164**, 119–23.

Lafferty, F.W. (1981) Primary hyperparathyroidism. Changing clinical spectrum, prevalence of hypertension, and discriminant analysis of laboratory tests. *Arch. Intern. Med.*, **141**, 1761–6.

Leedham, P.W. and Pollock, D.J. (1970) Intrafollicular amyloid in primary hyperparathyroidism. *J. Clin. Pathol.*, **23**, 811–8.

Legolvan, D.P., Moore, B.P. and Nishiyama, R.H. (1977) Parathyroid hamartoma. Report of two cases and review of the literature. *Am. J. Clin. Pathol.*, **67**, 31–5.

Leppla, D.C., Snyder, W. and Pak, C.Y. (1982) Sequential changes in bone density before and after parathyroidectomy in primary hyperparathyroidism. *Invest. Radiol.*, **17**, 604–6.

Linos, D.A., van Heerden, J.A., Abboud, C.F. and Edis, A.J. (1978) Primary hyperparathyroidism and peptic ulcer disease. *Arch. Surg.*, **113**, 384–6.

Ljungberg, O. and Tibblin, S. (1979) Peroperative fat staining of frozen sections in primary hyperparathyroidism. *Am. J. Pathol.*, **95**, 633–42.

Lloyd, H.M. (1968) Primary hyperparathyroidism: an analysis of the role of the parathyroid tumor. *Medicine (Baltimore)*, **47**, 53–71.

Löwhagen, T. and Sprenger, E. (1974) Cytologic presentation of thyroid tumors in aspiration biopsy smear: a review of 60 cases. *Acta Cytol.*, **18**, 192–7.

Mallette, L.E., Bilezikian, J.P., Ketcham, A.S. and Aurbach, G.D. (1974) Parathyroid carcinoma in familial hyperparathyroidism. *Am. J. Med.*, **57**, 642–8.

Marx, S.J., Attie, M.F., Spiegel, A.M. *et al.* (1982) An association between neonatal severe primary hyperparathyroidism and familial hypocalciuric hypercalcemia in three kindreds. *N. Engl. J. Med.*, **306**, 257–64.

Marx, S.J., Powell, D., Shimkin, P.M. *et al.* (1973) Familial hyperparathyroidism. Mild hypercalcemia in at least nine members of a kindred. *Ann. Intern. Med.*, **78**, 371–7.

Mayor, G. (1967) Chirurgie der Epithelkörperchen. Operationsindikation und Technik. *Langenbecks Arch. Klin. Chir.*, **319**, 212–4.

McGill, P.E., Grange, A.T. and Royston, C.S. (1984) Chondrocalcinosis in primary hyperparathyroidism. Influence of parathyroid activity and age. *Scand. J. Rheumatol.*, **13**, 56–8.

McGregor, D.H., Lotuaco, L.G., Rao, M.S. and Chu, L.L.H. (1978) Functioning oxyphil adenoma of parathyroid gland. An ultrastructural and biochemical study. *Am. J. Pathol.*, **92**, 691–712.

Mitrovic, D., Mazabraud, A., Ryckewaert, H. *et al.* (1967) Adénome parathyroïdien à cellules sombres et clairs. Deux aspects ultrastructuraux de la cellule claire. Etudes cytochimiques et au microscope électronique. *Arch. Anat. Pathol.*, **15**, 225–30.

Mixter, C.G., Keynes, W.M. and Cope, O. (1962) Further experience with pancreatitis as a diagnostic clue to hyperparathyroidism. *N. Engl. J. Med.*, **266**, 265–72.

Mühlethaler, J.P., Schärer, K. and Antener, I. (1967) Akuter Hyperparathyreoidismus bei primärer Nebenschilddrüsen-Hyperplasie. *Helv. Paediatr. Acta*, **22**, 529–57.

Nathaniels, E.K., Nathaniels, A.M. and Wang, C-A. (1970) Mediastinal parathyroid tumors: a clinical and pathological study of 84 cases. *Ann. Surg.*, **171**, 165–70.

Netelenbos, C., Lips, P. and van der Meer, C. (1983) Hyperparathyroidism following irradiation of benign diseases of the head and neck. *Cancer*, **52**, 458–61.

Nilsson, O. (1977) Studies on the ultrastructure of the human parathyroid glands in various pathological conditions. *Acta Pathol. Microbiol. Scand. [A]*, (Suppl. 263), 1–88.

Norris, E.H. (1947) The parathyroid adenoma. A study of 322 cases. *Int. Abstr. Surg.*, **84**, 1–41.

Obara, T., Fujimoto, Y., Yamaguchi, K. *et al.* (1985) Parathyroid carcinoma of the oxyphil cell type. A report of two cases, light and electron microscopic study. *Cancer*, **55**, 1482–9.

Ober, W.B. and Kaiser, G.A. (1958) Hamartoma of the parathyroid. *Cancer*, **11**, 601–6.

Ordoñez, N.G., Ibañez, M.L., Samaan, N.A. and Hickey, R.C. (1983) Immunoperoxidase study of uncommon parathyroid tumors. Report of two cases of nonfunctioning parathyroid carcinoma and one intrathyroid parathyroid tumor – producing amyloid. *Am. J. Surg. Pathol.*, **7**, 535–42.

Ordoñez, N.G., Ibañez, M.L., Mackay, B. *et al.* (1982) Functioning oxyphil cell adenomas of parathyroid gland: immunoperoxidase evidence of hormonal activity in oxyphil cells. *Am. J. Clin. Pathol.*, **78**, 681–9.

Palmer, J.A., Brown, W.A., Kerr, W.H. *et al.* (1975) The surgical aspects of hyperparathyroidism. *Arch. Surg.*, **110**, 1004–7.

Patten, B.M., Bilezikian, J.P., Mallette, L.E. *et al.* (1974) Neuromuscular disease in primary hyperparathyroidism. *Ann. Intern. Med.*, **80**, 182–93.

Polga, J.P. and Balikian, J.P. (1971) Partially calcified functioning parathyroid adenoma: case demonstrated roentgenographically. *Radiology*, **99**, 55–6.

Poole, G.V., Jr, Albertson, D.A., Marshall, R.B. and Myers, R.T. (1982) Oxyphil cell adenoma and hyperparathyroidism. *Surgery*, **92**, 799–805.

Purnell, D.C., Smith, L.H., Scholz, D.A. *et al.* (1971) Primary hyperparathyroidism: a prospective clinical study. *Am. J. Med.*, **50**, 670–8.

Randall, C. and Lauchlan, S.C. (1963) Parathyroid hyperplasia in an infant. *Am. J. Dis. Child.*, **105**, 364–7.

Rao, S.D., Frame, B., Miller, M.J. *et al.* (1980) Hyperparathyroidism following head and neck irradiation. *Arch. Intern. Med.*, **140**, 205–7.

Rapaport, D., Ziv, Y., Rubin, M. *et al.* (1986) Primary hyperparathyroidism in children. *J. Pediatr. Surg.*, **21**, 395–7.

Rhone, D.P. (1975) Primary neonatal hyperparathyroidism. Report of a case and review of the literature. *Am. J. Clin. Pathol.*, **64**, 488–99.

Roth, S.I. (1970) The ultrastructure of primary water-clear cell hyperplasia of the parathyroid glands. *Am. J. Pathol.*, **61**, 233–40.

Roth, S.I. (1971) Recent advances in parathyroid gland pathology. *Am. J. Med.*, **50**, 612–22.

Sahin, A. and Robinson, R.A. (1988) Papillae formation in parathyroid adenoma. A source of possible diagnostic error. *Arch. Pathol. Lab. Med.*, **112**, 99–100.

Sawady, J., Mendelsohn, G., Sirota, R.L. and Taxy, J.B. (1989) The intrathyroid hyperfunctioning parathyroid gland. *Mod. Pathol.*, **2**, 652–7.

Schantz, A. and Castleman, B. (1973) Parathyroid carcinoma. A study of 70 cases. *Cancer*, **31**, 600–5.

Schwindt, W.D. (1967) Multiple parathyroid adenomas. *JAMA*, **199**, 945–6.

Shane, E. and Bilezikian, J.P. (1982) Parathyroid carcinoma: a review of 62 patients. *Endocr. Rev.*, **3**, 218–26.

Shane, E. and Bilezikian, J.P. (1989) Parathyroid carcinoma. In *Textbook of Uncommon Cancer* (eds C.J. Williams, J.G. Krikorian, M.R. Green and D. Raghavan), Wiley, New York, pp. 763–71.

Shields, T.W. and Staley, C.J. (1961) Functioning parathyroid cysts. *Arch. Surg.*, **82**, 937–42.

Sicard, G.A. (1981) Parathyroid hyperplasia versus adenoma: clinical or histologic diagnosis. *Surgery*, **90**, 125–6.

Sieracki, J.C. and Horn, R.C., Jr (1960) Nonfunctional carcinoma of the parathyroid. *Cancer*, **13**, 502–6.

Sipple, J.H. (1961) The association of pheochromocytoma with carcinoma of the thyroid gland. *Am. J. Med.*, **31**, 163–6.

Sloane, J.A. and Moody, H.C. (1978) Parathyroid adenoma in the submucosa of esophagus. *Arch. Pathol. Lab. Med.*, **102**, 242–3.

Smith, J.F. and Coombs, R.R.H. (1984) Histological diagnosis of carcinoma of the parathyroid gland. *J. Clin. Pathol.*, **37**, 1370–8.

Snover, D.C. and Foucar, K. (1981) Mitotic activity in benign parathyroid disease. *Am. J. Clin. Pathol.*, **75**, 345–7.

Spiegel, A.M., Marx, S.J., Doppman, J.L. *et al.* (1975) Intrathyroidal parathyroid adenoma or hyperplasia. An occasional overlooked cause of surgical failure in primary hyperparathyroidism. *JAMA*, **234**, 1029–33.

Steiner, A.L., Goodman, A.D. and Powers, S.R. (1968) Study of a kindred with pheochromocytoma, medullary thyroid carcinoma, hyperparathyroidism and Cushing's disease: multiple endocrine neoplasia, type 2. *Medicine (Baltimore)*, **47**, 371–409.

Stephenson, H.U., Jr (1950) Malignant tumors of parathyroid glands; review of literature with report of a case. *Arch. Surg.*, **60**, 247–66.

Thompson, N.W., Eckhauser, F.E. and Harness, J.K. (1982) The anatomy of primary hyperparathyroidism. *Surgery*, **92**, 814–21.

Tisell, L-E, Hedman, I. and Hansson, G. (1981), Clinical characteristics and surgical results in hyperparathyroidism caused by water-clear cell hyperplasia. *World J. Surg*, **5**, 565–71.

Tisell, L-E., Carlsson, S., Fjälling, M. *et al.* (1985) Hyperparathyroidism subsequent to neck irradiation. Risk factors. *Cancer*, **56**, 1529–33.

Udén, P., Berglund, B., Zederfeldt, B., Aspelin, P. and Ljungberg, O. (1987) Parathyroid lipoadenoma: a rare cause of primary hyperparathyroidism. Case report. *Acta Chir. Scand.*, **153**, 635–9.

Utley, J.R. and Black, W.C. (1967) Hyperparathyroidism. A clinicopathologic evaluation. *Am. J. Surg.*, **114**, 788–95.

van Heerden, J.A., Weiland, L.H., ReMine, W.H. *et al.* (1979) Cancer of the parathyroid glands. *Arch. Surg.*, **114**, 475–80.

Verdonk, C.A. and Edis, A.J. (1981) Parathyroid 'double adenomas'. Fact or fiction? *Surgery*, **90**, 523–6.

Wang, C.A. (1971) Surgery of the parathyroid glands. *Adv. Surg.*, **5**, 109–27.

Wang, C.A. (1976) The anatomic basis of parathyroid surgery. *Ann. Surg.*, **183**, 271–5.

Wang, C.A. and Gaz, R.D. (1985) Natural history of parathyroid carcinoma. Diagnosis, treatment, and results. *Am. J. Surg.*, **149**, 522–7.

Wermer, P. (1963) Endocrine adenomatosis and peptic ulcer in a large kindred. Inherited multiple tumors and mosaic pleiotropism in man. *Am. J. Med.*, **35**, 205–12.

Wheeler, M.H., Williams, E.D. and Wade, J.S. (1987) The hyperfunctioning intrathyroidal parathyroid gland: a potential pitfall in parathyroid surgery. *World J. Surg.*, **11**, 110–4.

Wills, M.R. (1971) Normocalcaemic primary hyperparathyroidism. *Lancet*, **i**, 849–53.

Wolff, M. and Goodman, E.N. (1980) Functional lipoadenoma of a supernumerary parathyroid gland in the mediastinum. *Head Neck Surg.*, **2**, 302–7.

Wolpert, H.R., Vickery, A.L. and Wang, C-A. (1989) Functioning oxyphil cell adenomas of the parathyroid gland. A study of 15 cases. *Am. J. Surg. Pathol.*, **13**, 500–4.

Woolner, L.B., Keating, F.R., Jr and Black, B.M. (1952) Tumors and hyperplasia of the parathyroid glands. A review of the pathological findings in 140 cases of primary hyperparathyroidism. *Cancer*, **5**, 1069–88.

17 Other forms of hyperparathyroidism and related disorders

17.1 Secondary hyperparathyroidism

Secondary hyperparathyroidism is the result of a prolonged lowering of the plasma concentration of ionized calcium. It is usually a consequence of chronic renal disease or intestinal malabsorption. The depression of the calcium level stimulates all parathyroid glands to undergo hyperplasia. This phenomenon can be considered as a physiological compensatory reaction, leading to increased secretion of PTH and normalization of the serum calcium level (Breslau, 1987).

In renal insufficiency, phosphorus retention occurs and, as a result, there is a tendency to a low serum calcium level as an adjustment to the high serum phosphorus concentration. In addition, the renal conversion of 25(OH)D into its biological active form, $1,25(OH)_2D$, is impaired in renal disease. The resulting vitamin D deficiency decreases the intestinal absorption of calcium, and thus contributes to the reduction in serum calcium. Levels of circulating PTH are higher than those in patients with either primary hyperparathyroidism or secondary hyperparathyroidism without renal failure (Arnaud, 1973) and a considerable part of this PTH appears to be in the form of inactive degradation fragments (Freitag *et al.*, 1978). Recent evidence from *in vitro* studies with parathyroid cells obtained from uraemic patients, suggests that there is a shift in the set point for calcium-regulated PTH secretion. Thus, a higher serum concentration of ionized calcium may be necessary to suppress the release of PTH (Brown *et al.*, 1982). It is possible that the low levels of $1,25(OH)_2D$ seen in secondary hyperparathyroidism may contribute to the altered regulation of PTH secretion by calcium in this condition (Lopez-Hilker *et al.*, 1986).

Morphologically, secondary hyperplasia is similiar to primary chief cell hyperplasia (p. 332). All glands are enlarged, the degree of hyperplasia depending upon the duration and the degree of the hypocalcaemic stimulus. In the majority of cases the hyperplasia is diffuse and uniform, but pronounced nodularity may also be seen, especially in advanced and long-standing renal insufficiency. The cell pattern may vary and includes nodules with oxyphil cells and highly vacuolated chief cells. Altenähr (1972) pointed out that, ultrastructurally, increased amounts of glycogen and/or mitochondria are typical of cells in secondary hyperplasia. The number of lipid bodies is reduced (Altenähr, 1972; Nilsson, 1977). At the light microscopical level, this is shown as decreased amounts or the absence of lipid droplets in frozen sections stained with oil-red O.

17.2 Tertiary hyperparathyroidism

The term tertiary hyperparathyroidism refers to cases of severe renal disease or malabsorption with previous evidence of hypocalcaemia and secondary hyperparathyroidism, who later develop hypercalcaemia (St Goar, 1963; Davies et al., 1968). It is considered to represent an autonomous parathyroid hyperfunction that develops on the basis of a pre-existing secondary hyperparathyroidism. Today, it is most often seen after successful renal transplantation because of chronic renal insufficiency. In many cases, serum calcium levels return to normal within a few weeks to months and parathyroidectomy is rarely needed (Parfitt, 1982). In this situation the parathyroid glands may merely demonstrate a relative autonomy of function, being capable, after varying periods of time, of reverting to a normal state (Katz et al., 1969; Black et al., 1970; Schwartz et al., 1970). In a minority of cases, however, persistent or apparently permanent hypercalcaemia is observed. Often, the disabling symptomatology of hypercalcaemia in such cases, and its potential damage to the renal transplant, necessitates surgical intervention with subtotal parathyroidectomy.

All parathyroid glands are enlarged and show a diffuse or multinodular hyperplasia (Figs. 17.1 and 17.2). The parathyroid changes resemble those seen in secondary hyperplasia, but the glands tend to be larger with greater variety of cell types. The size of the glands is often impressive, with a 10- to 20-fold increase in the normal glandular weight in diffuse hyperplasia, and up to 40-fold increase in multinodular hyperplasia (Krause and Hedinger, 1985). Nodules may occasionally be composed exclusively of oxyphil cells (Misonou et al., 1987). Ultrastructurally, chief cells in parathyroid glands from patients with tertiary hyperparathyroidism show abundance of RER and secretory granules correlating with a high cellular secretory activity (Krause and Hedinger, 1985). Fat staining with oil-red O usually reveals complete absence of fat droplets in chief cells of the hyperplastic areas compatible with increased functional activity.

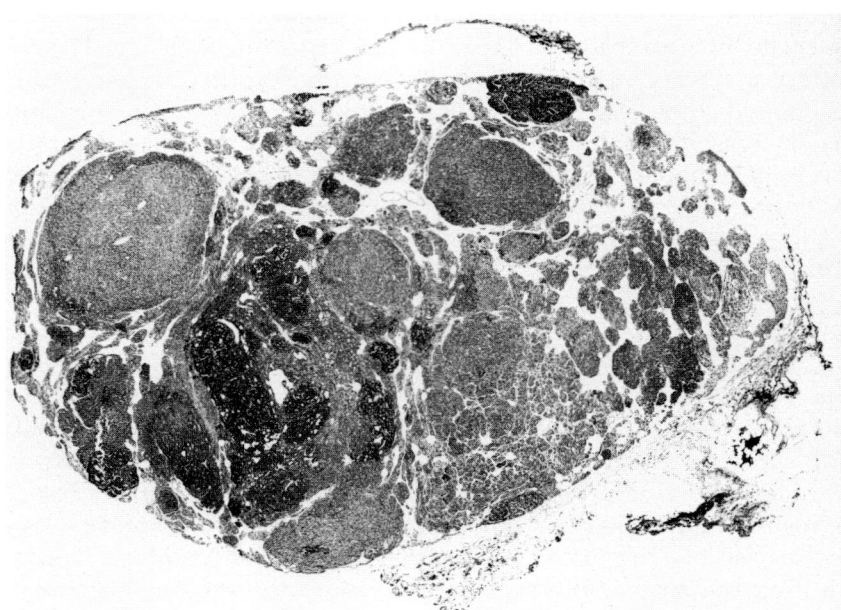

Figure 17.1 Tertiary hyperparathyroidism. The gland contains multiple small hyperplastic nodules. Intervening parenchyma shows some compact, probably diffusely hyperplastic areas; other parts are normal-looking (H&E, × 11).

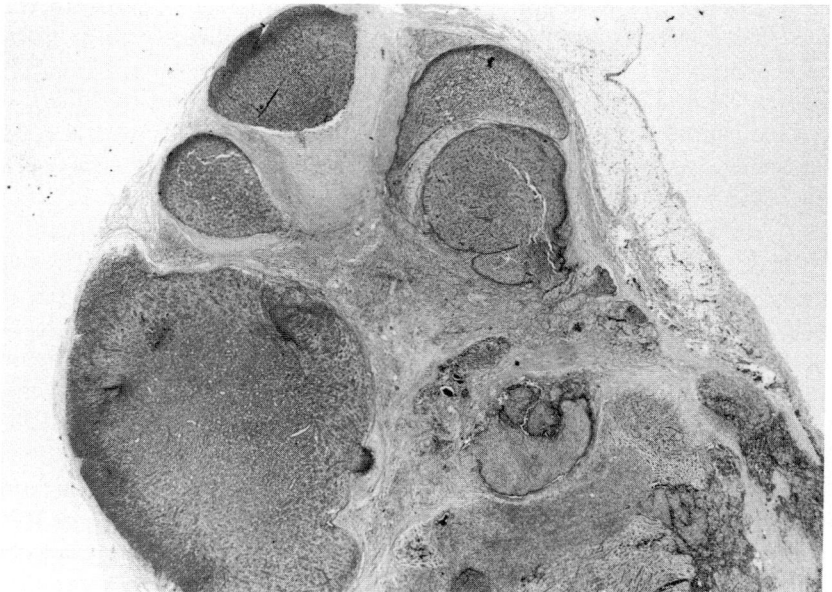

Figure 17.2 Parathyroid gland from a patient with tertiary hyperparathyroidism, showing advanced nodular hyperplasia and internodular fibrosis (H&E, × 9).

17.3   Hyperparathyroidism in multiple endocrine neoplasia syndromes

Primary hyperparathyroidism is a well-recognized finding in hereditary, and more exceptional sporadic, disorders with hyperfunction of two or more endocrine tissues. The two major complex endocrine syndromes that may involve the parathyroid glands are multiple endocrine neoplasia, type 1, and type 2. In type 1 (MEN 1), the parathyroids, pancreas and pituitary are involved. In type 2 (MEN 2), the patients present with medullary carcinoma of the thyroid gland and adrenal medullary and parathyroid disease (Steiner et al., 1968). Later, it was recognized that MEN 2 could be segregated into two variants, one without (MEN 2A) and the other with (MEN 2B) mucosal ganglioneuromatosis and other proliferative changes of the peripheral nervous system (Chong et al., 1975).

17.3.1   Multiple endocrine neoplasia, type 1

In its familial form, MEN 1 shows an autosomal dominant mode of inheritance. The patients present with hyperparathyroidism and hyperfunction of the pancreatic islets and/or anterior pituitary gland. Less common manifestations are adrenocortical tumours, follicular tumours of the thyroid gland, bronchial or intestinal carcinoids, and subcutaneous or intestinal lipomas.

Hyperparathyroidism is reported to occur most frequently, i.e. in 90–97% of cases, followed by pancreatic islet tumours in 30–80% of cases, and pituitary tumours in 15–50% of the patients (Brandi et al., 1987; Samaan et al., 1989). Expression of the syndrome often begins between 20 and 40 years of age. Hyperparathyroidism is generally present and may precede other manifestations of the syndrome. It is noteworthy that 1–18% of cases of apparently sporadic primary hyperparathyroidism are found to be associated with MEN 1, when other endocrine disorders are carefully sought (Fitzpatrick, 1989). In the majority of series, the incidence of renal impairment and bone disease is similar to that in sporadic primary hyperparathyroidism (Boey et al., 1975; Cornelis et al., 1979; Fitzpatrick, 1989; Samaan et al., 1989). However, cases of MEN 1 tend to have lower serum calcium levels compared to classic primary hyperparathyroidism (Fitzpatrick, 1989). The incidence of asymptomatic hypercalcaemia in MEN 1 ranges from 25% (Haff et al., 1970) to 71% (Christensson, 1976).

The parathyroid glands in hyperparathyroidism associated with MEN 1 show chief cell hyperplasia with gross and histological features similar to those in classic sporadic primary hyperparathyroidism (Chapter 16, p. 332). Although parathyroid adenomas have occasionally been reported in these cases (Samaan *et al.*, 1989), it would appear that most, if not all, cases are basically hyperplasia, which may sometimes involve the glands unevenly. The recurrence rate of hyperparathyroidism after surgery has been reported to be higher than in sporadic primary hyperparathyroidism (Cornelis *et al.*, 1979; Rizzoli *et al.*, 1985). Ths is probably because some cases are misdiagnosed as having adenomatous disease rather than hyperplasia, leading to removal of an insufficient amount of parathyroid tissue (van Heerden *et al.*, 1983). The surgical literature is reluctant to recommend parathyroidectomy unless significant hypercalcaemia or documented renal dysfuction or bone disease is present. But when present, it is important to identify all glands at the initial operation and perform a subtotal parathyroidectomy, or total parathyroidectomy with a fresh or cryopreserved autograft of parathyroid tissue (Clark and Duh, 1989; Fitzpatrick, 1989; Samaan *et al.*, 1989).

The familial form of MEN 1 appears to be associated with a specific chromosomal abnormality. By using molecular genetic techniques, such as cloned DNA sequences, which reveal restriction-fragment-length polymorphism, pancreatic tumours from MEN 1 patients have been shown to be associated with the loss of alleles on chromosome 11 (Larsson *et al.*, 1988). With similar techniques, Thakker *et al.*, (1989) have demonstrated that allelic deletions on chromosome 11 are also involved in the development of some of the parathyroid lesions in patients with this syndrome. The findings of allelic loss in some of the parathyroid lesions suggests a monoclonal cellular proliferation in these glands. The authors suggest that failure to demonstrate allelic deletions in other cases may reflect early events in the progression from hyperplasia to a neoplastic monoclonal proliferation. It can be speculated that the adenoma-like lesions which have sometimes been observed together with hyperplasia in patients with MEN 1 could possibly represent a progression from a hyperplastic state to a focal neoplastic lesion, i.e. an adenoma.

A close linkage has also been demonstrated between MEN 1 and the oncogene INT2, which encodes a protein of the fibroblast growth factor family (Dickson and Peters, 1987). Furthermore, Brandi *et al.* (1986) have shown that plasma from patients with MEN 1 contains a protein with parathyroid mitogenic activity, which has similarities to basic fibroblast growth factor.

17.3.2   *Multiple endocrine neoplasia, type 2*

Steiner *et al.* (1968) proposed the term 'multiple neoplasia, type 2' for a familial syndrome, which is also dominantly inherited and encompasses medullary thyroid carcinoma, phaeochromocytoma and parathyroid disease. This entity was later designated MEN 2A to separate it form a phenotypically distinct subgroup, referred to as MEN 2B, in which patients also had a characteristic appearance (thickened lips, marfanoid habitus) and mucosal ganglioneuromatosis, but usually no obvious parathyroid disease.

It is difficult to assess the true incidence of parathyroid disease in patients with MEN 2A; some patients who are normocalcaemic, nevertheless, are found to have hyperplastic parathyroid glands at operation for their medullary thyroid carcinoma. Hyperparathyroidism in MEN 2A is a significant problem less frequently than in MEN 1. The functional and morphological status of parathyroid glands in MEN 2A has been confusing and unpredictable. Thus a single, sometimes markedly, enlarged gland may be found and removed at the time of total thyroidectomy without evidence of abnormal serum calcium levels before or after operation. In other instances, parathyroid glands are grossly normal, but may show early histological changes of diffuse or nodular hyperplasia (Catalona et al., 1971; Block et al., 1975). Whereas most, if not all, patients with MEN 1 will eventually show clinical signs of hyperparathyroidism, this is a much less frequent event in MEN 2A, being present in some families but not in others (Steiner et al., 1968; Melvin et al., 1972; Keiser et al., 1973; Block et al., 1975; Sizemore et al., 1977; Gagel et al., 1988). When present, clinically evident hyperparathyroidism tends to be mild with only slightly elevated serum calcium levels. However, the glands of normocalcaemic patients with MEN 2A appear to be non-suppressible by calcium. This is in contrast to patients with MEN 2B in whom calcium infusion is reported to bring about a normal suppression of the serum PTH levels (Heath et al., 1976).

The surgical literature recommends that all patients with MEN 2A undergoing neck operation for medullary thyroid carcinoma should have all parathyroid glands identified, removing only grossly enlarged glands, unless hyperparathyroidism is clinically evident. However, once parathyroid disease is evident and produces complications of hypercalcaemia, a subtotal parathyroidectomy is necessary (van Heerden et al., 1983; Fitzpatrick, 1989).

MEN 2B patients generally lack clinical evidence of parathyroid disease, and parathyroidectomy for this disorder is not justified. The parathyroid glands appear structurally normal in childhood and through the mid teens. With increased age, however, they may show a mild chief cell hyperplasia (Carney et al., 1980).

17.4 Familial benign (hypocalciuric) hypercalcaemia

Familial benign (hypocalciuric) hypercalcaemia is a rare condition, which may be defined as a dominantly inherited disorder of calcium and magnesium metabolism, characterized by lifelong hypercalcaemia and hypermagnesaemia (Heath, 1989). It is not usually associated with clinical disease or ill health. The disorder may mimic mild primary hyperparathyroidism, but typically, subtotal parathyroidectomy fails to abolish the hypercalcaemia (Foley et al., 1972). A low calcium: creatinine clearance ratio suggests, although is not diagnostic of, this condition. Urinary calcium excretion is typically low relative to serum calcium levels. Serum PTH levels are usually normal or slightly elevated. In this respect, patients with familial benign hypercalcaemia differ from patients with primary hyperparathyroidism who have increased amounts of circulating PTH. The finding of normal serum PTH levels in combination with hypercalcaemia suggests an abnormality of parathyroidal responses to calcium.

On gross examination, the parathyroid glands are usually of normal size and microscopically, they are reported to be either normal (Law et al., 1984) or mildly hyperplastic (Thorgeirsson et al., 1981).

The underlying cellular and molecular pathogenesis of familial benign hypercalcaemia is obscure. It has been suggested that the kidneys are hypersensitive to PTH, leading to an increased reabsorption of calcium and magnesium in the renal tubules (Law et al., 1984). The patients have normal bone mineral density (Heath, 1989).

Neonatal primary hyperparathyroidism is a rare and life-threatening disease causing severe symptomatic hypercalcaemia, failure to thrive, and hypotonicity (Bradford et al., 1973). The disorder affects newborns in families with apparent familial benign hypercalcaemia (Jason et al., 1981; Marx et al., 1982; Fujita et al., 1983; Cooper et al., 1986). At operation, the parathyroid glands are diffusely hyperplastic. Microscopically, the chief cells appear highly vacuolated, reflecting the increased amounts of glycogen observed by electron microscopy (Garcia-Bunuel et al., 1974).

References

Altenähr, E. (1972) Ultrastructural pathology of parathyroid glands. *Curr. Top. Pathol.*, **56**, 1–54.

Arnaud, C.D. (1973) Hyperparathyroidism and renal failure. *Kidney Int.*, **4**, 89–95.

Black, W.C., III, Slatopolsky, E., Elkan, I. and Hoffsten, P. (1970) Parathyroid morphology in suppressible and nonsuppressible renal hyperparathyroidism. *Lab. Invest.*, **23**, 497–507.

Block, M.A., Jackson, C.E. and Tashjian, A.H., Jr (1975) Management of parathyroid glands in surgery for medullary carcinoma. *Arch. Surg.*, **110**, 617–24.

Boey, J.H. Cooke, T.J.C., Gilbert, J.M. et al. (1975) Occurrence of other endocrine tumors in primary hyperparathyroidism. *Lancet*, **ii**, 781–4.

Bradford, W.D., Wilson, J.W. and Gaede, J.T. (1973) Primary neonatal hyperparathyroidism: an unusual cause of failure to thrive. *Am. J. Clin. Pathol.*, **59**, 265–75.

Brandi, M.L., Aurbach, G.D. , Fitzpatrick, L.A. *et al.* (1986) Parathyroid mitogenic activity in plasma from patients with familial multiple endocrine neoplasia type I. *N. Engl. J. Med.*, **314**, 1287–93.

Brandi, M.L., Marx, S.J., Aurbach, G.D. and Fitzpatrick, L.A. (1987) Familial multiple endocrine neoplasia type I: a new look at pathophysiology. *Endocr. Rev.*, **8**, 391–405.

Breslau, N.A. (1987) Update on secondary forms of hyperparathyroidism. *Am. J. Med. Sci.*, **294**, 120–31.

Brown, E.M., Wilson, R.E., Eastman, R.C. *et al.* (1982) Abnormal regulation of parathyroid hormone release by calcium in secondary hyperparathyroidism due to chronic renal failure. *J. Clin. Endocrinol. Metab.*, **54**, 172–9.

Carney, J.A., Roth, S.I., Heath, H., III *et al.* (1980) The parathyroid glands in multiple endocrine neoplasia type 2b. *Am. J. Pathol.*, **99**, 387–98.

Catalona, W.J., Engelman, K., Ketcham, A.S. and Hammond, W.G. (1971) Familial medullary thyroid carcinoma, pheochromocytoma and parathyroid adenoma (Sipple's syndrome). *Cancer*, **28**, 1245–54.

Chong, G.C., Beahrs, O.H., Sizemore, G.W. and Wollner, L.B. (1975) Medullary carcinoma of the thyroid gland. *Cancer*, **35**, 695–704.

Christensson, T. (1976) Familial hyperparathyroidism. *Ann. Intern Med.*, **85**, 614–5.

Clark, O.H. and Duh, Q-Y. (1989) Primary hyperparathyroidism. A surgical perspective. *Endocrinol. Metab. Clin. North Am.*, **18**, 701–14.

Cooper, L., Wertheimer, J., Levey, R. *et al.* (1986) Severe primary hyperparathyroidism in a neonate with two hypercalcemic parents: management with parathyroidectomy and heterotopic autotransplantation. *Pediatrics*, **78**, 263–8.

Cornelis, B.H.W., Lamers, M.D., Paul, G.A.M. and Froeling, M.D. (1979) Clinical significance of hyperparathyroidism in familial multiple endocrine adenomatosis type I (MEA I). *Am. J. Med.*, **66**, 422–4.

Davies, D.R., Dent, C.E. and Watson, L. (1968) Tertiary hyperparathyroidism. *Br. Med. J.*, **iii**, 395–9.

Dickson, C. and Peters, G. (1987) Potential oncogene product related to growth factors. *Nature*, **326**, 833.

Fitzpatrick, L.A. (1989) Hypercalcemia in the multiple endocrine neoplasia syndromes. *Endocrinol. Metab. Clin. North Am.*, **18**, 741–52.

Foley, T.P., Harrison, H.C., Arnaud, C.D. and Harrison, H.E. (1972) Familial benign hypercalcemia. *J. Pediatr.*, **81**, 1060–7.

Freitag, J.K.J., Martin, K.A., Hurska, C. *et al.* (1978) Impaired parathyroid hormone metabolism in patients with chronic renal failure. *N. Engl. J. Med.*, **298**, 29–32.

Fujita, T., Watanabe, N., Fukase, M. *et al.* (1983) Familial hypocalciuric hypercalcemia involving four members of a kindred including a girl with severe neonatal primary hyperparathyroidism. *Miner. Electrolyte Metab.*, **9**, 51–4.

Gagel, R.F., Tashjian, A.H., Jr, Cummings, T. *et al.* (1988) The clinical outcome of prospective screening for multiple endocrine neoplasia type 2a. An 18-year experience. *N. Engl. J. Med.*, **318**, 478–84.

Garcia-Bunuel, R., Kutchemeshgi, A. and Brandes, D. (1974) Hereditary hyperparathyroidism: the fine structure of the parathyroid gland. *Arch. Pathol.*, **97**, 399–403.

Haff, R.C. , Black, W.C., III and Ballinger, W.F. (1970) Primary hyperparathyroidism: changing clinical, surgical and pathological aspects. *Ann. Surg.*, **171**, 85–92.

Heath, H.H., III (1989) Familial benign (hypocalciuric) hypercalcemia. A troublesome mimic of mild primary hyperparathyroidism. *Endocrinol. Metab. Clin. North Am.*, **18**, 723–40.

Heath, H., III, Sizemore, G.W. and Carney, J.A. (1976) Preoperative diagnosis of occult parathyroid hyperplasia by calcium infusion in patients with multiple endocrine neoplasia, type 2a. *J. Clin. Endocrinol. Metab.*, **43**, 428–35.

Jason, C., Arnaud, S.B., Harrison, M.R. *et al.* (1981) Primary hyperparathyroidism (HPT) in infancy. *Pediatr. Res.*, **15**, 504 (abstract).

Katz, A.I., Hampers, C.L. and Merrill, J.P. (1969) Secondary hyperparathyroidism and renal osteodystrophy in chronic renal failure. Analysis of 195 patients, with observations on the effects of chronic dialysis, kidney transplantation and subtotal parathyroidectomy. *Medicine (Baltimore)*, **48**, 333–74.

Keiser, H.R., Beaven, M.A., Doppman, J. *et al.* (1973) Sipple's syndrome: medullary thyroid carcinoma, pheochromocytoma, and parathyroid disease. *Ann. Intern. Med.*, **78**, 561–79.

Krause, M.W. and Hedinger, C.E. (1985) Pathologic study of parathyroid glands in tertiary hyperparathyroidism. *Hum. Pathol.*, **16**, 772–84.

Larsson, C., Skogseid, B., Öberg, K. *et al.* (1988) Multiple endocrine neoplasia type I gene maps to chromosome 11 and is lost in insulinoma. *Nature*, **332**, 85–7.

Law, W.M., Jr, Carney, J.A. and Heath, H., III (1984) Parathyroid glands in familial benign hypercalcemia (familial hypocalciuric hypercalcemia). *Am. J. Med.*, **76**, 1021–6.

Lopez-Hilker, S., Galceran, T., Chan, Y-L. *et al.* (1986) Hypocalcemia may not be essential for the development of secondary hyperparathyroidism in chronic renal failure. *J. Clin. Invest.*, **78**, 1097–102.

Marx, S.J., Attie, M.F., Spiegel, A.M. *et al.* (1982) An association between neonatal severe primary hyperparathyroidism and familial hypocalciuric hypercalcemia in three kindreds. *N. Engl. J. Med.*, **306**, 257–64.

Melvin, K.E.W., Tashjian, A.H., Jr and Miller, H.H. (1972) Studies in familial (medullary) thyroid caricnoma. *Recent Prog. Horm. Res.*, **28**, 399–460.

Misonou, J., Ishikura, H., Aizawa, M. and Ohira, S. (1987) Functioning oxyphil cell adenoma in a patient with secondary hyperparathyroidism. *Acta Pathol. Jpn*, **37**, 1357–66.

Nilsson, O. (1977) Studies on the ultrastructure of the human parathyroid glands in various pathological conditions. *Acta Pathol. Microbiol. Scand.* [A], (Suppl. 263), 1–88.

Parfitt, A.M. (1982) Hypercalcemic hyperparathyroidism following renal transplantation: differential diagnosis, management, and implications for cell population control in the parathyroid gland. *Miner. Electrolyte Metab.*, **8**, 92–112.

Rizzoli, R., Green, J., III and Marx, S.J. (1985) Primary hyperparathyroidism in familial multiple endocrine neoplasia type I. Long-term follow-up of serum calcium levels after parathyroidectomy. *Am. J. Med.*, **78**, 467–74.

Samaan, N.A., Ouais, S., Ordoñez, N.G. *et al.* (1989) Multiple endocrine syndrome type 1. Clinical, laboratory findings, and management in five families. *Cancer*, **64**, 741–52.

Schwartz, G.H., David, D.S., Riggio, R.R. *et al.* (1970) Hypercalcemia after renal transplantation. *Am. J. Med.*, **49**, 42–51.

Sizemore, G.W., Carney, J.A. and Heath, H., III (1977) Epidemiology of medullary carcinoma of the thyroid gland: a five-year experience (1971–1976). *Surg. Clin. North Am.*, **57**, 633–45.

St Goar, W.T. (1963) Comment to: Nicholas, G., Jr and Roth, S.I.: Case records of the Massachusetts General Hospital, case 29. *N. Engl. J. Med.*, **268**, 943–53.

Steiner, A.L., Goodman, A.D. and Powers, S.R. (1968) Study of a kindred with pheochromocytoma, medullary thyroid carcinoma, hyper-parathyroidism and Cushing's disease: multiple endocrine neoplasia, type 2. *Medicine (Baltimore)*, **47**, 371–409.

Thakker, R.V., Bouloux, P., Wooding, C. *et al.* (1989) Association of parathyroid tumors in multiple endocrine neoplasia type 1 with loss of alleles on chromosome 11. *N. Engl. J. Med.*, **321**, 218–224.,

Thorgeirsson, U., Costa, J. and Marx, S.J. (1981) The parathyroid glands in familial hypocalciuric hypercalcemia. *Hum. Pathol.*, **12**, 229–37.

van Heerden, J.A., Kent, R.B., III, Sizemore, G.W. *et al.* (1983) Primary hyper-parathyroidism in patients with multiple endocrine neoplasia syndromes. Surgical experience. *Arch. Surg.*, **118**, 533–6.

18 Miscellaneous lesions

18.1 Parathyroid cysts

Parathyroid cysts may develop on the basis of degeneration in hyperplastic or adenomatous glands and thus, are usually associated with primary hyperparathyroidism (Rogers et al., 1969; Fallon et al., 1982). The majority of cysts, however, are non-functional, and may represent retention cysts or remnants of the third or fourth branchial pouch. They are usually located in the neighbourhood of the lower thyroid poles, i.e. in the region of the lower pair of parathyroid glands. They may also be found in the upper part of the neck or in the mediastinum (Thacker et al., 1971; Calandra et al., 1983). Clinically they may be confused with thyroid cysts or nodules (Wang et al., 1972). Most cysts are about 4 cm in diameter, although occasional examples have been more than 10 cm, and may give rise to pressure symptoms (Gordon and Harcourt-Webster, 1965).

Grossly, the cysts usually have smooth surfaces and can easily be dissected free of the adjacent tissues. The lumen of the cyst is filled with a thin fluid, which is usually clear, but may be haemorrhagic or opalescent, and contains PTH (Silverman et al., 1986). The inner surface is lined by a single layer of chief cells which are usually vacuolated. Sometimes small islands of closely packed chief cells are embedded in the fibrous wall (Figs 18.1 and 18.2). Occasionally, oxyphil cells may be present. Lymphoid or thymic tissue has also been encountered in the cyst wall.

18.2 Amyloidosis

Amyloid deposits are common in the stromal tissue in parathyroid glands of patients with generalized amyloidosis both of the primary and the secondary type (Ellis and Mawhinney, 1984). Many of these patients also have chronic renal insufficiency, and secondary or tertiary hyperparathyroidism with diffuse or nodular parathyroid hyperplasia is then also present (Thiele et al., 1975).

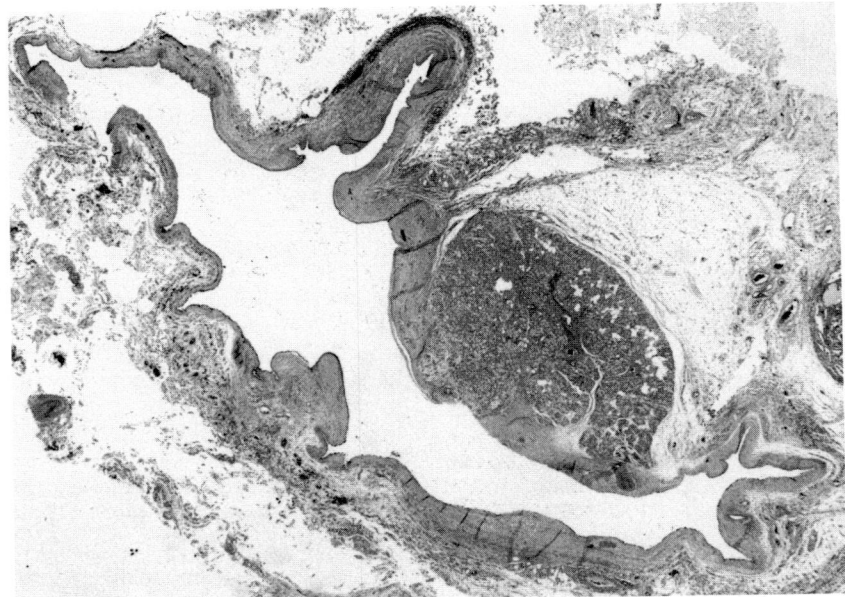

Figure 18.1 Parathyroid cyst from a patient with primary hyperparathyroidism. A nodule of compact parathyroid tissue is seen in the cyst wall (H&E, × 12,5)

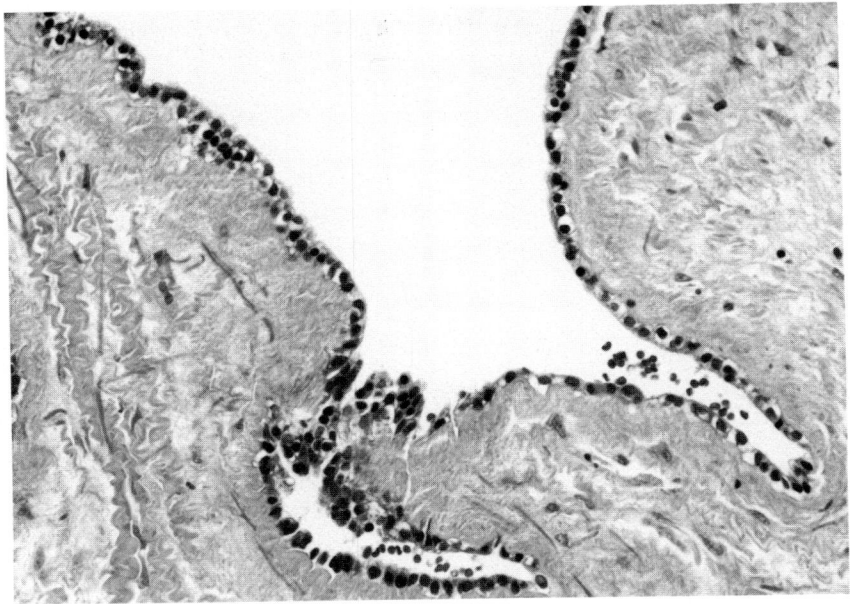

Figure 18.2 Same case as shown in Fig. 18.1. The cyst is lined by a layer of partly vacuolated chief cells (H&E, × 250).

References

Calandra, D.B., Shah, K.H., Prinz, R.A. *et al.* (1983) Parathyroid cysts. A report of eleven cases including two associated with hyperparathyroid crisis. *Surgery*, **94**, 887–892.

Ellis, H.A. and Mawhinney, W.H.B. (1984) Parathyroid amyloidosis. *Arch. Pathol. Lab. Med.*, **108**, 689–90.

Fallon, M.D., Haines, J.W. and Teitelbaum, S.L. (1982) Cystic parathyroid gland hyperplasia – hyperparathyroidism presenting as a neck mass. *Am. J. Clin. Pathol.*, **77**, 104–7.

Gordon, A. and Harcourt-Webster, J.N. (1965) Parathyroid cysts. A report of two cases. *J. Pathol. Bacteriol.*, **89**, 374–7.

Rogers, L.A., Fetter, B.F. and Peete, W.P.J. (1969) Parathyroid cyst and cystic degeneration of parathyroid adenoma. *Arch. Pathol.*, **88**, 476–9.

Silverman, J.F., Khazanie, P.G., Norris, H.T. and Fore, W.W. (1986) Parathyroid hormone (PTH) assay of parathyroid cysts examined by fine-needle aspiration biopsy. *Am. J. Clin. Pathol.*, **86**, 776–80.

Thacker, W.C., Wells, V.H. and Hall, E.R. (1971) Parathyroid cyst of the mediastinum. *Ann. Surg.*, **174**, 969–75.

Thiele, J., Ries, P. and Georgii, A. (1975) Spezielle und funktionelle Pathomorphologie der Epithelkörperchen in einem unausgewählten Obduktionsgut (589 Sektionen). *Virchows Arch. [A]*, **367**, 195–208.

Wang, C-A., Vickery, A.L., Jr and Maloof, F. (1972) Large parathyroid cysts mimicking thyroid nodules. *Ann Surg.*, **175**, 448–53.

Index

Adenolipoma
  parathyroid, *see* Lipoadenoma
  thyroid 126
Adenoma, *see* various types
Adipose tissue
  in parathyroid glands 299–300,
    301–2
  in thyroid lesions 126, 147, 149
Amyloid
  alkaline Congo red staining for
    28
  in intermediate type of thyroid
    carcinoma 228
  in medullary thyroid
    carcinoma 202–4, 219, 226,
    Plate 21
  in parathyroid adenoma 305, 327
Amyloidosis
  of parathyroid glands 360
Anaplastic carcinoma of the
    thyroid, *see* Undifferentiated
    thyroid carcinoma
Argentaffin cells
  in medullary carcinoma 13, 211
  in thyroid gland 13
Argyrophil staining reaction
  in medullary thyroid
    carcinoma 28, 207, 217
  in parafollicular cells 9, 28
  in paraganglioma of the
    thyroid gland 280
  in parathyroid adenoma 328
  in parathyroid cells 308, 315
Askanazy cell, *see* Oxyphil cell
Atypical adenoma
  of the thyroid gland 126, Plate
    12

Autoimmune thyroid disease, *see*
  Hashimoto's thyroiditis and
  Graves' disease

Basedow's disease, *see* Graves'
  disease
Biogenic amines, *see* Monoamines
Black thyroid 4

C-cells, *see* Parafollicular cells
Calcitonin
  in intermediate type of thyroid
    carcinoma 232
  in medullary thyroid
    carcinoma 216, 218, Plate 24
  in non-thyroidal neoplasms 226
  in parafollicular cells 10, 12, 18
  in poorly differentiated thyroid
    carcinoma 187, 212
  in ultimobranchial remnants 16
  in undifferentiated thyroid
    carcinoma 212, 216, 248
Carcinoembryonic antigen
  in medullary carcinoma 34,
    217, 223
  in oxyphil thyroid tumours 198
  in parafollicular cells 16, 34
Carcinoma, parathyroid, *see*
  Parathyroid carcinoma
Carcinoma, thyroid
  epidemiology of 138–9
  in childhood and adolescents
    290–4
  metastatic in thyroid gland 284–8
  radiation-induced 109, 139,
    290–3
  see also individually named types